T0205663

AGROECOSYSTEMS

Soils, Climate, Crops, Nutrient Dynamics, and Productivity

AGROECOSYSTEMS

Soils, Climate, Crops, Nutrient Dynamics, and Productivity

K. R. Krishna, PhD

Apple Academic Press Inc.
3333 Mistwell Crescent
Oakville, ON L6L 0A2
Canada

Apple Academic Press Inc.
9 Spinnaker Way
Waretown, NJ 08758
USA

©2014 by Apple Academic Press, Inc.

First issued in paperback 2021

Exclusive worldwide distribution by CRC Press, a member of Taylor & Francis Group

No claim to original U.S. Government works

ISBN 13: 978-1-77463-278-9 (pbk)
ISBN 13: 978-1-926895-48-2 (hbk)

This book contains information obtained from authentic and highly regarded sources. Reprinted material is quoted with permission and sources are indicated. Copyright for individual articles remains with the authors as indicated. A wide variety of references are listed. Reasonable efforts have been made to publish reliable data and information, but the authors, editors, and the publisher cannot assume responsibility for the validity of all materials or the consequences of their use. The authors, editors, and the publisher have attempted to trace the copyright holders of all material reproduced in this publication and apologize to copyright holders if permission to publish in this form has not been obtained. If any copyright material has not been acknowledged, please write and let us know so we may rectify in any future reprint.

Trademark Notice: Registered trademark of products or corporate names are used only for explanation and identification without intent to infringe.

Library of Congress Control Number: 2013949197

Library and Archives Canada Cataloguing in Publication

Krishna, K. R. (Kowligi R.), author
Agroecosystems: soils, climate, crops, nutrient dynamics, and productivity/K.R. Krishna, PhD.

Includes bibliographical references and index.
ISBN 978-1-926895-48-2
1. Anthropometry–Mathematical models–Data processing.
1. Agricultural ecology. 2. Crops--Nutrition. 3. Agricultural productivity. I. Title.

S589.7.K75 2013 577.5'5 C2013-906126-6

Apple Academic Press also publishes its books in a variety of electronic formats. Some content that appears in print may not be available in electronic format. For information about Apple Academic Press products, visit our website at **www.appleacademicpress.com** and the CRC Press website at **www.crcpress.com**

ABOUT THE AUTHOR

K. R. Krishna, PhD

K. R. Krishna received his PhD in Agriculture from the University of Agricultural Sciences, Bangalore, India. Retired from the International Crops Research Institute for the Semi-Arid Tropics (ICRISAT) in India, he has been a cereals scientist in India and a visiting professor and research scholar at the Soil and Water Science Department at the University of Florida, Gainesville, USA. Dr. Krishna is a member of several professional organizations, including the International Society for Precision Agriculture, the American Society of Agronomy, the Soil Science Society of America, the Ecological Society of America, and the Indian Society of Agronomy. He is author of several books, including *Maize Agroecosystem: Nutrient Dynamics and Productivity* and *Precision Farming: Soil Fertility and Productivity Aspects*, both published by Apple Academic Press, Inc.

CONTENTS

LIST OF ABBREVIATIONS

BGA	blue green algae
BMP	best management practice
BNF	biological nitrogen fixation
CEC	cation exchange capacity
CIFA	Centre de Investigacion y Formacion Agraria
CMS	cytoplasmic male sterile
CREC	Citrus Research and Experimental Centre
EOLSS	Encyclopedia of Life Support Systems
EONR	economically optimum nitrogen rates
ET	evapotranspiration
GDD	growing degree days
GLM	green leaf manures
HI	harvest index
ICARDA	International Centre for Agricultural Research in Dry Areas
ICID	International Commission on Irrigation and Drainage
ICRISAT	International Crops Research Institute for the Semi-Arid Tropics
INM	integrated nutrient management
IWMI	International Water Management Institute
LPM	liquid piggery manure
NSW	New South Wales
NTI	nutrient translocation index
NUE	nutrient use efficiency
PF	precision farming
RMW	rice mill waste
SAR	State Agency Recommendation
SOC	soil organic carbon
SOM	soil organic matter
SSA	Sub-Saharan Africa
SSNM	site-specific nutrient management
STCR	soil test crop response
WANA	West Asia and North Africa
WUE	water use efficiency

PREFACE

Historically, seeds for agricultural cropping were sown in the Fertile Crescent region of Middle Eastern Asia some 8–10,000 year ago. Since then, crops and human beings have evolved together. Several crop species were domesticated and cultivated during the Neolithic Period. In due course, cropping systems were refined and mildly intensified. During the medieval period, human ingenuity and man's inquisitiveness led him to migrate to different parts of the globe. This induced the spread of several crop species and facilitated formation of cropping belts across different continents. Soon, vast stretches of crops were developed and agronomic procedures were consistently refined to enhance productivity. Agroecosystems that were generated were either subsistent or intensively nurtured.

In the past century, agroecosystems were markedly dynamic with regard to crop species, genetic stocks, fertilizer/nutrient turnover, and productivity. Agroecosystems that supplied food grains and other products to large populations of humans and domesticated animals were preferentially intensified. Nutrient supply, crop genetic stocks, and facilitation of irrigation were the key factors that induced formation of productive agroecosystems in different parts of the world. Globally, we may encounter agroecosystems that are at different stages of evolution, development, sustenance, and productivity. Several of them could be still developing, others may have reached zenith, and yet others could be in process of deterioration or replacement. Such dynamics are induced by natural and man-made factors. Whatever may be the natural state of agroecosystems, we need to appreciate that each agroecosystem, small or large, has its say in terms of human nutrition and energetic functioning.

Agroecosystems that we have inherited and as we perceive them today are actually marvels created on earth's surface through human endeavor over a few millenniums. We have understood and accrued a large body of knowledge about them. Yet, we have too many things to investigate and perhaps control to our own advantage, of course, without deteriorating natural resources and our environment. Today, an entire posse of agroecosystems on the globe feeds over 7 billion human species and innumerable farm animals. We have to strive to modify these agroecosystems to feed a larger populace in the future.

Agroecosystems occupy a third of the global land surface. They are composed of a variety of soil types, crop species, and water resources. Major cereals such as wheat, maize, rice, few legumes, cash crops and plantations dominate the agrarian zones. Pastures and forage belts occur conspicuously in different continents.

The central theme of this book is agroecosystems that flourish in different continents of the world. Major focus is on the ingredients such as soils, climate, crops, nutrient dynamics, and productivity. There are three sections in this volume. The first section presents two chapters that support vast crop belts. In addition, there are 21 chapters that deal with the wide range of agroeocsystems that thrive on the globe. Detailed descriptions about crop species, their origin, spread, genetic aspects, agroclimatic requirements, expanse and productivity are available for 32 important crops. The second section presents principles that drive crop growth, nutrient dynamics, and ecosystematic functions within any agroecosystem. There are seven chapters that deal with various aspects of nutrient dynamics in an agroecosystem. Several natural and man-made factors regulate development, sustenance, and productivity of agroecosystems. The third section presents seven chapters that influence agronomic practices and factors such as soil microbes, organic matter, crop genetic nature, irrigation, weeds, and cropping systems that affect productivity of agroecosystems.

Over all, this book on agroecosystems provides detailed information on soils, crops, general principles that govern nutrient dynamics, and factors that affect productivity. It is useful to professors, researchers ,and students dealing with agriculture, ecology, and geography. It serves as a general reference book for all those interested in understanding our own nature and surroundings.

September 2013 **Kowligi Krishna**
Bangalore

ACKNOWLEDGMENTS

During preparation of this book, quite a number of specialists from agricultural research institutions and farming enterprises have offered research reports, publications, and photographs relevant to agronomic procedures adopted in different continents of the world. A few of them permitted me to use specific photographs in the chapters. I wish to acknowledge the following scientists/managers and agricultural research organizations for their contributions:

Mr. James Foxglove, Plow Creek Farms, Tiskalwa, Illinois, IL 61368, USA; Mr. David Nelson, Nelson Farms, Fort Dodge, Iowa, IA 50501, USA; William D. Emery, Erwin Orchards Inc., Silver Lake Road, South Lyon, Michigan, MI 48178, USA; Mr. B. Burgweger, B2 Farms LLC, Brooklyn, Wisconsin, WI 53521, USA; Dr. Lindsay Kennedy, Director-External Affairs, United Sorghum Checkoff Program, United States Department of Agriculture, Lubbock, Texas, 79403, USA; Dr. Kraig Roozeboom, Professor of Extension, Kansas State Extension Services, Kansas State University, Manhattan, USA, USA; Dr. Albert Quirogo and Dr. Alfred Bono, INTA, Anguil, Argentina; Carina Alvarez, Professor, Facultad de Agronomia, University of Buenos Aires, Av San Martin, Argentina.

International Crops Research Institute for the Semi-Arid Tropics, Patancheru 502324, Andhra Pradesh, India; Centro Internacional Tropicale Agricultura-CIAT, Cali, Columbia, EMBRAPA, Sao Paulo, Brazil; EMBRAPA, Campe Verde, Matto Grasso, Brazil; Citrus Research and Educational Centre, Lake Alfred, Florida, USA; Director, Austin Carey Forest Plantations, University of Florida, Gainesville, Florida, USA. International Centre for Agriculture in Dry Areas-ICARDA, Aleppo, Syria; Dr. Roger Atkins, Information Services, Rothamsted Experimental Station, Harpenden, Hertfordshire, England; Faculty and Administration at University of Agricultural Sciences, Bangalore 560065, Karnataka, India.

I wish to express my appreciation to several of my good friends and teachers of my high school, Gandhinagar High School, Kumara Park at Bangalore, India, whose reminiscence kept me encouraged and enthused in compiling this book rapidly for the publishers.

I wish to thank my wife Dr. Uma Krishna, and son, Sharath Kowligi.

PART I

AGROECOSYSTEMS OF THE WORLD

CHAPTER 1

AGROECOSYSTEMS: AN INTRODUCTION

CONTENTS

1.1 WHAT ARE AGROECOSYSTEMS?

Agroecosystems are unique cropping expanses that are well adapted to a particular geographic region. They exist as ecological entities and exhibit various ecosystematic functions and services. There are several ways of defining agroecosystems based on context, their composition and purpose. Definitions that highlight the geographic area, weather pattern, natural resources, cropping pattern and products are most common. One such definition states that an agroecosystem could be explained as an agricultural and natural resource system managed by human beings for the primary purpose of his food needs plus other useful products [1]. In certain situations, agroecosystem or cropping expanse could be located in different regions/continents and may be small or relatively larger in area. They often have common features that could be studied and analyzed collectively as an agroecosystem [2–5]. Some examples of typical agro-ecosystems are Wet land Rice Agroecosystem of South-east Asia, Temperate Wheat Cropping Zones of European Central plains, Dry lands of West Asia, Vertisol plains of South India that support Sorghum and Cotton expanses, Corn belt of North American Plains, Citrus Plantations of Florida and Sugar cane region of Western Gangetic Plains.

1.2 WHAT ARE THE INGREDIENTS OF AGROECOSYSTEMS

1.2.1 CROPS

It seems over 7,000 crop species are edible and congenial for use by humans. Since ages, they have been domesticated and cultivated in different regions of the world in small or large expanses. However, only 300 species are used consistently by us. Further, it is clear that during recent decades, humans have immensely depended on very few crop species. We have accentuated mere 30 domesticated crop species. They flourish as agroecosystems, in the plains, hills and coasts of different continents. Currently, these few crop species feed the entire human populace totaling 7 billion plus large number of domestic animals [1, 2]. It appears surprising yet it is true. Human preferences have been highly focused and narrow. Natural factors, diet preferences and economic advantages have literally restricted the crop species selected and cultivated by farmers. More interesting to note is the fact that carbohydrate requirement of large portion of human population and domesticated animals is answered by mere four or five cereals and few forage crops. Major cereals like Rice, Wheat and Maize together contribute carbohydrates to large fraction, over 90% human populace residing in Americas, Europe, Africa and parts of Asia. The Rice belt of South and South-east Asia supplies carbohydrates to over 2 billion humans. Following is a list of major cereals, legumes, sugar and fiber crops that are grown consistently in different regions of the world. They form cropping zones or agroecosystems of small or large size, in different regions and play vital role in sustaining humans and animals.

Cereals: Wheat, Rice, Maize, Sorghum, Barley, Pearl millet, Finger millet, Foxtail Millet, Prosomillet.

Legumes: Soybean, Chickpea, Lentils, Cowpea, Cajanus, Field bean, Peas, Green gram, Black gram, Lathyrus.

Oilseed Crops: Brassicas, Soybean, Sunflower, Canola, Groundnut, Sesamum, Corn, Olives.

Sugar crops: Sugarcane, Beet, Sweet sorghum.

Fiber Crops: Cotton.

Plantations: Citrus, Grapes, Mango, Banana, Apple, Palm, Coffee, Tea.

In a nutshell, energy, protein, fiber and even fuel requirements of human populace on earth is served by agroecosystems developed using very few crop species—about 4 or 5 cereals, some 5–8 legumes, oil seeds and a couple of sugar and fiber crops, plus few fruit crops. Perhaps, this situation could be modified to improve and diversify the crop species that fill the agrarian regions. The primary question is should we cultivate more crop species. Is it plausible? If so what are the advantages of diversification of cropping zones.

1.2.2 TERRAIN AND SOILS

Agroecosystems thrive on wide range of topographical conditions and soils that occur on the globe. Geographically, large plains, which hitherto supported wild prairies, have served the agricultural causes most ably. However, agroecosystems are spread all over the globe. Plateaus, undulated terrain, hills, high altitude mountain regions, valleys, low lands, swamps and flooded regions too support variety of cropping belts. Basically, crops adapt to wide range of topography, terrain and soil conditions. This aspect has been used shrewdly to develop, sustain and harvest products from agroecosystems. Globally, soils derived from wide range of parent material and fertility status have supported agricultural cropping. Soils that are weathered, rich in mineral nutrients and organic fraction have offered better crop productivity. Such fertile regions allow the development of highly intense cropping zones (e.g., Chernozems of Europe or Mollisols of Northern Great Plains). Large areas of moderately fertile Mollisols of Pampas in Argentina, Oxisols in the Cerrados of Brazil, Cambisols and Loess soils of European Plains, Calcareous Oxisols of Middle East Asia, Sandy Alfisols of West African Sahelian region, Inceptisols of Gangetic Plains, Alfisols and Vertisols of South India, Sandy soils and Laterites of South and South-east Asian coasts, Inceptisols of North China plains, Duplex soils, Red and Brown Earths of Australia, have all supported moderately productive agroecosystems. To a certain extent, soil productivity has governed the agricultural prowess of a particular region. Shallow soils, low in nutrient availability index and organic matter have allowed only agroecosystems of low productivity. The basic fact is that soil characteristics such as texture, structure, profile layers and depth, pH, redox potential, electrical conductivity, aeration, soil moisture holding capacity, organic matter content and mineral nutrient availability rather fertility *per se* have all influenced formation, productivity and perpetuation of agroecosystem in a given geographic location.

1.2.3 WATER

Water is a primary ingredient of any agroecosystem just like soils, nutrients and crop genetic stock. Atmospheric precipitation, ground water, rivers, lakes, irrigation dams, small reservoirs and wells are chief sources of water in any agroeosystem. Water resources have direct impact on composition of crop species, intensity of cropping and productivity of an agroeosystem. For example, creation of a dam across a river that augments irrigation can immediately induce farmers to change the cropping pattern and yield goals. Water influences many of the basic physiological manifestations of crops that fill the agroecosystem. Crop growth and yield formation are highly dependent on moisture status of soils. It seems clear that need for highly productive cropping belts has made water a more important and critical component [1]. On a wider scale, water resources have direct relevance to general vegetation in any part of the world. Aspects such as crop species, their mixtures, intensity of cropping, fluctuations in expanse and productivity of agroecosystems are highly correlated to soil moisture and precipitation patterns. Scanty precipitation coupled with lack of stored water resources has often induced formation of arid belts or dry land cropping zones. For example, Dry lands of Peru, Arid zone cropping belts of West Asia, North-west India, North-west China, Sudano-Sahelian zones in North and West Africa support only hardy crops able to adapt to low moisture conditions ranging from 250–450 mm annually. Water distribution pattern often has sharp and precise influence on crop species grown by farmers. For example, in West Asia, precipitation level of 350 mm allows wheat production, but a slight reduction by 50–80 mm a few miles away, restricts farmers to select barley since it thrives better in arid conditions. Semi-arid ecosystem created through low to moderate precipitation pattern (700–1100 mm annually) has allowed formation of arable cropping zones. Again, water reserves and precipitation pattern imparts sharp changes in the crop species that dominates the ecosystem. At 350–400 mm pearl millet dominates the Sahelian zone, but higher annual precipitation at 550–650 mm induces famers located just a few kilometers away to select sorghum and cowpea. As we traverse further to subtropics/tropics in Africa, farmers quickly change over to maize, rice, banana and cowpea-a crop mixture more productive than others. In South and South-east Asia, adequate water storage has allowed development of wetlands interspersed within large arable cropping zones. For example, rice and sugar cane cropping zones in South and South-east Asia occur within large tracts of arable cropping zones. Here again, fluctuations in water availability, creates marked changes in the cropping pattern and productivity of agroecosystems. Literally, farmers in any region have to match crop species and yield goals with several climatic parameters and this causes modifications to agroecosystems.

1.2.4 CLIMATE

Crops are highly versatile and adapt to a wide range of climatic conditions that prevail on the earth's surface. Globally, agricultural cropping occurs at varying intensities between 50° N to S. Of course, deep deserts, Arctic, Antarctic and high mountain regions with permafrost do not aid cropping. The region defined as 'Agrosphere' or area where agroecosystems flourish currently extends into a 3rd of earth's land area (13,500 m ha) [1, 2]. Crops negotiate fluctuations of climatic parameters that are common to various regions. For example, wheat, barley, oats adapt well to colder climatic conditions and are predominant in temperate agroecoregions, where in freezing temperatures, frost and squally climate occurs. Rice flourishes in tropical and subtropical wetland zones with irrigation support and high relative humidity. Hardy cereals like pearl millet, sorghum and finger millet thrive well in semiarid regions that are generally prone to drought, erratic precipitation patterns coupled often with hot temperature (30°–42°C). Legumes such as soybean and lentils adapt well to colder regions, while cowpea, pigeon peas flourish in tropics and subtropical climatic regions. Oil seed crops such as sunflower, groundnut, and brassicas relish warmer zones that experience moderate rainfall. Essentially, factors like crop duration, physiological requirement for photosynthetic radiation, temperature (degree days), moisture and precipitation pattern, soil depth and other relevant parameters are important. Despite most congenial conditions, extremes of a single or few climatic parameters can wipe a cropping ecosystem rather rapidly. Let us quote few examples. A season of drought can erase a large patch of pearl mille/cowpea intercrop in Sahel. Frost and cold bite literally restricts cereal crop yield in North America and Europe. Spread of cold front into North Florida has restricted citrus production to southern spodosol region of Florida. Droughts avoid sorghum and pigeon pea cultivation, therefore farmers replace it with cereal fodder and horse gram in South India. Farmers in West Asia switch over from wheat to barley if precipitation is relatively scanty.

1.2.5 NUTRIENTS

Nutrients occur in different chemical forms in soil, water, crop and atmosphere that are constituents of any agroecosystem. The principles of nutrient transformations and dynamics within agroecosystems are actually the centerpiece of this book. Major nutrients like C, N, P, K; secondary nutrients like Ca, Mg, S, and micronutrients like Zn, Fe, Mn, Mo, B, Co, Cl and their dynamics are vital aspects that partly regulate formation, perpetuation and productivity of each agroecosystem. Since crops acquire most of their nutrients from soil phase of the ecosystem, detailed knowledge about soil chemical conditions and nutrient availability are essential. Soil moisture plays perhaps the most vital role in regulating nutrient dynamics between soil and crop. We should note that all nutrients are absorbed in dissolved state (solute) and as such absorption rate of soil moisture has direct impact on nutrient acquisition by crop roots and its transport to above ground portion of agroecosystem. The nutrient input rates, accumulation pattern, loss via soil erosion, surface runoff, percolation and as harvested grains/forage influences nutrient dynamics in an agroecosystem rather markedly. Some of the above aspects have direct bearing on ground water quality.

The ingredients listed and discussed above are immediately pertinent to the development, perpetuation and productivity of agroecosystems. They are actually focused towards soil fertility and irrigation aspects of cropping systems. However, there are several other aspects or ingredients which agroecosystems encompass rather conspicuously. They are adjoining flora and fauna. For example, ever competing weeds, volunteers and indigenous plants species, antagonistic (pests) and useful (bees) insect species, birds, animals and soil microbes are important ingredients.

1.3 AGROECOSYSTEMS OF THE WORLD: EXAMPLES

The development of agroecosystems on earth is mediated through a complex interaction of several factors that operate at variable intensity and duration. Human ingenuity and his preferences are among the most important factors that influence development, expansion, perpetuation/extinction of agroecosystems. Several other factors such as natural resources—suitable topography, fertile soil, climatic factors, precipitation and soil moisture regimes, crop species—its adaptability, productivity and economic advantages also influence dynamics of agroecosystems around the world. Access to infrastructure, inputs and governmental legislation too affect agroecosystems. Through the ages, agroecosystems have fluctuated in terms of number, size, geographic location, cropping pattern and productivity. We can identify innumerable small, medium or large sized cropping expanses or agroecosystems. We should also realize that once developed and entrenched into a region, agroecosystems have their own impact on humans, surrounding flora and fauna. Following is a list of conspicuous agroecosystems that exist across different continents.

North America: The Wheat Agroecosystem of Great Plains in United States of America and Canada; Corn Belt of Northern Great Plains; Soybean Cropping zones of Midwestern USA; Cotton Belt of Central Great Plains; Peanuts in Southern USA; Citrus Plantations of Florida; Forestry regions of North-east USA, and Canada.

Central and South America: Maize/Legume/Vegetable cropping mixtures of Meso-America, Maize-Soybean Cropping belts of Cerrados and Matto Grasso of Brazil, Wheat belt of Pampas of Argentina, Maize cropping belt of Pampas of Argentina, Groundnut belt of South America.

Europe: Wheat belts of European Plains, Grape vines of southern France.

Africa: Cereal Agroecosystem of Sahelian West Africa, Pearl millet and sorghum expanses of North Africa, Groundnut belt of Senegal and adjoining countries of West Africa. Maize and legume crop mixtures of Central and South Africa. Sorghum in Southern and East Africa.

Asia: The Rice-Wheat Agroecosystem of Indo-Gangetic region, Pulse expanses of Central and South India, Sunflower belt of Vertisol regions of India, Groundnut belt of Gujarat and South India, Rice mono-cropping zones of Southern India Plains and Coasts, Spice crops of Western Ghats, Coffee growing region of Western Ghats, Tea belts of Eastern Indian hilly regions, Finger millet in Dry lands of South India, Sorghum belt of South Indian Vertisol region, Maize cropping zones of India. Rice Agroecosystem of South-east Asia, Legume intercropping zones of South-east Asia, Wheat and Maize belts of North-east China, Groundnut belts of Southern and Eastern China.

Australia: Wheat belts of Australia.

1.4 SALIENT FEATURES OF AGROECOSYSTEMS

Many of the following highlights and salient features of agroecosystems or crop production zones are drawn from Wood et al. [1]. Some of these salient features will allow us to appreciate the geographical relevance and importance of agricultural zones to humans.

- Satellite mediated surveys and ground mappings have shown that globally 28–30% of land mass is occupied by agricultural crops and pastures/forage grown to serve as food, fiber or fuel and to feed animals.
- Agricultural area cropped annually spreads into all continents and occupies 1.3 billion ha. It encompasses several large and small agroecosystems and patches of cropping zones. The agricultural cropping zones it seems has consistently increased by 2% each year and pastures by 0.3% over previous levels.
- Annual crops such as major cereals (Maize, Wheat, and Rice), legumes (Soybean, Cowpeas, Lentils etc.) and oil seed (Sunflower, Groundnut, Canola) occupy nearly 91% of cropped land. Perennial crops and plantations occupy the rest 9% of cropped zones of the world. Obviously, most agroecosystems are based on annual crops rather than perennial crops. The land surface indeed supports different crops each season based on rotations followed.
- Agricultural expanses that occur in the dry land or semiarid belts are mostly supported by natural precipitation patterns or via irrigation. There are of course large tracts that occur in torrential and high rainfall zones, in lowland flooded zones and regularly irrigated regions. Currently, about 300–320 million ha of agricultural expanses are under irrigation. They may receive protective irrigation or may entirely grow as irrigated crops. Globally, about 20% of cropped land is irrigated.
- Agroecosystems are distributed all across different agroclimatic regions. Satellite surveys suggest that 38% of global agroecosystems are found in the temperate regions, 38% in tropical zones and 28% in subtropical regions. A different assessment suggests that a fifth of crops thrive in semiarid and arid regions of the world. About one third of crops are grown in subhumid regions and a fifth in humid regions.
- The agroecosystems discussed in this book contribute about 94% of protein and 98% of carbohydrate requirements of humankind on earth. It may be valued at a few trillion dollars annually. Bizarre, yet true about 25% of global human population is preoccupied all through the year with agricultural procedures. Agroecosystems demand regular attention and human labor for upkeep and productivity. It seems about over 1.6 billion humans toil *directly* in fields or barnyards to produce food. A large fraction of human population depends on agriculture indirectly in agro-based industries and related activities. There are countries with only 7–10% population engaged in crop production, but there are others where 65–90% of population is entirely engaged in raising crops.
- Agroecosystems that occur in different parts of the world may encounter variety of different maladies related to soil, agroclimate, crops, disease pressure and natural calamities
- Agroecosystems thrive at different intensities of cropping, nutrient input schedules and productivity levels. Yet, the general trend is that many of the low-input cropping zones are being intensified at a regular pace. During recent years, expansion of agroecosystems has received reduced attention. However, intensification seems to have gained in momentum. The consequences of high input,

enhanced nutrient turnovers and productivity need to be assessed, perhaps simulated ahead and revised and re-revised to suit the farmer's economic goals and environmental standards.
- During recent years, greater effort has been bestowed to sequester relatively more C and adopt measures that lessen C loss from agroecosystems. It is estimated that agroecosystems garner and store about 18–24% of C in the above and below-ground portions.

KEYWORDS

- **cereals**
- **legumes**
- **oilseed**
- **pests**
- **plantations**
- **sugar crops**

REFERENCES

1. Wood, S.; Sebastian, K.; Scherr, S. J. Agroecosystems. A Pilot Analysis of Global Ecosystems. International Food Policy Research Institute: Washington D.C., USA, **2000,** pp. 185.
2. Krishna, K. R. Agrosphere: Nutrient Dynamics, Ecology and Productivity. Science Publishers Inc.: Enfield, New Hampshire, USA, **2003,** pp. 345.
3. Krishna, K. R. Peanut Agroecosystem: Nutrient Dynamics and Productivity. Alpha Science International Inc.: Oxford, England, **2008**, pp. 293.
4. Krishna, K. R. Agroecosystems of South India: Nutrient Dynamics, Ecology and Productivity. Brwonfialker Press Inc.: Boca Raton, Florida, USA, **2010**, pp. 565.
5. Krishna, K. R. Maize Agroecosystem: Nutrient Dynamics and Productivity. Apple Academic Press Inc.: New Jersey, USA, **2013**, pp. 345.

EXERCISE

1. Define Agroecosystem and mention different ingredients of Agroecosystems.
2. Mention at least 10 Agroecosystems encompassing field crops and plantations that spread across different continents.
3. Collect data about Expanse, Total Production and Productivity of at least 10 Agroecosystems.
4. Classify Agroecosystems based on Cropping intensity, Agroclimate and Geography.
5. Mention different factors that affect spread and productivity of Agroecosystems.
6. Mention at least 10 salient features of Agroecosystems.

FURTHER READING

1. Boehlen, P. J.; House, G. Sustainable Agroecosystems Management: Integrating Ecology, Economics and Society. CRC Press: Boca Raton, Florida, USA, **2009,** pp. 328

2. FAO. 2012, "World Agriculture: Towards 2015/2030, an FAO study." Food and Agricultural Organization of the United Nations, Rome, Italy.
3. Gliessman, S. E.; Rose Meyer, M. The conversion to Sustainable Agriculture: Principles, Processes and Practices. CRC Press, Boca Raton, Florida, USA, **2009,** pp. 380.
4. Hart, R. D. Agroecosystem Determinants. Winrock International, Morrilton, Arkansas, USA. **2012,** 1–25, pdf.usaid.goc/pdf.docs/ PNAAP019.pdf (July 11, 2012).
5. Krishna K. R. Agrosphere: Nutrient Dynamics, Ecology and Productivity. Science Publishers Inc.: Enfield, New Hampshire, USA, **2003,** pp. 343.
6. Lawrence, R.; Stinner, B. R.; House, G. J. Agricultural Ecosystems: Unifying Concepts. John Wiley and Sons. New York, **1984,** pp. 345.
7. Pearson, C. J. Field Crop Ecosystem (Ecosystems of the World). Elsevier Science Publishers Inc.: Netherlands, **1992,** pp. 576.
8. Wood, S.; Sebastian, K.; Scherr, S. J. Agroecosystems. A Pilot Analysis of Global Ecosystems. International Food Policy Research Institute: Washington D.C., USA, **2000,** pp. 185.

USEFUL WEBSITES

http://www.wri.org/wr2000 (July 14, 2012).
http://www.esa.org (August 14, 2012).
http://www.agronomy.org (July 10, 2012).
http://www.nrel.colostate.edu (October 3, 2012).
http://www.caryinstitute.org (October 10, 2012).

CHAPTER 2

SOILS OF AGROECOSYSTEMS

CONTENTS

Soil is a major component of all agroecosystems encountered across different continents of the earth. Soil forms the belowground portion and provides anchorage to crop. Soil acts as repository of moisture, mineral nutrients, organic matter, gaseous material and most importantly the soil microbes that mediate large number of nutrient transformations. Soil phase of any agroecosystem is a variegated system. Soils vary markedly for physicochemical properties, fertility and moisture held within the profile. Several types of soil that differ with regard to parent material, texture, structure, bulk density, pH, mineral nutrient distribution, organic matter and moisture-holding capacity are congenial to support cropping expanses. Soil fertility *per se* and ability to support cultivation of crops and productivity levels attained also vary in an agrarian region. Farmers may encounter different kinds of constraints to crop establishment, growth and yield formation. The number and types of soil maladies felt may also vary. Over all, soil phase almost decides the crop species, cropping pattern and productivity in a given agrarian belt.

2.1 SOILS OF AMERICAS

Agroecosystems in Americas thrive mostly on fertile soils classified as Udolls and Udalfs. Soil types, such as Mollisols, Aqualls, Ustolls, Fluvents, Aqualfs and Haplaquepts also support large expanses of crops (Table 1; Plate 1). Soils found in Great Plains are endowed with optimum level of organic fraction. Despite it, large amounts of crop residues are recycled. We should note that intensive cropping necessitates removal of massive quantities of nutrients into aboveground portion of ecosystem, mainly into grains/forage. Incessant cropping depletes soil fertility and diminishes SOM, therefore, irrespective of soil type; much of the cropping region is provided with inorganic and organic fertilizer replenishments. In New England zone, crops thrive on Udults. Soil productivity is held high through the use of fertilizers and farmyard manure. Crop yields are held high. For example, maize grown exclusively for forage may produce 25–30 t fresh forage/ha. In Southern Plains, crops thrive on Uderts and Ustolls. In the South-east of USA, (Georgia, Florida) crops are grown on Psamments, Quartzipsammets, Ultisols and Aquults. Piedmonts of Georgia, Carolinas and Virginia also support large-scale farming.

South American agroecosystems thrive on Mollisols (Ustolls), Aridisols (Argids,), Ultisols (Udults), Alfisols (Ustalfs) and Entisols (Psamments) [1]. Soils in Argentina comprise of eight different orders of soil taxonomy, but Mollisols are most common in Rolling Pampas [2]. Entisols and Aridisols are other soil types used for crop production. The *Gran Chaco* region of South America supports crop production on Haplustolls and Hapludolls. Soil erosion and runoff could be rampant in the agroecosystem. As a consequence, soil organic carbon gets reduced in the topsoil. Annually, 4–5% soil-N and 20% SOC could be lost from the soil profile due to erosion. Hapludolls are light textured, loamy and fertile. Soil fertility status is deemed as moderately good. Therefore, extraneous supply of nutrients is limited. Fertilizer inputs to crops cultivated after soybean, such as maize or sorghum is relatively low. Nitrogen supplements to cereals grown after peanuts or other legumes are usually small. Soils in the Cordoba region are sandy loams with organic matter content ranging from 1–3%. They are low in available-P, but high in K. Major soil-related constraints to nutrient recovery are soil erosion, loss of organic matter and soil crusting [3–5]. Incessant cultivation of soil without proper nutrient replenishment has caused severe imbalances in availability of nutrients to plant roots.

Soils of Western Chaco region are loamy but slightly acidic or neutral in upper horizons. Soils are suitable for cereal cropping as well as intercropping it with leguminous trees like *Prosopis*. Soils are generally fertile, but water is limiting.

PLATE 1 Surface views of Mollisols found in Kansas (above) and Iowa (below) region of the Great Plains that supports production of Maize, Wheat and Sorghum, in addition to legumes such as Soybean. Soils are assessed rapidly for pH and fertility using Precision Instruments. *Source:* Veris Technology, Kansas, USA; Nelson Farms, Fort Dodge, Iowa, USA).

The Cerrados of Brazil is a vast expanse of savanna vegetation and agricultural enterprise. It has predominantly a drier sandy or loamy stretch of soil. Highly weathered Oxisols are predominant covering up to 47% of Cerrado region (*see* Table 1). Other major soil types encountered are Ultisols (15%) and Entisols (15%). These soil types exhibit serious limitations with regard to agricultural cropping and are classified as low in fertility and mineral contents. The Oxisols are acidic and low in available N,

P and K. Secondary nutrients such as Ca, Mg and S are also deficient. Micronutrients like Zn, B, Cu and Mo are found in concentrations less than threshold for maize/soybean production [6, 7]. Major constraint to crop production on these soil types is the high Al content. It leads to P-fixation and reduction in fertilizer efficiency. High Al content actually affects rooting and nutrient acquisition by crop roots. Since most soils are highly acidic with pH 4.5 to 5.5, correction using lime is almost mandatory in any region of Cerrado. Liming essentially increases soil pH and at the same time corrects Al toxicity. It is useful to apply Dolomitic lime because it contains both Ca and Mg. Firstly, pH gets corrected plus it adds Ca and Mg that are usually deficient in soils of Cerrados. Actually, Al/(Ca + Mg) ratios in soils are carefully assessed before planting. On virgin soils, a large basal doze of lime is applied to correct pH and Al toxicity. One way of improving soil P status is to apply fertilizer-P in quantities slightly more than that required for crop growth and yield formation. It helps in satisfying P fixation and then allows some to accumulate in soil, so that P deficiency does not get expressed repeatedly. Similar principles could be adopted in building N and K reserves in the cropping belt.

TABLE 1 Descriptions of Major Soil types encountered in various Cropping zones of the World.

Corn Belt of USA: Soil Types – Ustolls and Udolls, Kastanozems (Kastanozems – WRB; Ustolls-US; Kastanozem-FAO):

Kastanozems are found in steppes or prairies. They have thick dark brown top soil. Kastanozems are chemically rich with saturated bases. Soil reaction (pH) may be slightly alkaline. Soil aggregates in a Kastanozem is susceptible to erosion through wind and water. Kastonezems are commonly found in the North American plains and in Pampas of Argentina [8]. Potentially, Kastenozems are fertile soils. Yet, crop production may get constrained due to paucity of water. Irrigation needs to be regulated since it could induce salinization. Adequate tillage and repeated fallows are common methods to avoid deterioration of soil fertility. Nutrient turnover rates are naturally high. Crop productivity is high. Chernozems are also encountered in Northern plains of USA. These soils are rich in organic matter, nutrients and biotic component. Several of the soil physicochemical properties are congenial for crop production. Soil aggregates are stable and relatively resistant to wind and water erosion. They possess fertile surface horizon, thus support crop production excellently.

Cerrados of Brazil: Soil Types – Luvisols, Alisols, Oxisols, Laterites:

Cerrados of Brazil is a large stretch of semiarid plains in the eastern half of the country. It supports large expanses of dry land cereals, legumes and pastures. Maize and soybean are important crops grown mostly on Alisols, Ultisols and Oxisols.

Udalfs (Luvisols-WRB; Udalfs-US; Luvisol-FAO):

Luvisols are characterized by surface horizons depleted of clay fraction and nutrients. They exhibit subsurface accumulation of high clay activity. Most traits of Luvisols favor crop production particularly if it is well drained. Luvisols contain minerals and organic fraction sufficient to support cropping. However, Drystic Luviosls may need amendments with lime to improve fertility. They are sometimes associated with Argids, especially in dry land belts. When situated closer to water sources, these soils are associated with gleyic or stagnic luvisols.

Alisols (Alisols-WRB; Ultisols-US; Alisols-FAO):

Alisols have brown colored surface horizon. Alisols possess high clay activity and Al exchange capacity. Alisols are less weathered compared to Lixisols or Nitisols. Usually, they are well drained but deficient in plant nutrients. Al toxicity in surface horizons is very common. Soil erosion is rampant. Crop productivity is generally low on Alisols. Nutrient buffering capacity is low; thus fertilizer and lime supply are necessary.

Ferralsols or Oxisols (Ferralsols-WRB; Oxisols-US; Ferralsol-FAO):

Oxisols are highly weathered mineral soils rich in sesquioxides and well suited for cropping. They have a ferralic horizon extending from 30–200 cm depth. Such ferralic horizon results from intensive weathering (ferralitization). The clay fraction is dominated by low activity clays. The silt and sand are constituted by highly resistant goethite, gibbsite and hematite. The physical aspects are quite congenial for cropping. Yet, they have low nutrient reserves and high exchangeable Al and Mn. Major constraints to cropping are acidity and Al toxicity. These factors actually result in improper rooting and low P uptake. Chemical fixation of P gets accentuated due to high Al^{++} and Mn^{++} activity. Liming and higher dosages of P fertilizers are usually recommended. Split application of nutrients improves crop growth and yield.

Pampas of Argentina: Soil Types – Mollisols (Ustolls), Ultisols, Entisols, Aridisols (Argids):

Mollisols: For details about soil characteristics see Ustolls under Corn Belt of USA.

Ultisols: For details about soil characteristics see Alisols under Brazilian Cerrados.

European Plains: Soil Types – Udalfs (Luvisols); Cambisols; Fluvisols; Chernozems:

Cereal production in the European plains occurs on soil types classified as Cambisols, Luviols and Fluvisols in Western Europe; on Podzols and Black Earths in Central Europe and Russia; and on Calcixeralfs in Mediterranean region of Europe.

Chernozems: (Chernozems-WRB; Udolls-US; Chernozem-FAO):

Chernozems are most commonly encountered in cropping zones of Europe including Russia. Soil physicochemical properties like bulk density, porosity, water holding capacity, cation exchange capacity (CEC), high SOM and neutral pH are congenial. Chernozems are blackish soils rich in organic fraction and prominent calcareous subsurface. Chernozems possess deep, very dark gray, humus and nutrient rich surface horizon. They are supposedly few of the best soil types for agricultural cropping [8, 9]. The biotic fraction of soil is well stabilized and this results in surface horizon as thick as 2 m. Therefore, it supports good rooting of most crops, leading to high amounts of nutrient absorption. The soil aggregates are relatively more stable than those encountered in Kastanozems. Soil aggregates easily withstand wind and water erosion.

Udalfs (Luvisols-WRB; Udalfs-US; Luvisol-FAO):

Descriptions for Luvisols or Udalfs are available under Cerrados of Brazil.

Podzols (Podzols – WRB; Spodosols-US; Podzol-FAO):

Podzols are quite frequently used for cultivation of cereal crops in Eastern European and Russian plains. They are acid soils with blackish-brownish-reddish subsoil with Illuvial Fe-Al-organic compounds. The name podzol means soil is with subsurface that has appearance of ash due to strong acid aided bleaching. It has a spodic Illuvial horizon. Podzols could be sandy or loamy. Coarse textured soils obviously have low water retention capacity. The CEC is moderate or low depending on SOM content. The C: N ratio ranges from 20 to 25. Podzols of Eastern European plains could be infertile. They support moderate crop yield, despite fertilizer application. This is attributed to sandy texture, low water retention and low soil fertility status [8].

Solonetz (Solonetz-WRB; Several soil orders-US Solonetz-FAO):

Agroecosystems in Ukraine, Russia and parts of Eastern Europe thrive on Solonetz or saline humic soils. Solonetz possess subsurface horizon with clays rich in sodium content. Actually, inherent salt contents, nature of parent material and ground water together induce formation of solonetz or saline/alkaline soils. High Na content and pH are major constraints to crop production. Usually, salt tolerant cultivars are preferred.

West African Savannas and Guinean Zones: Soil Types – Sandy Oxisols, Ultisols and Alfisols

For descriptions on Oxisols see under Brazilian Cerrados.

West Asia and North Africa: Soil types – Aridisols, Cambisols

The cropping regions of West Asia are endowed with Entisols, Mollisols, Vertisols, Inceptisols and Aridisols. North Africa is dominated by Inceptisols, Lithosols or Shallow soils, Entisols and Aridisols.

Aridosol (Arenosols-WRB; Psamments-US; Arenosol-FAO):

Aridosols are common in the dry land cropping belts of West Asia. Aridosols are rich in salts. Such salt accumulation can impair nutrient recovery leading to depressed growth. Major plant nutrients are very deficient. Boron toxicity is quite frequently reported in West Asia. Crops need both fertilizer supplements and irrigation in order to reach moderate level of productivity (1.5–2.5 t grain/ha).

Cambisols: (Cambisols-WRB; Inceptisol-US; Cambisol-FAO):

Cambisols possess horizons that indicate moderate weathering. They are difficult to characterize because of large variations in physical and chemical traits. They could be sandy, silty or clayey depending on parent material. In general, Cambisols are good for cropping, particularly when they are rich in bases. The acid and coarse textured Cambisols could be improved by fertilizer and organic manure addition.

Southern Indian Plains: Vertisols and Alfisols, other soil types are Inceptisol and Entisol
Vertisol (Vertisol-WRB; Vertisol-US; Vertisol-FAO):

Agroecosystems of South India occupy large tracts of Vertisols, especially in North Karnataka and Andhra Pradesh. They are commonly called Black cotton soil or *Regur*. Vertisols are deep, black colored, clayey (30%) soils. Clay minerals like montomorillonite, smectite and illite predominate. These Vertisol stretches in India are fertile and productivity is relatively high. They show remarkable swell and shrink characteristics. Vertisols show tendency to crack during drought or paucity of moisture. Nutrient and moisture buffering capacity is relatively high. Soil becomes very sticky during rainy season. Therefore, agronomic procedures like tillage, planting and interculture could become difficult or impossible during early kharif season. Post rainy crops are possible because of better moisture storage capacity.

Alfisol (Acrisol-WRB; Alfisol US; Alfisol-FAO):

South Indian agroecosystems thrive on Alfisols that are commonly called Red soils. Alfisols are deep, either loamy or gravely at times with relatively higher Fe content. Soils are rich in mica, quartz, feldspar and hematite. Moisture holding capacity is moderate or low. Tillage induces hard pan and loss of SOC. As such, SOM is low owing to high soil temperature, oxidized state and semiarid environment. Loss of surface soil and nutrients through erosion, seepage and percolation could be high. Moisture infiltration rates are high due to sandy/gravely nature of Alfisols.

South-east Asia and China: Soil types – Alfisols, Entisols, Inceptisols and Ultisols:

Alfisol (Acrisol-WRB; Alfisol US; Alfisol-FAO):

Descriptions are provided under South Indian plains.

Alisols (Alisols-WRB; Ultisols-US; Alisols –FAO):

See under Brazilian Cerrados

Cambisols: (Cambisols-WRB; Inceptisol-US; Cambisol-FAO):

See under West Asia and North Africa

Soils of Australia: Soil Types – Red and Yellow Ferralsols

Agricultural soils of Australia are classifiable as Quartzipsamments, Calciorthids, Palexeralfs, Haplaustalfs, Paleusterts, Chromousterts, Torrerts and Duplex soils (RhodoXeralfs, Natustalfs, and Paleustalfs). Black Earths of Queensland, Red and Yellow Earths of Western Australia and New South Wales are also used for crop production.

Ferralsols (Ferralsols-WRB; Oxisols-US, Ferralsols-FAO)

Ferralsols possess a ferralic horizon at some depth-say 20–30 cm from surface. The ferralic horizon is derived from long and intensive weathering commonly termed as *Ferralitization.* High ambient temperatures and rainfall induces formation of ferralic horizon. The Ferralsols have a weakly expressed soil structure. The clay fraction is constituted by low active clays. The Ferralsols are amenable for maize production. Ferralsols are less prone to erosion. Chemically, Ferralsols have variable CEC that is dependent on pH. Nutrient supply capacity decreases with cultivation. Liming becomes essential under acidic conditions. The Red Ferralsols of Queens land has been incessantly tilled and cropped. Tillage has induced loss of soil structure and fertility. Above it, organic matter recycling tends to be low causing loss of SOC. Periodic droughts have accentuated loss of soil fertility and quality.

Source: Refs. [1, 6, 8, 10–13].

Note: WRB = World Reference Base; US = United States Soil Taxonomy; FAO = Food and Agricultural Organization's soil taxonomy. Soils types described above are relevant to specific agroecosystems. They do not apply to entire agrarian regions of a continent.

2.2 SOILS OF EUROPEAN PLAINS

Soils found in European plains support a wide array of crops and cropping belts. Major soil types encountered are Cambisols in Britain and France, Luvisols in Germany, and Fluvisols in Netherlands (*see* Table 1). These soils are arable and suited for agricultural crop production. In the Central Europe, major soil types encountered are Loess soils, Podzols, Light Brown Steppe soils, Black Earths and Sandy soils. Chernozems are predominant in Russian and Eastern European plains. In Russia, nearly 48% of soil is classifiable as Chernozems, 14% as Podzols, 10% as Solonetz and rest as Grey soils. Chernozems are frequent in Ukraine, Bylorussia, Lithuania and Estonia. In the Czech Republic and adjoining regions Chernozems, Fluvisols and Luvisols are encountered [14]. In the Mediterranean region of southern Europe, calcareous soils with sediments are used to produce crops. Argillic Cambisols, Alfisols and Entisols are most common in the Mediterranean Europe (*see* Refs. [1, 8, 13]).

2.3 SOILS OF WEST ASIA AND NORTH AFRICA

Soils in West Asia support crops like wheat, maize, barley, lentils, berseem, etc. Crop production occurs mostly under low input subsistence farming conditions. In general, cropping regions of North Africa (Algeria, Sudan, Egypt, and Libya) possess Alfisols (Ustalfs, Psamments). Agricultural soils in Egypt are congenial for production of several types of cereals, legumes, oilseeds, vegetable and forage crops. Crops such as wheat, maize, lentils and peanuts get rotated on moderately fertile soils that are invariably deficient in organic matter. Further, in newly reclaimed zones, organic matter content is <1%. Such sandy soils are calcareous with high $CaCO_3$ content. In other parts of North Africa, crops grow on Aridisols, Entisols, Inceptisols and Vertic Inceptisols (e.g., in Morocco). The productivity of soils in North Africa depends much on fertilizer replenishments, crop rotations and most importantly irrigation. Xerolls (Xerorthents) and Cambisols are frequently encountered in the Middle-east, especially in Syria, Iran, Iraq and Israel (Table 1). They are calciferous soils. Actually, such gypsiferrous soils are common in the semiarid regions of West Asia. These soils exhibit deficiency of major nutrients and organic matter. Deficiency of micronutrients like Zn and B could also be discerned.

2.4 SOILS IN CROPPING REGIONS OF WEST AFRICA

Soils that support crop production in West Africa are mostly sandy and classifiable as Paleudalfs, Tropequents, Eutropepts and Dystrpepts (*see* Table 1; Plate 2). Soil erosion and nutrient loss could be rampant in regions where subsistence farming is practiced. As such, heavy precipitation events can cause considerable loss of nutrients from the crop fields. Carbon sequestration methods are important, since inherent SOC is low (0.5 to 1%) in most of these sandy soils. Soil pH ranges from 5.5 to 7.8. Soil N content is to be replenished using chemical fertilizers and organic manures. Total and available P pools too are meager to support a high yielding crop. Phosphorus replenishment should satisfy chemical fixation as well crop's need. Soil K might be optimum, yet incessant cropping could deplete this element swiftly.

PLATE 2 Sandy Oxisols found in Southern Niger, West Africa.

Note: Sandy soils are prone to wind erosion and loss of nutrients via percolation, seepage and emissions. Soil moisture retention is very low owing to low organic matter content. Loss of C as CO_2 is accentuated due to high soil temperatures and microbial activity.

Source: Dr K R Krishna, ICRISAT, Hyderabad, India.

2.5 SOILS OF INDIAN SUB-CONTINENT

Agricultural cropping zones thrive on a wide range of soil types found in the Indian subcontinent. In the Indo-Gangetic plains, agroecosystems occupy Inceptisols, Entisols and Alluvial soils. Rice-wheat, maize-wheat or oilseed-wheat rotations are most common in the Indo-Gangetic plains. In the Coasts, sandy soils support production of several different crop species. The Vertisols (or Black Cotton soils) found in Central and Southern plains of India are deep, clayey (montomorillonite) and rich in plant nutrients. They are slightly alkaline in reaction, and CEC is high. Therefore, nutrients are buffered better and moisture holding capacity is again high. They exhibit swell-shrink characteristic, and hence drought spells may affect soil structure and rooting. The productivity of crops is relatively high in the Vertisol belt. It ranges from 3–4 t grain plus 6–8 forage/ha. Alfisols, more commonly known as Red soils or Red Sandy loams, are wide spread in South Indian plains. They are well suited to lowland rice as well as arable crop production. Dry land cereals are usually intercropped on Red soils with variety of legumes, oilseeds and vegetables. Major share of grain production in South Indian plains is accomplished during rainy season (kharif).

Southern Indian agroecosystems thrive on Vertisols, Vertic types of Entisols and Inceptisols (Plate 3). To a good extent, agroecosystems flourish on Alfisols, especially Red Alfisols found in Karnataka and Tamil Nadu. Lateritic soils are also used to cultivate crops, but it is sparse and confined to west coast. Several crop species grow excellently on black cotton soils that are clayey due to higher fraction of montomorillonite clay (Vertic soils). Crops grown here tolerate soil pH of 6.0 to 8.5, and mild levels of salinity and alkalinity. Over all, soils are deemed low in fertility, especially for major nutrients-N, P, K and micronutrients Zn and Fe [15–17]. Incessant cropping has resulted in depletion of micronutrients.

Rice agroecosystem of South India thrives on Ustalfs (Troporthents) and Usterts (*see* Ref. [1]). Rice is cultivated on several types of soils. A large portion of rice agroecosystem in Andhra Pradesh region actually thrives on deltaic alluvial soil. These are derived from riverine or marine deposits. They are deep, medium to fine textured and neutral or alkaline in reaction. Soils used for paddy cultivation are relatively more fertile. Paddy cultivation in Godavari region occurs on soils with smectite (40%) clay minerals, mica (20%), traces of vermiculite and high CEC [18]. Soils in East coast of Indian Peninsula are sandy. They may possess high salt concentration due to salinity and alkalinity. Therefore, these soils require irrigation to flush excessive salts. Major soil types encountered in the rice belt of Tamil Nadu are Alfisols, Vertisols, Red laterites and Coastal sandy soils. In Karnataka, rice belt occurs on Red Sandy loams, Lateritic soils and Vertisols (Black soils). In Kerala, rice is cultivated predominantly on lateritic and alluvial soils. Rice productivity is sustained at higher level through nutrient inputs [19]. Soil pH preferred by rice plant ranges from 5.0 to 8.0 [20]. Soil acidity is a severe problem that appears in locations with acid lateritic soils. Soil pH in these areas is low at 4.3 to 5.5, which is uncongenial for rice farming. Much of the acidity is ascribed to nonexchangeable acidity. Such pH dependent acidity is due to kaolinite clay and high sesquioxides. Lateritic soils are prone to leaching, yet may show high accumulation of Fe. Rice crop responds to P inputs, since P availability is below critical limit in most areas.

PLATE 3 *Top:* A Typical Red Alfisol field at Hebbal Agricultural Experimental Station, University of Agricultural Sciences, Bangalore in Karnataka, South India. Cereals are produced on such Red Soil strips during kharif season. They are often intercropped or rotated with legumes, mainly Pigeonpea, Cowpea, Field bean and Horse gram. *Bottom:* A Vertisol field (Black Cotton Soil) prepared for sowing near Bijapur in the South Indian Plains.
Source: Dr Krishna, Bangalore.

2.6 SOILS OF SOUTH-EAST ASIA

Soil types encountered in Chinese agrarian zones are Alfisols, Entisols, Inceptisols and Ultisols. In the North-west, Aridosols and Orthents are used to grow crops. In the North-east, crop production occurs on iso-humisols with long history of cropping. In Shandong and nearby regions, crops are grown on gravely soils derived from weathered granites, shale and gneiss. In Jiangxi, Hennan and Hubei province cropping proceeds on red and yellow acidic soils. In Myanmar, major soil types encountered in the maize cropping zones are Fluvisols, Gleysols, Lithosols, Cambisols, Andisols, Vertisols and Luvisol. Rice is the major cereal, yet maize is found intercropped with legumes like black gram or peanuts.

In Vietnam, crops are grown on Ferralsols that are acidic in reaction and possess strong P adsorption traits. As an example, soils used for peanut production in Vietnam may contain about 1 to 1.5% total C, 0.1.0.15% total N and 2.8 to 4 mg available P/ kg soil. In parts of Mekong delta, crops are cultivated on heavy soils that are high in organic matter (5–6%).

Based on FAO-UNESCO classification of soils, Myanmar possesses five different soil groups. They are: 1) Fluvisols, 2) Gleysols, Lithosols, 3) Cambisols, Andosols, 4. Vertisols, Luvisols and Acrisols and 5) Ferrosols. Based on a combination of precipitation patterns and soil types, 10 different agroecological zones have been identified in Myanmar. Rice is the major crop in wet zones. In Indonesia, Alfisols with low fertility status occupy most of cereal planting area. Soils used for crop production are neutral to alkaline, with Fe deficiency [21]. Both medium heavy and light soils are used to grow crops. Alfisols show deficiencies of both major (N, P, K, S) and micronutrients especially Fe, Zn and Mn [22].

2.7 SOILS USED FOR CROP PRODUCTION IN AUSTRALIA

Soil types used to produce cereals like wheat and maize are generally coarse textured Quartzi-Psamments or Torri-Psamments. Calciorthids, Palexeralfs or Haplustalfs termed as earths are also good for cereal production. Cereals are also cultivated on Duplex soils (Rhodoxeralfs, Natustalfs and Paleustalfs) [23]. In Western Australia, cereal crops thrive on duplex and sandy soils. These are generally low in organic-C and have restricted soil fertility status. Application of fertilizers is important on such soil types. The Brown soils of New South Wales, Victoria and South Australia are used mainly to produce wheat, but maize is also cultivated for grain and forage, although in relatively small areas. The Black Earths of Queensland supports large tracts of cereal production. In addition, Australian cereal belt contains Yellow and Red Earths, especially in Western Australia and New South Wales. In North-west Australia, Vertisols rich in exchangeable-K have supported cereal production. However, recent reports suggest that incessant cropping has led to depletion of K. Potassium deficiencies have been observed [24].

Most crops in Australia tolerate soils with pH from 5.5 to 7.0. Soils that are highly acidic need liming to correct their reaction. Deep red soils that are strongly acidic in reaction can be found scattered in Queensland. These soils were originally forested but currently are used for field crop production such as peanuts, maize, sugarcane etc. The initial fertility level of these soils has declined due to incessant cropping. Currently, peanut crop grown on these soils respond to addition of P, K and Mo. Despite such limitations, these soils are supposedly very productive in Australia [25]. Red earths and yellow earths are inherently low in fertility, low in N, P and micronutrient content. Peanut cultivated on these soils respond well to good management. Chocolate and Brown soils in New South Wales (NSW) are moderately acidic to neutral in reaction. These are intensively farmed zones, especially in some pockets of north east of NSW, wherein a variety of crops including peanuts may be grown.

KEYWORDS

- **Aqualfs**
- **Aqualls**
- **Chernozems**
- **Ferralsols**
- **Fluvents**
- **Haplaquepts**
- **Luvisols**
- **Mollisols**
- **Podzols**
- **Ustolls**

REFERENCES

1. Brady, N. C. *Nature and Properties of Soils.* Prentice Hall of India: New Delhi, **1995,** 576.
2. Diaz, R. A.; Graciela, O.; Maria, M.; Travasso, I.; Rafeal, O.; Rodgriguez, A. A. Climatic change and its impact on the properties of Agricultural soils in Argentinean Rolling Pampas. *Climate Research* **1997,** *9,* 25–30.
3. Pedelini, R. Peanut Production in Argentina. *Proceedings of American Peanut Research and Education Society* **2002,** *30,* 60.
4. Garbulsky, M. F.; Deregibus, A. Argentina: The Country, Pasture and Forage Resource. http://www.fao.org/waicent/faoinfo/agricult/agpc/doc/counprof/ argentina.html **2005,** 1–23 (September 20, 2012).
5. Riveros, F. The Gran Chaco. Technical paper on Gran Chaco. Crop and Grassland Service. Food and Agricultural Organization: Rome Italy, **2005,** 1–42.
6. Sanchez, A. *Properties and Management of Soils in the Tropics.* Wiley: New York, **1978,** 372.
7. Lopes, S. Soils under Brazilian Cerrado: A success story in Soil Management. *Better Crops International* **1996,** *10,* 8–14.
8. Beyer, L. Soil Geography and Sustainability of Cultivation. In: Soil fertility and Crop Production. Krishna. K. R. (Ed.) Science Publishers Inc.: Enfield, NH, USA, **2002,** 33–63.
9. Blume, P. Soil Formation, Soil Taxonomy and Soil Geography. In: Text Book of Soil Science. Schachtschable Blume. Brummer, G.; Hartge, K. H.; Schwertmann, U. (Eds.) Springer Verlag, Stuttgart, Germany, **1998,** 373–400.
10. Gerpacio, R. The Maize Economy of Asia. International Maize and Wheat Center, Mexico. www.cimmyt.org/ Research/Economics/map/pdfs/ ImpactsAsia_Chapter1.pdf. **2002,** 1–15 (October 4, 2012).
11. FAO-AGLW. Water Management Group Maize. Food and Agricultural Organization of the United Nations. Rome, Italy, **2002,** 1–7.
12. Krishna, K. R. Agrosphere: Nutrient Dynamics, Ecology and Productivity. Science Publishers Inc.: New Hampshire, USA, **2003,** 68–104.
13. Krishna, K. R. Maize Agroecosystem: Nutrient Dynamics and Productivity. Apple Academic Press Inc.: Point Pleasant, New Jersey, USA, **2013,** 342.

14. Maly, S.; Sarapalka, B.; Krskova, M. Seasonal variability in Soil-N Mineralization and Nitrification as influenced by N fertilization. Rostlinna Vyroba **2002,** *48,* 389–396.
15. Fink, A.; Venkateswaralu, J. Chemical Properties and Fertility Management of Vertisols. Transactions of the 12 International Congress of Soil Science, – Symposia on Vertisols. Indian Society of Soil Science, New Delhi, **1982,** 61–79.
16. Kanwar, J. S. Managing Soil Resources to meet the Challenges to Mankind. Transactions of 12 International Soil Science Congress-Plenary Session. Indian Society of Soil Science, New Delhi, **1982,** 1–32.
17. Virmani, S. M.; Sahrawat, K. L.; Burford, J. R. Physical and Chemical properties of Vertisols and their Management. Transactions of 12 International Congress on Soil Science, Symposia on Vertisol. Indian Society of Soil Science. New Delhi, **1982,** 80–93.
18. Reddy, R.; Raju, A. Integrated Nutrient Management for Rice. 18 World Congress of Soil Science, Philadelphia, USA, http://crops.confex.com/crops/wc2006/ techprogram/P15736.htm. **2006,** 1–2 (July 12, 2012).
19. Rao, T. N. Rice Productivity in Kerala: Challenges and Strategies. Potash and Phosphate Institute, Gurgaon, India, www.ppi-ppic.org/ppiweb/sindia.nsf. **2006,** 1–5 (July 12, 2012).
20. KissanKerala, Soils of Kerala. http://kissankerala.net/kissan/kisancontents/soil.jsp **2006,** 1–7 (August 12, 2012).
21. Taufiq, A. Direct and indirect effect of ZK-Plus on a planting pattern of Groundnut-Soybean on a calcareous soil. Penelitian Pertanian Tanaman Pangan, Indonesia **2002,** *20,* 12–18.
22. Taufiq, A. Macro and Micronutrient status of Alfisol soil and fertilizer optimization of Groundnut. Penelitian Pertanian Tanaman Pangan, Indonesia, **2000,** *19,* 81–90.
23. Perry, M. C. Cereal and Fallow/Pasture System in Australia. In: Field Crop Ecosystem. Pearson, C. J. (Ed.) Elsevier, Amsterdam, Netherlands, **1992,** 451–481.
24. Carter, M. A.; Singh, B. Response of Maize and Potassium dynamics in Vertisols following Potassium fertilization. http://www.regional.org /au/au/asssi/supersoil2004/s13/ oral/1603_carterm.htm **2004,** 1–12. (August 12, 2012).
25. CSIRO, Soils of Australia. Year Book of Australia. Canberra, **2002,** 1–6.
26. Tiwari, K. N. Future of Plant Nutrition Research in India. Indian Journal of Fertilizers **2006,** *2,* 73–98.

EXERCISE

1. Describe a typical Soil profile found in Agrarian zones. Mention all the horizons.
2. Mention the three major soil classification systems adopted by agricultural experts.
3. Mentions major soil types encountered in each continent. Describe soil characteristics of at least 10 different soil types.
4. Describe the major features of soils found in at least five different agroecosystems of the world, such as Wheat belt in Great Plains, Corn Belt in North America, Rice-Wheat in Indo-Gangetic Plains; Soybean cropping zones of Brazilian Cerrados, Maize/soybean intercrops in Pampas and Wheat belt Northern Plains of China.

FURTHER READING

1. Brady, N. C. In *Nature and Properties of Soils*. Prentice hall of India Ltd.: New Delhi, **1975,** pp. 576
2. Carter, M. R.; Gregorich, E. G., In *Soil Sampling and Methods Analysis.* CRC Press, Taylor and Francis Company: Boca Raton, Florida, USA, **2006,** pp. 85.
3. Huang, P. M. In *Hand Book of Soil Science: Resources Management and Environmental Impacts*. CRC Press: Boca Raton, Florida, USA, **2011,** pp. 830.

4. Krishna, K. R., In *Soil Fertility and Crop Production.* Science Publishers Inc.: New Hampshire, USA, **2002,** pp. 556.
5. Lal. R., Encyclopedia of Soil Science. Marcel Dekker, New York, **2003,** pp. 2060.
6. Power, J. F., In *Soil Fertility for Sustainable Agriculture*. CRC Press: Boca Raton, Florida, USA, **1997,** pp. 384.

USEFUL WEBSITES

http://www.fao.org/nr/land/soils/soil/en (October 14, 2012).
ftp://ftp.fao.org/agl/agll/docs/guidel_soil_descr.pdf (October 14, 2012).
http://www.sssa.org (October 14, 2012).
http://soils.usda.gov (October 14, 2012).

CHAPTER 3

WHEAT AGROECOSYSTEM

CONTENTS

3.1 INTRODUCTION

Seeds for global wheat agroecosystem were sown by the early prehistoric farming communities thriving in the *Fertile Crescent*, comprising the present day nations such as Syria, Turkey, Jordan, Iraq and Iran. Some of the earliest excavations that contained wheat seeds occurs in Northern Iraq. Egyptians grew wheat varieties in the Nile valley during 5000 B.C. Wheat grains were traced in the archeological sites located in the Indus valley that date back to 3000 B.C. Wheat reached the North American main land along with European migrations that occurred during 1600s. Currently, global wheat agroecosystem is a conglomerate of cropping belts that occur all across different continents. It encompasses several types of wheat like emmer, durum, einkorn, spelt and a large number of different genotypes. Researchers at the International Maize and Wheat Centre, Mexico actually demarcate global wheat into several mega-environments based on water resources, soil fertility and crop productivity. They also stress on human population that each wheat mega-environment supports in that region [1].

The global wheat cropping expanses extend into areas located between 50° N latitude to 50° S latitude. The wheat cropping zones thrive in regions that differ markedly with regard to topography, soil, agroclimate and productivity. Generally, wheat culture is predominant in temperate and subtropical zones. Tropical zones too support wheat production. Globally, wheat agroecosystem spreads across 232 m ha. It contributes 682 m t grains with an average productivity of 3.23 t grain/ha. Wheat is a staple carbohydrate to a large populace, nearly 2.3 billion people living in different continents on earth (Table 1). Global wheat agroecosystem is actually a conglomerate of several wheat-farming zones traceable in different continents. The Great Plains of North America, Pampas in Argentina, European Plains, West Asian Fertile Crescent area, Indo-Gangetic Plains, Plains of North-east China and South-east Australia, all support wheat production in large scale. Wheat crop also thrives under wide range of agroclimatic conditions such as cool temperate, subtropics, Mediterranean, semiarid and dry lands. Wheat production occurs from sea level to mountain regions with altitude up to 2,000 m.a.s.l. Wheat crop also tolerates different temperature regimes ranging subzero to 35°C. Regarding sowing season, it is said wheat varieties suited to all longitudes are available. Wheat is sown and/or harvested almost daily in some part of the world.

TABLE 1 Wheat: Origin, Classification, Nomenclature and Uses—A summary.

Wheat is native to *Fertile Crescent* region of West Asia. Wheat was domesticated in West Asia. It later spread to other agrarian zones in Southern Europe, Northern Plains of Europe, and Russia. Wheat reached North America during medieval times through the Spanish voyagers and conquerors. English settlers spread wheat into vast plains in North America. Wheat was introduced to Indian plains during 2nd millennium B.C. through human migration from Asia Minor. At present wheat cultivation occur in all continents.

Classification

Kingdom-Plantae, Order-Poales, Family Poaceae, Genus-*Triticum*, Species *T.aestivum* or *T. durum.*

$2n = 14$

There are several species of wheat grown and consumed by humans. Cultivated wheat species can be grouped as follows:

Hexaploid wheats ($2n = 6n = 42$): *Triticum aestivum* is the most commonly used bread wheat, Another hexaploid wheat known as Spelts or *Triticum spelta* is used feebly. Sometimes it is called as subspecies of common bread wheat.

Tetraploid Wheats ($2n = 4n = 28$): *Triticum durum* is widely cultivated wheat in West Asia. *Triticum dicoccum*, also known as Emmer wheat is also common in many regions of the World.

Diploid Wheat ($2n = 14$): *Triticum monococcum* is the wild species of cultivated variants. Einkorn wheat was domesticated almost at the same period as Durum wheats.

Wheat is also identified and classified based on common use and morphogentics. For example in World markets following wheat types are easily recognized.

Durum Wheat: These are light colored and used to prepare semolina flour, pasta. They are high in protein content

Hard Red Spring Wheat: Hard grains which are brownish and used to prepare bread.

Hard Red Winter Wheat: Brownish grains with high protein content and used frequently for making bread.

Soft Red Winter Wheat: Grains are soft low in protein and cultivated in temperate climate.

Hard White Wheat: Grains are hard and light colored. They are medium in protein content and adapted well to temperate climate.

Soft White: Grains are soft and light colored, very low in protein content. They are useful in making pies, cakcs and pastry.

Major Germplasm centers for Wheat are available at International Maizc and Wheat Centre, Mexico and in several other nations. Wheat Germplasm is maintained separately in many European nations based on their individual requirements. In India, Wheat Research Centre at Karrnal in Haryana, and a few Agricultural Universities situated in the Indo-Gangetic belt do maintain wheat germplasm. Beijing Agricultural University in Beijing, China also maintains wheat germplasm.

Nomenclature

Bulgarian-Pshenihsa; Croatian-Psenica; Czech-Psenice; Danish-Havede; Dutch-Tarwe, Finnish-Vehne; French-Leble; Geman-Lebble, Weizen; Hindi-Gehun; Italian-Grano; Norwegian-Hevete; Polish-Pszenice; Portuguese-Trigo; Romanian-Gran; Russian-Pshenithse; Spanish-Trigo; Swedish-Vee; English-Wheat; Hindi-Gehun; Kannada-Godhi, Telugu-Godumulu, Tamil-Kothumai.

Uses of Wheat

Wheat is mainly consumed as food for human populations. Wheat grains are cooked into a variety of foodstuffs in different geographic regions. Food items differ based on ethnicity of populace, cooking facilities, wheat varieties available, economic considerations, palatability and human preferences. Wheat grains are easily dried, stored and transported across long distances. Wheat grains possess gluten protein, which enables the dough to be enleavened. This process allows wheat dough to be mended into excellent bread and related products. There are several types of wheat grains used by human populations across different regions. Europeans and Russians use wheat more commonly than other communities. They draw nearly 30% of their daily calorie requirements from wheat products. In United State of America, hard red spring and hard red winter wheat's are used to prepare bread. Hard white wheat is also common in Central Plains of North America. Other wheat types common to Americans and Europeans are soft red winter and soft white types. Durum wheat is consumed in North America and Mediterranean regions. Durum wheat is used prepare sphagetti, macaroni and crispies. Soft red winter wheat's are used to make biscuits, pastry and flours. Wheat is also used to make pancakes, pizzas, puddings, noodles, pie crusts, muffins, bread rolls, cakes and waffles. Wheat straw and other byproducts from processing industries are used to feed farm animals. Wheat grains are used in industries to derive fermented products like alcohol, malts, starch, wheat bran oil, and gluten.

Nutritive Value of Wheat Grains (per 100 g crude grains):

Energy-1360 k cal, Carbohydrate 51.8 g, Dietary fiber 13.2 g; Fat 9.72 g; Protein 23.15 g; Thiamine 1.88 mg; Riboflavin 0.5 mg; Niacin 6.8 mg; Pantothenic acid 0.05 mg; Vitamin B6–1.3 mg; Folate 281 ug; Ca 39 mg; Fe 6.26 mg; Mg 239 mg, P 842 mg, Zn 12.3 mg.

Source: http://www.nal.usda.gov/fnic/foodcomp/search/; Krishna, 2003.

3.2 WHEAT AGROECOSYSTEM OF THE WORLD

3.2.1 WHEAT CROPPING BELTS OF NORTH AMERICA

The Wheat Agroecosystem of North American Great Plains extends into landscapes in Southern Canada and Great Plains region within United States of America (Fig. 1). It extends into 20.2 m ha. During the past decades, this wheat belt has provided wheat grains and related products to human population within the North America, as well as those situated away in other continents. Major soil types that support wheat production in Great Plains are Mollisols (Ustolls and Udolls); Alfisols (Ustalfs); Aridisols (Argids). Mollisols of Northern plains are relatively rich in nutrients and organic matter. They are supposedly among best suited to raise wheat crop. The resources and technology for production of wheat, for example, crop genetic stocks, tillage intensity, fertilizer supply, nutrient dynamics and productivity have kept changing rapidly since 1950. Currently, wheat production in Great Plains is relatively intensive. Farmers impinge large quantities of inorganic nutrients via chemical fertilizers, organic manures and other amendments. Of course, net removal of nutrient via grain and forage harvest is proportionately high. Wheat monocrops are predominant in the Northern United States of America and Canada. The conventional wheat-fallow with stubble mulch has been in vogue for several decades. Farmers adopt several wheat-based sequences, intercrops and cropping mixtures. Typically, intensive wheat-based cropping sequences followed in this region of globe are wheat-corn-fallow; wheat-sorghum-fallow; wheat-proso millet-fallow and wheat-soybean [2, 3] (Plate 1). Cumulative weather data from over 80 years suggests that precipitation received during growing season in the Great Plains ranges from 95 mm to 290 mm. The annual flux of moisture through precipitation events is 150–600 mm. According to Peterson [3], tillage variations and precipitation use efficiency are important aspects. Tillage systems commonly adopted by farmers in North America are conventional-shallow tillage using disks, maximum tillage, minimum or restricted tillage and stubble mulching. Zero-tillage practices are preferred by most farmers, since it reduces loss of soil carbon. Timing of fallows is import, since it has direct relevance to soil moisture storage and refurbishment of soil fertility.

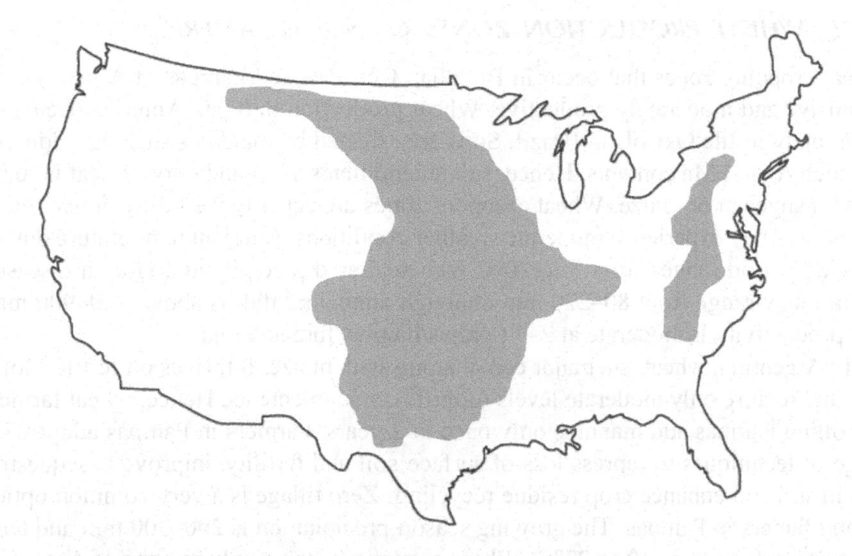

FIGURE 1 Wheat cultivation zone in the Great Plains and other adjoining areas within USA.

PLATE 1 Wheat fields in Northern Great Plains.
Note: Wheat production in Wisconsin and other states are accomplished in large sized farms of over 10,000 ha. They are highly mechanized with GPS-guided Precision Farming instrumentation. Fertilizer-based nutrient inputs match the higher yield expectations. Intensive farming techniques lead to high nutrient turnover in the soil profile.
Source: Nelson Farms, Fort Dodge, Iowa; B2 Farms LLC, Brooklyn, Wisconsin, USA.

3.2.2 WHEAT PRODUCTION ZONES OF SOUTH AMERICA

Wheat cropping zones that occur in Brazilian Cerrados and Pampas of Argentina are expansive and moderately productive. Wheat production in South America occurs on moderately fertile Oxisols in Brazil. Soils are affected by maladies such as acidic pH and high Al and Mn contents. Hence, soil amendments are mandatory. Wheat is rotated with soybean or maize. Wheat cropping zones are actually well distributed within Cerrados. They experience moderate weather conditions. Ambient temperature during growing period ranges from 8 to 20°C with scattered precipitation. Growing season rainfall may range from 80–220 mm although annual rainfall is above 700–900 mm. The productivity is moderate at 2–3 t grains/ha plus forage 5 t/ha.

In Argentina, wheat is a major cereal along with maize. It thrives on fertile Mollisols that require only moderate levels of fertilizer supplements. Hence, wheat farmers in Rolling Pampas add manures only once in 2 years. Farmers in Pampas adopt wide range of techniques to repress loss of surface soil and fertility, improve C sequestration in soil and enhance crop residue recycling. Zero tillage is a very common option among famers in Pampas. The growing season precipitation is 200–300 mm and temperatures ranges from 10 to 23°C. Wheat grown near the southern cone of Argentina experiences extreme cold. Hence, in this area, cold tolerant wheat genotypes are preferred. The productivity of wheat in Pampas ranges from 2–4 t grain/ha plus forage (Plate 2). Wheat-soybean-fallow and wheat-sorghum-wheat-fallow are common rotations practiced in the Pampas of Argentina [4–6].

PLATE 2 Ripened wheat ready for harvest in the Argentinean Pampas.
Source: Dr Alvarez, Department of Agronomy, University of Buenos Aires.

3.2.3 WHEAT IN EUROPEAN PLAINS

The European wheat belt is demarcated into Northern Europe, Western Europe, Central Europe, Russia and Eastern Europe, Southern and Mediterranean Europe. Major wheat production zones in Northern and North-western Europe extends into countries such as United Kingdom (Plate 3), Ireland, Denmark, Netherlands, Belgium, Germany and France. It thrives predominantly on Cambisols and Luvisols. Wheat production in the European Central plains extends from 41° 41′ and 54° 54 N' latitude and between 9° 54′ to 28° 37′ E longitude. The Central European wheat belt extends into countries such as Poland, Bulgaria, Hungary, Czechoslovakia, Romania and parts of Eastern

Germany. Wheat belt in Central Europe is supported by Chernozems, Podzols and Loess soils. Within the former Soviet Union, wheat cropping zone extends into 188 m ha located between 45° and 56°N latitude. Wheat production in the Ukrainian and Russian plains is accomplished mostly on fertile Chernozems. The region encompassing Ukraine and Volga territory is known as *Bread basket of Russia and Eastern Block.* The wheat belt of southern Europe enjoys Mediterranean weather pattern. It extends into regions within France, Spain, Portugal, Greece and Italy. Cambisols, Alfisols and Entisols are encountered more frequently in this region. The weather pattern across the entire European wheat belt differs and perceptibly wide fluctuations occur with regard to various parameters. In the Northern Europe, growing season temperature ranges between 3° and 15°C. *Note:* Mean daily temperature during July can reach 20°C. Mean monthly precipitation during cropping season ranges from 60 mm to 95 mm/month. The total rainfall in Western European wheat production is 550 to 950 mm. Major cropping systems encountered within the European Wheat Agroecosystem are Wheat (continuous); wheat-fallow; wheat-grain legumes; wheat –tobacco sugar beet etc.

PLATE 3 Long-term manurial trial at Broaldbalk in England.
Note: Such annual field trials determine and rank the response of wheat genotypes. There are several wheat yield evaluation centers all across in European nations. Wheat genotypes released from them literally determine the crop genetic composition and genotypes that dominate the wheat agroecosystem of European plains. We should expect proportionate changes in the nutrient dynamics as and when genotypes with greater nutrient use efficiency or those needing high amounts of manure are selected by researchers/farmers for large-scale production.
Source: Roger Atkins, Information Department, Rothamsted Experimental Station, Harpenden, England.

3.2.4 WHEAT IN WEST ASIA AND AFRICA

Wheat production zones are well distributed in the dry lands of West Asia and North Africa (WANA) (Plate 4). Wheat crop often encounters cold temperatures during the season. The temperature during cropping season ranges from +5 to 25°C. Wheat cropping occurs mostly in the relatively wetter areas with annual precipitation 450–600 mm. Whereas, drier zones with 300 mm precipitation supports barley and intercrops like lentils or vetch. In fact, wheat or barley cultivation zones follow the precipitation pattern and soil fertility resources rather stringently. Soil types that support wheat production in West Asia are Cambisols, Xerosols and Lithosols. Calcareous Oxisols with moderate levels of P deficiency and a consistent requirement for fertilizer-N also support wheat cropping. The organic matter content of soils found in this region is low at <1%. Hence, repeated addition of organic manure is needed. Farmers in West Asia prefer to grow durum wheat. Productivity of rain fed wheat crop is obviously low at 1.5 t grains/ha. Wheat genotypes with ability to negotiate intermittent drought spells better and still offer good harvest are selected and released into ecosystem. Such wheat genotypes literally hold sway over the nutrient dynamics in the individual fields and entire cropping zone (Plate 4). It actually depends on the extent to which each genotype dominates and influences nutrient dynamics.

PLATE 4 Wheat on CalciXeralfs found in Turkey.
Source: Information Division, ICARDA, Aleppo, Syria.

Wheat production in Southern Africa confines to areas with cooler climate or those with subtropical climatic conditions. Wheat cropping zones extend into countries like South Africa, Zambia, Zimbabwe, Namibia, etc. Wheat grown here thrives predominately on low fertility sandy Alfisols or Inceptisols. The wheat belt is filled with an assortment of genotypes that adapt well to subtropics and to low fertility soils. Thanks to specific breeding programs of International Maize and Wheat Center, Mexico, that has generated several genotypes that suit Southern and Eastern African conditions (Plate 5). Over all, productivity of wheat in Southern Africa has varied depending on soils, irrigation and fertilizer supply. The average grain yield is 2–3 t/ha under moderate fertilizer supply and less than a ton, if grown under subsistence rain fed situations.

PLATE 5 Wheat expanses in Kenya.
Source: http://www.pecad.fas.usda.gov/photo_gallery/images/photo_gallery/eafrica/kenya_wheat9.jpg

3.2.5 WHEAT CROPPING ZONES OF SOUTH ASIA

The wheat belt in South Asia that includes the famed Indo-Gangetic plains is among the intensely cultivated zones of the world. It stretches into 28 m ha. Average productivity ranges from 2.5–2.8 grain/ha. The cropping zones of Indo-Gangetic plains could be subclassified into (a) Indus plains covering Punjab, Sind and Deltaic plains of Pakistan; (b) Trans-Gangetic Plains covering Punjab, Haryana and parts of Rajasthan in India; and (c) Upper-Gangetic Plains, Middle-Gangetic Plains and Lower-Gangetic Plains. Wheat is sown generally during early part of winter using stored moisture. The wheat season begins around November and lasts till April. Here, precipitation during wheat season ranges from 200–300 mm. The wheat crop itself may require 550–600 mm to support growth and yield formation. The water use efficiency of wheat grown in Indo-Gangetic plains ranges widely from 10 kg/ha/mm to 140 kg/ha/mm. The Indo-Gangetic wheat belt thrives on Mollisols and Inceptisols in western region (Pakistan). Soils in Gangetic plains are classifiable as Mollisols, silty Inceptisols, sandy loams, and alluvial soils. Farmers adopt restricted or no tillage systems and integrated nutrient management procedures. Soils are moderately fertile, yet repeated cropping depletes soil nutrients. Hence, inorganic fertilizer and FYM supplements are required. The cropping pattern of the Indo-Gangetic plains has been dynamic during past century. Initially, during first half of twentieth century, it was all an arable belt with Wheat-Oil seed (rape) or Wheat-legume-fallow. However, since 1980s this belt is popularly known as Rice-Wheat agroecosystem [7]. Rice-wheat fallow or Rice-wheat-legume is the most common cropping sequence adopted currently. The productivity of wheat ranges from 3.0 to 3.5 t grain/ha plus forage.

3.2.6 WHEAT IN CHINA

The Chinese wheat agroecosystem is large and spans into 24.2 m ha, with an average productivity of 4.75 t grain/ha. The annual contribution to global wheat harvest ranges from 115–120 m t depending on precipitation pattern and demand. Wheat farming zones spread into many geographic regions within China (Plate 6). It occupies the riverine zones around major rivers like Yangtze and Huang He, alluvial plains, hilly tracts, southern and eastern coasts and dry lands in the North-west. Wheat belt thrives on Inceptisols, Alfisols, Alluvial soils, Coastal sandy soils, lateritic soils of hilly zones and Mollisols in the North-east. Wheat agroecosystem of China extends into areas with moderately tropical climatic conditions in the South and East, semiarid and dry climate in the North-west and cool temperate climate in the North-east. The North-east China supports an intensive wheat-cropping zone. Here, wheat grain yield ranges from 5 to 6 t/ha plus forage. The dry North-west China supports only subsistence wheat cropping area. The productivity is low at 1–2 t grain/ha.

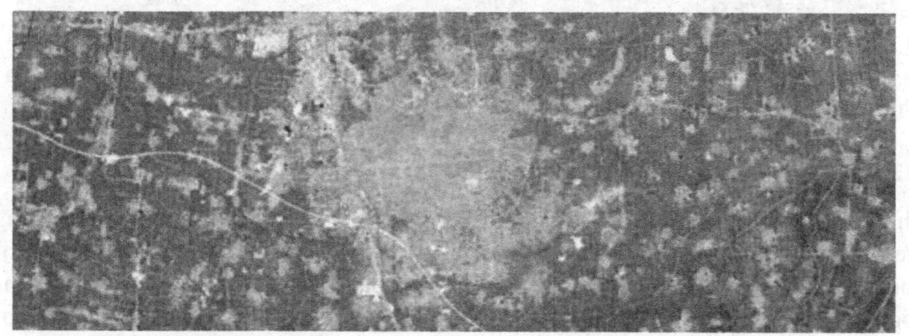

PLATE 6 A satellite picture of Wheat growing region of Southern Henan in China.
Note: Darker zones indicate ripened winter wheat and lighter region shows spring-sown seedlings (May, 2004). Remote sensing of wheat belt allows the Chinese Agricultural Researchers to assess the crop growth, its nutritional status and forecast harvest level.

Source: www.fas.usda.gov/remote/china_countrypage.jpg

3.2.7 WHEAT CROPPING AREAS OF AUSTRALIA

The Australian wheat belt got initiated with arrival of European settlers during 18th century. At present, wheat is grown in all states of Australia, except Northern Territory. Wheat cropping expanses cover about 4.5 m ha in Western Australia, 3.5 m ha in New South Wales, and about 3.m ha together in Victoria, Queensland and Tasmania. The wheat zone in southern parts of the country receives about 275–450 mm precipitation. The rainfall peaks during April to October. Water loss due to surface flow, erosion and seepage could be considerable in the vast wheat expanses. Terminal drought is common in regions with erratic precipitation pattern. The average temperature during wheat cropping season fluctuates between 9–21°C. Mid-season temperature can dip to as low as 8°C. Major soil types encountered within the Australian wheat belt are Quartzipsamments, Torripsamments, Earths (Calciorthids), Clays (Paleusterts) and Duplex soils (Rhodoxeralfs). Both Red and Yellow earths are used to produce wheat. Most common cropping systems followed in the Australian wheat belt are Wheat-wheat, wheat-fallow, wheat-lupin, wheat-fallow-sorghum, wheat-fallow-soybean [8, 9].

3.3 PRODUCTIVITY OF WHEAT BELTS

As stated earlier, currently wheat belt spreads into all continents. It is sown and harvested all through the year in some location on the surface of earth. Wheat agroeocsytem is most pronounced in Canada, United States of America, European nations, West Asian countries, India, China and Australia. During 2010, top wheat producers were China 115 m t; India 81 mt; United States of America 60 m t; Russia 42 m t; France 38 m t; Germany 24 m t; Pakistan 23 m t; Canada 23 m t; Australia 22 m t and Turkey 19 m t. Global wheat harvest during 2010 was 651 m t at 3.1 t/ha productivity. Some of the best productivity figures reported is 8.9 t grain/ha [10].

KEYWORDS

- **Fertile Crescent**
- **Triticum aestivum**
- **Triticum dicoccum**
- **Triticum durum**
- **Triticum monococcum**
- **Triticum spelta**

REFERENCES

1. CIMMYT, People and Production affected by Wheat within each Wheat mega-Environment, targeted by Wheat CRP. International Centre for Maize and Wheat. Mexico. Wheatatlas.cimmyt.org. **2011,** 1–3 (July 7, 2012).
2. Westfall, D. G.; Havlin, J. L.; Hergert, G. W.; Raun, W. R. Nitrogen Management in Dry land Cropping Systems in the Great Plains: A Review. Journal of Production Agriculture **1996,** *9,* 1992–199.
3. Peterson, G. A.; Schlegel, A. J.; Tanaka, D. L.; Jones, O. R. Precipitation Use-Efficiency as affected by Cropping and Tillage Systems. Journal of Production Agriculture **1996,** *9,* 181–186.
4. Hall, A. J.; Vivella, F.; Trapani, N.; Chimenti, C. The effects of water stress and genotype on the dynamics of pollen-shedding and silking. Field Crops Research **1992,** *5,* 349–363.
5. Galantini, J. A.; Landricini, M. R.; Iglesias, J. O.; Maglierina, A. M.; Rosell, R. A. The effects of Crop rotation and fertilization on Wheat productivity in the Pampas of Argentina. Soil and Tillage Research **2000,** *53,* 137–144.
6. Gonzalez-Montaner, J. H.; Madonni, G. A.; DiNapoli, M. R. Modeling grain yield and grain yield response to Nitrogen in Spring wheat in the Argentinean Southern Pampas. Field Crops Research **1997,** *51,* 241–252.
7. Velayutham, M. Sustainable productivity under Wheat-Rice cropping system-Issues and Imperatives for Research. In: Sustainable Soil productivity under Rice-Wheat System. Biswas, T. D.; Narayanswamy, G. (Eds.) Indian Society of Soil Science Bulletin No 18, **1997,** 1–6.
8. Perry, M. C. Cereal and Fallow/pasture systems in Australia. In: Field Crop Ecosystems. Pearson, E. J. (Ed.) Elsevier, Amsterdam, **1992,** 451–481.

9. Anderson, C. C.; Fillery, I. J.; Dunin, F. X.; Dolling, J.; Assesng, S. Nitrogen and Water flows under pasture, wheat and lupin-wheat rotation in deep sand in Western Australia. Australian Journal of Agricultural Research **1998,** *49,* 345–361.
10. FAOSTAT, Wheat Statistics. Food and Agricultural Organizations of the United Nations. Rome, Italy, http://www.faostat.org **2010** (October 12, 2012).
11. Krishna, K. R. **2003**, Agrosphere: Nutrient Dynamics, Ecology and Productivity. Science Publishers Inc.: Enfield, New Hampshire, USA pp. 348.

EXERCISE

1. Write an essay on origin, centers of origin and genetic diversity. Mention classification of Wheat.
2. Discuss Wheat Agroecosystems of Great Plains of North America and Pampas of Argentina with regard to Natural Resources and Productivity.
3. Mention major wheat-based cropping systems followed in Indo-Gangetic Plains and North-east China.
4. Write an essay on Wheat Grain Production trends during past 50 years in USA, and Europe.
5. List the top 10 wheat grain producing countries.
6. Discuss the soil characteristics of major wheat belts found in Americas.

FURTHER READING

1. Gooding, M. J.; Davies, W. P. In: *Wheat Production and Utilization.* CAB International: Oxfords, English, **1997,** pp. 352.
2. Regmi, A. P. In: *Effects of long-term application of Mineral Fertilizers and Manure on Rice-Rice-Wheat system.* International Maize and Wheat Centre (CIMMYT): El Batan, Mexico, **2003,** pp. 134.
3. Woolston, J. E. In: *A 1000 Peer-Reviewed Journal Articles from CIMMYT and its Collaborators-1966–2002.* International Maize and Wheat Centre: El Batan, Mexico, **2003,** pp. 23–49.
4. Zawail, R. M. Y.; El-Desouky, S. A.; Abd El-Haleem Khedr, Z. M. In: *Wheat Productivity using Biofertilizers and Anti-oxidants.* LAP Lambert Academic Publishing Inc.: New York, **2011,** pp. 232.

USEFUL WEBSITES

http://apps.cimmyt.org/english/wps/publs/Catalogdb/catalog.cfm?data=4&monitor=3 (October 14, 2012).

http://www.nue.okstate.edu/Crop_Information/World_Wheat_Production.htm (August 26, 2012).

http://www.cimmyt.org (October 14, 2012).

http://www.fao.org/docrep/006/yr4011e/yr4011e0 s.htm (October 14, 2012).

CHAPTER 4

MAIZE AGROECOSYSTEM

CONTENTS

4.1 GLOBAL MAIZE AGROECOSYSTEM: AN INTRODUCTION

Maize is among three foremost cereal crops preferred by human race. It serves as staple carbohydrate for a large population in the Americas and elsewhere in Asia and Africa. It is an excellent forage source. It is a versatile crop species and forms large cropping belts in all continents. Its cultivation extends from 50° N to 50° S latitude. The maize agroecosystem thrives on variety of soil types, weather patterns and economic zones. The Northern Great Plains of USA, supports a large expanse of maize crop. Here, it is grown intensively using large amounts of inorganic fertilizers, gypsum and FYM in about ten states of Northern Great Plains. Collectively, this agroecosystem is referred as *Corn Belt of United States of America*. This stretch of maize contributes nearly 60–70% of total maize produce of United States of America. The *Cerrados* of Brazil is another region where maize is a dominant crop on the vast plains. Here, it is often rotated or intercropped with soybean. The Oxisols of Cerrados are moderate in fertility and exhibit Al toxicity. Hence, maize yield is moderate despite repeated adoption of soil amelioration techniques. Maize is a sought after forage crop in the European plains. Fertile soils and high fertilizer-based nutrient supply induces massive production of forage reaching 30–35 t biomass/ha. In West and Central Africa, tropical maize is a staple cereal to many tribes. It is grown in mixtures with beans/legumes and bananas. Maize culture is localized to Nile region in North-east Africa. In Egypt, precipitation trends and irrigation resource are key factors to maize cropping and its productivity. Maize cropping zone is gradually increasing in the Southern Indian Plains and Hills. It is being preferred over other dry land crops due to its ability to offer greater quantity of food grains to humans plus forage. In North-east China, maize is cultivated intensively using relatively larger dosages of fertilizers and FYM. Maize rotated with soybean dominates the North China agrarian zone. Maize is also cultivated as an important rainy season crop in many other parts of South-east Asia. Maize belt in Australia offers both forage grains. Its expanses are confined to Queensland, New South Wales and few other locations.

Globally, maize production strategies vary. Maize belts differ in terms of composition of genotypes, intensity of cultivation and productivity (Fig 1). In the USA, and European agrarian zones, maize belt is highly intensive. In order to sustain high grain and forage productivity, appropriately high nutrient and irrigation inputs are practiced (Table 1). Maize agroecosystem is only moderately intense with regard to nutrient supply and grain productivity in many Asian countries. Maize agroecosystem is held at subsistence levels of nutrient and water supply in Sahelian, Guinea Savanna and other regions of Africa. Maize productivity under such semiarid or dry land conditions is low. The ability to thrive in different environments is attributable to highly versatile nature of maize plant.

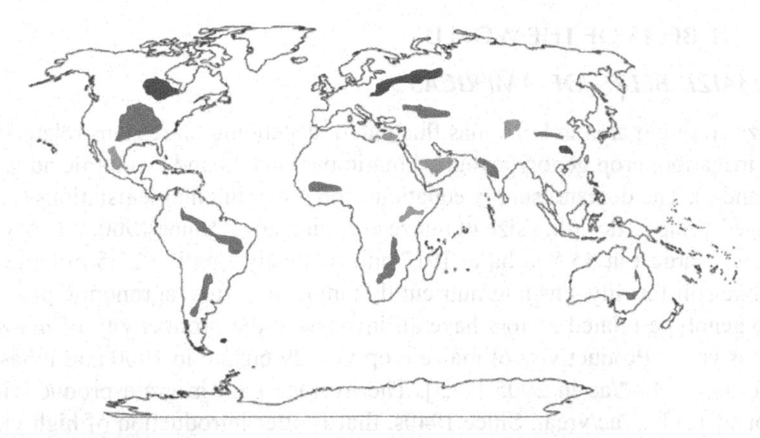

FIGURE 1 The Maize Agroecosystem of the World.
Note: Deep dark areas represent Intensive Maize Cropping zones (6–8 t grain/ha); Moderately dark areas indicate moderately intense production zones (2.5–6.0 t grain/ha). Lighter area indicates subsistence, low input maize cultivation zones (0.5–2.5 t grain/ha).
Source: Ref. [1].

TABLE 1 Maize Agroecosystem – A classification based on Intensity of Cropping and Productivity.

Nature of Maize Agroecosystem	Productivity	Countries/Region
Low Productivity Maize Farming Regions Nutrient supply is low 20-100 kg NPK ha^{-1} Low seeding rate, rain fed, dry lands	0.5–.2 t grain ha^{-} 5–8 t forage ha^{-1}	Bangladesh[1], Botswana, Burkina Faso Chad, India, Nepal, Pakistan, Mali, Mauritania, Togo, Senegal, Sierra Leone, Congo, Ghana, Niger, Saudi Arabia, Libya,
Moderately Intense Maize Belts Nutrient supply exceeds (200 kg ha^{-1} NPK)	2.5–6.0 t grains ha^{-1}	China, Thailand, Korea, Vietnam, Argentina, Hungary, Serbia, Russia
Rain fed or Irrigation restricted to Critical stages	10–15 t forage ha^{-1}	Mexico, Brazil, Venezuela, Syria, Uzbekistan, Morocco Slovenia, Kazakhstan, Turkey
Intensive Mono-Cropping Regions of Maize High nutrient inputs (300 kg ha^{-1} NPK)	7–11 t grains ha^{-1}	USA (Corn Belt), Canada, France, Italy, Austria, Spain, Belgium

Source: Refs. [1–4].
Note: Maize ecosystem is intense and highly productive in North America and European plains. Moderately intense cropping is confined to Asia, Mesoamerica and Tropical Africa. Maize ecosystem is subsistent and least productive in Sahelian West Africa and other semiarid regions.

4.2 MAIZE BELTS OF THE WORLD

4.2.1 MAIZE BELTS OF AMERICAS

The maize cropping area in USA, has fluctuated, depending on reasons related to soil fertility, irrigation, crop genotype, agroclimatic parameters and economic advantages (Figs. 1 and 2). The demand/supply equations and governmental legislations too must have played a role in deciding size of maize cropping zone. Since 1900, the maize belt in USA, was largest at 45.5 m ha in 1917 and relatively small at 27.5 m ha in 1969. Factors like soil fertility, ensuing nutrient dynamics in farms, agronomic procedures and crop genotype related factors have all influenced the productivity of maize during various years. Productivity of maize crop was 29 bu*/ac in 1900 and it has since increased to 143 bu*/ac in 2005 [2, 5]. The average gain in grain productivity has been around 1.0 bu*/ac/yrear. Since 1940s, that is after introduction of high yielding genotypes and chemical fertilizers, gain in grain yield has jumped to 1.8 bu*/ac/year (*One bushel = 27 kg grains).

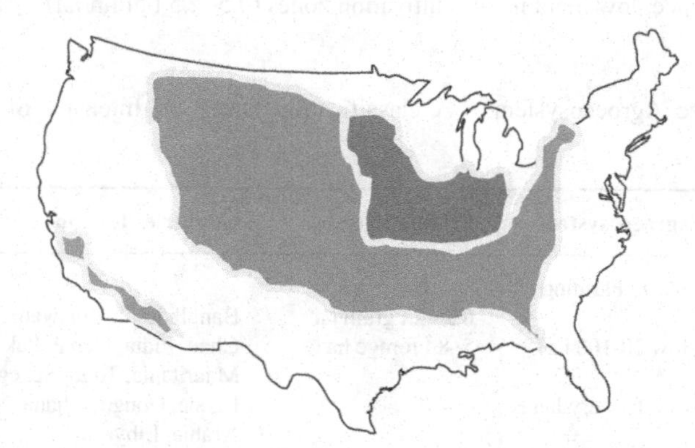

FIGURE 2 The Intensive Maize Cropping Zone of USA, – 'The Corn Belt' (Darker area).
Note: Maize is grown in many other states of USA, (lighter area), in addition to 'Corn Belt States,' wherever agro-climate is congenial.

Maize grain productivity has been enhanced through manipulation of several factors like crop genotype, soil fertility and irrigation. In USA, maize agroecosystem was first impinged with nitrogenous and phosphorus fertilizers in early 1940. This aspect affected the soil nutrient dynamics within the agroecosystem. It actually helped the farmers in matching nutrient demands of the crop with supply, based on the yield goals that were envisaged. Improved soil fertility status, it seems allowed maize farmers to plant maize seedlings closely. Further, they achieved a better crop stand by increasing planting density from 12,000–14,000 plants/ac in 1900 to over 27,000 plants/ac in 1940s. Also, traditional maize/small cereal grain intercrops got replaced by monoculture of corn alternated with soybean [6]. This led to the development of an intensive maize-growing region in Northern part of USA. This region is more commonly known as 'American Corn Belt' (Fig. 2). The states that support intensive maize production are aptly called 'Corn States.' They are Wisconsin, Minnesota, Missouri, Michigan, Illinois, Iowa, Ohio, Indiana, also parts of North Dakota, South Dakota, Nebraska, Kansas (Plate 1), Arkansas, Tennessee and Kentucky. Obviously, maize dominates the agricultural horizon in the 'Corn belt.' Therefore, it has greatest impact on the nutrient dynamics, ecosystematic functions and agricultural productivity of the region. The six states in 'Corn belt,' namely Iowa, Illinois, Nebraska, Minnesota, Indiana and Ohio account for 82% of maize produced in USA, annually. During 2007, national maize acreage was 35 m ha. Iowa is the top maize producing state in USA. Maize belt in Iowa extended into 5.2 m ha. It contributed 22% of annual grain harvest of USA. We ought to note that maize cropping zones in USA, do extend beyond 'Corn Belt' into many other states. Overall, 60% of maize produced in USA, is used to feed farm animals, about 20% is exported, remaining portion is used for human consumption and industrial uses, like starch production [7].

PLATE 1 A view of Maize fields in Kansas State of USA.
Source: Prof Kraig Roozeboome, Extension Service, Kansas State University, Manhattan, Kansas, USA.

The maize agroecosystem in Canada is large and spreads across southern region of the country. It occupies about 1.3 m ha and contributes 9.5 m t grains annually. The productivity is high at 7.82 t grains/ha. It is attributable to chemical fertilizer supply and irrigation. Such high biomass generation allows for ample residue recycling and C sequestration in soil [8, 9].

Currently, Mesoamerican maize belt thrives in regions endowed with rich vegetation and tropical or subtropical climate. Mexican maize farmers often clear forests and practice 'slash and burn' or 'shifting cultivation.' Maize is cultivated along with beans, gourds and other vegetables. Therefore, here maize thrives predominantly as an intercrop. Maize is an important crop for inhabitants of Andes and its slopes. As such, they specialize in cultivating maize on steep mountains by adopting terrace farming. For example, Guatemalan highland supports maize production through terrace formation. In Central Mexico, maize cropping zones evolved into *Chinampas* or floating gardens. It seems *Chinampas* allowed intensification of maize farming in this region. *Chinampas* is a unique maize farming technique. It was standardized much before Spaniards entered Mesoamerica. Currently, maize farming stretches into entire Mesoamerican agricultural zone. Maize is a dominant cereal among the crop mixtures grown in Central America. During past 30 years, Mexican farmers have been cultivating maize and Macuna in rotation, mainly to stabilize soil fertility, its quality and to obtain steady grain yield [10]. The maize belt extends into 7.7 m ha in Mexico and contributes 3% of global maize harvest (21 m t). The productivity of maize fields is about 3.1 t grains/ha plus forage. Maize cultivation spreads into 600,000 ha in Guatemala, 300,000 ha in Honduras and 130,000 ha in Cuba.

The Argentine maize belt is concentrated in the Pampas (Fig. 3). Maize belt of Argentina is large and spreads into 2.8 m ha. Annually, Argentine maize agroecosystem produces 19.5 m t grains, which is equivalent to 3% of global harvest. The productivity of Argentinean maize belt is relatively high at 6.5 t grain/ha plus forage. Brazil supports a large expanse of maize cropping zone (Fig. 3). It is mainly concentrated in the Cerrados. Maize belt also extends into other regions like Amazonia. The Brazilian maize belt spreads into 12.4 m ha. It accounts for 8.5% of global maize agroecosystem in terms of area. Productivity of Brazilian maize agroecosystem is relatively low compared to those in USA, or China. It ranges from 3.2–3.7 t grains/ha plus forage. Annually, about 51 m t maize grains are harvested from Brazilian maize growing regions [11]. In the Andes region of South America, white maize is most preferred. It is cultivated in high altitude regions ranging from 1200–3600 m.a.s.l. Maize encounters abiotic stresses like cold temperature and drought. Yet, the grain yield may range from 2–3 t/ha. Farmers tend to produce much of their maize under low input systems owing to environmental vagaries. On fertile soils, if grown under high input systems, these highland maize cultivars (white floury types) yield 9–12 t grain/ha [12].

FIGURE 3 Maize Cropping Zones of Cerrados of Brazil and Pampas of Argentina.
Source: Ref. [1].

4.2.2 *MAIZE IN EUROPE*

The European maize belt occurs between 5°W to 140°E longitude and 38°N to 55°N latitude. It experiences wide variations in agroclimatic conditions. The European maize belt is vast and occupies regions that are predominantly cold temperate, moderately cold temperate and Mediterranean. In the cold temperate plains, mean temperature during crop season fluctuates between 3°C and 15°C. However, maximum temperature can reach 20–25°C. Diurnal variations do affect maize-based cropping systems. Day length period varies from 8.5 h to 15 h. Mean monthly precipitation ranges from 95–110 mm during crop season. Total rainfall is around 550 mm in Western European countries like Spain and France. It is around 950 mm in German plains. Maize crop experiences cool Mediterranean climate in Southern European agricultural zones. Evapotranspiration of moisture from the ecosystem is low. The temperature fluctuates between 7°C and 18°C during cool season (January–March) and between 24°C and 28°C during June to September. Maize cropping in Russia and Ukraine depends much on frost-free period. The annual precipitation in Southern European plains (Spain, Portugal, Southern France, and Greece) ranges from 500–1000 mm. Relative humidity ranges from 30 to 60% during the cropping season. Maize belt in Europe thrives mostly on plains, where in altitude ranges from 300–500 m.a.s.l. Maize culture also occurs at higher altitudes in the fringes of Alps. In the Hungarian plains, maize is sown in April. Sowing date depends on soil fertility and temperature. A delay in sowing maize by one month leads to a loss of 20–25% grain yield. For optimum nutrient recovery and fertilizer use efficiency, maize should be sown by April 10 to May 5. The heat threshold for most maize hybrids sown is 10°C. However, there are genotypes that germinate at lower temperature of 6–8°C [13]. Maize culture is prominent in France. It occupies 1.7 m ha and contributes 13.2 m t grains annually, which is equivalent to 2% of global maize harvests. The productivity is markedly high at 7.8 t grains ha. The German maize growing regions extend into 0.47 m ha. Maize is intensively cropped and thus productivity ranges from 7.2 to 7.7 t grains/ha. In Romania, maize belt extends into 3.0 m ha and productivity is 4.3 t grains/ha plus forage. Ukraine supports maize belt that extends into 2 m ha on the plains and contributes 6.5 m t grains annually. Productivity is moderate at 3 t grains/ha. Hungary in Eastern Europe supports a large maize belt of 1.2 m ha. Productivity is moderate at 3.94 t grains/ha. The annual grain harvest is about 4.5 m t grains plus forage. The Italian maize belt spreads into entire country. It is predominant in agricultural zones of Northern Italy. Italy is actually a major maize-producing nation in Southern Europe. Among European nations, Italy contributes 2% of global maize harvest (14.3 m t/yr). The maize agroecosystem in Italy occupies 1.46 m ha. Again, productivity is rather high due to intensive farming techniques that employ high amounts of nutrient and water supply. Spain has maize belt of 0.47 m ha that thrives under intensive cropping system. The productivity is rather high at 9.5 t grains/ha plus forage. Maize crop meant for silage is prominent in Europe. In Northern Europe it is rotated with pasture grasses and grown mostly as rain fed crop with no irrigation. In Hungarian plains of Central Europe, maize for silage is rotated with wheat. In Italy and other Southern European regions, maize for silage is rotated with wheat and soybean [14].

4.2.3 MAIZE IN WEST ASIA AND NORTH AFRICA

Maize cultivation in North Africa and West Asia proceeds at subsistent to moderately intense levels. Maize agroecosystem in North Africa and West Asia varies enormously with regard to intensity of cropping and nutrient supply. In Morocco, Algeria and Libya, maize is cultivated with low input using subsistence farming methods. It yields 0.8 to 2.4 t grains/ha. In Lebanon and Syria, maize production is moderately intense. Farmers in these countries harvest 3–4.5 t grains/ha. The productivity of maize is highly dependent on soil fertility and nutrient dynamics. Maize crop yields 12 t grains/ha plus forage in countries like Israel. Here, maize cultivation occurs on sandy soils found in plains, undulated landscapes and coasts. The agroclimate of maize growing regions is highly variable. The variation in precipitation affects cropping systems. In the Mediterranean region, temperature during crop season ranges from 5°C to 18°C. Cold winters with <5°C and slightly hotter summer with 25°C are a clear possibility during maize culture.

4.2.4 MAIZE AGROECOSYSTEM OF AFRICA

In West Africa, maize culture is pronounced in wetter savanna and guinea regions. Maize cultivation is sparse in Sahelian zone, because of relatively lower levels of precipitation (350–600 mm/yr). In addition, moisture-holding capacity of sandy Oxisols is rather meager. Instead, Sahel is often occupied by drought tolerant pearl millet or sorghum or cowpeas. Maize cultivation zone extends between 3°N – 10°N latitude and 10°W to 10°E longitude. Maize culture is often confined to rainy season, lasting from June/July to October. However, in wet tropical zones of Southern Nigeria, Ghana, Cameroon, Gabon and other countries, maize culture proceeds in all the three cropping seasons. Maize is an important cereal in entire tropical Nigeria. Its planting time seems arbitrary and at times improper choice of dates results in yield depreciation. Most often, it is sown between April and June in the three major agroecological regions for maize production, namely Southern Savanna, Southern Guinea Savanna and Sudano-Savanna (*see* Ref. [1]). Early planting is generally aimed at harnessing precipitation and inherent soil fertility more efficiently [15]. Average temperature ranges from 22–30°C, but it could reach 38–40°C in summer. Relative humidity is high (60%) in Coastal plains as well as in forested zones. Diurnal variations are small (12–14 h) owing to closeness of this region to equator. Plants reach maturity faster because thermal requirements are satisfied rather quickly.

Maize is mostly intercropped with legumes, cassava or oilseeds to improve land use efficiency. It is cultivated from sea level to 500 m.a.s.l. Maize is an important intercrop in Central and East Africa. It is cultivated on plains as well as on mountainous terrain, at altitudes ranging from 0 to 1000 m.a.s.l. Major maize producing countries are Kenya, Tanzania, Congo, Central African Republic, Ethiopia and Somalia. The tropical climate supports maize culture all through the year. Precipitation levels are high at 2,000 to 2,500 mm annually. Soils are moderately fertile and rich in organic matter content. Relative humidity during rainy season is high at >80%.

Maize is an important cereal crop in Southern African agricultural zones that extend from 15°S–35°S to 10°E–40°E. Southern African maize belt thrives on semiarid plains, coasts, river valleys, undulated terrains and mountains. It occupies regions that experience tropical, subtropical and temperate agroclimate. It is preferred as mono-crop and equally so as an intercrop with vegetables/legumes. Small stretches of continuous maize are also found in southern Africa. Major countries that produce maize are South Africa, Angola, Botswana, Zimbabwe, Mozambique and Zambia. Maize culture extends from sea levels to 1,000 m.a.s.l. The maize agroecosystem thrives on sandy soils and drier climate in regions closer to or on the fringes of Kalahari. Precipitation in the semiarid region is low or moderate ranging from 400–800 mm annually. Growing season precipitation of 250–400 mm is distributed mostly during January to March. In South Africa, maize cropping extends into temperate climatic zones. Precipitation pattern allows only one crop during rainy season. Therefore, post rainy and summer crops need supplemental irrigation. The dry regions with low relative humidity are predominant. Mean annual temperature ranges from 10°C to 28°C, but maximum temperature reaches 36–40°C. The white maize is an important cereal Southern Africa. It is grown both under rain fed and irrigated conditions. Maize is a preferred crop in the flood plains of Namibian rivers. Maize planted in August/September is harvested during February next year. Maize is also planted during December/January and harvested by June /July [16].

4.2.5 MAIZE IN SOUTH AND SOUTH-EAST ASIA

In India, maize cropping zones stretch from 8° to 32°N latitude and 74° to 96°E extending into 23 states, but it is a major crop in only few states like Rajasthan, Bihar, Karnataka, Andhra Pradesh and Gujarat. Karnataka and Andhra Pradesh are the main corn producing states in South India. Majority of maize belt thrives under dry land conditions that persist in Southern Indian plains. Moderately fertile Vertisol and Red Alfisol regions of South India support vast stretches of maize, mostly as mono-crop in rotation with legumes or cotton. Maize cultivation is sparse in Himalayan region, yet it is harvested regularly in Sivalik Mountains at altitudes around 2,500–3,000 m.a.s.l. Similarly, Western Ghats supports mixed farming that includes maize. Here, maize grows under evergreen tropical conditions at altitudes 1,500 m.a.s.l. The maize ecosystem spreads all through the humid coastal plains of South India (Plates 2, and 3). The maize improvement project in India identifies at least four different agro-ecoregions that support its production in significant quantities (*see* Ref. [17]). They are:

Himalayan Zone: It is a temperate zone with elevations above 600 m.a.s.l. Crop duration is relatively longer.

North-west Plains: This region is characterized by wet and arid tropics. Moisture stress at flowering and grain is common. Soils are sandy or alluvial and need fertilizer replenishment.

North-east Plains: This region is characterized with hot and humid weather. Soils are loamy and alluvial. Nutrient deficiencies need correction through fertilizer application.

Peninsula Region: It is characterized by tropical and subtropical climate in the plains and hill zones. The coastal area supports cultivation of maize throughout the year. Soils used for maize production in South India are sandy Alfisols, Vertisols and Inceptisols. They experience deficiency of major nutrients. Micronutrient deficiencies are sporadic.

Source: Refs. [1, 17, 18].

Maize is an important cereal in several nations of South-east Asia. The Nepalese maize belt can be demarcated into at least four different ecoregions. Maize is an important cereal in the 'Tarai region.' Here, maize is intensively cultivated on the flood plains. Maize cropping also extends into mountainous regions up to 1,000–3,000 m.a.s.l. Maize is grown on terraced fields. Maize is mostly sown during rainy season (kharif) spanning from June/July to November. In the Tarai region, maize meant for fodder is grown in summer. There are many small rivers that support maize production on their banks. On riverbanks, maize is rotated after a rice crop.

PLATE 2 A Maize crop grown on moderately fertile Red Alfisol of South Indian Plains.
Note: Maize is an important arable crop that thrives both under irrigated and rain fed dry land conditions. Inorganic fertilizer and Farm Yard Manure inputs are regulated based on soil fertility status, nutrient dynamics and yield goals.
Location: A Farm near Doddaballapur in Bangalore District, India (13.17° N latitude and 77.30° E longitude).
Source: Ref. [17].

The Indonesian maize production zones are spread in a mosaic covering the entire length and breadth of the country. Much of the maize crop is cultivated in *Tegalan* (rain fed dry land) or raised lands within flooded wetland region. Maize is also grown under tidal swamp conditions called *Surjan.* Tidal maize is more frequent in Java. Actually, rice is grown in standing water in sunken beds, but maize is planted on raised beds close by.

The Philippines maize belt extends mostly into subtropical zones. Maize thrives predominantly as an intercrop with legumes or vegetable. It is grown in valleys and terraces. Rice is the major cereal in low lands. Maize is planted in uplands prior to onset of monsoon. Maize is also grown in post-rainy and summer seasons. It seems, in Philippines, traditional distinction of maize production systems is not based on agroclimatic characteristics. It is easier to demarcate maize belt based on grain types. Mostly, maize with white colored grains is sown to regions with poor soil fertility and low productivity. Whereas, yellow colored grain types are confined to high soil fertility and high input zones. Here, maize is mostly used for forage and starch production (*see* Ref. [1]; Plate 3).

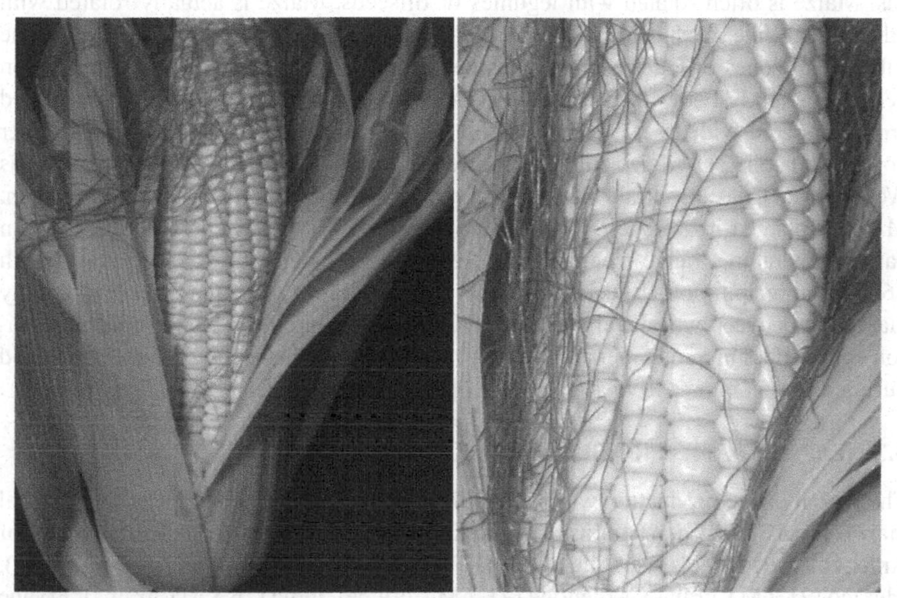

PLATE 3 A frequently preferred Maize genotype (Dent Corn) in Asia, Americas and Africa.
Note: Healthy grains on cobs that supply carbohydrates.
Source: Dr K.R. Krishna, Bangalore, India.

The maize agroecosystem extends into almost all provinces of China. It spreads from 26° N to 50°N latitude and 72°E to 130°E longitude. Maize belt in China extends into a range of topographic and agroclimatic conditions. It extends into plains, plateau and mountainous regions. Maize is grown on hills, slopes and terraces at altitudes reaching 1,500 m.a.s.l. Maize culture extends into low lands where generally rice predominates. The temperate region in the North and North-east China supports a large expanse of maize. It contributes nearly 66% of Chinese maize. Tropical and subtropical climate available in southern China is also congenial for maize production [19, 20]. Here, the mean temperature during maize cropping season ranges from 11°C to 20°C. Severely, low temperatures too occur reaching as low as −5°C in Northern China. Obviously, length of frost-free period is important. It ranges from 130–330 days depending on region. Maize crop is cultivated both in rainy and post rainy season. In the intensive agricultural zones of North-east China, maize mono-cropping occurs throughout the year. In the tropical regions of South-east Asia, especially in Vietnam, Indonesia, Malaysia and Burma, it is sown immediately after first rains during June.

The temperature is warm at 18°C–22°C and relative humidity is high at 70–80%. The maize belt that thrives on plains, undulated landscapes and mountains in the above countries may receive 1200–1700 mm annually.

4.2.6 MAIZE CROPPING ZONES OF AUSTRALIA

Australian maize belt is relatively small if compared with those encountered in Americas. Maize is often rotated with legumes or oilseeds. Maize is actually rotated with other cereals like sorghum or wheat. Maize-cotton rotation is gaining acceptance in the subtropical zones of Queensland. It has been reported that maize grown after cotton performs slightly better than continuous maize. Maize is relatively a shallow rooted crop compared with trees. It extracts soil moisture and nutrients better from upper horizons of soil [21]. Maize cultivation spreads into most parts of Queensland such as Wet Tropical Coast, Central Queensland, Darling Downs, South Burnett and Moreton. Maize is planted in October/November and reaches harvest in 100–120 days. Mean daily temperatures range from 20°C to 32°C. Maize adapts to dry land zones with <600 mm precipitation, especially in New South Wales in Australia. Maize is usually planted in between wheat stubbles under zero tillage conditions. Maize cropping season commences in September and ends by mid-January. The net precipitation received during a season ranges from 190–240 mm [1].

4.3 EXPANSE AND PRODUCTIVITY OF MAIZE BELTS

The global maize production was 817 million in 2011. Nearly one half of total global maize, about 332 million t grains equivalent to 47.2% was harvested in United States of America. The People's Republic of China (163. 9 MT grains), Brazil (51.6 MT grains), Mexico (21.7 MT grains), Argentina (21.7 MT grains), India (16.8 MT grains), France (13.1 MT grains), Indonesia (12.4 MT grains), Canada (10.5 MT grains) and Italy (9.9 MT grains) are other important maize producing nations [22]. Annually, during the past decade, USA, contributed 35–42% global maize grain harvest; China contributed 15%, European countries together about 14%, Brazil 8.4% and India 3%. About 60% of global maize cultivation zones occur within six countries. Continent wise, North America contributed 45%, South America 10%, Europe 12%, Asia, 26% and Africa 7% of maize grain harvest. Total area (m ha) and Productivity (t grains/ha) of maize farms in top 10 countries are as follows:

Country:	USA	China	Brazil	Mexico	India	Nigeria	Indonesia	S. Africa	Rumania	Philippines	Global
Area (m ha)	29.8	25.4	12.4	7.7	7.5	4.4	3.4	3.2	3.0	2.7	159
Yield (t/ha)	10.2	5.4	3.7	3.1	2.1	0.9	3.2	3.0	4.3	2.2	5.12

Source: Refs. [1, 3, 23].

The productivity of maize crop in Africa is 1.5 t grains/ha plus forage, depending on harvest index. Maize farms in West Africa that are kept under low input and meant for subsistence, produce a mere 0.7 to 1.5 t grain/ha [24] (Table 2). The average productivity of maize crop in South America is 3.0 t/ha. In most parts of South Asia, including India it is 1.7 to 2.3 t grains/ha. Maize productivity is about 2.4 t grains/ha in China and Fareast. Maize grain yields are relatively higher in the Corn Belt and other regions of North America, including southern Canadian plains. Maize is an excellent forage crop. Maize is also grown exclusively for forage. Maize yields 20 t fresh biomass (stalks and leaves), which is suitable for ensilage. The forage productivity depends on irrigation and soil fertility conditions. In dry lands, forage productivity is about 18–29 t/ha, but under irrigated conditions it reaches 42 t/ha. Maize is mostly a forage crop in few of the European countries. In European plains and loess region, maize forage yield fluctuates between 35 to 50 t/ha. For example in France it is 40 t forage/ha, in United Kingdom it is 25–35 t/ha in Eastern European region, again it ranges from 25–35 t forage/ha. The grain yield in European agrarian zones productivity ranges from 5.5 to 7 t/ha [1, 24].

TABLE 2 Yield Potential and Current Yield of Maize crop raised in different parts of the World.

Region	**Yield Potential (Current Yield)** t grain/ha	
	Low or Moderate Soil fertility	**High Soil fertility**
Fareast and South-east Asia 5.0 (2.5)	8.0 (3.0)	
South Asia	5.0 (1.7)	7.0 (2.6)
West Asia and North Africa	5.0 (3.2)	7.5 (3.2)
Sub-Saharan Africa	5.0 (0.6)	7.0 (2.5)
Latin America and Caribbean	6.0 (1.1)	10.0 (4.0)
Corn Belt of USA®	6–7 (5.5)	12.0 (10.0)

Source: Refs. [1, 23, 25–27].

KEYWORDS

- **Cerrados**
- **Chinampas**
- **kharif**
- **Tarai**
- **Tegalan**

REFERENCES

1. Krishna, K. R. Maize Agroecosystem: Nutrient Dynamics and Productivity. Apple Academic Press Inc.: Toronto, Canada, **2013,** 345.
2. FAOSTAT, Maize Statistics. Food and Agriculture Organization of the United Nations. Rome, Italy, www.faostat.fao.org **2005** (August 23, 2012).

3. FAOSTAT, Maize Statistics. Food and Agricultural Organization of the United Nations. Rome, Italy, www.faostat.org **2010** (July 20, 2012).
4. OKSTATE, World Wheat and Maize Production. http://nue.okstate.edu/Crop_information/World_Wheat_ Produtcion.htm **2009,** 1–11. (July 12, 2012).
5. Larson, W. E.; Cardwell, Y. B. History of US Corn Production. http://www.mindully.org/Farm/US-Corn-Production1999.htm **1999,** 106. (September 2, 2012).
6. Allmaras, R. R.; Wilkins, D. E.; Burndisde, O. C.; Mulla, D. J. Agricultural Technology and adoption of Conservation Practices. In: The State of Site Specific Management for Agriculture. Agronomy Association of America, Madison, Wisconsin, USA, **1997,** 99–158.
7. OSU, Zea: Introduction. http://www.gramene.org/species/zea/maize_intro.html **2009,** 1–3 (August 20, 2012).
8. Lal, R.; Kimble, J. M.; Follet, R. F.; Cole C. The potential of US Crop land to sequester carbon and mitigate the greenhouse effect. Slleping Bera Press, Ann Arbor, Michigan, MI, USA, **1998,** 335.
9. Amos, B.; Arkebauer, T. J.; Doran, J. W. Soil Surface Fluxes of Green House gases in an Irrigated Maize-based Agroeocsystem. Soil Science Society of America Journal **2005,** *69,* 387–395.
10. Aguillar-Jimemez, C. E.; Tolon-Becerra, A.; Lastra-Bravo, X. Assessment of the maize (*Zea mays*)-Macun (*Macuna deerringiniam*) Agroecosystem. American Journal of Agricultural and Biological Sciences **2010,** *7,* 186–193.
11. IPNI, Maize Planting and Production in the World. International Plant Nutrition Institute, Norcross, Georgia, USA, http://www.ipni.net/ppiweb/nchina.nsf/$webindex/3FA09D72EC945–883482573BB0030EB2A **2008,** 1–8. (July 17, 2012).
12. Beck, D. Research on Tropical Highland Maize. http://www.cimmyt.org/Research/Maize/results/mzhigh99–00mrhigh99–00_res.pdf **2000,** 1–12. (May 12, 2012).
13. Racz, F.; Illes, O.; Pok, I.; Szoke, C.; Zsubori, Z. Role of sowing time in Maize Production. www.date.hu/acta-agraria/2003–11i/racz.pdf **2003,** 1–6 (August 20, 2012).
14. Vasileiadis, V.P., Sattin, M., Otto, S., Veres, A., Ban, R., Pons, X., Kudsk, P., Van der Weide, R., Czembor, E., Moonen, A.C., Kiss, J. Crop protection in European Maize-based cropping systems: Current Practices and Recommendations for Innovative integrated Pest Management. *Agricultural Systems, 104,* 533–540.
15. Kamara, A. Y.; Ekeleme, F.; Chikoye, D.; Omogui, L. O. Planting date and cultivar effects on grain yield in dry land corn production. *Agronomy J.* **2009,** *101,* 91–98.
16. NAB, Grain controlled crops-White Maize. http://www.nab.com.na/white_maize.htm **2007,** 1–2 (June 2012).
17. Krishna, K. R. Agroecosystems of South India: Nutrient Dynamics, Ecology and Productivity. BrownWalker Press Inc.: Boca Raton, Florida, USA, **2010,** 543.
18. IKISAN, Maize: Crop Technologies. http://www.IKISAN.com/lomks/ ap_maizeCropTechnologie s.shtml. **2007,** 1–9.
19. Dowswell, Paliwal, C. R.; R. L.; Cantrell, R. Maize in the Third World. Westview Press, Boulder, Colorado, USA, **1996,** 275.
20. Pray, C.; Rozelle, S.; Huang, J. Country Case Study in China. In: *Maize Seed Industry.* Morris, M. L. (Ed.). Lynne Reinner Publishers: USA, **1998,** 319.
21. Devereux, A. F.; Fukai, S.; Hullugale, N. R. The effects of Maize rotation on Soil quality and Nutrient availability in Cotton-based cropping. www.reginal.org.au/au/asa/2008/concurrent/plant_nutrition/5815_devereuxaf.htm **2008,** (May 15, 2012) 1–5.
22. FAO, Food and Agricultural Organization of the United Nations: Economics and Social Department-Statistics Department. www.faostat.fao.org **2008.**

23. Ping, J. L.; Ferguson, R. B.; Dobermann, A. Site-Specific Nitrogen and Planting Density Management in Irrigated Maize. Agronomy Journal **2008,** *100,* 1193–1204.
24. Ofori, E.; Kyei-Baffour, N.; Agodzo, S. K. Developing effective climate information for managing rained crop production, in some selected farming centers in Ghana. Proceedings of the School of Engineering Research. Accra, Ghana, **2004,** 1–18.
25. Duvick, D. N.; Cassmann, K. G. Post Green revolution trends in yield potential of temperate Maize in the North Central USA.; Crop Science **1999,** *39,* 1622–1630.
26. Egli, D. B. **2008,** Comparison of Corn and Soybean Yields in the USA-Historical trends and future prospects. Agronomy Journal 100 S78–79.
27. Ofori, E.; Kyei-Baffour, N. Agrometeorology and Maize Production in Africa. www.wmo.int/pages/prog/wcp/agm/gamp/documents/chap13C-draft.pdf **2008,** 1–27. (July 23, 2012).

EXERCISE

1. Classify maize-cropping zones of different continents based on Productivity levels into low, medium and high.
2. Mention the states that constitute 'Corn Belt of United States of America.'
3. List the top 10 maize producing countries. Write about soils, agro-climate and productivity pertaining to each nation.
4. Mention various Industrial uses of maize.
5. Discuss in detail origin and centers of genetic diversity of maize.

FURTHER READING

1. Bennetzen, J. E.; Hake, S. C. In *Hand Book of Maize: Genetics.* Springer Verlag Inc: Heidelberg, **2009,** pp. 800.
2. Krishna, K. R. In *Maize Agroecosystem: Nutrient Dynamics and Productivity.* Apple Academic Press Inc.: Toronto, New Jersey, **2013,** pp. 341.
3. Panda, C. S. In *Maize Crop Science.* Riddhi International: New Delhi, **2010,** pp. 210.
4. Wayne Smith, C.; Javier Betran, E. C.; Runge, A. In *Corn: Origin, History, Technology and Production.* John Wiley and Sons: New York, **2004**, 976.

USEFUL WEBSITES

http://www.cymmit.org
www.agronext.iastate.edu/corn
corn.agronomy.wisc.edu
http://cropwatch.unl.edu/web/soils/home

CHAPTER 5

WET LAND RICE AGROECOSYSTEM

CONTENTS

5.1 INTRODUCTION

Rice is an important cereal crop of the world. It serves as a major food grain to nearly 2 billion people, equating to one third of global population. Globally, rice is grown on 161.4 m ha. Its production in 2009 amounted to 704.4 MT paddy, i.e., 476 t milled rice [1] at an average productivity of 4.2 t/ha. Asians produce and consume nearly 95% of the global rice. The rice agroecosystem of the world thrives on variety of soils and climatic conditions experienced in over 115 countries situated between 53° N to 35° S. China, India and other South-east Asian nations support a very large *Rice Agroecosystem.* Rice belt of India extends into 44.6 m ha and contributes about 86 m t of rice grains annually. Historically, domestication of rice and its culture for regular consumption as carbohydrate source occurred in north-east of Indian subcontinent. Rice was associated with Chinese and Indian civilizations since 6000 years [2]. Archeological samples of rice from Chalcolithic period were estimated between 6570 and 4530 B.C. [3, 4]. Rice cultivation began in Gangetic plains around 2000 to 1500 B.C. Rice culture spread to South India around 1400 B.C. Archaeological studies on Neolithic sites in North Karnataka and Andhra Pradesh, in India indicate that rice cultivation occurred in these locations around 1500 B.C. [5]. Fuller et al. [5] believe that rice may have also been independently domesticated in South India around 1st millennium B.C. Randhawa [6] states that rice spread into Southern Indian plains during Iron Age.

Regarding rice types, *Indica* rice spread into Malaysia and Indonesian archipelago around 2000 to 1400 B.C., and into Sri Lanka around 543 B.C. It also traversed to Coastal South China and Yangtze river valley. Grains of *Hsien* or *Indica* rice found in east China were estimated to be 5,000 years old [7]. The *Javanica* rice originated in Asia and moved into Taiwan, Philippines and Indonesia between 2000 and 1000 B.C. [8] (Table 1). Roschevez [9] has stated that rice cultivation might have started in India by 2800 B.C.

Archeological remains from Harappan culture indicate that cultivation of rice was in vogue by 2500 B.C. The *Rigveda* suggests that divergent rice types existed in the area comprising Assam, Yunan area, Indo-China and Malayan peninsula [10]. Reports suggest that *O. sativa indica* was domesticated in Himalayan zone and *O. sativa japonica* in Southern China. Rice is said to have originated in Assam, a north-east state in India and Yunnan in China [11]. China is one of the centers of origin of rice. Rice cultivation was practiced in China during Sheng Nung period (2700 B.C.) [12, 13]. Asian rice spread into Korea in 1030 B.C. [14]. Rice moved into Japan via Korea. Actually, rice cultivation began in Japan around 1000 B.C. corresponding to late *Jomon* period [15]. Some estimates indicate that rice was introduced into Japan around third century B.C. from China. Romans learnt about rice and its cultivation through the expedition of "Alexander the Great" to India during 327 to 324 B.C. Rice was introduced into Europe through several routes. Rice seems to have moved from Persia to Egypt during first century B.C. Rice cultivation spread from Greece or Egypt to Spain. Italy and Balkans received rice sometime during 13th to 18th century A.D. It is said that Turks brought rice from India to Balkans. Russians received rice from both Korea and China in the east as well as from Persia via Caspian Sea in early 1770s [7].

Rice cultivation began in USA, during early 17th century. Initially, rice plantings were made in Virginia, but its cultivation stabilized by 1690 A.D. with regular cropping in South Carolina. Later, its cultivation spread into low land areas of Mississippi, Louisiana, Texas, Arkansas and parts of Florida around Everglades.

Regarding the West African rice, *O. glaberrima* was domesticated from *O. barthii*. The primary center of origin is supposedly in Niger delta. The secondary center of origin for African rice exists around Guinean coast [16] (Table 1). The West African rice *O. glaberrima* exhibits relatively less diversity owing to its shorter history and restricted cultivation zones. Cultivars of *O. glaberrima* are confined to savannas in West Africa.

TABLE 1 Rice: Botany, Classification, Nomenclature, Types and Uses—A summary.

Botany and Classification

Rice belongs to division Magnoliophyta; class Liliopsida; order Poales; family Poaceae; tribe *Oryzae* and genus *Oryza*. The genus *Oryza* includes 20 wild and cultivated species. Commonly grown rice or Asian rice is known as *Oryza sativa* L. It is commercially grown and consumed in 112 countries. Where as, *Oryza glaberrima* Stued is West African rice. Wild species of rice is widely distributed in humid and sub tropics of Asia, Africa, Central and South America and Australia (Kipple and Ornelas, 2006). Wild species have both diploid ($2n = 24$) and tetraploid ($2n = 4\times = 48$). Cultigens are diploid. It is believed that *O. glaberrima* in Africa and *O. sativa* evolved from perennial wild species. Their parallel evolution is as follows:

Africa: *O. longistaminata* > *O. barthii* > *O. glaberrima*

Asia: *O. rufiipogon* > *O. nivra* > *O. sativa*

Nomenclature

Indian languages: *Arisi* in Tamil, *Akki* or *Batha* in Kannada, *Biyam* in Telugu, *Ari* in Malayalam, *Chawal* in Konkani *Chawal* in Hindi, Punjabi, *Chawal* in Orriyan, *Chawal* in Bengali, *Chawal* in Assamee, *Chawal* in Rajasthani, *Chawal* in Marathi, Other languages: *Rice* in English, *Fan* in Chinese, Ar-*ruzz* in Arabic, *Riz* in Russian, Arroz in Spanish, *Riz* in French, Arrouz in Portuguese; *Oryza* in Greek; *Oriza* in Latin; *Riso* in Italian; *Reis* in German.

Types of Rice

Evolution and selection process during past centuries has led to development of three types of rice. The differentiation of rice varieties is based on amylose content. *Indica* rice is high in amylose. Its grains are cooked fluffy. *Japonica* types are low in amylase. These are cooked sticky and made suitable to be eaten with chopsticks. *Javanica* types are intermediate in amylose content and stickiness. Rice is also classified as long, medium or short-grained varieties. Flavored rice types are also common.

Uses

In general, demand for rice and its products partially decide intensity and expanse of rice agroecosystem. Of course, rice is a staple cereal and demand for it is almost inelastic in South-east Asia. In South Asia and Fareast, rice cultivation depends to a large extent on various uses of rice grain and straw. Rice is used in several ways. There is wide range of food preparations. Rice varieties are highly variable in their culinary characteristics. South Asians use different varieties such as long grain, short grain, white polished or brown red, basmati or jasmine, parboiled etc. Majority of the rice is consumed as white rice after being cooked in boiling water or in a pressurized cooker. Actually, white rice is derived after removing bran and impurities. Brown rice needs longer cooking time and is supposedly richer in vitamins, especially vitamin B. Rice puddings (*Idly*) prepared from fermented rice plus black gram mix is a favorite breakfast item among Southern Indians. Similarly, pancakes (*dosas*) prepared from fermented rice dough are relished by Southern Indians. Preparations from rice are popularly consumed during rituals and festivals in southern India. Rice preparations are consumed during Major festivals like Pongal in Tamil Nadu, Onam in Kerala, Huthri in Coorg, Shankranthi in Karnataka and Andhra Pradesh.

Other Uses

Husks and panicles of rice are used for bedding, fuels and building board. Charred rice grains are useful as filtration material and in preparing charcoal briquettes. Rice bran oil is used in cooking. It also has anticorrosive properties useful in emulsions and paints. Rice straw is a good animal feed. Straw is useful raw material in paper industry. Rice straw is a good substrate for mushroom production. Toughened rice straw is useful in preparing ropes and thatching/roofs for hutments. Rice extracts, oils and starch are important base materials for several types of cosmetics. Rice oil is an important component in sun care products that absorb UV rays and in air conditioners. Rice proteins are known to nourish good hair growth. Rice is cholesterol-free hence useful medicinally. Rice bran is rich in fiber that assists reduction of fats in gut. Rice fiber aids better digestion. Oil from rice is rich in vitamin E and vitamin E group of compounds. It has antioxidant qualities. Rice based syrups are components in medicinal preparations. In India, rice is called the *wonder grain* or *blessed grain* because it adorns most of our families, blesses our ceremonies and decorates our homes.

Nutritive Value (100/g grain)

Energy–365 k cal; Carbohydrates–80 g; Dietary fiber–1.3 g; Fat–0.66 g; Protein–7.13 g; Water–11.61 g; Thiamine–0.07 mg; Riboflavin–0.1 mg; Niacin–1.6 mg; Pantothenic acid–1.0 mg; Vitamin B6–0.16 mg; Ca–28 mg; Fe–0.8 mg; Mg–1.0 mg; P–115 mg; K–115 mg; Zn–1.0 mg.

Sources: http://en.wikipedia.org/wik/rice.html; http://www.plantcultures.org.uk/plants/rice_otheruses.html (November 22, 2012); http://www.plantcultures.org.uk/plants/rice_western_medicine.html; Refs. [1, 17–21]; http://www.nal.usda/fnic/foodcomp/search. (November 22, 2012).

5.2 SOILS, AGROCLIMATE AND WATER RELATIONS

5.2.1 SOILS OF RICE AGROECOSYSTEM

The rice growing regions that occupy the swamps and delta region of Mississippi in Lousiana, those in adjacent states thrive on sandy Ultisols, Inceptisols and Alluvial soils. In Florida, rice-growing zones of Everglades survives on Histosols, Spodosols and Ultisols. The spodosols are richer in nutrients and organic matter.

The West African rice belt spreads into areas with Coastal sandy soils, riverine alluvial soils, sandy Oxisols and Alfisols. Soil fertility is relatively low and needs periodic replenishment of nutrients through fertilizers and organic manures. Rice cropping zones in Southern and Eastern Africa thrive on Inceptisols, Alfisols and Coastal soils. Soils are moderately fertile. Tropical condition induces loss of soil organic carbon and nutrients.

Much of the South Indian rice agroecosystem thrives on Ustalfs (Troporthents) and Usterts [22]. A large portion of rice agroecosystem in Andhra Pradesh actually thrives on deltaic alluvial soil. These are derived from riverine or marine deposits. They are deep, medium to fine textured and neutral or alkaline in reaction. Soils used for paddy cultivation are relatively more fertile. Major soil types encountered in the rice belt are Alfisols, Vertisols, Red laterites and Coastal sandy soils. In the Gangetic Plains, rice agroecosystem flourishes on Inceptisols and Alluvial soils.

5.2.2 PRECIPITATION, WATER REQUIREMENTS AND WATER USE EFFICIENCY

Historically, rice was first cultivated as a rain fed crop, suited better to low lying areas where rain water could be retained for a length of time [17]. Usually, such areas were marshy, flood-free sites on the banks of river and lake. However, through time, irrigated rice that yielded better gained in importance. Much of the Indian rice belt lies in semiarid plains. The rice-growing region depends predominantly on monsoon rains and irrigation facilities. The Indo-Gangetic Rice belt receives 700–900 mm precipitation annually. Yet precipitation could be erratic, hence irrigation supplement is common. In Southern India, Tamil Nadu region receives between 940 to 1,000 mm rainfall annually. About 325 mm is contributed by South-west monsoon and 470 mm by North-east monsoon. The rice belt in Kerala receives relatively greater amounts of precipitation during the year. The average annual precipitation is 2823 ± 409 mm. The precipitation pattern is almost stable and dependable. The rice cropping zone of Andhra Pradesh receives precipitation through monsoons. The summer monsoon lasts from June to September and contributes about 65% (600 mm) of total precipitation. The winter monsoon lasting from October to December supplies remaining 35% precipitation. On an average, rice regions here receive 1,000 mm. Annual precipitation in the rice growing zones of Karnataka varies depending on geographic location. In the plains, it is drier and annual precipitation may not exceed 900 mm. Whereas, in Western Ghats it reaches as high as 2,300 mm per year.

The water requirement of rice crop varies slightly with geographic location. Water level in the low land paddy fields are maintained at 1.5 cm at transplantation and increased gradually to 5 cm until maximum tillering. Water is drained off the field about 15 days prior to harvest. Critical stages are maximum tillering and heading periods. For summer rice, 5 cm irrigation every two days after disappearance of pounded water is essential. Where water resources are limited phasic stress is practiced. It reduces consumption of water by 25–30%, but grain formation is reduced by at least 5–10% compared to normal [23].

The water requirement of a rice crop or sequences involving it is higher than other cereals. A rice-rice sequence uses as much as 230 to 240 cm per sequence. Whereas, rice sequenced with legumes or groundnut requires 125 to 140 cm per sequence. The water use efficiency (WUE) is better, if rice is sequenced with a legume or oil seed (sunflower or groundnut) or dry land crop. The WUE obtained through rice-legume or rice-oilseed sequences ranges from 78 to 97 kg/ha cm. Whereas, WUE is least at only

46 kg/ha cm for rice-rice intensive cropping sequence. The WUE could be significantly improved by intercropping rice with legumes such as black gram or green gram

5.2.3 RICE-BASED CROPPING SYSTEMS

Rice farmers practice several different combinations of cropping sequences. Crop sequences such as Rice-Wheat-Fallow and Rice-Wheat-short season legume or Rice-Wheat-vegetable are common in Indo-Gangetic Plains. Regarding South India, for example, Rice-Rice-Gingelly/groundnut intercrop; Rice-rice-pulse-groundnut-pulse; rice-rice-pulse are popular in coasts [24]. Rice monoculture, rice-rice-sesamum or rice-rice-pigeonpea are popular with farmers in Thanjavur district and other Coastal districts of Tamil Nadu in India. Traditional rotations that involve rice are rice-rice-black gram, maize-rice-black gram, maize-rice-sun hemp, rice-rice-sun hemp, maize-rice-sesame and rice-banana.

Rice is followed by an intercrop of pulses such as pigeonpea, cowpea, green gram or black gram with chilies in the Vertisol plains of South India. Rice-rice monoculture is also common, especially in southern districts of Karnataka. Farmers in Kerala adopt several combinations of rice-based rotations. Two most common rotations involving rice are rice-rice-vegetable or pulse or oilseeds or green manure and rice-rice-fallow [23].

Rice monoculture dominates the lowland flooded ecosystem of Andhra Pradesh, especially where irrigation facilities are commensurate. Rice-rice-green manure or rice-rice-fallow is a preferred rotation in Nagarjuna sagar command area as well as delta zones of Godavari and Krishna [25]. There are many variations of rice-based cropping sequences followed by farmers in this part of South India. A few examples are as follows: Rice-Indian Mustard, Rice-green gram, Rice-soybean, Rice-black gram, Rice-groundnut, Rice-sunflower, Rice-maize and Rice-rice mono cropping [26, 30].

5.3 EXPANSE AND PRODUCTIVITY

Rice agroecosystem dominates the agrarian landscape of South and South-east Asia. Lowland ecosystem is large and contributes most to global rice harvest. The upland paddy is important in the arable and dry tracts. Major rice producing nations of the world are China (29.9 m ha; 191 mt/yr), India (44.1 m ha; 131 mt/yr), Indonesia (12.8 m ha; 64 mt/yr), Bangladesh (11.5 m ha; 45 mt/yr), Thailand (10.9 m ha; 31.6 mt/yr) and Vietnam (7.4 m ha; 38 mt/yr) [27].

5.3.1 RICE IN CHINA AND SOUTH-EAST ASIA

Rice agroecosystem is historically more than 6,000 years old. It has expanded slowly commensurate with enlargement of agrarian zones and demography. The intensity of rice production, crop genotype, soil fertility and yield obtained has all altered to different extents through the ages. The nutrient cycling and recycling procedures adopted by Chinese farmers have played an important role in regulating the rice agroecosystem. The rice agroecosystem extends into all agrarian zones in China (Plate 1). Rice occupies predominantly irrigated zones. It is cultivated intensively in the Huang He and Yangtze riverine zones. Riverine irrigation in the South-west China is another important rice production area. It extends into 31 m ha. The productivity of rice is high due to large inputs of fertilizer-based nutrients and organic matter, in addition to timely irrigation. The rice agroecosystem is also sown mostly with high yielding cultivars. During recent years, high yield hybrids with better harvest index have replaced the rice agroecosystem. The productivity of rice fields in China is 6.6 t grain/ha plus 12–15 t forage. The total rice production in China is 195 m t/yr [20, 31]. About 2% of Chinese rice production is contributed by upland paddy fields.

Major soil types that support Chinese rice belts are Inceptisols, Oxisols, Alfisols and Alluvial soils. Rice crop is transplanted throughout the year across the nation, wherever weather parameters are matching and congenial, offering high yield goals. Irrigated, lowland rice requires about 1200–1800 mm per season, which needs to be met via riverine irrigation, ground water or precipitation. Flooded agroecosystem in China and other South-east Asian nations includes a conspicuous, floating fern called Azolla. The azolla is a fern that harbors nitrogen fixing blue green algae. Hence, its presence in the flooded paddy fields adds to soil-N. On an average, 20–40 kg N/ha is added per season. Azolla adds to soil organic carbon when its biomass is incorporated. The succulent fern decomposes rapidly releasing nutrients into soil. *Azolla pinnata* and *Azolla sp* are common in Asian lowland rice ecosystem. Rice cultivation involves flooding of soil that generates anaerobic conditions in soil. It has consequences on soil microflora and physicochemical process. However, in many locations, rice is rotated with an arable crop that restores oxidative nature of soil profile.

PLATE 1 Terrace farming of Rice in South-east Asia, particularly Philippines (left) and China (right).

Note: Terrace farming offers a unique dimension to nutrient dynamics during rice culture. Firstly, plots are small. Large-scale tillage is totally unsuitable. Soil erosion, loss and carry-over of nutrients/water from upper to lower terrace need specific soil management procedures. Nutrient supply has to tailored after due consideration to nutrients that leach or add up from time to time.

Source: International Rice Research Institute, Manila, Philippines.

Rice is rotated with crops like maize, wheat, legumes and vegetables. Relay cropping systems that allow rice fields to be intercropped initially under arable (no-flooding) conditions are possible. For example, it is possible to cultivate an arable legume sown ahead of paddy transplantation. Paddy fields could be flooded after harvest of the first crop, which is generally a short duration crop. Therefore, rice agroecosystem in China, or anywhere else in South-east Asia, literally is a conglomerate of continuous paddy (flooded) and arable crop that is grown in rotation or even as intercrop.

5.3.2 RICE IN INDIA

Rice is an important cereal crop of India. The rice agroecosystem in India is vast and spans into 45 m ha. During 2009, total rice production in India was 148 m t annually. It contributes 42% of total food grains produced in the country. The productivity of rice is around 3.7–4.0 t grains/ha in the Indian subcontinent. The rice agroecosystem in India consumes about 40% of chemical fertilizer used annually by the nation. It consumes relatively high amounts of nutrients and water. Nutrient supply and turnover rates are held high to achieve better productivity. The rice-wheat system of Gangetic belt is a vast stretch that annually generates food grains for over 550 million. The Southern Indian Rice agroecosystem extends into states such as Andhra Pradesh, Tamil Nadu, Karnataka, Kerala, Pondicherry and Andaman-Nicobar islands. It extends into at least 8.5 m ha. Rice farming is intense in this zone, owing mainly to demand and farmer's desire for higher harvests. Intensification has been achieved using high yielding varieties or hybrids, inorganic fertilizers and irrigation. Since past 15 years, rice hybrids are gaining in area with in the Asian lowland agroecosystem. It is mainly due to better grain and forage yield (Plates 2 and 3).

PLATE 2 Paddy field in the Alfisol plains of South India.
Note: Rice genotypes thrive on Red Alfisols irrigated intermittently allowing a longer stretch of dry arable conditions in paddy fields. It improves water use efficiency and reduces water requirements.
Source: Main Research Station, Gandhi Krishi Vignana Kendra, University of Agricultural Science, Bangalore, India.

Rice is primarily cultivated in irrigated wetlands available in plains, coasts and hilly tracts. Uplands with relatively lower precipitation levels are also used. The rice agroecosystem includes intense, moderately intense and subsistence farming zones. Obviously, productivity of rice varies enormously based on water supply levels, nutrient inputs and cultivation practices. Overall, agroclimate that prevails during rice cultivation immensely influences productivity. For example, rice grown under irrigated conditions occupies nearly 46% of total rice area in India and productivity ranges from 2.8 to 3.5 t/ha. Whereas, upland paddy cultivated under favorable conditions yields 1.5 t/ha, but under drought conditions grain formation is limited to 0.6 t/ha. Upland rice occupies about 14% of rice zone. Rain fed low lands too contribute large portion of rice grain produce. Rain fed lowland constitutes about 29% of total rice area in India. On an average, low lands that are rain fed may yield 1.2–1.5 t/ha. Coastal wetlands constitute only 2–4% of rice area in South India. Here, productivity of rice is not high but restricted to 1.0 t/ha [4].

PLATE 3 A stall canvasing for high protein rice genotypes at a Farm Exhibition, GKVK Agricultural Station, Bangalore, South India. Rice genotype with enhanced protein concentration in the grain is preferred.
Source: Krishna, 2011

5.3.3 RICE IN WEST AFRICA

Rice is an important cereal in parts of West African coastline countries and riverine zones. The West African rice belt extends into 2.56 m ha and contributes 4.1 m t grains to the total global harvest [29]. The rice species that flourishes in this ecosystem is *O. glaberrima*. It is known to cook into a good starchy, paste and the population relishes it as a carbohydrate source. Rice is dominant crop in parts of Liberia, Ghana, Ivory Coast, Gambia and Senegal. The sandy Oxisols that are slightly low in organic matter content, support rice cultivation in swampy conditions. The flooded ecosystem is mostly supplied with stored water collected via precipitation. Matching precipitation pattern with rice transplantation is important in this zone. It helps in improving precipitation use efficiency. Supplemental irrigation from river and canals are possible in some areas of Senegal, Liberia, Ghana and Nigeria. Paddy transplantation is usually done around June/July at the onset of rains and harvested by October/November, depending on the variety. Soils are moderately fertile and need periodic replenishment of chemical fertilizers and FYM. Annual supply of NPK fertilizer may fluctuate between 140–200 kg/ha, depending on yield goals. The rice belt in West Africa is therefore moderately intense. Paddy is often rotated with legumes such as cowpea, macuna or vegetables.

5.3.4 RICE IN NORTH AMERICA

Major rice production zones of North America are actually scattered in the southern states of USA, mainly Alabama, Texas, Lousiana, Mississippi and Florida. The North American rice belt contributes 11.3 m t grains annually. Louisiana and Florida the most important regions that support flooded rice ecosystem. The Mississippi–Missouri swamps in the delta region and Everglades in South-Central Florida support sizeable production of rice. The red rice is the most preferred rice variety in Southern United States of America. Major soil types that support rice production in USA, are the Ultisols, Spodosols and Coastal sandy soils. These soils are moderately endowed with nutrients and soil organic matter and need replenishments based on nutrient depletion trends and yield goals. The rice season begins during April/May and lasts till October. Rice is harvested well before the onset of cold spell in the southern states.

KEYWORDS

- **Alfisols**
- **Alluvial**
- ***Azolla pinnata***
- ***Dosas***
- ***Idly***
- **Inceptisols**
- ***O. glaberrima***
- **Oxisols**
- **Troporthents**
- **wonder grain**

REFERENCES

1. FAO. FAOSTAT agricultural data. http://faostat.fao.org. Rome, Italy. **2006** (October 5, 2012).
2. Kumar, T. T. History of Rice in India: Mythology, Culture and Agriculture. Gian Publication, New Delhi, **1988,** 242.
3. Vishnu-Mittre, B. Discussions in Early History of Agriculture. Philosophical Transactions of Royal Society of London. **1976,** *275,* 141.
4. Sharma, S. K.; Gangwar, K. S.; Pandey, D. K.; Tomar, O. K. Increasing Productivity of Rice-based systems for Rainfed Upland and Irrigated areas of India. Indian Journal of Fertilizers **2006,** *2,* 29–40.
5. Fuller, D. Q.; Korisettar, R.; Venkatasubbiah, C.; Jones, M. K. Early plant domestications in Southern India; Some Preliminary Archaeobotanical results. Vegetational History and Archaeobotany **2004,** *13,* 115–129.

6. Randhawa, M. S. A History of Agriculture in India. Indian Council of Agricultural Research, New Delhi, **1980,** 243.
7. Lu, J. J.; Chang, T. T. Rice in its temporal and spatial perspectives. In: Rice Production and Utilization. Luh, B. S. (Ed.). Westport, Connecticut. USA, **1980,** 1–74.
8. Chang, T. T. The Ethnobotany of Rice in Human Civilization and Population expansion. Interdiskiplinary Science Reviews **1988,** *12,* 63–69.
9. Roschevez, R. J. A contribution to the knowledge of Rice (in Russian) with English summary). Bulletin of Applied Botany, Genetics and Plant Breeding. Leningrad, Russia, **1931,** *27,* 1–133.
10. Manasala, K. A new look at Vedic India. http://asiapacificuniverse.com/pkm/vedicindia.html. **2006,** 1–17 (October 10, 2012).
11. Zhimin, A. Origin of Chinese Rice cultivation and its spread East. http://http-server.carleton.ca/~bgordon/rice/ papers/zhimin99.htm **1999,** 1–11 (September 2, 2012).
12. Chang, T. T. Crop History and Genetic conservation in Rice- a case study. Iowa State Journal of Research **1985,** *59,* 405–455.
13. Ting, Y. Chronological studies of the cultivation and distribution of rice varieties *Keng* and *Sen* (in Chinese with English summary). Sun Yetsen University Agronomy Bulletin **1969,** *6,* 1–32.
14. Chen, W. H. Several problems concerning the origin of Rice growing in China (in Chinese). Agricultural Archeology **1989,***1,* 173–183.
15. Akazawa, T. An outline of Japanese Prehistory. In: Recent Progress of Natural Sciences in Japan. Anthropology **1983,** *8,* 1–11.
16. Harlan, J. R. Genetic Resources of some Major Field crops in Africa. In: Genetic resources in plants-their exploration and conservation. Frankell, O. H.; Bennet, E. (Eds.) Philadelphia, Pennsylvania, USA, **1973,** 19–32.
17. Kipple, K. F Ornelas, K. C. Rice. The Cambridge World History of Food. http://www. cambridge.org/us/books/kiple/rice.htm **2006,** 1–22 (October 20, 2012).
18. Londo, J.; Chiang, Y, Hung, K.; Chiang, T.; Schaal, B. A. Phylogeography of Asian wild rice, *Oryza rufipogon*, reveals multiple independent domestications of cultivated rice, *Oryza sativa*. Proceedings of National Academy of Sciences of USA, **2006,** *103,* 9578–9583.
19. IRRI, Rice in China. www.irri.org/partnerships/contents_profiles/ asia_oceania/india/rice_ch-china. **2012b,** 14 (October 12, 2012).
20. FAOSTAT, Rice statistics. Food and Agricultural Organization of the United Nations. Rome, Italy. http://www.faostat.org **2008** (October *10,*2012).
21. Krishna, K. R. Rice Agroecosystem of South India. In: Agroecosystems of South India: Nutrient Dynamics, Ecology and Productivity. BrownWalker Press Inc.; Boca Raton, Florida, USA, **2010,** 565.
22. Brady, N. C. US Soil Taxonomy. In: Nature and Properties of Soils. Prentice Hall of India. New Delhi, **1990,** 623.
23. Kisssankerala, Crop Information: Rice (*Oryza sativa*). http://www.kissankerala.net/kissan/kissan/kissancontents/rice.htm **2006,** 1–13 (October 10, 2012).
24. TNAU, [15]N Fertilizer Studies. Soil Science Laboratory, Department of Soil Science and Agricultural Chemistry, Tamil Nadu Agricultural University, Coimbatore, TN, India http://www.tnau.ac.in/scms/SSAC/Res/ SACN151.htm **2003,** 1–6.
25. ANGRAU, Integrated Nutrient Management. http://www.angrau.net/research-management.htm **2006,** 1–6 (July 10, 2012).
26. Avil Kumar, K.; Reddy, N. V.; Sadasiva Rao, K. Profitable and energy-efficient Rice (*Oryza sativa*)-based cropping systems in Northern Telangana of Andhra Pradesh. Indian Journal of Agronomy **2005,** *50,* 6–9.

27. FAOSTAT, Rice Statistics. Food and Agricultural organization of United Nations, Rome, Italy, http://www.fao.org. **2010.**
28. IRRI, Rice in India. www.irri.org/partnerships/contents_profiles/asia_oceania/india/rice_in_india. **2012a,** 12 (October 12, 2012).
29. CILLS, West Africa's Rice Production To Decline 12% In 2011–12. Oryza News, 2012, *4,* 1. http://Oryza.Com/Rice News/14868.html.
30. Rao, T. N. Rice Productivity in Kerala: Challenges and Strategies. Potash and Phosphate Institute, Gurgaon, India, www.ppi-ppic.org/ppiweb/sindia.nsf **2006,** 1–5 (August 12, 2012).

EXERCISE

1. Mention names of wild and domesticated species of Rice.
2. Mention different types of Rice used in India, Fareast and West Africa.
3. Delineate the Rice growing regions of the World on a Map.
4. Mention at least 10 currently cultivated varieties of rice.
5. Mention the top 10 nations that contribute to global rice harvest.
6. Mention the various crop rotations followed in different continents that include rice.
7. What is hybrid rice? Mention its potential yield in China and India.

FURTHER READING

1. Bouman, B. A. M.; Hengsdijck, H.; Hardy, L.; Bindralan, P. S.; Toung, J. P.; Ladha, K. In *Water-wize rice Production.* International Rice Research Institute: Los Banos, Philippines, **2002,** 356.
2. Dowling, N. G.; Greenfield, S. M.; Fischer, K. S. In *Sustainability of Rice in the Global Food System.* International Rice Research Institute: Manila, Philippines, **1998,** pp. 404
3. Floresca, J. P.; Alcanatara, A. J.; Lamug, C. B.; Rapera, C. L.; Adall, C. B. Assessment of Ecosystem Services of Low land Rice Agroecosystem in Echagua, Isabela, Philippines. *J. Environ. Sci. Manag.* **2009,** *12,* 140–173
4. Jena, K. R.; Hardy, B. Advances in Temperate Rice Research. International Rice Research Institute, Los Banos, Philippines, **2012,** pp. 105.
5. Kush, G. S. Breaking Yield Frontier of Rice. *Geo J.* **1996,** *35,* 329–332.
6. Mohamed Sah, S. A.; Mohamed Shah, A. S. R.; Mohamed, C. S. In *Rice Agroecosystem,* USM Press: Buku, Malaysia, **2009,** pp. 242
7. Timsina, J.; Buresh, R. J., Dobermann, A.; Dixon, J. In *Rice-Maize system in Asia: Current Situation and Potential.* CIMMYT-IRRI: Manila, Philippines, **2011,** pp. 242.
8. Xie, F.; Hardy, B. In *Accelerating Hybrid Rice Development.* International Rice Research Institute: Manila, Philippines, **2009,** pp. 687.

USEFUL WEBSITES

http://irri.org/index.php?option=com_k2&view=itemlist&layout=category&task=category&id=408&Itemid=100301&lang=en (October 10, 2012).
http://irri.org/index.php?option=com_k2&view=item&id=11368%3Anmrice&lang=en (October 14, 2012).

CHAPTER 6

SORGHUM AGROECOSYSTEM OF ASIA, AFRICA AND AMERICAS

CONTENTS

6.1 INTRODUCTION

Seeds for a sorghum agroecosystem on earth were sown some 4000 years ago, by the Neolithic humans that inhabited Eastern African highlands. Ethiopians were earliest to domesticate sorghum landraces and use it as staple food. Botanical surveys have indicated that, greatest genetic diversity for genus *Sorghum* is found in the region comprising Nile belt of Southern Egypt, parts of Chad, Sudan and Ethiopian Highlands. Genetic separation of *S. bicolor*, the cultivated species, is said to have taken place in these areas during early Neolithic age. Domesticated sorghum spread from Ethiopia to other locations in East Africa, then to South and West Africa. Domesticated sorghum occurred in West Africa around 3000–2000 B.C. (Table 1). West African locations in the upper reaches of river Niger could be considered as yet another or secondary center of origin for sorghum.

Kaoliang was the term used for sorghum on the silk route between China and India. Sorghum spread into China and Fareast via trade routes. The Chinese mainland received amber sorghums (*sorghos*) that are useful as forage and in making syrup. Sorghum was also referred as *Samshu* (*fiery spirit*) in ancient China. Archaeological studies indicate that sorghum was cultivated in Arabia and Yemen in particular, around 2500 B.C. Assyrians cultivated sorghum as early as 700 B.C. [1]. In the Mediterranean countries, it was called *Juar-I Hindi*, indicating that it was derived from India.

Like many other crops, sorghum too moved into Americas during mid-19th century via slave trade and merchants. Its cultivation became prominent in Southern plains, especially in Texas, Kansas and Nebraska during early 1900s. Guinea corn and chicken were exported from West African ports to North America and Brazil. Earliest of evidences indicate that Europeans cultivated sorghum around first century A.D.

Sorghum might have reached the Indian peninsula around 1500 to 1000 B.C., via trade routes from Arabian coast in East Africa to Southern Indian ports. Merchants using these trade routes introduced sorghum and other crops into South India. Sorghums derived from both North-east African coast and those from Mozambique ports are traceable in Southern Indian peninsula. Archaebotanical investigations suggest that domesticated sorghum was used by humans in Neolithic settlements of Southern India (1500 B.C.), especially in North Karnataka and Andhra Pradesh [2, 3]. Historical reports suggest that by 1000 B.C. Sorghum grains were regularly produced for food by Southern Indians. However, certain evidences indicate that true domesticated *S. bicolor* existed in semiarid zones of India, Oman and Yemen much earlier.

Sorghum was an important cereal crop grown in ancient South India. During ancient period that is between 1st to fifth century A.D. human transit and development of new settlements induced rapid spread of sorghum cultivation into several locations in South India. The expanse and intensity of its cropping improved markedly, much later, during medieval times, especially in Northern Karnataka and Andhra Pradesh. Sorghum was among major cereals grown and consumed during medieval period from 13th to 17th century [4]. During modern times, from 18th to twentieth century, sorghum cultivation spread into all three southern states. However, sorghum competed with cereals like rice, finger millet, foxtail millet, little millet and kodo millet as a source of carbohydrates. Rotations and intercrops involving sorghum were fairly standardized during modern period. Productivity of sorghum grown in Karnataka and Tamil Nadu ranged from 15–16 bushels/ha (1.0 Bu 27.5 kg grains) [5].

During early 1900s, sorghum agroecosystem further stabilized and spread into hitherto uncultivated zones in South India. It occupied between 2–3 m ha, but its productivity was low 0.2 t to 0.3 t grains/ha. During second half of twentieth century, sorghum belt experienced drastic changes in plant genotype and cropping procedures. Hybrids (CSH series) and high yielding composites (SPV series) were introduced. The National Center for Sorghum headquartered at Hyderabad, has so far disseminated over 30 high yielding varieties and hybrids that tolerate abiotic and biotic stress [6]. Further, Sorghum belt was also intensified using fertilizers and irrigation. Consequently, productivity improved to 0.87 t/ha. At present, sorghum cultivation is predominant in semiarid tracts of South India. Several types of sorghum are grown depending on local agro-climate, adaptability and yield goals (Table 1). Mostly, hybrids and improved composites dominate the sorghum belt in South India. International Crops Research Institute for the Semi-arid Tropics (ICRISAT), situated at Patancheru in South India is the world's largest germplasm repository for genus *Sorghum*. It currently holds over 37,000 accessions of sorghum. A large number of improved varieties, composites and hybrids have also been developed and disseminated into other continents from this agricultural center [7].

TABLE 1 Sorghum: Botany, Nomenclature and Uses—A summary.

Botany

Kingdom–Plantae; Division–Magnoliophyta; Class–Liliopsida; Order–Poales; Family–Poaceae; Sub family–Panicoidea; tribe–Andropoganeae; Genus–*Sorghum*; Species–*S.bicolor*

There are at least 30 species of Sorghum. The progenitor of cultivated sorghum (*S.bicolor*) is said to be *S. arundinaceaum*, which is an annual found in the tropics of Central and West Africa. *Sorghum halepense* and the perennial *S. propinquum* found in Indian subcontinent are also considered as progenitors of *S. bicolor*.

Sorghums are subdivided into four subgroups, namely Grain Sorghums (e.g., Milo); Grass Sorghums (pasture and hay), Sweet Sorghums (also called Guinea corn) and Broom corn (brooms and brushes). However, Harlan and de Wet [8] identify five basic races Guinea, Caudatum, Kafir, Durra, Bicolor and their hybrid races. They have grouped sorghums into three main sections:

Sorghum halepense are rhizomatic species of European origin that diffused into India.

Sorghum propinquum are species of South India and Srilanka that diffused into East Asia.

Sorghum bicolor is the cultivated species found in Africa and Asia.

$2n = 20$

Nomenclature of Sorghum in different Languages:

America/English-*Milo*; French-*Sorgho*; Spanish-*Sorgo*; Italian-*Sorgo*; German-*Mohrenhirse*; Arabic-*Durra*

Persian: *Juar-I-hindi*; Chinese-*Kaoling*

African Languages: Sudanese-*Durra*; Ethiopean-*Bachanta*; West African Nations -*Guinea corn*; East African nations—*Mtama, Kafir corn*

Indian Languages: Hindi-*Jowar*; Gujarati-*Jowar*; Marathi-*Javar*; Kannada-*Jola*; Telugu-*Jonna*; Tamil-*Cholam*; Malayalam-*Cholam*

Uses

At present, sorghum is a staple cereal in parts of Africa, Asia, Southern USA, Meso-and Southern America. Sorghum products are also consumed by Chinese and other South-east Asians. It is a staple diet in South India, especially in parts of North Karnataka and Andhra Pradesh. Sorghum flour is used to prepare Roti (*Bhakri*). Sorghum can be used to extract edible oil, starch and dextrose paste. It is a major source of carbohydrate and good base material for several types of alcoholic beverages. Sorghum porridges are also common in South Indian households. Sweet sorghum also called *Sorgo* is grown for its juicy stalks to extract sweet syrups, to make molasses, jiggery, etc. Sorghum is used as base material for syrups. Sorghum is an important carbohydrate source in alcohol fermentation industry. Sorghum is also used as a good forage crop. Sorghum stover is a valuable animal feed. For example, it is used as fodder for cattle. Sorghum stover (stems) is also used to make silage, brooms, rough fiber, bedding, fencing and roofing. Sorghum stalk is used as a fuel in villages, especially to cook food.

Nutritive Value (192/g grains)

Calories 651 k cal; carbohydrates-143 g; Fat-6.3 g; Protein-21.7 g; Vitamins A, B6, Folate, B12, C, D and K all found in negligible quantities. Thiamine 0.5 mg; Riboflavin-0.3 mg; Niacin-5.6 mg; Polyunsaturated fats-2.6 g; Total Omega fatty acids-125 mg; Minerals: Ca-53.8 mg; Fe-8.4 mg; P-551 mg; K-672 mg; Na—11.2 mg; Sterols-nil; Water 17.7%. Ash-3%

6.2 SOILS, AGROCLIMATE AND CROPPING SYSTEMS

6.2.1 SOILS THAT SUPPORT SORGHUM AGROECOSYSTEM

The Sorghum growing region of North America is confined mostly to Central and Southern United States of America. Here, it thrives on Ultisols and Mollisols. Sorghum meant for forage is often restricted to soils with low fertility. In Brazil, moderately productive Oxisols rich in Al and Fe salts support sorghum production. Sorghum thrives with moderate productivity on Inceptisols and Mollisols of Pampas of Argentina. Lateritic soils found in the fringes of Andes region and hilly tracts allow sorghum production. In Europe, soil types such as Cambisols, Chernozems, Loess and sandy Oxisols are used for sorghum production. Sandy Oxisols and Alfisols with deficiencies of N and P are used for sorghum culture in West African Sahel and Sudanian region. These soils are also low in organic matter. Sorghum productivity is low since it is often grown for subsistence and under low input systems. Again, much of the sorghum belts in Southern and Eastern Africa thrives on low fertility Alfisols and Inceptisols. Sandy coasts also support sorghum production. The Indian sorghum belt thrives on Vertisols, Vertic types of Entisols and Inceptisols. To a good extent, sorghum crop flourishes on Alfisols, especially Red Alfisols. Sorghum crop grows excellently on Black Cotton soils that are clayey due to higher fraction of montomorillonite clay (Vertic soils). Sorghum tolerates soil pH of 6.0 to 8.5, and mild levels of salinity and alkalinity. Lateritic soils are also used to cultivate sorghum, but it is sparse and confined to few districts of Karnataka and couple of them in Kerala. The post-rainy crop of sorghum is largely confined to Vertisols, since they are endowed with better nutrient and moisture buffering capacity.

6.2.2 AGROCLIMATE OF THE SORGHUM BELT

Topographically, sorghum agroecosystem thrives from sea level to 2300 m.a.s.l.. Sorghum grows equally well in plains, hills, undulated terrain and coastal regions. Sorghum crop is well adapted to tropical conditions and it withstands relatively higher temperatures during the growing period. It tolerates drought better compared to other cereals such as wheat or rice. Sorghum grows even in dry zones where annual rainfall is about 400 mm, but thrives reasonably well in regions with annual rainfall between 400 to 600 mm. Sorghum can be successfully grown till maturity in areas wherever annual rainfall ranges from 600 mm to 1000 mm. Long-term trials across different continents have shown that sorghum needs 500 mm moisture per season. The critical stages for irrigation are early vegetative growth, flowering and seed filling period. During a year, water balance in sorghum fields is governed by at least 4 different components. The surface drainage and runoff loss may account for 25–28%, deep percolation 9–10%, evapotranspiration (ET) during monsoon 24–28% and ET during post monsoon 40–42% [9]. Sorghum seeds germinate at temperatures as low as 7 °C to 10°C although optimum for germination is >15°C and that for crop growth is 26° to 30°C. Sorghum is a short day plant. Flowering is hastened by short days and delayed under long days [7,10].

6.2.3 SORGHUM-BASED CROPPING SYSTEMS

Sorghum is a versatile crop that adapts exceptionally well to variations in seasons, weather and cropping patterns that are commonly encountered in agrarian zones. Sorghum is cultivated as monocrop or intercropped with crops like cotton, soybean or legumes in the Southern plains of USA. Relay cropping is also practiced. Farmers in Brazilian Cerrados prefer to grow sorghum for grain in rotation with legumes, cotton or vegetables. Sorghum for forage is grown in mixed pastures. In the Pampas of Argentina, sorghum-groundnut, sorghum wheat or maize, sorghum-sunflower are most frequently adopted rotation. Sorghum groundnut intercrops improve land use efficiency [11]. In Africa, sorghum is grown in mixtures adopting several different combinations. In the large tracts of sub-Sahara, it is grown as a subsistence crop, mostly as monocrop or along with legumes. Sorghum-based rotations practiced in semiarid South India are dependent on various factors related to soil fertility, local agro-climate, crop genotype, economic constraints and human preferences. Two-year rotations are commonly practiced on Vertisols and Alfisols of Southern India. Usually, cotton-sorghum-cowpea, sorghum-groundnut or sorghum–legume sequences are preferred on Alfisols. On Vertisols, cotton-sorghum-groundnut sequence seems congenial in terms of nutrient dynamics. Sorghum mono-cropping is also practiced on Vertisols of Peninsular India [12].

6.3 EXPANSE AND PRODUCTIVITY

As stated earlier, seeds for an expanded sorghum-based agroecosystem were sown long ago, at least some 4000 years ago. Sorghum agroeocsystem actually thrives on demand for grains and other products by over 1.2 billion people. During recent years, higher productivity per unit land and shifts in human preferences to other cereals has induced a proportionate shrinkage in sorghum agroecosystem. At present, sorghum occupies wide range of agro-climates in different continents. Globally, it is grown between 40°N to 40°S of equator (Fig. 1). It is a major food crop in Africa and Asia. It is cultivated mostly for fodder in Europe and Americas. Globally, sorghum is cultivated on 44 m ha. It is cropped in over 99 countries located in different continents. During 2009, the global production was 58 m t. Currently, global sorghum agroecosystem spreads into 63 m ha, across different continents. The sorghum agroecosystem is largest in Africa. About 59% of world sorghum area is in Africa. Asian countries occupy 25% of world sorghum area. North and Central America covers 11% of sorghum area and 4% is in South America. The developing countries in Asia and Africa are more dependent on sorghum. They contribute more than 70% of total sorghum production in the world. North and Central America produces 21% of sorghum and 6% is produced in South America. Major producers of sorghum are USA, (10.2 mt), India (7.8), Nigeria (8.3 mt), China (7.8 mt), Mexico (5.6 mt), Sudan (4.3 mt), Ethiopia (1.2 mt) and Argentina (2.1 mt) [7, 13]. Sorghum agroecosystem in North America is mostly confined to southern United States of America (Plates 1 and 2). It produces 10.2 m t grains. It is grown for grains, forage and fuel (alchohol). The meso-American sorghum belt is scattered into all countries. Sorghum production occurs both under subsistence mixed-cropping conditions and in large farms as intensively grown monocrop. The productivity of sorghum is moderate (1.8–2.2 t/ha). The Cerrados of Brazil supports a large and expansive sorghum agroecosystem. It also extends into other regions of Brazil, such as Amazonia and Matto Grasso. The average productivity of Brazilian sorghum belt is 1.2 t/ha). Sorghum is one of the major cereals of Argentina. Its cultivation spreads into entire Pampas region. Sorghum agroecosystem extends into 1.80 m ha with an average productivity of 1.3 t/ha.

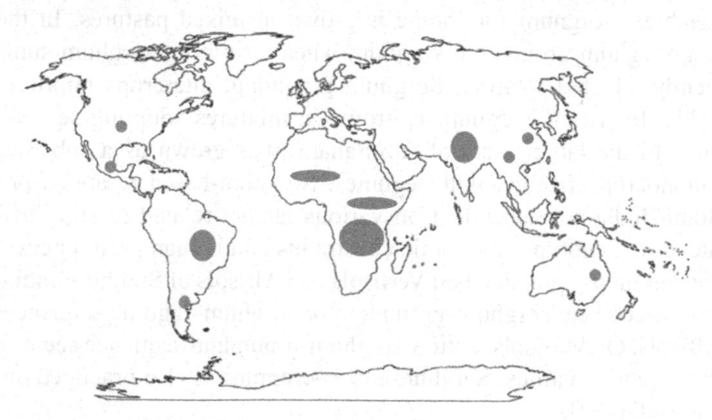

FIGURE 1 Sorghum Cropping Zones of the World.

PLATE 1 Sorghum crop with ripened heads and ready for harvest.
Note: Field is located near Lubbock, in the North Texas, USA.
Source: Kennedy, 2012, Sorghum *checkoff* Program, Lubbock, Texas, USA.
http://www.sorghumcheckoff.com/sites/default/files/09–16–08_NM_193.jpg http://www.sorghumcheckoff.com/sites/default/files/09–16–08_NM_193.jpg

PLATE 2 A high productivity Forage Sorghum field near Lubbock in North Texas.
Note: Such forage sorghum is supplied with moderately high dosages of chemical fertilizers, mainly N, P, K and FYM. Nutritious forage is essential to support cattle and other animal stock. Crop residue recycling is marginal, since much of the soil nutrients garnered into plant tissue reaches animal farms and is not recycled efficiently. Note that forage sorghum genotypes are planted closely. Therefore, rooting density and underground biomass too are expectedly higher.
Source: Kennedy, 2012, Sorghum *Check off* Program, Lubbock, North Texas, USA.
http://www.sorghumcheckoff.com/sites/default/files/2008–07–09_3.jpg http://www.sorghumcheckoff.com/sites/default/files/2008–07–09_3.jpg

Sorghum is grown in several countries of North Africa, spanning from Morocco in the West of Sahara to Sudan and Egypt in the east. Algeria, Tunisia, and Sudan are major sorghum producers in this region. The West African Sahel supports a large expanse of sorghum. It also extends into tropical areas of this region. Nigeria is a major sorghum producing nation in this region. Sorghum is a staple diet to a large section of people in Sub-Saharan countries. The sorghum belt, here extends into 4.6 m ha. The productivity of subsistence farms is low at 1–1.2 t/ha. Sorghum belt in Africa occupies the semiarid, fringes of Kalahari Desert, plains, hills and coastline. Major sorghum producing countries are Namibia, Zimbabwe, Botswana, Zambia, Malawi, Tanzania, and Kenya.

Sorghum agroecosystem of Asia is moderately intense, but much of the area still falls within low soil fertility, rain fed zone. China, India, Pakistan, Saudi Arabia, Thailand and Vietnam are major sorghum producing countries in Asia. Sorghum is an important carbohydrate source for large population in Peninsular India. The sorghum agroecosystem in India is mostly confined to dry lands of peninsular states. It is sown on 9.2 m ha, about 5.0 m ha during kharif and rest 4.2 in the rabi season. Annually, it constitutes 21% of global sorghum production. During recent years, both areas sown to sorghum and total grain production during kharif season have declined. The decline in area was actually compensated by higher grain productivity per unit area. Hence, during past decade, total grain sorghum harvests have plateaued instead of steeply declining [14]. Maharashtra contributes 48% of India's annual sorghum production (4.8 m t). Sorghum is a staple cereal to large fraction of rural population residing in the semiarid region of Southern Indian states such as Karnataka, Andhra Pradesh and Tamil Nadu. Currently, sorghum agroecosystem of South India occupies 3.1 m ha and annual production of grains is about 2.96 m t. Productivity of sorghum sown in kharif has improved enormously from 500 kg/ha in 1960s to 1000 kg/ha at present in the three major states namely Andhra Pradesh, Karnataka and Tamil Nadu. According to Dar [15], such an increase in sorghum grain productivity has allowed a portion of agricultural land to be released for cultivation of other crops. Obviously, sorghum cropping belt has shrunk marginally in South India. Sorghum is a highly preferred cereal in North Karnataka and Andhra Pradesh (Plate 3). It is a predominant cereal on Vertisol plains of South India. The productivity of sorghum in southern states of India fluctuates depending on factors like, soil fertility, precipitation pattern, season and location. It ranges between 0.5 t/ha and 2–2.5 t/ha.

PLATE 3 Sorghum genotypes get evaluated for grain/forage production in different locations. Sorghum response to chemical fertilizer and bio-fertilizers is important in the semiarid regions of Southern India. Above field trial depicts sorghum cultivated near Bhavanisagar project, near Coimbatore, Tamil Nadu, South India.
Source: Ref. [10], ICRISAT, Hyderabad.

Sorghum cropping zones are easily traceable in many of the South-east Asian nations. It is grown for both grains and fodder. The Chinese agrarian zone supports sorghum production. Its cultivation is mostly confined to Western and Southern parts of / China. Sorghum cultivation extends into 6.3 m ha.

KEYWORDS

- **Bhakri**
- **Germplasm**
- **Juar-I Hindi**
- **Kaoliang**
- **Samshu**
- **Sorghum**
- **Sorgo**

REFERENCES

1. Undersander, D. J.; Smith, L. H, Kaminski, A. R.; Kelling, K. A.; Doll, J. D. Sorghum-Forage. Alternative Field Crops Manual. http://www. hort.purdue.edu/newcrop /afcm /forage. html. **1990,** 1–7 (June 6, 2012).
2. Fuller, D. Q.; Korizettar, R.; Venkatasubbaiah, C.; Jones, M. Kearly. Plant domestications in Southern India: Some preliminary Archaeobotanical results. Vegetation History and Archaeobotany . **2004,** *13,* 115–120.
3. Fuller, D. Q. Ceramics, Seeds, and Culinary change in Prehistoric India. Antiquity **2005,** *79,* 761–777.
4. Morrison, K. D. Fields of Victory. Manoharlal Munshiram Publishers: New Delhi, **2000,** 201.
5. Buchanan, F. A Journey from Madras through the countries of Mysore, Canara and Malabar. W.; Bulmer and Company, Cleveland row, St James, London, **1807,** *1,* 1–370; *2*, 1–510; *3*, 1–440. (in three volumes).
6. NRCS, Sorghum: A perspective. National Center for Sorghum Research. Indian Council for Agricultural Research, Hyderabad, India. http://www.nrcsorghum.res.in/perspective. asp **2001,** 1–3 (June 15 2012).
7. CGIAR, Sorghum. http://www.cgiar.org/ompact/research/sorghum.html **2004,** 1–3.
8. Harlan, J. T.; De Wet, J. M. J. A simplified classification of cultivated Sorghum. Crop Science **1972,** *12,* 172–176.
9. Venkateswarulu, J. Rainfed Agriculture in India. Indian Council of Agricultural Research, New Delhi, **2003,** 566.
10. Stigter, K.; Brunini, O. Agrometeorology and Sorghum Production. http://www.agrometeorology.org/ fileadmin /insam/repository/gamp_chapt13 g.pdf **2007,** 1–24 (June 20, 2012).
11. Krishna, K. R. Agrosphere: Nutrient Dynamics, Ecology and Productivity. Science Publishers Inc.: New Hampshire, USA, **2003,** 348.
12. Krishna, K. R. Agroecosystems of South India: Nutrient Dynamics and Productivity. BrownWalker Press Inc.: Boca Raton, Florida, USA, **2010,** 565.
13. FAOSTAT, Sorghum statistics. Food and Agricultural Organization of the United nations, Rome, Italy. **2010** (June2 2012).

14. ICRISAT, Sorghum nutrition for needy. International Crops Research Institute for the Semi-arid Tropics. *SATrends*-Monthly News letter, **2006,** *69,* 2–4.
15. Dar, W. Sorghum yield doubled in India Thanks to ICRISAT-ICAR partnership. http://www.icrisat.org/media/2001/media24.htm. **2001,** 1. (May 19, 2012).

EXERCISE

1. Describe the Sorghum growing regions of World and mark thcm on a map.
2. Mention names of wild and domesticated species of Sorghum.
3. Mention names of different types of Sorghum.
4. Discuss production trends of Sorghum in India and West Africa.
5. Discuss various uses of sorghum.
6. Mention few sorghum based rotations practiced in South India.

FURTHER READINGS

1. Bantilan, C. S.; Gowda, C. L. L.; Reddy, B. V. S.; Obilana, A. B.; Evenson, R. E. *Sorghum Genetic Enhancement: Research Process, Dissemination and Impacts.* Research Report. International Crops Research Institute for the Semi-Arid Tropics. http://oar.icrisat.org/1180/. **2004,** pp. 288.
2. Krishna, K. R. Agroecosystems of South India: Nutrient Dynamics, Ecology and Productivity. BrownWalker Press Inc: Boca Raton, Florida, USA, **2010,** pp. 565
3. Wayne Smith, C.; Richard A. Frederiksen, Sorghum: Origin, History, Technology, and Production. Wiley Series in Crop Science, Wiley Inc.: New York, **2000,** pp. 840.

USEFUL WEBSITES

http://www.eolss.net/ebooks/Sample%20Chapters/C10/E5-02-01-04.pdf
http://www.sorghumcheckoff.com
http.www.icrisat.org
http://intsormil.org
http://www.icrisat.org/crop-sorghum.htm

CHAPTER 7

FINGERMILLET CROPPING ZONES OF ASIA AND AFRICA

CONTENTS

7.1 INTRODUCTION: DEVELOPMENT OF A FINGER MILLET AGROECOSYSTEM IN ASIA AND AFRICA

Finger millet is widely grown in South Asia and Africa. It was domesticated, initially in the highlands of Eastern Africa [1–3]. Later, it spread into tropics of Central and Eastern Africa. Finger millet (*E.coracana subsp coracana*) was regularly cultivated by people inhabiting Southern Indian Neolithic settlements, around 2nd millennium B.C. [4–7]. Seed and plant material attributable to *E. coracana* were obtained from Neolithic phase-3 sites near Hullur in North Karnataka [4]. There are suggestions that finger millet was exported from Africa to Southern Indian Peninsula between 2000–1500 B.C. During Medieval times, finger millet found greater acceptance among farmers in South India. During 18th and 19th centuries, finger millet cultivation spread rapidly into southern Indian plains. According to Buchanan [8], farmers in Southern Karnataka grew at least three types of finger millet. They were called *Cari* (black), *Kempu* (red) and *Hulupuria*. All three types of finger millets were equally productive. They yielded about 23–24 bushels/ac (1 bushel = 27.5 kg grains). The *hulupuria* types were prone to high dehiscence at the time of harvest. Finger millet was rotated with other cereals like Jola (Jowar) and Tovary (pigeonpea) on sandy (Marulu) and Doray soils. Black soil or Erray bhumi were highly fertile and best suited. Over all, a section of Southern Indians depended on finger millet for carbohydrates and used its stover to feed animals (Table 1). Naturally, it found easy acceptance among farmers. As a result, finger millet agroecosystem spread into large portions of dry land zones of South India [9, 10].

Globally, finger millet cultivation zone is confined to 20°N to 20°S. Finger millet cultivation is conspicuous in Eastern and Southern African nations. It is an important cereal crop in Ethiopia, Somalia, Kenya, Tanzania, Zaire, Zambia and Uganda. Finger millet is a major cereal in Uganda. Finger millet in Uganda extends into 0.4 m ha. Finger millet belt extends into south as far as Zambia, Rwanda and Mozambique in Africa. In Southern Africa, finger millet is an important back-up cereal food in times of drought and famine. Overall, the African finger millet zone extends into 1.2 m ha. *E coracana* is also traceable in Australia and warmer regions of Japan. However, its area of cultivation in these countries is negligible.

Historically, finger millet belt in India has shrunk in area during past 5 decades. The Finger millet is an important dry land crop of Southern States of India. Its cultivation is predominant in Karnataka, Andhra Pradesh and Tamil Nadu. In Southern India, finger millet occupies 2.6 m ha and contributes 3.6 m t grains annually. It is staple cereal to large section of rural population in Karnataka, Tamil Nadu and Andhra Pradesh (Plate 1). The spread of the finger millet belt seems to depend to great extent on the food habits and economic advantages related to finger millet farming [10]. In the Himalayas, its cultivation is scattered but it spreads until 30°N.

7.1.1 FINGER MILLET CROPPING ZONES BECOME HIGHLY PRODUCTIVE IN SOUTH INDIA: GENETIC STOCKS

As stated earlier, major share of global finger millet grain/forage is contributed by Southern Indian plains. Farmers, in this agrarian zone have actually cultivated numerous cultivars/hybrids of finger millet. The composition of cropping zone with regard to genetic stocks of finger millet has kept changing with time, depending on farmer's preferences and resources. The genetic potential of finger millet genotype with regard to ability to withstand drought, disease and soil fertility related constraints have been important. Obviously genetic improvement programs have played crucial roles in sustaining and improving productivity of finger millet cropping zones.

PLATE 1 Top: An expanse of Finger Millet just before heading. Finger Millet belts thrive in the Dry plains of Southern India. It is a prominent cereal during rainy season (kharif) since it withstands the vagaries of weather and rainfall pattern of monsoon period during June to November/December. During recent years, farmers have preferred to supply fertilizers and cultivate high yielding drought tolerant genotypes. Cultivars such as Poorna, Hamsa, Indaf series and GPB series dominate the Alfisol zones of Southern Indian states such as Karnataka, Tamil Nadu and Andhra Pradesh. Finger millet carves its own niche in this region mostly as a monocrop or as an intercrop with legumes like Dolichos, Horse gram, Pigeon pea or sometimes with other cereals such as Sorghum (Jowar) or Maize. In a nutshell, the finger millet cropping zone has been among the best bets under low input, sustainable farming conditions of Southern Indian plains. Middle row: Finger millet at heading stage and heads. Bottom: Finger Millet (GPU series), grown as sole crop under irrigated condition on Red Alfisols of South India. Above crop is supplied with Inorganic fertilizers (80N:30P:30K kg/ha) plus 5 t FYM. Such a crop often yields 2.0–2.5 t grains/ha and 5–8 t forage/ha, depending on cultivar.
Source: Regional Experimental Station, GKVK, UAS, Bangalore, India.

PLATE 2 Finger Millet/Brassica Intercrop.
Note: Seeds of intercrops were broad caste. They were not sown in rows. Finger millet is amenable to cultivation under variety of combinations with other crops. Farmers cultivate it in mixtures with legumes and oil seed crops. The above example depicts finger millet/mustard mixed crop at a location near Ghati Subramanya, near Doddaballapur, Bangalore India 13.17°N latitude and 77.3°E longitude. The two crops mature at different times during late kharif and allow better land use efficiency. Mixed cropping also ensures farmers with a certain degree of economic insurance in case precipitation pattern is uncongenial to main crop of finger millet. Finger millet expanses grown as intercrop, relay crop or broadcasted mixtures are easily traceable in interior regions of Southern Indian plains.
Source: Dr KR Krishna, Bangalore, India.

Earliest systematic effort to select and genetically improve finger millet was initiated around 1914 by Coleman [11] at Hebbal Farm near Bangalore in South India. Finger millet strains that were generally better than commercial mixtures (e.g., Dodda ragi) and produced 25% more grains were developed. These selections (strains) of finger millet ripened uniformly. During that period commercial mixtures were erratic, in ear head bearing pattern, ripening and yield. Coleman [11] has documented that due to selection for uniformity, better ear traits and number of grains, productivity of ragi strains improved by 2.0 q/ha. It increased from 4.6 q/ha to 6.8 q/ha. Around 1940s, Narasimhan [12] developed cultivars like K1. Several other crosses such as *RO-889* (Majjige ragi *x* Ro324) were developed during 1950s at Hebbal farm, near Bangalore in India. For example, varieties such as *Aruna, Udaya* and *Purna* dominated the finger millet belt in South India, right until mid-1970s, especially in Karnataka. Finger millet cultivars such as AKP and VZM series developed at Anakapalle, spread into different parts of Andhra Pradesh. Similarly, improved varieties like *Co7, CO_2, PLR-1 and Co8* found acceptance with farmers in Tamil Nadu. The grain productivity of all of these new strains was higher. Obviously, introduction of these newly developed cultivars altered the nutrient dynamics of the cropping zone. Firstly, they needed better soil fertility and water resources. Therefore, extraneous nutrients were supplied in the form of inorganic fertilizers in addition to organic manures [10]. *All India Co-coordinated Research Project on Finger Millet* was initiated in 1963 with major activities at Hebbal, Bangalore. It aimed at standardizing agronomic procedures and enhancing genetic potential of Eleusine. The soil fertility trials showed that grain yield of finger millet could be enhanced to 3–4 t/ha by manipulating nutrient dynamics in the field. The development of crosses between Eleusine strains from Indian and African cropping belts is an important achievement. It markedly changed the genetic composition of finger millet agroecosystem of South India. The *Indaf* varieties were resistant to diseases, sturdy, withstood drought and most importantly they responded excellently to fertilizers. The potential grain yield jumped from 1.5 or 2.5 t/ha to 4–5 t/ha due to introduction of Indaf varieties beginning in 1975 [13]. At present, finger millet genotypes such as Indaf-15 and GPU-28 fetch 5 t grain/ha plus 10–12 t straw/ha under high fertility condition.

TABLE 1 Botany, Classification, Nomenclature and Uses of Finger Millet—A summary.

Eleusine coracana is cultivated in diverse ecological regions. An African highland race is cultivated in Ethiopian highlands and Central Africa. A low land race occupies various locations in Africa and India. It is a predominant crop in South India, mainly Karnataka. An Indian race with origin in North-east India is localized in Himalayan tracts. The African high land races are said to be most primitive, from which the low land versions now grown in Africa and India were derived. Natural selection was the major cause for evolution of major races of *E.coracana* [1]. At present, different types of *E.coracana* and its subspecies are cultivated in Karnataka and adjoining states of South India. Therefore, considering its relative abundance and diversity of ecotypes available, this ecoregion comprising Karnataka, Andhra Pradesh and parts of Western Tamil Nadu could be considered a Secondary Center of Origin of Finger millet. Domesticated *E.coracana* was grown during Neolithic age, around 2nd millennium B.C. by Southern Indians. Since then, its cropping zone has expanded steadily in Southern Indian dry lands. University of Agricultural Sciences, GKVK, campus, Bangalore is a major repository for *Eleusine* germplasm. The collection contains accessions from both Africa and Asia. International Crops Research Institute for the Semi-arid Tropics (ICRISAT) at Hyderabad, India also maintains set of useful germ plasm of *Eleusine sp.* Totaling 5949.

Botany and Classification

There are two subspecies of African Finger Millet. The wild species is *E.coracana subsp africana* and cultivated species is *E. coracana subsp coracana.* The wild subspecies *africana* is almost similar to the wild subspecies found in India but with minor variations. In India, wild species is *E.coracana subsp indica.* It also called Indian goose grass, and the cultivated species, once again is *E.coracana subsp coracana* (Plate 1).

Wild Species Cultivated Species

In Africa *E.coracana subsp africana* ▶ *E.coracana subsp coracana*

In India *E.coracana subsp indica* ▶ *E.coracana subsp coracana*

Kingdom–Plantae, Division-Magnoliophyta; Class-Liliopsida; Order-Poales; Family Graminae (Poaceae); Subfamily: Chloridoideae; Genus-*Eleusine*. The genus Eleusine comprises 9 species including diploids and tetraploids. *Eleusine coracana* is believed to be an allotetraploid ($2n$=36) derived by hybridization of *E. indica* and an unknown diploid.

Cultivated species: *Eleusine coracana* Gaertn. The cultivated species is a tetraploid ($2n$ = 36).

Synonyms: *Cynosurus coracanus L.*

Nomenclature

Indian languages: Ragi-Kannada; Ragulu, Chodalu, Sodee, Soloo, Tamidelu-Telugu; Kaeppni, Kelvaragu, Kevir, Kapai-Tamil; Raagi, Moothari-Malayalam; Nagli, Maruva, Mandua-Hindi; Nachini-Marathi; Koddo-Nepalese; Nacchini, Kurakkan in Sinhalese

African languages: Uphoko-Zulu; Mpogo-Pedi; Ulezior wimbi-Swahili; Mawe or Malesi-Malawi; Mbege-Tanzania; Dagussa-Amharic/Sod; Bulo-Uganda; Tokuso or Barankiya in Oromo-Ethiopia; Rukweza or mazavole or njera or poho-Zimbabwe; Bule or Amale or Bawale or Lupoko or Kambale-Zambia; Ceyut in Beri-Sudan; Goose grass or Osgrass-Afrikans; Tialabon-Arabic

European languages: Finger hirse-German; Petit Mil or Eleusine cultive-French; Finger millet, Birdsfoot millet, Coracana millet-English; Vogel geirst-Dutch

Uses

Whole grains are usually ground and used to make porridges, puddings, cakes and pancakes. A common food for people in Karnataka Andhra Pradesh and Parts of Tamil Nadu in South India is known as *Mudde* or Ragi Sankati. Ragi dosa, ragi roti, ragi balls, biscuits, crispys and bread are common preparations made in Southern states. Grains are sometimes soaked and fermented foods are also prepared. Beverages made of malted ragi are popular in South India. Katti is a special dish prepared from ragi powder by Keralites in Idukki and other districts. Husk from finger millet is used in underground storage, as packaging material in pillows and cushions. Finger millet straw is a good fodder, plus it is used in preparing beds for animals. Finger millet seed is used to prepare tonic, astringents and syrups. It is used to treat billousness and hepatitis. The juice from ragi leaves has diaphoretic and diuretic effects. Folklore states that ragi is good to cure diseases like measles, pleurisy, leprosy and small pox.

Nutritional Value

On an average, finger millet seeds contain following ingredients in 100 g DW:

370 calories; water-negligible; Protein 7.6 g; Fat-1.5 g; Carbohydrate-88 g; Ash-2.2 g; Minerals-Ca-410 mg; P-290 mg; Fe-12.6; plus traces of Mg, Na, K, Zn

Vitamins: Vitamin A-0.48 mg; Thiamine 9B1)-0.33 mg; Riboflavin (B2); 21 mg; Niacin-1.2 mg; Cyanocobalamine (B12)-nil; Vitamin C-nil.

Source: [1, 3, 9, 10, 14–17]; http://www.nandyala.org/mahanandi/archives/category/flourpindi/ragi-flour.htm; http://en.wikipedia.org/wiki/finger_millet.htm; http://dacnet.nic.in/millets/10year_finger.htm
http://www.icrisat.org/crop-fingermillet.htm

7.2 SOILS, AGROCIMATOLOGY, WATER USE AND CROPPING SYSTEMS

7.2.1 SOILS

The finger millet zones of Central and Eastern Africa thrive on several different types of soils. The soil fertility status and constraints to crop production are important criteria to consider. In Uganda, finger millet thrives on Alfisols and Xerorthents. Udults, Udalfs and Ustalfs found in Tanzania, Botswana and Kenya also support good crop of finger millet. Finger millet thrives on sandy or loamy arable soils. It needs porous soils, since its growth is affected, if water logged conditions prevail. In Southern India, it is grown on Red Alfisols, Vertisols, Vertic Inceptisols and lateritic soils. Finger millet is most common on Red Sandy Loams or Alfisols of South Karnataka, parts of Andhra Pradesh and Tamil Nadu. The Red Alfisols are deep, moderately fertile and rich in K content. These soils are low in organic fraction, N and P contents. Phosphate fixation capacity is generally high ranging from 20–40% of fertilizer-P added. Soil and nutrient loss due to erosion is also frequently observed. The optimum soil pH is 6–7, but it withstands soil with pH ranging from 5.5–8.2. Repeated supply of N, P and organic manure is essential.

7.2.2 AGROCLIMATOLOGY

Finger millet is a preferred cereal in dry and semiarid tracts of South India. Finger millet thrives well under tropical temperatures ranging from 25°C–38°C. In South India, it grows well when the average temperatures fluctuate between 18°C–35°C. The preferred average temperature is 27°C. However, it also tolerates cool temperatures that occur at higher altitudes in South India and Himalayas. At high altitudes, it thrives in low temperatures ranging from 8°C–17°C. Finger millet is a short day plant. Optimum day length for flowering and grain formation is 12 h. Regarding altitude, its cultivation is mostly confined to altitudes between 500 and 1200 m.a.s.l. In the Himalayas, its cultivation occurs up to 2000 m.a.s.l. However, both in East Africa and India its cultivation is predominant on plains and undulated landscapes that are situated at 300 to 1000 m.a.s.l.

Field trials to fix appropriate sowing date perhaps began during 1930s. Overall, they showed that during kharif season seeds could be dibbled between June and August 1st. The July planted crop used soil nutrients best and productivity ranged from 820–930 kg/ha [9]. Rabi crop could be initiated during October and continued till February. Summer or hot weather crop initiated during December performed better than one started in February. In Andhra Pradesh, *Pedda panta* (kharif) is grown from August/September to November/December. The *Punasa* or hot weather season lasts from May to September. The cool season crop from December to March is called *Pyru*. Transplanting 2–3 week-old seedlings also establishes a finger millet crop. There are reports that transplanting increases productivity of kharif crop, but it is usually on par with dibbling seeds during rabi season. During recent years, crop season for finger millet production in Karnataka is as follows: The main crop spans from July/August to December. A cool season crop may begin in December. If irrigation is available, summer crops last from February till June. Over all, finger millet is amenable for cultivation in all three seasons throughout the year. The African finger millet farmers begin seeding to coincide with rains during June for a crop that lasts till October. Summer crop lasts from January till May. It is highly dependent on soil moisture status and irrigation facilities.

7.2.3 WATER RELATIONS

Finger millet is a dry land crop. It is often grown as rain fed crop without any reasonable supplemental water supply. In South India, its cultivation is possible in areas receiving 500–1000 mm precipitation annually. It is preferred in areas prone to low and erratic precipitation patterns. Since it withstands intermittent drought, it is a staple cereal for subsistence farming communities in drought prone areas. In relatively wetter areas, sorghum or pearl millet replaces finger millet.

Finger millet is mostly grown as rain fed crop in South India. Hence, water requirements and water use efficiency (WUE) are important parameters to be considered while deciding the genotype. The consumptive water use ranges from 29–35 cm and WUE from 40–46 kg grain cm water. Application of P is known to improve water use and its efficiency by 15–20% over control [18]. Several other reports state that finger millet requires 400–450 mm water, either through precipitation, irrigation or stored moisture [19, 20]. Irrigation is required during 3 phases of crop development, namely vegetative, reproductive and ripening. Supply of moisture is crucial during early vegetative phase especially initial 2–3 weeks. The critical stage for irrigation is panicle initiation, flowering and grain filling. The WUE of finger millet genotype is a crucial factor in dry lands, since water resources are scarce. We should note that at 350–400 mm per season, finger millet is one the most efficient users of water, compared with other dry land cereals like sorghum (500 mm) or pearl millet (400–500 mm). According to some estimates a rain fed crop of finger millet requires at least 310 mm to produce optimum yield (4100 kg grain/ha). The WUE of finger millet is relatively high at 13.4 kg grain mm, compared to 8.0 kg grain ha/mm for sorghum and pearl millet [19].

Rainfall pattern is an important factor that decides the size and intensity of African finger millet belt. Basically, a moderately yielding finger millet crop requires 550–600 mm water per season. The Central African finger millet belt generally receives 900–1700 mm annually. The tropical condition may induce higher degree of evapotranspiration. In most parts of Central and Eastern Africa, finger millet thrives as rain fed crop. The rainy season lasts from June to October. Finger millet cultivation also proceeds during post rainy and summer seasons, depending on irrigation facilities.

7.2.4 CROPPING SYSTEMS

Finger millet monocrops grown under rainfed conditions are most common in drier areas of Eastern Africa. Tropical Central Africa supports scattered regions of finger millet/legume intercrops. Finger millet is grown as mixed crop with cassava, plantain and vegetables. Finger millet is most commonly rotated with legumes such as cowpea or pigeon pea in Eastern Africa. Most commonly practiced crop mixtures in South India that involve Finger millet are:

With legumes: Finger millet/*Dolichos*; Finger millet /Pigeonpea; Finger millet/Black gram; Finger millet/castor

Finger/Cowpea; Finger millet/Horse gram; Finger millet/Soybean

With Cereals: Finger millet/Maize; Finger millet/Foxtail millet; Finger millet/Jowar; Finger millet/Little millet

With other species: Finger millet/Brassicas; Finger millet/mustard (Plate 2)

(Commonly adopted row ratios are 5:1 finger millet: legume species).

Finger millet is grown as a mono-crop, season after season in some places within Karnataka, in India. In this area it is an intensely cultivated crop. Finger millet -fallow rotation is also common. Finger millet is usually rotated with crops like sorghum, cotton, gingely, pearl millet and tobacco. It is also rotated with spices like chilies and coriander or vegetables like onions or sweet potato [9]. There are indeed several combinations of crop rotations practiced in South India that include finger millet. They all aim at maximizing land use, improving soil fertility status, maintaining optimum nutrient dynamics and better productivity of finger millet. Taller crop is usually grown in fewer rows during mixed cropping. Whereas, less robust and small statured legumes like cluster bean or black gram or green gram is often sown in higher row ratios. In the Southern districts of Karnataka, finger millet/*Dolichos lablab* is a preferred crop mixture. *Dolichos sp* being a legume adds to soil-N status. Finger millet is also compatible with horticultural species. It is grown as under story crop with perennial fruit trees like mango, sapota or coconuts.

7.3 EXPANSE AND PRODUCTIVITY

Finger millet is an important cereal in the dry lands of South India. The finger millet expanse in South India extends into 1.2 m ha in Karnataka, 0.15 m ha in Andhra Pradesh and 1.30 m ha in Tamil Nadu. During past 3 decades finger millet expanse in South India has experienced shrinkage in area. It is attributable to preference to major fodder producers such as maize. During 2005–2010, average production of finger millet belt was 3.1 m t grains/*y*. The productivity of finger millet crop has improved from 1.35 t grains to 1.82 t grain/ha. The enhanced productivity is attributable to both improved genotype and fertilizer/irrigation. The Ugandan finger millet thrives mostly on subsistence farming techniques. Yet there are areas that support finger millet on fertile soils and with high inputs. Productivity of finger millet grown in Central and East Africa ranges from 1.2–2.4 t grain/ha [10].

KEYWORDS

- **Cari**
- **Dodda ragi**
- **Eleusine coracana**
- **Hulupuria**
- **Kempu**
- **Pedda panta**

REFERENCES

1. Hilu, K. W.; De Wet, J. M. J. Evolution in *Eleusine coracana subsp coracana* (Finger Millet). American Journal of Botany **1976,** *63,* 1311–1318.
2. Hilu, K. W.; De Wet, J. M. J.; Harlan, J. R. Archaeobotanical studies of *Eleusine coracana subsp coracana* (Finger millet). American Journal of Botany **1979,** *66,* 330–333.
3. Hilu, K. W.; Johnson, J. L. Systematics of *Eleusine* geartn. (Poaceae: Chloridoideae): chloroplast DNA Total evidence. Annals of the Missouri Botanical Garden **1997,** *84,* 841–847.
4. Fuller, D.; Korizettar, R.; Venkatasubbaiah, C.; Jones, M. K. Early plant domestications in Southern India: Some preliminary archaeobotanical results. Vegetational History and Archaeobotany **2004,** *13,* 115–129.

5. Misra, N. Prehistoric Human Colonization of India. http://ww.ias.ac.in/jbiosc/nov2001/491.pdf **2001,** 1–57.
6. Vishnu-Mittre. Ancient Plant Economy at Hullur. In: Prehistoric cultures of the Thungabadhra Valley (Hullur excavations) Nagraja Rao M. S. (Ed.). Karnataka University, Dharwar, **1971,** 1–9.
7. Devraj, D. V.; Shafer, J. G.; Patil, C. S.; Balasubramanya, K. The Watgal excavations: an interim report. Man and Environment **1995,** *20,* 57–74.
8. Buchanan, F. A Journey from Madras through the countries of Mysore, Canara and Malabar. W.; Bulmer & Company, Cleveland Row, St James, London, **1807,** *1*, 1–370; *2*, 1–510; *3*, 1–440. (in three volumes).
9. Rachie, K. O.; Peters, L. A. The Eleusines: A review of the world literature. International Crops Research Institute for the Semi-arid Tropics, Patancheru, India, **1977,** 178.
10. Krishna, K. R. Finger millet Agroecosystem of South India. In: Agroecosystems of South India. Brown Walker press Inc.: Boca Raton, Florida, USA, **2010,** 546.
11. Coleman, L. C. The cultivation of Ragi in Mysore. Bulletin of the Department of Agriculture Report. Mysore, **1920,** 1–152.
12. Narasimhan, M. J. Early maturing Ragi. Mysore Agriculture Calendar **1940,** 20–21
13. UAS Research report-RAGI (Finger millet). http://uasbng.kar.nic.in/ragi.htm **2006a,** 1–4 (April 30, 2012).
14. Baker, R. D. Millet Production- Guide A-414. http://lubbock.tamu.edu/othercrops/docs/nmsumilletprod.htm **2006,**1–5.
15. Morris, R. Plants for a future: *Eleusine coracana.* Creative commons. http://www.pfaf.org/database/plants.php?Eleusine+indica.htm **2004,** 1–9 (October 10, 2012).
16. UAS, Package of Practices for Karnataka. University of Agricultural Sciences, Bangalore and Dharwad, **2006b,** 285.
17. Vietmeyer, N. D. Lost Crops of Africa; Volume 1; Grains-Finger Millet. The National Academy of Sciences Press, Washington, USA, **1996,** 39–40.
18. PPI-PPIC Potassium in Tamil Nadu Agriculture. Potash and Phosphate Institute, Norcross, Georgia, GA, USA. http://www.ppi-ppic.org/ppiweb/sindia.nsf/$webindex/051db5ececa5af 6765256f0–100457. **2006,** 1–3 (April 10, 2012).
19. Venkateshwarulu, J. Rain fed Agriculture in India. Indian Council of Agricultural Research, New Delhi, **2003,** 507.
20. IKISAN, Irrigation. http://www.IKISAN.com/links/ap_irrigation.shtml, **2007,** 1–14.

EXERCISE

Mention the names of Wild ancestors and domesticated species of Finger Millet grown in Asia and Africa.

List the countries in Africa and Asia and the specific regions that support Finger millet agroecosystem.

Mentions major cropping systems practiced in Africa and Asia that include Finger Millet.

Write an essay on geographic area and productivity of Finger Millet in Africa and Asia.

Write about agronomic procedures involved during finger millet production. Focus on soil tillage, management, planting density, fertilizer supply, irrigation and grain yield.

List major finger millet varieties grown in India during the past 50 years. State nutrient supply schedules for as many cultivars.

FURTHER READING

1. Board on Science and Technology for International Development 19996 Lost Crops of Africa. Volume 1. Grains. The National Academies Press: Washington D.C. USA, 408.
2. Kallimuthu, R.; Kumaran, V.; Nallasamy, T. System of Crop intensification in Finger Millet. Lap Lambert Academic Publishing Inc.: **2012,** pp. 108.
3. Premavalli, K. S. Finger Millet: A valued cereal. Nova Science Publisher Inc.: New York, **2012,** pp. 180.

USEFUL WEBSITES

http://www.icrisat.org/crop-fingermillet.htm

CHAPTER 8

MINOR CEREALS AND MILLETS

CONTENTS

Millets are small seeded grass species belonging to family Poaceae. There are several genera and species of millets that are well spread all over the agrarian regions of the world. Millets are preferred species in harsh environments, regions with low fertility soils and unpredictable rainfall pattern. Millets are of great value in the semiarid and arid regions of the world. They are particularly accepted in areas with short growing season that occurs in regions afflicted with either cold and frost or dry hot spells. Millets usually complete their life cycle within the short congenial period. Millets suit the economically less endowed farmers situated in different continents. They are perhaps the best bets for subsistence farming conditions. However, they may also be equally good preferences to make in regions endowed with fertile soils, congenial agroclimate and high fertilizer inputs. In tune with the context of this book, we should note that millets, almost each species, forms a small or large scattered or uniformly even cropping belt-in other words agroecosystems. These millet species have carved out a niche for themselves under the given agroclimatic condition and natural resources. As stated above, there are indeed many species of millets used by humans for grains, forage and several other purposes related to industrial uses, etc. Pearl millet (*Pennesitum glaucum*); Finger millet (*Eleusine coracana*); Foxtail millet (*Setaria italica*); Proso millet (*Panicum miliaceum*); Kodo millet (*Panicum romosum*); Little millet (*Panicum miliare*); Japanese millet (*Echinochloa frumentaceaea*); Tef (*Ergrostis tef*) and Rye (*Secale cereale*) are few examples. Considering the brevity of this volume and importance of the crop at present, only two minor cereal species and three millet species have been dealt in this chapter.

BARLEY (*HORDEUM VULGARE*)

8.1 INTRODUCTION

Barley is among the oldest of cereal grains used by human beings for their sustenance and to satisfy energy needs. Historically, barley was supposedly domesticated in the *Fertile Crescent* encompassing regions within Syria, Southern Turkey, Eastern Iraq, Israel and Western Iran [1]. Domesticated barley known as *Hordeum vulgare* is derived from its wild ancestor known as *Hordeum spontaneum*. Archeological sites indicate that both barley and wheat were used by prehistoric agrarian societies of West Asia. Barley grains have been traced in the archeological sites at Jarmo, in Iraq dating 6500 B.C. Early and middle Neolithic migration introduced barley into several other agrarian regions in North Africa, Europe, and East Asia. Archeological analysis at Nubia and other locations on the banks of Nile, suggests that barley was cultivated in Egypt since 3000 B.C. [2]. Egyptians spread barley into many regions within North Africa and Southern Europe. Ancient Greeks and Romans used barley as a regular diet for their militaries [1]. Archeological samples in the Western European plains especially in Spain, France and Germany indicate use of barley during 5th millennium B.C. Barley was an important grain source to populace in Southern Russia and Caucasus. Barley cultivation was well established in Scotland and Nordic region during 1–8th century A.D. Barley grains and forage was used by ancient civilizations of Indo-Gangetic plains as early as 3000 B.C. Historical sites of Korea and Japan suggest cultivation of barley by 100 B.C. It seems barley spread to China during 2nd millennium B.C. There are other reports suggesting that barley was cultivated as early as 1500–850 B.C. in the Korean peninsula. Barley was introduced into North American plains through Spanish explorers and conquerors. Barley has been in regular cultivation in the New England region since early 17th century [1].

Barley, it seems was introduced into Latin America by Columbus during his second sojourn to Hispaniola. Spanish merchants and warriors planted barley at Isabella in Puerto Rico in 1493. Like most other crops, barley cropping zones spread rapidly in the New World during medieval period. By 1530, it was grown in the Santa Fe region of Argentina, and Sancti Spiritus [3]. Later during mid-1500s, barley was introduced into Andean region. Regular research to improve the genetic stock in barley belts began in 1912 in Uruguay. Currently, barley is grown both for forage and grains in almost all nations of South America (Table 1).

TABLE 1 Barley—Origin, Distribution, Classification, Nomenclature, Uses and Nutritional aspects—A summary.

Barley was one of the earliest cereals to be domesticated in the 'Fertile Crescent' area comprising present day Syria, Iraq, Palestine, Israel and Lebanon. Barley was cultivated since prehistoric times in West Asia, North Africa, South America, Russian Plains and South Asia. Analysis of world barley germplasm for nuclear genes indicates that there are three germplasm pools. First one is derived from Eastern Africa (Eritrea and Ethiopia) and South America (Ecuador, Peru and Chile). Genetic analysis has shown that South American germplasm was originally derived from East Asia and share a common genetic base. The second genetic pool is from Caucasus (Armenia and Georgia), and the third group belongs to Central Asia, West Asia and East Asia. Studies on subspecies of barley suggest that cultivated species includes two rowed and six rowed barley [4].

Classification

Kingdom-Plantae; Order-Poales; subfamily Pooidae; tribe-Triticae; Genus *Hordeum*; Species-*H, vulgare*

$2n = 14$

Wild barley-*Hordeum spontaneum* and then *H. agriocrithon*

Nomenclature

Barley-English; Jau-Hindi; Satu-Marathi; Barley-Kannada; Barli Arisi-Tamil; Joba-Bengali

Uses

The two row barley with higher protein is useful as feed. The six-row barley that has relatively higher sugar continent is preferred in malting and beer preparation. The six row is barley commonly used in producing American lager Beer. Barley is used many traditional foodstuffs in West Asia and Europe. For example, barley soup is consumed during Ramadan in Arabia. Cholent or Hamin is a traditional stew for Jews during Sabbath. In Eastern Europe barley is used to prepare wide range of soups and stews. In Northern Britain isles it is converted into beer and used. In India, barley flour is used to prepare porridges known as Ambli and malted drinks.

Nutritional Aspects

Nutritional value per 100 g grains is as follows: Energy-352 kcal; Carbohydrates-77.8 g; Sugars-0.8 g; Dietary Fiber-15.6 g; Fat-1.2 g; Protein-9.9 g; Thiamine-0.2 mg; Riboflavin-0.1 mg; Niacin-4.6 mg; Pantothenic acid-0.3 mg; Vitamin B6-0.3 mg; Folate-23 μg; Ca-29 mg; Fe-2.5 mg; Mg-79 mg; P-280 mg; K-280 mg; Zn-2.1 mg.

Source: [2, 4];
http://www.icarda.org/; http://en.wikipedia.org/w/ondex.php?title=Barley&oldid=51091086.

8.2 SOILS, AGROCLIMATE AND CROPPING SYSTEMS

Barley is well adapted to different types of soils encountered in temperate countries. The barley cropping zones thrive well on Chernozems found in North America, Eastern Europe and Russian plains. Cambisols of Northern Europe too support barley production zones. It grows equally well on low fertility Xeralsols found in West Asia. Usually, soils with low fertility and regions with relatively lower levels of precipitation, around 300–400 mm supports barley crop in West Asia. Barley cropping zone thrives well on Inceptisols and Alluvial soil found in the Indo-Gangetic belt. Barley is grown on sandy soils in the Northern Plains of China. Barley withstands alkalinity in soils but it is susceptible to acidity in soils.

Barley prefers temperate or subtropical climate during growing season. It matures in 110–120 days depending on heat units received and onset of frost. Barley has excellent cold tolerance. Barley is relatively a short duration crop. In the arctic zone it ripens in 60–70 days. In the Canadian plains it is sown in May and harvested by September end. Barley is drought tolerant. It is grown in dry regions of North America and in Sub Sahara. Barley is a shallow rooted cereal and exploits soil nutrients and moisture available in the upper layers of soil. Barley production zones also extend into deserts where it negotiates salty and drought prone soils better than other crop species. Barley is a rain fed crop and as such fertilizer supply is commensurately low. In the North American plains barley meant for forage receives 30–60 kg N/ha depending on SOM content of the field and soil NO_3 test values [5]. Barley meant for grains is supplied with 60–80 kg N/ha, again depending on SOM content and soil NO_3 test values. If the previous crop is a legume, then N input to barley is regulated based on N derived from legume. High inputs of fertilizer-N may lead to lodging and reduce grain harvest. Production of beer and malting may get affected if the crop is supplied excessive quantities of fertilizer-N.

Barley mono-cropping zones are predominant in the drier tracts of West Asia and North Africa. However, barley is often rotated or intercropped with other crop species in this region. Barley-fallow seems least productive, but is preferred in regions with scanty and erratic precipitation pattern. Barley-barley continuous cropping is moderately productive and is perhaps most common in West Asia. Barley is rotated with variety of legumes that add to soil-N fertility. For example, in Syria and Cyprus, barley-vetch is preferred because it offers better grain and forage harvest. Actually, barley-vetch or barley-peas rotations offer better quantities of protein source to West Asian farmers. Barley-medic, barley-continuous and barley are rotations encountered in Northern Iran and Central Asian Republics [6]. Barley-legume rotations provided with 30 kg N and 80 kg P/ha have performed best under the given rainfall pattern.

8.3 EXPANSE AND PRODUCTIVITY

Barley agroecosystem is relatively small compared to other major cereals like wheat or maize. Yet, its cultivation spreads into over 100 countries. It spreads into entire temperate region, but the crop itself is grown scattered. Currently barley is an important cereal crop in many of the temperate countries situated closer to arctic regions. Barley cultivation is most pronounced in Russia, followed by Western Europe (Germany, France, and Spain), Turkey in Middle East and Australia. Globally, barley production is 151.7 m t/yr. Barley grain production in top 10 nations are as follows: Russia 17.9 m t/yr; Ukraine-12.9 m t/yr; France-12.9 m t/yr; Germany-12.3 m t/yr; Canada-9.5 m t/yr; Australia-9.5 m t/yr; Turkey-7.3 m t/yr, United Kingdom-6.8 m t/yr and United States of America-5.0 m t/yr [7].

In the Canadian Prairies, barley is best suited as both grain and silage source. Usually smooth-awned varieties are cultivated in this region of North America. Barley varieties meant for forage that are predominant in Canadian plains are Ranger, Westford and Hawkeye. In Canada, barley-cropping zone is wide spread in all agrarian regions. During 2005 to 2010, the expanse of agroecosystem has fluctuated between 7.2 to 10.1 m ha annually [8]. The two-row barley type constitutes 70% of Canadian barley agroecosystem. About 10% area is covered by six-row barley and 15% area by forage barley. In general, barley-cropping zones have shrunk marginally during the past 5 years. The reasons attributed are many. Alternative crops and new variations in rotations that do not favor barley are in vogue. Farmers are not inclined to invest in barley since they prefer high input large scale farming not so congenial for barley. Barley is sometimes treated as specialty crop for malting hence its choice gets restricted. During recent years demand for forage barley has declined. In some regions barley is replaced with soybean because it withstands wet weather much better than barley. Barley is an important crop in Western Canada. It is supplied with fertilizers. Usually, 65–70 kg N/ha· 45 kg P/ha and 33 kg k/ha is applied for a crop with yield goal of 1.5 t grain and 8 t forage.

In USA, barley is grown predominantly for forage. The barley growing regions extend into North Dakota, Minnesota, Montana, Idaho, Washington and California. Winter barley is grown in the Southern Plains of North America. It extends into Nebraska, Oklahoma, Georgia and North Carolina. Barley production improved steadily from 3.9 m bushels in 1950 to 6.2 m bushels in 1985, but later it declined gradually to 2.1 m bushels in 2011 [9] (*Note:* One bushel = 28.5 kg grains).

In South America, barley-cropping zones extend into most regions. Barley is grown for forage and grains. Barley cropping spreads into plains in the Cerrado region of Brazil to hilly tracts in Colombia and other Andean countries. The barley crop adapts to 1500 m.a.sl in Bolivia and up to 3800 m.a.s.l in Chile. Barley production is predominant in Brazil, Argentina, Peru, Bolivia, Equador, Colombia and Chile. During past two decades, barley cropping zone has extending into 120,000–145,000 ha in Peru, 80,000–100,000 ha in Bolivia, 40,000–60,000 ha in Equador and 20,000–40,000 ha in Colombia. The average productivity of barley crop grown in different regions of South America ranges from 1,500–2,000 kg grain/ha. In Argentina, barley meant for forage was predominant during 1950s and extended into 0.8–1.0 m ha. Currently, malting barleys are preferred in the Argentine Pampas [10].

Wheat is the major cereal in the Eastern European and Russian plains. Barley is the second most important crop in this region. Russia has consistently produced 16 m t grains/yr during the past. Area of the barley-cropping belt in Russia has fluctuated between 9–12 m ha. The barley belt is dominant during spring. Nearly 95% of barley area is composed of spring barley. As stated earlier, Russia and Ukraine are top barley producing regions of the world. During past decade, Russia has contributed 12–18% and Ukraine contributed 9–14% of global harvest of barley grains [11]. Barley is grown in the Arctic regions of Russia. Barley agaroecosystem is relatively intense and wide spread in the Russian plains and Eastern Europe. Barley thrives on Chernozems and sandy podzols in the Russian plains. It is a short duration crop with ability to mature before the onset of frost and cold. The productivity is dependent on soil type and its inherent fertility. Yet, on an average, barley yields 1.2 t grain/haplus forage in the Russian plains.

Barley is an important grain crop in Mediterranean and subtropical regions of West and South Asia. It is cultivated wherever agrarian enterprise is a possibility. Barley agroecosystem often thrives better in regions with relatively lower rainfall, where wheat does not perform well. Reports from ICARDA, Aleppo, Syria suggest that in regions with 400 mm rainfall wheat is sown, but at slightly lower levels of precipitation < 350 mm barley takes over as the prime-cropping belt. Productivity of barley belt in West Asia depends on rainfall and soil fertility measures if any. Barley mono-cropping with barley-barley sequence is common in arid belts, where other combinations of rotations may fail to be remunerative. Fertilizer and organic manure supply is essential, if yield goals are revised up words. Protective irrigation is necessary. Rain fed barley produces 0.8–1.5 t grain/ha and 6 t forage.

Agrarian regions in Ethiopia, Morocco, Algeria and South Africa support barley-cropping zones of different intensities. Barley production in other African regions like Egypt, Kenya or Lesotho is feeble. The end use of barley crop may also vary based on geographic region, soil fertility status and yield goals. Ethiopia has the largest barley-cropping belt in Africa. Barley is a fairly important cereal in certain portions of cooler Southern African regions. Barley grain harvests are used in malting and beer making. It is also used as forage crop. Western Cape region contributes 73% barley cropping area and production of South Africa. During 2009, South African barley agroecosystem stretched into 75000 ha with a total grain produce up to 224,000 tons. The barley cropping zone in South Africa has stayed constant in terms of area. Its expanse has fluctuated marginally between 74,000 and 77,000 ha [12].

Barley cropping belt in India is relatively small and has fluctuated in expanse between 75,000 and 100,000 ha. Barley belt occurs in states such as Punjab, Rajasthan, Madhya Pradesh, Uttar Pradesh and Haryana. Barley cropping is sparse in Utharakund, Himachal and Bihar. Rajasthan and Uttar Pradesh together constitute 70% of barley agroecosystem. Barley productivity on Inceptisols of Indo-Gangetic plains and dry sandy loams has been moderate at 1.9 t grain/ha and 8 t forage/ha. Annually, the

Indian barley cropping belt contributes 1.2 to 1.5 million tons to global harvest [13, 14].

The Chinese barley cultivation zones are predominant in the Northern dry region that receives scanty precipitation. The Chinese barley-cropping zone has shown a steady decline in area and has shrunk from 3.1 m ha in 1960 to 0.75 m ha in 2008. It is clearly attributable to preference to wheat and soybean. The total barley production in China was 3.2 m t annually with an average productivity of 3.92 t grain and 10 t forage/ha. To a certain extent decrease in barley production area has been offset by improvement in productivity from 1.18 t grain to 3.29 t grain/ha.

The Australian barley belt has grown from 1.1 m ha in expanse during 1960 to 4.35 m ha in 2010. The barley grain productivity has increased from 1.1 m t annually in 1960 to 1.87 m t in 2010. To a large extent, improvement in barley harvest from 1.42 m t annually to 7.4 m t in 2010 is attributable to enlargement of barley agroecosystem in Australia and New Zealand [7, 14].

KEYWORDS

- **Echinochloa frumentaceaea**
- **Eleusine coracana**
- **Ergrostis tef**
- **Panicum miliaceum**
- **Panicum miliare**
- **Pennesitum glaucum**
- **Secale cereale**
- **Setaria italic**

REFERENCES

1. Newman. C. W.; Newman, R. K. A brief History of Barley foods. Cereal Foods World **2006,** *51,* 1–7.
2. Palmer, S. A.; Moore, J. D.; Clapham, A. J.; Rose, P.; Allaby, R. G. Archaeogenetic evidence of Ancient Nubian Barley evolution from six to two row indicates local adaptation. The McDonald Institute for Archaeological Research, University of Cambridge, Cambridge, United Kingdom, **2009,** 1–13.
3. Arias, G. Mejoramiento Genetico y Producion de Cebada Cervecera en America el sur. Regional Office of the Food and Agricultural Organization for Latin America and Caribbean. **1995,** 160.
4. Jilal, A.; Grando, S.; Henry, R. J.; Lee, L. S.; Rice, N. F.; Hill, H.; Baum, M.; Ceccaelli, S. Genetic diversity of ICARDA's worldwide barley landrace collection. Genetic Resources and Crop Evolution. http:/dx.doi.org/10.1007/s10722-008-9322-1. **2008,** 1–9 (September 6, 2012).
5. Davis, J. G.; Westfall, D. G. Fertilizing spring-seeded small grains. Colorado State University Extension Service Bulletin **2009,** 1–6.

6. Al-Ajlouni, M. M.; Al-Ghazwi, L. A.; Al-Tawaha, A. B. Crop rotation and fertilization effect on Barley yield grown in Arid conditions. Journal of Food, Agriculture and Environment **2010,** *8,* 869–872.
7. FAOSTAT. Barley Production Statistics. Food and Agricultural Organization of the United Nations. Rome, Italy, **2010** (September 5, 2012).
8. Therrien, M. C. Barley production and cultivar development trends for Western Canada. http://www.Canadabarley.htm **2012,** 1–5 (September 6, 2012).
9. North Dakota Barley Council. The USA, Barley Production. North Dakota Barley Council, Fargo, USA, **2012,** 1–3.
10. Capettini, F.; Cataneo, M.; Geman, S.; Pando, G. L.; Minella, E.; Pieroni, S.; Rivadeneira, M.; Zamore. Barley enhancement in Latin America in the last 20 years: A story of success. http://www.03_ABTS-07_Capettini-barley_in_latin-america.pdf **2011,** 1–24 (September 6, 2012).
11. Lindeman, M. Russia: Agricultural Overview. United States Department of Agriculture, Beltsville, Maryland, USA, **2003,** 1–8.
12. Department of Agriculture South Africa, Barley: Market value chain profile 2010–2011. http://www.nda.agric.za/docs/ AMCP/BarleyMVCP2010–2011.pdf. **2011,** 1–23 (September 5, 2012).
13. Multi Commodity Exchange Ltd, Barley. http://www.mcxindia.com/sitepages/ contract-Specification.aspx? production=Barley.htm **2010,**1–4 (September 5, 2012).
14. Tiwari, V. Growth and Production of Barley. Encyclopedia of Life Support Systems. http://www.eolss.net/sample-chapters/c10/E1–05A-11–00.pdf **2010,** 1–18 (September 5, 2012).

EXERCISE

1. Demarcate Barley growing regions of the World.
2. Mention names of Wild type and Cultivated species of Barley.
3. Discuss the Total grain/forage production and productivity trends of Barley in Russia and Ukraine.

FURTHER READING

1. Ullrich, S. E. Barley: Production, Improvement and Uses. John Wiley and Sons: New York, **2010,** pp. 500.
2. Elfson, S. E. Barley: Production, Cultivation and Uses. Nova Publishers Inc.: New York, **2011,** pp. 335.

USEFUL WEBSITES

http://www.icarda.org
http://www. usda.gov

OATS (*AVENA SATIVA*)

8.1 INTRODUCTION

It is believed that domestication and spread of oat culture occurred much later than wheat or barley. Historical evidences accumulated so far indicate that earliest culture of oats occurred in Asia Minor and Southern Europe [1]. Archeological remains from locations in Egypt suggest that oats were consumed by people living during 12th dynasty that is around 2000 B.C. Archaeological sites near Alps in Switzerland prove that oats were regularly cultured during Bronze age. Oats reached North American main land during early 1600s. Domesticated oats were cultivated on Elizabeth islands in Massachusetts during mid-1700s. The westward movement of oat culture into Mississippi valley became rapid during 1870s [1] (Table 1).

TABLE 1 Oats: Origin, Classification, Nomenclature and Uses—A summary.

The Oat agroecosystem is worldwide in distribution. Oat cultivation, its expanses and production are more pronounced in temperate regions. The hexaploid wild oat *A.sterilis* is supposedly closely related to *A.sativa*. The genetic analysis shows that *A sterilis* is wide spread in Fertile Crescent. Wild relatives of oats are also traceable in Central Asia. The exact origin of domesticated Oat seems still unclear. Archaeological remains of cultivated oat show that its farming was in vogue in Egypt by 2000 B.C.

Classification

Kingdom-Plantae; Order-Poales; Family Poaceae; Genus-*Avena*; Species-*A.sativa*

$2n = 42$

Nomenclature

Yan mai-Chinese; Oat-Malay; Obveoc-Mangolian; Tagalog-Philipino; Yen Mach-Vietnamese; Haver-Afrikanse; Tagji-Albanian; Shufan-Arabic; Jyulaf-Azerbaijani; Shibolet shual-Hebrew; Yulaf-Persian; Obec-Bulgaria; Civda, Catalan; Zob-Croatian; Oves-Czech; HavreDanish; Oats-English; Kaura-Finnish; Hafer-German; Zab-Hungarian; Aveno-Irish; Avena-Italian; Auzas-Latvian; Havre-Norwegian; Owies-Polish; Ovaz-Romanian; Oves-Slovania; Havre-Swedesh; Ovec Ukrainian

Uses

Oats produce both grains for human consumption and forage for livestock. Products such as flakes, flour and grits made from oats are most commonly consumed as breakfast items. Oat breads, cakes and other pastries are common to people in North America and Europe. Oats are used in ice creams and other dairy products. Oats are useful in preparing food items that should contain higher quantities of fibers. Pastures with oat and legume mixtures are predominant in Northern Europe. Oat straw is an important item use for bedding and feeding cattle. Fermented oats are also popular. Furfural derivatives from oat are used in industries. Oat bran is used in medications that aim at reducing cholesterol levels.

Oat malts are consumed by people in temperate climates. Oats have good medicinal value. Oat products are prescribed for several types of ailments including exhaustion, mild sickness, flu, skin care etc. Oat leaf decoctions are good for skin ailments. Hydrolized oat protein is good as hair tonic. Oat starch is used to replace talcum in facial products.

Nutritional Aspects

100 g of oat grains contain 12.1 g water and 3.6 g ash; other ingredients found are Protein-5 g; Carbohydrates-24 g, Dietary Fiber 3.5 g; Starch 2.7 g; Sugar 2.1 g; Fat-2.4 g; Vitamin A-170 IU; Vitamin C 52 μg; Vitamin E-96 μg; Vitamin K-0.22 μg; Folate 38 μg; Vitamin B6-163 μg; Minerals: Ca-25 mg; Fe-3.0 mg; Mg-29 mg; P-80 mg; K-140 mg; Na33 mg; 1.0 mg Zn; 1.4 mg Mn; 81 μg Cu; 2.1 μg Se.

Sources: [1–4]; http://nutrition.indobase.com/articles/oats-nutrition.php (September 9, 2012); http://www.factanddetails.com/world. php? itemid=1581&catid=54 subcatid=343.htm (September 9, 2012).

8.2 SOILS, AGROCLIMATE AND CROPPING SYSTEMS

8.2.1 SOILS

Oat culture is wide spread in Europe and North America. It adapts to colder temperate climates better. However, farmers in regions experiencing Mediterranean climate do cultivate oats. Oat production is conspicuous in the temperate regions or Russia, Canada, United States of America, Nordic regions like Finland, Sweden, Norway, Northern British Isles and cooler regions of Central European Plains. Oats adapt to different soil types and conditions. They are cultivated on Chernozems, Podzols, Peaty soils and Silty loams in the European regions and on Mollisols and Ultisols in North America. Slightly dry Calci-Xerosols of Mediterranean support moderate grain and forage yield of oats. Oats are cultivated as intercrops and included in a variety of cropping systems. In North America, oats are cultured both during winter and spring season. In the Canadian plains, oats are cultivated for grazing and silage. It may not make excellent silage like barley, yet it is useful to farmers. Smooth awned varieties are preferred. Cultivars like Bell, Baler and Foothills are common in the Canadian plains. Oat cultivation is encountered more frequently in Minnesota, Wisconsin, Iowa, South Dakota and Michigan. In the Russian plains, oats withstands soil acidity. The oat belt of Russia thrives well on loamy or sandy loams or peaty soils. Oats respond well to fertilizer inputs. Farmers aiming at higher yield usually supply N, P and K based on soil test values. For example, farmers in Manitoba, Canada add 20–30 kg N, 12–15 kg P_2O_5, 7 kg K_2O and 7 kg/ha after confirming soil nutrient levels through chemical tests.

PLATE 1 Oat crop at flowering and seed-fill stage.
Source: Krishna, K. R., Bangalore, India.

8.2.2 AGROCLIMATE AND CROPPING SYSTEMS

Oats are wide spread in the temperate regions of the world. Oats require relatively lower amount of thermal heat units to mature. Oat production is pronounced between 35°–65°N latitude. Oats are mostly sown in spring, but autumn planting is in vogue in hilly tracts. Oats are also preferred in regions with severe winters, for example in Canada, Russia and North America [2]. Oat is an annual grass. Oat varieties grown during spring usually have les tillers. The stem terminates in an inflorescence which flowers and bears seeds. In Russia, oat varieties planted in the North need 14 hr daylight but those sown in south may need only 12 hr to grow and mature. As stated earlier, oats need relatively lower degree-days to mature. Oat varieties mature in 75–120 days in the Russian planes. Early varieties are preferred, if one wishes to avoid frost. Oats can withstand low temperatures of 3°C–4°C during germination. About 12°C is needed for seed germination. The total heat units needed are 1,300 for early varieties and 1,500 for late varieties. Oats need 1,100–1,300 heat units during vegetative phase. Oats do not grow and survive at high temperatures of 38°–40°C.

8.3 EXPANSE AND PRODUCTIVITY

Oat agroecosystem spreads into different continents. Oat belts occur as monocrops. They are rotated with other cereals or legumes. Oat is cultivated in larger areas in Russia, Canada, United States of America, Ukraine, Poland, Australia, etc. [4].

Globally, the expanse of oat cropping zone was 12.02 m ha in 2008. During 2008, the oat belt was 0.18 m ha in Africa, 0.58 m ha in Asia, 7.46 m ha in Europe, 2.16 m ha in North America, 0.66 m ha in South America, 0.93 m ha in Australia and New Zealand. Russian Federation supports the largest oat agroecosystem (3.5 m ha). It is followed by Canada at 1.42 m ha, USA, 0.74 m ha, Australia 0.92 m ha and Poland 0.55 m ha. In Asia, China has a large oat belt at 0.3 m ha. The average productivity of oat is 2.2 t grains/ha. However, continent wise, North America and Europe have better grain yield at 2.2–2.3 t/ha, followed by South America and Asia at 1.9 t grain/ha and Africa at 1.1 t grain/ha. The inherent soil fertility of the geographic region, fertilizer supply, irrigation facility, and agroclimate are the major reasons for variations in grain productivity. Globally, total oat grain production was 25.4 m tons during 2008. Europe contributed over 70% of the global oat produce at 16.5 m ton grain. It was followed by 5.2 m t grain by North America, 1.3 m t by South America, 1.1 m t by Asia and 0.25 by African continent [3]. Oats are also grown exclusively for fodder. During 2006, North American forage oat occupied 1.8 million ha, out of which Canada supported 800,000 ha of forage oats [5].

The vast temperate plains of Russia and Eastern Europe support oat production. The agroecosystem is large and annual production is relatively high. However, during past three decades, total production of oat grains has decreased from 12.2 million tons in 1987 to 5.3 million tons annually in Russia alone. Farmers' preference to wheat, maize and alternate cropping system seems to have caused this decline in oat cropping area. Oats are grown mainly in the dry northern region in China. Other areas with temperate climate such as Tibet also support oats. The Chinese oat production has hovered constantly around 0.6–0.8 m t grain annually. During certain years such as 1997 and 2006 it has dipped to 0.4 m t owing to agroclimate, farmer's preferences and economic advantages.

Oats is grown in different parts of Australia, wherever temperate agroclimate allows normal productivity. Major oat producing states are South Australia, Victoria and New South Wales. Relatively smaller oat producing areas occur in Western Australia, Tasmania and Queensland. It is preferred as both grain and forage crop [6]. In Australia oats grown for forage is mostly exported. Western Australia, South Australia and Victoria export large quantities of oat forage to nations like Korea and Japan. During 2010, 0.5 million t hay was exported from Australia, out of which half the quantity was contributed by Western Australia.

KEYWORDS

- ***A. sativa***
- ***A. sterilis***
- **agroclimate**
- **agroecosystem**

REFERENCES

1. Gibson, L.; Benson, G. Origin, History and Uses of oat (Avena sativa) and Wheat (*Triticum aestivum*). Http:// www.agron.iastate.edu/couses/agron212/readings/ oat_wheat-history.htm **2002,** 1–5 (September 9, 2012).
2. Kang, J. S. **2008,** Oats. Department of Agronomy, Punjab Agricultural University, Ludhiana, Internal Report 1–12.
3. Tiwari, V. Growth and Production of Oat and Rye. Encyclopedia of Life Support Systems, **2010.**
4. Gashkova, I. 2009 Interactive Agricultural Atlas of Russia and Neighboring Countries. http://www.agroatlas.ru/en/content/cultural/Avena-sativa/ **2010,** 13 (September 8, 2012).
5. Fraser, J.; McCartney, D. Fodder oats in North America. Food and Agricultural Organization of the United Nations, Rome, Italy, **2007,** 1–13 http://www.fas.org/docrep/008/yr57656e/yr5765e07.htm (September 9, 2012).
6. Alfonso, Y. *Avena sativa*. http://ausgrasse2.myspecies.info/content/avena-sativa. **2012,** 1–3 (September 9, 2012).

EXERCISE

1. Describe global Oat producing regions with regard to soils, agroclimate and productivity.
2. Mention the names of wild type and cultivated species of Oat.
3. Mention various uses of Oats.

FURTHER READING

1. Murphy, D. E. Oats: Cultivation, Uses and Health Effects. Nova Publishers Inc.: New York, **2011,** pp. 167.

USEFUL WEBSITES

http://www.icarda.org
http://www.usda.org

PEARL MILLET (*PENNESITUM GLAUCUM*)

8.1 INTRODUCTION

Pearl millet is a staple cereal crop in the agrarian regions of West African Sahel, Southern Africa, North West India and parts of Deccan in South India. It is a hardy cereal that withstands harsh environmental conditions and yet offers sizeable grain and forage harvest to subsistence farmers in Africa and Asia. Archaebotanical studies suggest that pearl millet is native to West African Sahel where it was domesticated [1]. This region comprising upper reaches of river Niger and Sahelian zone that spreads up to Chad, seems to be the primary center of diversity for pearl millet. Pearl millet spread to southern African semiarid regions on the fringes of Kalahari and up to Kenya through human migration, perhaps during late Neolithic age. Then, it spread to Indian subcontinent. Pearl millet reached many regions in South-east Asia, where currently it is grown more for forage. Archaeological studies suggest that pearl millet was cultivated as a staple cereal in the region comprising present day Mali, Burkina-Faso, Northern Nigeria, Niger and Cameroon during 2 millennium B.C. Evidences from Southern Indian Plains suggest its use as food grains during 1800–1200 B.C. [2, 3]. Pearl millet is also cultivated in southern European dry regions. Pearl millet is a forage crop in Kansas, Texas and Oklahoma in North America (Table 1).

TABLE 1 Pearl Millet: Origin, Botany, Classification and Uses of Pearl Millet in Africa Indian Subcontinent—A summary.

Pearl millet was domesticated for human consumption by the per-historic population that resided in the Upper reaches of river Niger in West Africa. The primary center of origin is in the West African Sahelian region. North-west India comprising the states of Rajasthan, Punjab, and Gujarat is considered yet another center of origin for pearl millet. Pearl millet spread to other parts of African continent such as Southern African semiarid regions, North Africa, and West Asia through human migration, trade routes and conquests. Currently, pearl millet is grown for either for grain and/or forage in many parts of Brazilian Cerrados, North American plains, southern Europe, West Asia, Indian subcontinent and dry regions of China. However, major cultivation zones of pearl millet are concentrated in West Africa, Southern Africa, North-west and Southern India.

Botany and Classification

Kingdom-Plantae; Order-Poales; family Poaceae; Subfamily Panicoideae; Genus-*Pennesitum*; Species-*P. glaucum*

$2n = 14$

Wild type: *Pennizetum violaceum*

The genus *Pennesitum* has 80 species that are spread all through the tropics in Africa and Asia.

Synonyms: *Pennizetum americanum; Pennizetum spicatum; Pennizetum typhoides*;

The pearl millet germplasm is maintained in different regions of the world. International Crops Research Institute for the Semi-arid tropics, at Patancheru in South India is the largest repository

---

of Pearl millet germ plasm lines. They preserve over 24,000 accessions of pearl millet. There are other germ plasm centers for pearl millet. For example in Coastal Plains Experimental Station at Tifton, Georgia, Pakistan Agricultural Research Center, Rawalpindi, ICRISAT Sahelian Centre, Niamey, Niger, International Plant Genetic Resources Institute, Station at Ouagadougou in Burkina Faso, ORSTOM in Senegal.

Nomenclature in Different Continents

Bajra-Hindi; Bajri-Rajastani; Kambu-Tamil; Gantillu-Telugu; Sajje-Kannada; Kambam-Malayalam; Maiwa-Hausa; Mordo-Ngomo; Mar-Fyer; Amar-Ninzo; Mwan-Ningye; Mer-Sur; Mayi-Nupe (Upper Volta); Mara-Mbala (Banut region); Maar-Mbat (Bantu region); Mahangu-Namibia; Mexeiora-Mozambique; Sanio or Gero-Zarma; Zembwe-Botswana; Mhunga-Zimbabwe; Mwele-Swahili; Dro'o-Tunisian; Dokhn-Yemeni; Petit mil-French; Massango Liso-Polish; Milheto-Portuguese (Brazil); Bulrush Millet-English (Australia).

Uses

Pearl millet grain is a staple carbohydrate source to people in West Africa, India and other locations that experience semiarid climate. Pearl millet powder is used to prepare rotis, bread, pancakes and porridges. Pearl millet is used as an admixture in cereal flours. Pearl millet grits are common food items in India and West Africa. Pearl millet stem and leaves are good forages to farm animals. Pearl millet porridges/pastes called To' in West Africa and Ambli in India are staple diets to poorer populations in these regions.

Nutritive value: Protein-11.5 %; carbohydrate-77%; Fat-4.7%; Dietary fiber-9.7; Ash-1.6%; Ca-36 mg 100/g; Fe-9.6 mg 100/g; Vitamin A 22 IU 100/g; Lysine-3.1 g 100-g.

Compared with carbohydrate rich rice, millets are preferred as food grain for people afflicted with diabetes because millet grains release sugars slowly and for longer duration. Hence, millet eaters are less prone to diabetes.

Sources: [3, 4]; en.wikipedia.org/wiki/Pearl_millet; www.prota.org

8.2 SOILS, AGROCLIMATE AND CROPPING SYSTEMS

Pearl millet is a versatile crop that withstands vagaries of soil moisture and nutrient availability slightly better than many other crop species given similar conditions. In West Africa, pearl millet thrives on sandy Oxisols, low in available N and P. The soils are mildly acidic. Soil water retention is insufficient since sandy soils allow easy percolation down the profile. Pearl millet is also grown on sandy loams and loamy soils in parts of Africa. In Eastern Africa, pearl millet is grown on Lixisols and Alfisols. In North-west India, pearl millet thrives on sandy soils. Its cultivation also spreads into Inceptisols and silty Mollisols of Gangetic Plains. In Gujarat, Central India and parts of Maharashtra, pearl millet occupies Vertisols and drought-affected belts. In South India, pearl millet is predominantly grown on Red Alfisols and sandy loams rich in Fe salts. The productivity of these soils is moderate. In China and other South-east Asian nations, pearl millet thrives on Inceptisols and Alluvial soils. In USA, pearl millet belt occupies low productivity Ultisols and Mollisols found in the Kansas, Oklahoma and Texas. Pearl millet is mostly grown as forage crop in the Canadian plains. Here, it thrives on light textured soils and sandy loams.

The West African pearl millet belt experiences semiarid climate with a short rainy season that lasts from June to September. The average precipitation in this pearl millet belt ranges from 600–900 mm annually. Much of the precipitation occurs during June to September. Pearl millet farmers use the precipitation efficiently by planting early with the onset of the rainy season. Yet, this region is prone to periodic drought spells that pearl mille crop has to negotiate. The tropical conditions allow the crop with full complement of heat units (2,400–2,600 degree days). Pearl millet tillers profusely and puts forth biomass rather rapidly during 60–70 days from seedling stage, tillering to grain formation. The relative humidity of pearl millet cropping zone is usually high during rainy season. Pearl millet responds to fertilizer input. The sandy soils are poor in SOM content, hence FYM supply along with fertilizers produces remarkable increases in grain and forage harvests. Wind erosion could be rampant in the Sahelian zone. Loss of topsoil and fertility could reduce crop growth and grain yield. Hence, application of mulches surrounding the seed hills and application of FYM is essential to sustain the cropping belt. Wind brakes made of agroforestry trees is very useful. Tree species such as *Acacia albida, A. holosericae, Bauhinia rufescence, Zisipus mauritania* (Plate 1).

PLATE 1 Pearl Millet monocrop grown at ICRISAT's Agricultural Experiment Station, Patancheru, near Hyderabad, India. *Note:* Pearl millet genotypes that yield higher amounts of grain and fodder, resistant to diseases like Downy Mildew and pests are selected and spread into different regions of Peninsular India and Plains.
Source: Krishna, ICRISAT, Hyderabad, India.

The Indian pearl millet agroecosystem too thrives during warm, rainy season that lasts from June/July to September/October. The temperature during crop season averages 24°C–32°C. However, pearl millet crop withstands up to 38°C during the season. The Rainy season brings in 400–500 mm during a season. Annual precipitation of this region may hover around 750–950 mm. Pearl millet is a photosynthetically efficient crop and requires 12–14 hr sunlight during rainy season. Pearl millet is provided with fertilizer and FYM based on yield goals. Otherwise, a large portion of pearl millet belt is grown as subsistence crop without much fertilizer inputs.

8.2.1 CROPPING SYSTEMS

Pearl millet mono-crops occur in the Sahelian region of West Africa. Pearl millet is also intercropped with legumes such as cowpea, groundnut and vegetables. Post rainy crop requires assured irrigation and is not very common in West Africa. There are indeed many reports that deal with field evaluations of pearl millet-legume rotations and intercrops. Continuous pearl millet/legume intercrops have served the Sahelian farmer best in terms of soil fertility, biomass accumulation and grain yield [5]. In North-west India, Pearl millet is grown as monocrop or as intercrop with legumes such as guar, cowpea, groundnut, pigeonpea etc. In the Deccan region, pearl millet is grown as intercrop with pigeonpea, cotton, oilseeds such as sunflower or canola. Pearl millet belt in South India is filled up with monocrops or intercrops of pearl millet and legumes/oilseeds. Pearl millet is grown as a catch crop during summer (Plates 2 and 3).

PLATE 2 Pearl millet crop near Gabore in Southern Niger.
Note: Pearl millet crop establishment depends on sand shifts and timely precipitation. Nutrient recovery pattern depends immensely on soil moisture content. Pearl millet hills are planted with wide spacing. It allows profuse rooting and improves nutrient/moisture absorption. Neem (*Azadirechta indica*) trees are a source of shade and organic matter to soil.
Source: Krishna, ICRISAT, Hyderabad, India.

PLATE 3 Pearl Millet at ICRISAT Experimental farm, Sadore, near Niamey in Niger.
Note: Generally, pearl millet landraces and cultivars grown in Sahel are taller and profusely tillering. Pearl millet thrives on sandy Oxisols of West African Sahel.
Source: Krishna, K.R. ICRISAT, Hyderabad, India.

8.3 EXPANSE AND PRODUCTIVITY

Globally, pearl millet cropping zone extends into 14 m ha in Africa and another 14 m ha in Asia. It also spreads into smaller areas in North America and Europe. Pearl millet agroecosystem is predominantly encountered inn semiarid and dry agroclimatic regions of the world. The total production of pearl millet grains has averaged 22 m t. Pearl millet dominates the agricultural landscape in Sahelian West Africa, parts of Southern African semiarid regions and North African dry regions. Major pearl millet producing nations in Africa are Nigeria that contributes 32% of global grain harvest, Niger 8%, Burkina Faso 3%, Uganda 3% and Sudan 2%. The average productivity of rain fed pearl millet grown in West African Sahel is rather low at 700–800 kg/ha plus 5 t forage. Evaluation at ICRISAT's Experimental Farm at Sadore near Niamey in Niger has shown that, on an average, pearl millet monocrop may produce 1.0 t grain/ha. An intercrop of pearl millet with cowpea yields 0.94 t pearl millet grain plus 0.08 t cowpea grains/ha. Pearl millet/ Stylosanthes intercrop produces 0.78 t and 0.03 t grain/ha and pearl millet/*S. hamata* produces 0.96 t plus 0.02 t grain/ha respectively (Garba and Renard, 1991).

Pearl millet cultivation areas are also well spread in Southern and Eastern Africa. Cultivars such as KAT/PM-1, 2 and 3 that mature in 2.5–3 months are dominant in the Kenyan pearl millet growing regions. These varieties thrive well from 0 to 1500 m.a.s.l. They offer gray colored grains and produce 0.7–1.1 t grain/ha. Varieties such as Okoa and Shibe are common in pearl millet regions of Tanzania. They are well spread from 0–1300 m.a.s.l. The productivity is relatively better at 2–2.5 t grain/ha.

Asia produces 43% of total global pearl millet harvest. In Asia, India harbors the largest pearl millet agroecosystem that extends into 9.8 m ha. The total production is 9.4 m t grain. Pearl millet belt of China extends into 1.1 m ha and produces 2.2 m t grains annually. The productivity of pearl millet in China is relatively high and hovers around 1.9 to 2 t grain/ha. Pakistan has 343,000 ha under pearl millet and produces 193,000 t grains annually. Myanmar grew pearl millet on 253,000 ha and harvested 150,000 t grains during 2008.

The productivity of pearl millet belt in India has steadily improved from just 350 kg ha grain/ha in 1950 to 950 kg grain/ha in 2008 [6]. Pearl millet is well spread into the North-west semiarid and dry areas on the fringes of Thar Desert. It is also major cereal in adjoining states like Gujarat and Maharashtra. Nearly 52% of pearl agroecosystem in India is filled up with hybrids and rest has cultivars, composites and local landraces. In India, pearl millet is mostly consumed by middle or low-income group people. It is among the best bets to farmers in subsistence farming regions of North-west and South India. Rajasthan, Gujarat, Haryana and Maharashtra are the major pearl millet producing regions in India. Together, they constitute 76% of total pearl millet area in India. The average productivity of cropping zone in this region ranges from 970 kg to 1252 kg grain/ha [7, 8].

Pearl millet cropping zone that occurs in Southern Canada is meant mainly for forage. Pearl millet is grown for forage in Saskatchewan, Ontario and Quebec. Pearl millet is supposedly a high protein alternative to several other cereals silages. The total biomass production ranges from 8–12 t/ha with 16–24 % crude protein [9].

Pearl millet is palatable forage to farm animals. The pearl millet forage crop belt in USA, stretches into 1.5 m ha annually [10, 11]. Actually, pearl millet forage can be grown in most of the states of United States of America [12]. Pearl millet is a forage crop in the Northern Great Plains especially in states such as North Dakota [13]. Here, it grows tall between 6 and 10 ft. in height. Pearl millet tillers profusely and forms large biomass that could serve as a good feed to farm animals. Ratooning pearl millet meant for forage has been profitable. Pearl millet forage yield reported for areas in North Dakota range from 2–7.6 t/ha depending location, soil type and fertilizer inputs. In the dry land, annual forage production of pearl millet hybrids ranged from 1.9 to 2.5 t/ha. Pearl millet is not grown as a major crop in Brazil. However, its cultivation has spread considerably in some parts of Cerrado region. Initial indications suggest that pearl millet is a good intercrop with soybean [14].

KEYWORDS

- **Alfisols**
- **Inceptisols**
- **Lixisols**
- **Mollisols**
- **Oxisols**
- ***Pennizetum violaceum***
- **Vertisols**

REFERENCES

1. D'Andrea, B.; Klee, M.; Casey, J. Archaebotanical evidence for Pearl Millet (Pennesitum glaucum) in Sub-Sahara in West Africa. Antiquity **2001,** *75,* 341–348.
2. Blench, R. Archaeology, Language and the African past. AltaMira Press, Lanham, Maryland, USA. **2006,** pp. 361.
3. Blench, R. The contribution of vernacular names for Pearl Millet to its Early History in Africa and Asia. Proceedings of RIHN symposium on'Small Millets in Africa and Asia'. Tokyo, Japan, **2010,** 1–10 http://www.rogerblech.info/RBOP.htm (September 11, 2012).
4. Taylor, J. R. N.; Emmambux, M. N. Traditional African Cereal Grains-Overview. http://www.sik.se/traditionalgrains/review/oral%20present%20.PDF%20files/Taylor.pdf **2008,** 1–6 (September 8, 2012).
5. Garba, M.; Renard, C. Biomass production, yields and Water use efficiency in some pearl millet/legume cropping systems at Sadore, Niger. Proceedings of Workshop on Soil Water Balance in the Sudano-Sahelian Zone. IAHS Publication Niamey, Niger. **1991,** *199,* 431–439.
6. ACIPMIP, All India Coordinated Pearl Millet Program, Mandor, Rajasthan, India Internal Report. **2008,** 1–2.
7. Agropedia, Area and Distribution of Pearl millet. http://vasat.icrisat.org/crops/pearl_millet/pm_production/htmlm3/ index.html **2009,** 1–4 (September 9, 2012).

8. Basavaraj, G.; Parthasarathy Rao, Bhagavatula and Ahmed, W. Availability and Utilization of Pearl Millet in India. SAT eJournal. Ejournal.icrisat.org **2010,** *8*, 1–10 (September 9, 2010).
9. AERC, Canadian Forage Pearl millet. Agriculture Environmental Renewal Canada Inc.: **2005,** 1–5 http://www.aerc.ca/ forageparlmillet.html (September 9, 2012).
10. Myers, R. L. Pearl Millet: A new Grain crop option for sandy soils or other moisture limited conditions. Jefferson Institute, Columbia, Missouri, USA. http://www.hort.purdue.edu/nw-crop/articles/ji-millet.html **1999,** 1–3 (September 9, 2012).
11. Jefferson Institute, Pearl Millet: Overview http//www.jeffersoninstitute.org/pearl_millet.php **1999,** (December 3, 2012).
12. Hannaway, D. B.; Larson, C. Pearl Millet (*Pennizetum americanum*). http://www.forages.oregonstate.edu /php/ fact_sheet_print_grass.php?SpecID=34&use=Soil. **2004** (September 8, 2012).
13. Sediviec, K. K.; Schatz, B. G. Pearl Millet: Forage production in North Dakota. http://www.ag.ndsu.edu **1991,** 1–6.
14. ICRISAT, Brazilian farmers get a boost from Sahel. SAT Trends-ICRISAT News Letter **2001,** *9*, 1–4.

EXERCISE

1. Describe geographic regions of origin of Pearl millet. Mention the areas for Primary and Secondary Centers of Genetic diversification
2. Mention top 10 pearl millet producing nations. Include total production, area and productivity for 2010.
3. Mention various uses of Pearl millet.

FURTHER READING

1. Andrews, D. J.; Kumar, A. Pearl millet, for food, feed and forage. Advances in Agronomy. **1992,** *48*, 90–139.
2. Fuller, D. Q. Indus and Non-Indus agricultural traditions: New Perspectives from the Field. Weber, A.; Belcher, W. R. (Eds.). Lanham Press: New York, **2003,** 343–396
3. Khem Singh, Gill. Pearl Millet and Its Improvement. Indian Council of Agricultural Research, Krishi Anusandhan Bhavan, Pusa, New Delhi, **1991,** pp. 483.
4. Rachie, K. O. Millets: Importance, Utilization and Outlook. International Crops Research Institute for the Semi-arid Tropics. Patancheru, India, **1975,** pp. 148.

USEFUL WEBSITES

http://www.infonet-biovision.org-Millet.htm (August 31, 2012)
http://www.wholegrainscouncil.org/(October 12, 2012)
http://intsormil.org/ (October 12, 2012)
http://www.icrisat.org (November 20, 2012)

FOXTAIL MILLET (*SETARIA ITALICA*)

8.1 INTRODUCTION

Foxtail millet was grown in China in the prehistoric period around 5000 B.C. Today, it is an important cereal fodder crop in drier parts of North China. Archeological evidences indicate that foxtail millet cultivation was practiced by 2000 B.C. in European plains and by 600 B.C. around Turkey and other West Asian regions. Foxtail millet growing regions were prominent in the Southern Europe, West Asia, India and China during middle ages. Foxtail millet was an important grain crop during Roman period. However, gradually it got replaced by large grain cereals such as wheat, maize and sorghum during recent period of history. Obviously, impact of foxtail millet as a large cropping zone or agroecosystem must have faded commensurate with its cultivation trends, through the ages. It seems foxtail millet was introduced into Great Plains of North America by European travelers during mid-1700s. At present, foxtail millet is a widely distributed crop. It is grown in many agrarian regions of the world. Its cultivation is currently in vogue in the Middle Europe, Mediterranean, Asia Minor, Iran, the Himalayas, Mongolia, China, Southern Asia, North America. In the territory of the former USSR, it is cultivated in regions of Volga, Middle Dnepr, Black Sea Coast, Crimea, Southern Ural; Western and Eastern Siberia [1]. In Northern Russia, foxtail millet zones extend up to 50°N latitude (Table 1).

TABLE 1 Foxtail Millet: Botany, Classification, Nomenclature and Uses—A summary.

Foxtail millet was probably domesticated and grown in China by 5000 B.C. Perhaps it was domesticated simultaneously in other regions such as West Asia. Currently, Foxtail millet agroecosystem that thrives on marginal soils and regions with relatively low precipitation extends into agrarian regions of North America, Europe and Asia. Its cultivation is not pronounced in Africa. In Europe, foxtail millet regions gave way for wheat and maize during medieval period.

Botany and Classification

Kingdom-Plantae; Order-Poales; Family-Poaceae; Subfamily Panicoidea; Genus-*Setaria*; Species *S.italica*

$2n = 18$; Wild types/ancestors are distributed in Europe, Mongolia, China and North American Plains. Foxtail millet germplasm has been preserved in several countries. In India, foxtail millet genotypes are maintained at Universities and Agricultural Research Institutes under All India Coordinated Small Millet Improvement Project. The International Crops Research Institute for the Semi-arid Tropics at Patancheru near Hyderabad, South India holds a fairly large collection of foxtail millet germ plasm lines numbering over 1,535. The Chinese Agricultural University campus at Beijing holds over 25,300 accessions and University of Agricultural Sciences, Bangalore has 1,300 germ plasm lines of Foxtail millet.

Nomenclature

Foxtail millet, Italian Millet or German millet-English; Panis, millet des oizeaux, millet dc italie-French; Painco, milho PaincoItalian; Xiao Mi-Chinese; Tinai-Tamil; Navane-Kannada Korra-Telugu; Kangni-Hindi; Thina-Malayalam.

Uses

Powdered grain of foxtail millet is used in preparing noodles common in China. Foxtail millet grains are used brewing alcoholic beverages. Grains are cooked directly and consumed as cereal food. Porridges are prepared from flour. Foxtail millet flour is also used to prepare pancakes.

Foxtail millet based pastures are common in many parts of the world. The crop should be young and leaves succulent. A grown up crop may show up bristles that affects feeding by livestock. Mature crops may have diuretic effect on animals. Foxtail millet has laxative effect on horses and is not used as sole forage for many farm animal species. Nitrate accumulation in plant tissue is a problem, if a well-fertilized crop is used as forage.

Nutritional Aspects

Foxtail millet is nonglutinous and has an abundance of Manganese, Iron, Phosphorus, Vitamin B and Tryptophan.
Nutritive value: Following constituents occur in 100 g edible seeds of foxtail millet:
Water-12 g; 351 Kcal energy; 11.2 g Protein; 4 g Fat; 63.2 Carbohydrates; 6.7 g Crude Fiber; 31 mg Ca; 2.8 mg Fe; 0.6 mg Thiamine; 0.1 mg Riboflavin; 3.2 mg Niacin.
Essential amino acids in 100 g powdered Foxtail millet grains: 103 mg Tryptophan; 233 mg Lysine; 296 mg Methionine; 706 mg Phenylalanine; 328 Threonine; 728 Valine; 1764 Leucine and 83 mg Isoleucine.

Source: Ref. [2]; http://www.icrisat.org/crop-foxtailmillet.htm; http://www.agroatlas.ru/en/content/related/ Setaria_italica/; http://www.prota4u.org/protav8.asp?h=M4&t=Setaria,italica&p=Setaria+italica

8.2 SOILS, AGROCLIMATE AND CROPPING SYSTEMS

8.2.1 SOILS

Foxtail millet cropping belts thrive well both in mountainous regions and plains [2]. Foxtail millet grown for seed and/or forage is suitable for production on Dark Brown, Black, Dark Grey and Grey soils found in the Canadian Plains. Foxtail millet is grown on marginal soils, commonly referred as brown or dark brown soils of low productivity. Such soils are commonly traced in Southern Canada, North Dakota, Minnesota, Nebraska and Colorado. Foxtail millet sown in Saskatchewan and Alberta may receive 25–70 kg N, 20–35 kg P, 0–35 kg K and 0–20 kg S depending on soil type and soil test values.

Alfisols, Xerasols and Calci-Xerals found in West Asia support relatively larger patches of foxtail millet. Soil fertility is low but it sustained through effective residue recycling procedures. In Kenya, foxtail millet thrives on shallow sandy soils that are infertile, where many other crops may just fail to reach flowering/grain formation.

Foxtail millet is grown on Inceptisols and Alluvial soils found in the Indo-Gangetic plains. Low fertility Mollisols are also used in the upper India. Its cultivation is scattered in the Vertisol belts of Central India. Foxtail millet is grown on low fertility Alfisols (red sandy soils) of South India.

8.2.2 AGROCLIMATE AND WATER REQUIREMENTS

In the Canadian plains, foxtail millet is grown for forage. It is cultivated as a short season cereal and usually harvested in about 60–90 days, depending on heat units received by the crop. Generally, foxtail millet is sown after onset of warm period that occurs in early June in Saskatchewan, Alberta and Manitoba. Its sowing coincides with first rains of the season. Foxtail millet grows up to 1.5 m in height. Late maturing crops with medium thick stems and leafy are preferred as forage. Foxtail millet reaches vegetative stage during July–August. It puts forth leafs using each spell of rainfall and remains in vegetative phase for longer durations. It is a shallow rooted cereal, yet it tolerates drought and erratic precipitation pattern. Foxtail millet grown in North America offers golden yellow grains good for consumption by farm animals and birds. Weeds, especially volunteer cereals, grassy species and wild millet may affect foxtail millet growth and establishment. Hence, grassy weeds are usually thoroughly removed before planting seeds. Sometimes, seeding is delayed up to last week of June to overcome weed infestation.

Foxtail millet is an important small-grained cereal grown in United States of America. It is primarily a forage crop, but seeds are usually meant for birds. Foxtail millet is grown in fallows or soils with constraints related to fertility and irrigation. Foxtail millet is a preferred catch crop or short season crop in North Dakota, Nebraska, and Colorado. Foxtail millet was introduced recently into Colorado, during early 1990s, mainly to produce forage and grain for birds, during the season summer fallows or on wheat stubbles. Foxtail millet has also been used for grazing as pasture grass, but ratoons are slow to grow and not efficient.

In West Asia, foxtail millet crop sown immediately after the main cereal or lentil is provided with irrigation to achieve rapid germination of seeds and to protect the crop in the face long spell of dryness. Irrigation efficiency is important; hence, furrow irrigation is adopted [3].

Foxtail millet is a grown as a hardy drought tolerant crop in the southern Indian plains. It withstands warm temperatures that occur during summer. The growing season temperature fluctuates between 22°C–36°C. Water requirements range from 350–450 mm. Foxtail millet is usually sown on residual moisture. Therefore, inherent soil moisture, sporadic rains during post-rainy season and protective irrigation, if any sustains the crop.

Foxtail millet is among preferred small grain cereals in the Northern dry regions of China. Here, it negotiates drought and marginal soil fertility and still offers farmers with acceptable grain/forge yield [4]. Studies on leaf senescence pattern and photosynthetic efficiency indicate that, tolerance of foxtail millet genotypes to water deficits is related to antioxidant metabolism and accumulation of certain secondary metabolites [4].

8.2.3 CROPPING SYSTEMS

Foxtail millet is a short season crop that suits to fill the summer fallows that occur after harvest of wheat or maize in the states such as Nebraska, Minnesota and Colorado. Computer-based simulations suggest that shallow rooted foxtail millet removes soil moisture and nutrients still left in soil after wheat. Wheat-foxtail millet sequences actually improve precipitation and fertilizer use efficiency [9].

In East Africa, foxtail millets rotated with legumes such as pigeonpea or cowpea. It is intercropped with other cereals such as sorghum or maize. Foxtail millet is also suitable as post rainy crop that effectively uses residual moisture.

Foxtail millet is grown in southern Indian plains on marginal soils or as rotation crop in sequence with dry land cereals/legumes. It is a good species to develop planted fallows, since it is hardy, drought tolerant and produces useful grains/forage to the farmers. Foxtail millet is often grown during post-rainy or summer season in the Gangetic plains. It serves to produce small quantities of grains and forage during the short fallow season. Foxtail millet is grown as a mono crop or as intercrop with legumes or oil seed crops. Foxtail millet is rotated with short season legumes that support biological N fixation. For example, in the Alfisol regions of Telangana plains in South India, green gram-foxtail rotations are provided with small amounts of fertilizers and grown under rain fed conditions. Green gram improves soil-N fertility. Recycling of its residues further adds N, P, K and organic matter to soil. Therefore, foxtail millet that follows the legume grows rapidly on residual moisture and produces moderate harvests of cereal grain and forage during post rainy season [6] (Plate 1).

PLATE 1 A Foxtail Millet field at Grain-fill stage.
Source: Krishna, K.R. Bangalore

8.3 EXPANSE AND PRODUCTIVITY

Foxtail millet or Italian millet is grown on marginal soils with low fertility traceable in different continents. Currently, it is a minor cereal grown in Asia, Europe and Americas. Its cultivation is feeble in Africa. Since it tolerates drought, low soil fertility and organic matter, it is commonly grown in regions where other major cereals may fail or become nonviable economically. Foxtail millet does not form a major cropping expanse, but occurs as sporadically scattered patches. Globally, foxtail millet production is around 5.0 m t annually and is equivalent to 18% of total grain production by all millet species together. The productivity of foxtail millet ranges from 0.8 t −1.8 t grain/ha depending on water supply. In China irrigated fox tail millet crop produces 1.8 t grain and 11 t fresh forage or 3.5 t hay.

In the Great Plains of North America, foxtail millet is cultivated for forage and seeds that could be used as animal feed. Its productivity is relatively low, since it is often grown on low fertility soils that experience different kinds of maladies. The productivity ranges from 6–8 t/ha forage and 0.8–1.5 t grain/ha. The foxtail growing regions also extend into southern regions of USA.

Foxtail millet grown in North America is often supplied with 25–30 kg N and 8–10 kg P/ha to support satisfactory growth and formation of 6–10 t forage [7].

Foxtail millet is produced on the Calci-Xerasols of West Asia, mainly for grains/forage. Foxtail millet is good as second crop after wheat or barley grown in the dry regions of Turkey, Syria, Northern Iran and Iraq [3]. In Iran, Foxtail is grown in the semiarid regions, mostly under rain fed conditions, but with facility for protective irrigation.

According to Obilana [8], foxtail millet is not an important millet species in the semiarid regions of Africa. However, it is grown in small patches in humid tropics of Uganda and Central African Republic. Here, it thrives on well-drained sandy Oxisols. The productivity is not high, but depends on inherent soil fertility, residue recycling procedures and fertilizer supply, if any. Small areas of foxtail millet occur in coasts and hill regions of Kenya. Cultivar such as KAT/FM-1 fills the foxtail millet growing landscape. It is a preferred cultivar in semiarid regions, in low areas and up to 1500 m.a.s.l. Foxtail millet cultivars stay in the field for 3 months and produce 0.7–0.9 t grain/ha. Cultivar Lanet/FM-1 is more common in higher altitudes with cold climate. Foxtail millet cultivation is relatively sparse in southern African agrarian regions. It is grown in Zimbabwe, Zambia, Mozambique, South Africa and Lesotho.

Foxtail millet cultivation is scattered all around the southern Indian plains, hills and coastal region. Several cultivars are grown by Famers. Cultivars that are predominant in different states of South India are TAU43 and TNAU 196. Cultivars like SR51, PRK1 and PS4 are more common in Rajasthan and Gangetic Plains. Most of the foxtail cultivars currently grown in semiarid regions of India are bold seeded and golden yellow in color. The productivity of these cultivars ranges from 1–1.8 t grain plus 6–8 t forage under normally prescribed fertilizer inputs and irrigation [9].

Major foxtail millet producing regions in China are Hopei, Horank and Sanshi. It is grown in large patches in Northern dry regions of China. The productivity is relatively low owing to low input and unfavorable agroclimate. About 0.7–1.2 t grain and 5–8 t forage/ha is produced per crop [2].

KEYWORDS

- **Calci-Xerasols**
- **foxtail millet**
- **marginal soils**
- **pigeonpea**

REFERENCES

1. Smekalova, N. Interactive Agricultural Atlas of Russia and neighboring countries. http://www.agroatlas.ru/en /content/related/Setaria_italica/ **2009,** 1–3 (September 3, 2012).
2. Baltensperger, D. D. Foxtail and Proso Millet. In: Progress in New Crops. Janick, J. (Ed.). ASHS Press, Alexandria, Virginia, USA, **1996,** 182–190.
3. Heidari, H. Foxtail millet (*Setaria italica*) mother plants exposure to deficit and Alternate furrow irrigation and their effects on seed germination. Annals of Biological Rcscarch **2012,** 3, 2559–2564.
4. Dai, H.; Shan, C.; Wei, A.; Yang, T.; Sa, W.; Feng, B. Leaf senescence and photosynthesis in Foxtail millet (*Setaria italica*) varieties exposed to drought conditions. Australian Journal of Crop Science **2012,** 6, 232–237.
5. Saseendran, S. A.; Nielsen, D. C.; Lyon, D. J.; Ma, L.; Felter, D. G.; Baltensperger, D. D.; Hoogenboom, G.; Ahuja, L. R. Modeling responses of dry land spring triticale, proso millet and foxtail millet to initial soil water in the high plains. Field Crops Research **2009,** *113,* 48–63.
6. Yakadri, M.; Thatikunta, R. Production potential and Economics of Green gram-Foxtail millet crop sequence as influenced by Nitrogen and Phosphorus application. Agricultural Science Digest **2004,** *24,* 77–78.
7. Manitoba Forage Council Inc.: Golden German Millet production. Manitoba Forage council, Manitoba, Canada. Mbforagaecouncil.mb.ca **2003,** 1–5 (August 31, 2003).
8. Obilana, A. B. Overview: Importance of Millets in Africa. Intentional crops Research Institute for the semiarid tropics, Patancheru, India. **2005.**
9. AICSMIP, A. Crop Improvement: Recently released varieties of Small Millets. All India Coordinated Small Millets Improvement Project (AICSMIP). ICAR, New Delhi, **2011,** 1–2.

EXERCISE

1. Mention the names of Wild and Cultivated species of Foxtail millet.
2. Mention the soil types and agroclimatic conditions that support Foxtail millet production.
3. Mention top 3 Foxtail millet producing nations and include Production and Area for 2010.

FURTHER READING

1. De Wet, J. M. J.; Oestry-Stidd, L. L.; Cubero, J. H. Origin and Evolution of Foxtail Millet. Journal de Agriculture Traditionelle et de Botanique Applique **1979,** *26*, 53–64.
2. Malm, R. N.; Rachie, K. O. *Setaria* millets. A review of world literature. University of Nebraska, Lincoln, USA, **1971,** pp. 133.
3. Rahayu, M.; Jansen, P. C. M. *Setaria italica* (Foxtail millet). In: Plant Resources of South-east Asia. Grubben, G. J. H.; Partohodjon, S. (Eds.). No 10. Cereals. Backhuys Publisher Inc. Leiden, Netherlands, **1996,** 127–130.

USEFUL WEBSITES

http://www.cnseed.org/ugandan-foxtail-millet-setaria-italica.htm (August 31, 20212)
http://www.infonet-biovision.org-Millet.htm (August 31, 2012)
http://www.icrisat.org/crop-foxtailmillet.htm (August 31, 2012)
http://www.wholegrainscouncil.org/ (August 31, 2012)

PROSO MILLET (*PANICUM MILIACEUM*)

8.1 INTRODUCTION

Historical records suggest that proso millet was grown in Caucasus and the Northern plains region of China during 2nd millennium B.C. (Small, 2012). Proso millet was domesticated independently in Transcaucasia and China about 7000 years ago. Proso millet spread into different agrarian regions through human migration. Currently, it is grown extensively in Indo-Gangetic Plains, Russia, Ukraine, Romania, Turkey and adjoining regions in Syria and Israel. Proso millet has been cultivated in Northern China, Manchuria and Mongolia since prehistoric period. Even today, it is a staple cereal for some of the local tribes in Korea and Mongolia.

Over 20 millet species are traced in the Canadian prairies. Among them, proso millet is an important one, grown mostly in the Central Prairies and Ontario. Proso millet grows rapidly reaching 1–2 m hight. It produces small grains that are whitish, gray or brown in color. Proso millet cultivation is also in vogue in North Dakota, Nebraska and Colorado in USA. Proso millet served as an important cereal grain to human population in Europe, Russian and Central Asian plains region, during prehistoric period [1]. Proso millet is not an important millet in the African continent [2]. In South Asia, especially India, proso millet is a useful cereal grain crop. Proso millet cultivation has been in vogue in Egypt and Arabia since prehistoric times. Proso millet was grown as a principal small grain cereal in Asia, Africa and Europe during Middle Ages. It was gradually replaced by wheat during 19th century. Currently, proso millet farming is confined to low fertility and drought prone regions of Russian plains, China, Indian Sub-continent, West Asia and North America (Table 1).

TABLE 1 Proso Millet: Botany, Classification, Nomenclature and Uses—A summary.

Proso millet was domesticated by human tribes in Transcaucasia and Northern China during Neolithic period. It has been serving as cereal food for humans since 5000 years. Proso millet agroecosystem spreads into different continents. Currently, it thrives mostly in regions with low soil fertility and scanty precipitation. It thrives well in temperate and subtropics. It is equally adapted to semiarid and dry belts of Asia. The proso millet agroecosystem extends from 0 to 2,500 m.a.s.l. Proso millet thrives on wide range of soil types with differing texture and soil moisture holding capacity.

Botany and Classification

Kingdom-Plantae; Subkingdom-Angiosperms; Order-Cyperales; Family-Poaceae; Genus-Panicum; Species-*P. miliaceaum*

$2n = 36$

Wild Progenitor–Unknown; Wild weedy relatives occur in Central Asia, Mongolia, Xinjiang and Russian plains

The International Crops Research institute for the Semi-arid tropics at Patancheru near Hyderabad, in South India preserves over 840 germ plasm lines of proso millet.

Nomenclature

Proso millet, Broom corn millet, White millet, Hog millet-English; Echte Hirse-German; Miglo-Italian; Le Millet Commun-French; Proso-Russian; Mawele-Swahili; Siao mi-Chinese; Barri-Hindi; Kodra-Marathi; Varigulu-Telugu; Panivaragu-Tamil; Baragu or Harka-Kannada; Kodo-Bengali.

Uses

Proso millet grains are fed to farm animals, poultry and caged birds. Proso millet serves as forage. It is used as hay and silage. Proso millet flour is used in making roti, pancakes and bread. Proso millet grits are used in making porridges and other food items. In Northern China and Mongolia, proso millet powder is mixed with water and the dough is shaped into ball and steamed. Proso millet is a staple cereal for some sections of people in Northern Korea. In India, proso millet powder is used to prepare chapathis.

Nutritive Value of Seeds (dry weight basis):

Crude protein-12.0 %; Crude fiber 8.0%; Fat-4%; Total digestible Nutrients-75%; Energy-1500 K calories/lb; Ca-0.05%; Phosphorus-0.3 % B-complex vitamins: Thiamine-3 mg/lb; Niacin-10.5 mg/lb; Riboflavin-1.7 mg/lb; Pantothenic acid 5 mg/lb; Amino acids: Lysine-0.23%; Methionene-0.29%; Theonine-0.40% and Tryptophan-0.17%.

Sources: Ref. [3]; http://en.wikipedia.org/w/index.php?title=Proso_millet&oldid=495432253; http://www.hort.purdue.edu/newcrop/afcm/millet.html; http://www.icrisat.org/crop-prosomillet.htm

8.2 SOILS, AGROCLIMATE AND CROPPING SYSTEMS

8.2.1 SOILS

Proso millet cropping zone thrives well on Mollisols commonly used to grow wheat or maize. The residual soil fertility and moisture plays vital role in crop growth and productivity. Low fertility Ultisols and Alluvial soils are also used to grow Proso millet. Proso millet receives only small dosages of fertilizers commensurate with yield goals. It is supplied with 30–40 kg N, 20 kg P and 5 t/ha FYM for a yield goal of 6–10 t forage/ha and 8–1.2 t grain/ha. Proso millet thrives well on soils with neutral or acidic reaction with pH 6.0. It is shallow rooted crop and grows well as intercrop with deep rooted species such as cotton, sunflower or maize. On the Mollisols of North America, proso millet is supplied with30–40 kg N/ha depending soil NO_3-N content Phosphorus inputs are based on available soil-P levels and it ranges from 17 to 20 kg P_2O_5/ha. Irrigated proso millet may be supplied with 20–30 kg K_2O based on DTPA extractable K levels [4]. In East Africa, proso millet cultivation zones are sparse and traced mostly in regions with sandy Alfisols or coastal alluvium.

Proso millet cultivation proceeds on several types of soils found in Indo-Gangetic plains and Peninsular India. Marginal soils are most often allocated to proso millet. Major soil types such as Mollisols, Inceptisols, Vertisols, Alfisols, Alluvial soils and sandy coastal soils are all suited for cultivation of proso millet in South Asia.

8.2.2 AGROCLIMATE

Proso millet is sown with onset of warm season in the Great Plains. It is sown by June/July in Nebraska, Colorado and Kansas. In North Dakota, proso millet is sown in May and harvested by mid-September [3]. Early planting avoids cold and frost damage. In the Southern region of USA, for example, in Florida and Georgia, dove fields or proso millet is sown in June and harvested by September end. It needs 100–110 days to mature [5]. Proso millet requires 70–90 days to mature. Crop duration depends on number degree-days and heat units received. Proso millet takes about 70–90 days to mature, if grown in Northern Plains of America. Time needed for maturity depends on hybrids or cultivars used, geographic region, seeding period, and most importantly the available degree-days. In the Canadian plains a shortage of degree-days extends maturity proportionately. Proso millet is susceptible to early frost or extreme temperatures that might occur at flowering. The short duration between 3rd leaf and 6th leaf stage is said to be critical for full expression of growth potential and yield formation.

Proso millet is grown in rotation with wheat in many regions of North America. Here, it grows using stored moisture. Tillage after wheat harvest may induce loss of soil moisture and organic matter, hence plowing is often avoided before sowing proso millet. Soil moisture retention is a key aspect during wheat-proso millet rotations. Proso millet requires relatively less water and has low transpiration ratio since it is a C_4 cereal crop. A proso millet crop grown in the Great Plains may transpire 270 mm water to reach maturity. Its requirement for water is relatively much less compared with 530 mm by wheat, 560 by cotton or 300 by sorghum. Proso millet also tolerates drought and erratic precipitation pattern, but it needs protective irrigation during critical stages like flowering and grain fill [6]. Proso millet grows erect and reaches 30–100 cm in height. It tillers feebly. Proso millet tillers at 430 growing degree days (GDD). Flowering is initiated at 750 GDD. Proso millet thrives well from 0–2500 m.a.s.l. Proso millet is grown in the mountainous regions of Russia that are 1200 m.a.s.l. and in the India Himalayas it is found grown at 2500 ma.s.l. [7]. Proso millet is an efficient converter of water to biomass and it accumulates more biomass in grains compared to several other millets [7].

8.2.3 CROPPING SYSTEMS

Proso millet is among the preferred crops to follow wheat in the Central Great Plains of USA. It is sown into wheat stubbles with least disturbance to soil. Proso millet that follows wheat or maize is usually sown with no-tillage or minimum tillage. Specific rotations followed in the Great Plains are wheat-proso millet-fallow; wheat-corn-proso millet-fallow; wheat-corn-proso millet forage. Among the various rotations practiced that include proso millet, corn followed by wheat and then proso millet seems to be popular in the region [6]. Proso millet is preferred because it reduces water requirement of the crop sequence and halts loss of soil moisture. In parts of Florida, proso millet is sown first in June as a sole crop and it is followed by a mixture of maize/proso millet and legume during next season. Simulations using computer models suggest that proso millet is an excellent species to shorten or fill the fallow period that occurs after harvest of wheat [8]. Analysis of rooting pattern and water depletion using

RZWQM2 model suggests that proso millet effectively exploits residual moisture, still available in the upper layers of soil after wheat season.

In the Indian subcontinent, proso millet serves to fill the fallow period and improve both precipitation and nutrient use efficiency. Proso millet is grown in rotation with major cereals like wheat or maize. The short fallow period that occurs after a cereal-legume rotation is used to sow and reap proso millet. Proso millet is actually a versatile short season species, hence it adapts itself to several combinations of intercropping, relay cropping and sequences. Proso millet is often intercropped with legume or oil-seeds.

8.3 EXPANSE AND PRODUCTIVITY

Proso millet cultivation extends into several states in the Northern and Central regions of Great Plains. It also extends into Southern and south-eastern states like Georgia and Florida. The proso millet growing regions occur as rotation crops with wheat or maize. It is also intercropped with maize or cotton. The productivity is low and is based on fertilizer and irrigation supply. Proso millet yields 10 t forage plus 1.0–1.5 t grain/ha. In North Dakota, the proso millet growing regions are filled with cultivars such as Cerize (red grain), Early Bird, Rize, Snowbird, Sunup, Huntsman, Minsum and Horizon. These genotypes are short or medium statured, early or medium duration crops with small to medium sized seeds [3]. Most of these cultivars grown in North Dakota produce 1.3–1.8 t grain/ha plus forage. In Florida and Georgia, proso millet is predominately grown to feed animals and birds. Such a crop may yield 1.0 t grain plus forage.

Proso millet cultivation is in vogue European plains from west coast of Spain to Southern Russia. It is also cultivated in agrarian regions of Southern Europe such as Italy, Romania, Hungary and Czeckoslavia. Proso millet offers marginal grain yield reaching 1–2 t/ha and forage.

Small areas of proso millet are traceable in coasts and hill tracts of Kenya in East Africa. Its cultivation spreads from 0–2400 m.a.s.l. It is preferred in regions with low precipitation. Incidentally, proso millet has one of the lowest water requirements among cereals. It can grow up to maturity in Kenya, in areas with average rainfall of 200–450 mm annually. About 30–40% of rainfall should occur during growing period. Currently, Kenyan proso millet growing region is filled predominantly with variety-KAT/PRO-1. It thrives from 0–2000 m.a.s.l., offers cream or white colored grains and has a yield potential of 0.7 t grain/ha. It is a drought tolerant cultivar, hence thrives successfully when other crops begin to perish.

Proso millet cultivation though sparse, it is well spread out into Indo-Gangetic plains. It is mostly an intercrop or summer crop or one that fills up fallows. Proso millet cultivation proceeds from 0–3000 m.a.s.l in the Himalayan region. Proso millet cultivation is well scattered in the southern India plains. In Tamil Nadu, Panivarugu-CO4 is popular among farmers. It is mainly a forage/grain crop meant for poor people. Proso millet serves as good forage crop in Tamil Nadu [9]. Cultivars such as GPUP 8, GPUP21 and CO-6 are common in millet growing regions of Karnataka and adjoining areas in Tamil Nadu [10]. Proso millet grain yield is relatively low in the dry lands of South India. Farmers harvest about 0.8 to 1.4 t grain/ha and forage. Expansion of proso millet regions depends on summer fallow period and irrigation resources. Proso millet is often allocated to low fertility zones or those under reclamation, hence it limits expression of full potential for grain and forage production. Intercropping proso millet more frequently with other high input crops will enlarge its cropping zone and improve productivity.

KEYWORDS

- **Mollisols**
- **proso millet**
- **RZWQM2 model**
- **warm season**

REFERENCES

1. Small, E.; Millets, The Canadian Encyclopedia.htm **2012,** 1–2 (August 31, 2012).
2. Obilana, A. B. Overview: Importance of Millets in Africa. Intentional crops Research Institute for the semiarid tropics, Patancheru, India. **2005.**
3. Berglund, A.; Proso, D. R. Millet in North Dakota. North Dakota State University Agricultural Extension Service bulletin. http://www.ag.ndsu.edu **2007,** 1–3 (August 31, 2007).
4. Davis, J. G.; Westfall, D. G. Fertilizing spring-seeded small grains. Colorado State University Extension Service Bulletin **2009, 1**–6.
5. Giuliano, W.; Selph, J. F.; Hodges, K.; Wiley, N. Dove Fields in Florida. Institute of Food and Agricultural Science, University of Florida, Gainesville, USA, **2012,** 1–7.
6. Croissant, R. L.; Peterson, G. A.; Westfall, D. G. Dry land Cropping Systems. Colorado State University Extension Services Bulletin, Fort Collins, USA. **2012,** 1–5.
7. Baltensperger, D. D. Foxtail and Proso Millet. In: Progress in New Crops. Janick, J. (Ed.). ASHS Press, Alexandria, Virginia, USA, **1996,** 182–190.
8. Saseendran, S. A.; Nielsen, D. C.; Lyon, D. J.; Ma, L.; Felter, D. G.; Baltensperger, D. D.; Hoogenboom, G.; Ahuja, L. R. Modeling responses of dry land spring triticale, proso millet and foxtail millet to initial soil water in the high plains. Field Crops Research, **2009,** *113,* 48–63.
9. TNAU, Millets. Department of Millets. Tamil Nadu Agricultural University, Coimbatore, India 1–2 TNAU Agritech Portal **2012,** (August 31, 2012).
10. AICSMIP, Crop Improvement: Recently released varieties of Small Millets. All India Coordinated Small Millets Improvement Project (AICSMIP). ICAR, New Delhi, **2011,** 1–2.

EXERCISE

1. Describe Proso millet growing regions of the World.
2. Describe Agronomic procedures, Fertilizer inputs and Water requirements for Proso millet.
3. Mention at least five different cropping sequences involving Proso millet.
4. Describe various uses of Proso millet.
5. Mention the Nutritive value of forage derived from Proso millet.

FURTHER READING

1. Baltensperger, D.; Lyon, D.; Anderson, R.; Holman, T. J.; Stymiest, C.; Shanahan, J.; Nelson, L.; DeBoer, K.; Hein, G.; Krull, J. Producing and Marketing Proso Millet in the Higher Plains. Institute of Agriculture and Natural Resources, University of Nebraska-Lincoln, Nebraska, **1995,** pp. 1–22.
2. Lyon, D. J.; Bargener, P. A.; DeBoer, K. L.; Halverson, R. M.; Hein, G. L.; Hergert, G. W.; Holman, T. L., Nelson, L. A.; Johnson, J. J.; Nleya, T.; Krull, J. M.; Nielson, D. C.; Vigil, M. F. Proso Millet in the Great Plains. University of Nebraska-Lincoln, Nebraska, USA, **2008,** pp. 19.

USEFUL WEBSITES

http://www.wholegrainscouncil.org (September 25, 2012)
http://www.icrisat.org/crop-prosomillet.htm (September 25, 2012)

CHAPTER 9

SOYBEAN PRODUCTION ZONES

CONTENTS

9.1 INTRODUCTION

Soybean is native crop of China and adjoining South-east Asian regions in Japan, Korea and Taiwan. Soybean culture began in North America during mid-1700s. It seems soybean was introduced into Georgian plains by sailors returning from China. During early 1900s, soybean cultivation spread rapidly. Soybean farms and companies supported by major business houses and philanthropic organization started expanding rapidly in the Northern Plains of USA. Regular use of soymilk and ice creams made of soy powder and extraction of vegetable oil induced expansion of soybean belt in the Great Plains during 1880s. Farmers in Mexican agrarian region started growing soybean in 1877. Soybeans reached Caribbean Islands during 1760s, but remained a small cropping belt. Soybeans were introduced into Argentina in 1882 [1]. Soybean was introduced into European plains by British colonialists during mid-1700s. Soybean reached Central Europe and Transcaucasia during 1867 [2]. In Africa, Egyptians received soybean during 1850s. The soybean agroecosystem spread into Australia with its introduction by seafarers and explorers during 1770s [3]. Soybean companies that used soy powder became prominent in Australia during early 1800s. South-east Asian famers grew soybeans during 1300s. Soybeans were introduced into Indonesian archipelago during early fourteenth century. However, in India, soybean farming began only during 1660s with its introduction by Dutch travelers [4].

TABLE 1 Soybean: Origin, Classification, Nomenclature and Uses—A summary.
Soybeans were domesticated in China and Fareast. Its cultivation spread into agrarian zones of different continents rapidly during mid-1700s. Soybean cropping zones were intensified during 19th century. It is driven by greater demand for its protein filled grains, oil and other products.

Botany and Classification

Kingdom-Plantae; subkingdom-Angiosperms; Order-Fabales; Family Fabaceae SubFamily Faboideae; Genus-*Glycine*; Species-*G. max*

Wild Type is known as *Glycine soja*. It is traced in South-eastern Russia, China, Japan, Korea, and Taiwan. There are 16 wild species of Glycine traced in its native region.

Synonyms: *Glycine angustifolia; G. gracilis; G. hispida; G soja*

$2n = 40$

Nomenclature

European languages: Czech-Soja; Dutch-Sojaboon; German-Sojabohne; Italian-Semia de Soia; Portuguese-Fava de Soja; Russian-Soevic; Spanish-Soja; Swedish-Sojabona; Hindi-Soya; Marathi-Soybin; Kannada-Soya bean; Tamil-Soya Payiru; Bengali-Gari kalai; Chinese-Dodou; Japanese-Diazu.

Uses

Globally, soybeans are used in many ways in addition to foods. Nearly 85% of soybean is used as grains, extracted vegetable oil and processed soymeal that is rich in protein. Soybeans are also used as fresh vegetables. Soybean seeds are fried and consumed with salt and pepper as snacks. A wide range of snacks are prepared using soybean flour. Soybean four is fermented

---

---

to prepare a popular food item called *Tofu*. Soybean pastes and seasonings such as *kinako, edmame, natto* and *miso* are common in many countries of Asia and America. Soybean milk is an important food item in North America. Soybean powder and milk is used in infant foods. Lecithinated soybean is popular in North America. Soybean meal derived after extraction of oil is rich in protein and is used as animal feed. Soybean fodder is used to feed farm animals. It's recycling aids in improving soil organic matter content and soil-N. Soybeans are used in several types of industries. Mainly, they are used in vegetable oil production, protein food manufacture, paint industry, preparation of phenols, plastics, medicines etc.

Nutritive Value (g 100/g grains)

Energy-446 k cal; Carbohydrates 30.2; sugars 7.33; Dietary fiber-9.3; Fat-19.93; Saturated Fat-2.88; Monounsaturated fat-4.04; Polyunsaturated fat-11.25; Protein-36.94; (Tryptophan-0.59 mg; Threonine-1.76 mg; Isoleucine-1.97 mg; Leucine-3.30 mg; Methionine-0.55 mg; Cystine-0.65 mg; Phynylealanine-2.12 mg; Tyrosine-1.54 mg; Valine-2.02 mg; Arginine-3.13 mg; Histidine-1.09 mg; Alanine-1.92 mg; Aspartic acid-5.12 mg; Glutamic acid-7.84 mg; Glycine-1.88 mg; Proline-2.39 mg; Serine-2.36 mg) Water-8.55 g; Vitamin A 1.0 μg; Vitamin B6-0.377; Vitamin B12-nil; Choline-115.6 mg; Vitamin C 6.0 mg; Vitamin K-47 μg; Calcium-277 mg; Iron-15.70 mg; Magnesium-280 mg; Phosphorus 704 mg; Potassium-1797 mg; Sodium-2 mg Zinc-4.8 mg

---

Source: USDA Nutrient Database-http://www.nal.usda.gov/fnic/foodcomp/search; Refs. [1–4].

Soybean is an annual herbaceous plant that grows to 1.0–2.0 m in height (*see* Table 1; Plates 1, 2, and 3). The plant is hairy. It bears trifoliate leaves with three or four leaflets. Leaves usually senesce by the time pods are mature. Flowers are found in the axil of leaf. They vary in color. The pods are hairy, borne in clusters of 3–5 pods per axil and each of the pods has 3–4 seeds. Soybean seeds vary in color. The seed coat color could be yellow, brown, blue, green, mottled or black. The seed coat is hard. Soybean cropping zones thrive in subtropical and humid tropical climates. It grows with in regions with man temperature of 18°C−32°C. Ambient temperature below 18°C may affect growth rate, unless it is a cold tolerant genotype. Soybean crop needs 80–120 days to reach maturity

Soybean agroecosystem is worldwide in distribution. Soybeans flourish on variety of soil types and fertility regimes encountered in different continents. Largest of soybean cropping zones that occur in North America thrive on Mollisols and Ultisols. In Brazilian Cerrados, soybean grows on sandy Oxisols that are acidic and deficient in nutrients such as N and P. Here, soils are prone to high P fixation. Soybean intercrops found in Argentina occur on low fertility Mollisols. In the European region, soybean crop adapts to cold Cambisols, Podzols and Chernozems. In Asia, soybean belts occur on Inceptisols, Black clay loams, Red sandy loams, Oxisols and even Laterites. Soils that support soybean could be deficient for N. Soybean is a legume that derives N from atmosphere via symbiotic association with Rhizobium. Therefore, in highly fertile zones of North America and even in other regions, soybeans are supplied with relatively lower levels of fertilizer-N or many a times nil. Soybean, in fact contributes 20–80 kg

N/ha per season to soil-N pool. Soils with micronutrient nutrient deficiencies may affect crop growth and yield formation. Hence, foliar sprays or soil application of micronutrients is necessary, especially Fe and Mo.

9.2 SOYBEAN CROPPING ZONES OF THE WORLD

Soybean is a dominant legume in the Northern Plains of United States of America. Soybean cultivation is prominent in Iowa, Illinois, Indiana, Ohio and Missouri. It extends into over 3.2 m ha area. The productivity of soybean is high owing to lavish supply of nutrients and water. Soybean is most often intercropped and /or rotated with wheat or maize. In addition to the intensive cropping within the Mid-west region, soybean is cultivated in other regions of USA, and Canada (Plate 1; Fig. 1). Soybean cropping zone thrives mostly on fertile soils. Soybean derives a sizeable amount (20–40 kg/ha) of N from atmosphere via symbiosis with Bradyrhizobium. Soybean is often sown in field with restricted tillage or even zero tillage, immediately after the harvest of previous crop. Soybean seedlings are often protected using mulches. Mulching reduces soil erosion and maintains soil temperature during cold spell. Soybean needs 550–600 mm water per season.

PLATE 1 Large scale soybean planting using GPS-guided planters. Soybean is planted in rotation with Maize in many of the Mid-west states and Great Lake region. Precision farming techniques adopted in soybean cropping zones is altering the soil nutrient dynamics. Precision farming imparts accuracy during fertilizer and water supply.
Source: Mr David Nelson, Nelson Farms, Iowa.

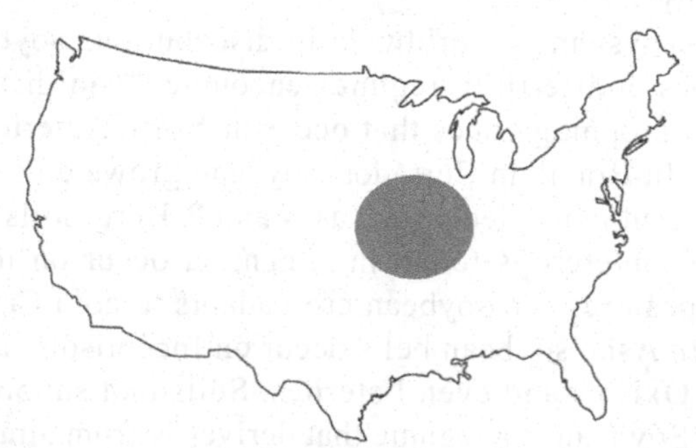

FIGURE 1 Soybean cultivating zone of United States of America.
Note: Soybean is rotated or intercropped with wheat or maize in the Great Plains zones. Soybean mono-crops are also frequently encountered. Soybean producing states in USA, are Illinois, Indianan, Iowa, Kansa, Michigan, Minnesota, Ohio, Arkansas and Tennessee.

Soybean cultivation has been in vogue in South America since 1880s. It is primarily preferred for oil and meal that is good to feed animals. The expansion of soybean cropping zones in Brazil was actually prompted by the high demand for legume grains created by erratic legume production trends in other parts of the world, such as USSR. During 1970s, soybean belt in Brazilian Cerrados reached 5 m ha [5] (Plate 2). During recent years, between 2008 and 2010, it fluctuated around 4.7 m ha. The Brazilian soybean agroecosystem, especially in the Matto Grosso region experienced rapid increase in productivity from 0.8–1.0 t grain/ha to 3.5 t grain/ha, due to fertilizer and irrigation supplements. The Brazilian soybean season coincides with rainy period. Nearly 80% of precipitation occurs during the growing season. Technologies that improve drainage and avoid water stagnation are essential while embarking on large scale soybean cropping. The correction of soil maladies such as Al and uncongenial pH using gypsum is almost always required. Soil fertility is low in the Cerrados. Usually, a slightly higher amount of fertilizer-P is supplied to the field, to overcome soil-P fixation capacity. Nearly, 80% of farmers in the soybean cropping zones adopt zero-tillage concepts, along with cover crops. Cover crops reduce loss of soil nutrients and weed growth. Cover crops, mostly grasses like *Pennesitum*, *Brachiara* or a legume such as *Cajanus cajan* are planted [6]. Conventional tillage requires use of heavy disk plows, turning of topsoil and ridging. This procedure involves cost to farmers and induces perceptible loss of soil organic carbon. It creates highly oxidized soil conditions for root growth. Planting time of cover crops seems crucial; otherwise it may lead to rapid depletion of nutrients in the topsoil. Brazilian soybean is exported in large amounts to Europe and Africa. Soybean is also used in wide range of preparations by South Americans. It supplies both fat and protein to them.

PLATE 2 The famed Soybean expanses of Brazilian Cerrados.
Source: EMBRAPA, Sao Paulo, Brazil.

PLATE 3 Top: A Soybean field at GKVK Agricultural Experimental Station, Bangalore, South India. It is a preferred rotation crop in the dry arable plains of South India. Bottom Left: Soybean Pods; Right: grains. Soybean offers grains with relatively higher protein content ranging from 28–32% on dry weight basis.
Source: Krishna, K. R., 2010.

Soybean is cultivated as a source of protein and fat in India. Soybean cropping is mostly confined to Central Plains and Southern Peninsular region of India (Plate 3). Annually soybean farming in India extends into 1.1 m ha. Soybean agroecosystem thrives on Vetisols and Alfisols that are moderately fertile and need periodic replenishment of lost nutrients. Soybean is grown both as rain fed and irrigated crop. Farmers often grow soybean under assured rains or with protective irrigation facilities. Nutrient supply is restricted to a primer doze of N and full dose of P and K. A good share of N requirement of soybean cropping zone is derived via BNF. Soybean is grown as intercrop with cereals such as maize or sorghum in the arable cropping zones of Central and South India. It is rotated with maize, cotton or vegetables. Soybean offers nutritionally rich fodder. The productivity of soybean in India is moderate and fluctuates between 1.5–2.0 t grain/ha. Each crop of soybean adds to soil N and C fertility. Soybean powder is used to supplement milk.

Soybean cropping zones are well distributed in China. They are most intense in the North-easterm Plains and major river valleys such as Hung He and Yangtze Yang. Soybean culture is feeble and subsistent in the north-west dry regions. Soybean production is a preoccupation the south-east, where its productivity is high. The productivity of soybean in China ranges from 0.8 t/ha in subsistent low input regions of North-west to 3.5–4 t/ha in the plains region.

Soybean cropping is relatively feeble in Australia. During recent years, its area and production is increasing. The spread of soybean cropping zones encounters competition from canola that dominates oilseed sector in Australia. Soybean cropping extends into parts of New South Wales, Northern Victoria and Southern Queens land. Currently, Australian agrarian zones produce 35,000 t/ha. A large share of soybean requirement is actually imported. Hence, there is scope to enlarge soybean area in Australia.

9.2.1 SOYBEAN PRODUCTION

Globally, soybean production during 2010 was 249 million ton. The average productivity was 2.5 t grain/ha. About 75% of global soybean harvests are derived from agroecosystems of three countries namely USA, (91 mt), Argentina (68.5 mt) and Brazil (53 mt). The productivity of soybean farms in the top three regions was 3 t grain/ha. Productivity of soybean is relatively high in Northern Great Plains, Cerrados and Pampas. The grain yield potential of soybean is still higher in some regions. It may reach up to 10.7 grain/ha. China produced 9.8 mt grains and India 7.4 m t during 2010. Several other countries such as Canada (4.3 mt), Uruguay (1.8 mt), Ukraine (1.7 mt) and Bolivia (1.6 mt) also possess large soybean belts with appreciable productivity [7].

KEYWORDS

- **Bradyrhizobium**
- ***edmame***
- ***Glycine soja***
- ***kinako***
- ***miso***
- ***natto***
- ***Tofu***

REFERENCES

1. SoyInfo Centre. History of Soybeans and Soybean foods in South America (1882–2009) http://www. soybeaninfocenter.com/books/132 **2011a** (October 2, 2012).
2. SoyInfo Centre History of Soybeans and soybean foods in Africa. http://www.soybeaninfocenter.com/books/134 **2011b** (October 2, 2012).
3. SoyInfo Centre History of soybeans and Soybean food in Australia http://www.soybeaninfocenter.com/books/138 **2011c** (October 2, 2012).
4. SoyInfo Centre History of soybeans and Soybean food in Australia http://www.soybeaninfocenter.com/books/140 **2011d** (October 2, 2012).
5. MCFFEY, M.; BAUMEL, P.; WISNER, R. 2000 "BRAZILIAN SOYBEANS—WHAT IS THE POTENTIAL?" *AGDM NEWSLETTER, 10,* 5–7.
6. Pacheco, L.; Petter, F. A. Benefits of cover crops in Soybean Plantation in Brazilian Cerrados. Research Report of the federal University of Piuf, Brazil, **2010,** 67–93.
7. FAOSTAT, Soybean Statistics. Food and Agricultural Organization of the United Nations. Rome, Italy. http://www.faostat.fao.org/site/339/default.aspx **2011** (October 23 2012).

EXERCISE

1. Discuss the Origin, Centers of Genetic diversity and spread of Soybeans.
2. Mention top 10 countries with regard to Soybean Area, Production and Productivity.
3. Mention the various uses of soybean.
4. Mention at least 10 food items prepared using Soybeans.
5. Mention at least 10 most popular Soybean cultivars grown currently in North and South America, along with Potential yield.

FURTHER READINGS

1. Ng, T. In: *Soybean: Applications and Technology*. Intech Publications: USA, **2011,** pp. 402
2. Singh, G. In: *The Soybean: Botany, Production and Uses.* CAB International: England, **2010,** pp. 494

USEFUL WEBSITES

http://www.rirdc.gov.au/programs/established-rural-industries/pollination/soybean.cfm
http://www.encyclopedia.com/topic/soybean.aspx
http://www.soybeans.umn.edu/pdfs/FieldBook.pdf

CHAPTER 10

LENTIL CROPPING BELTS

CONTENTS

10.1 INTRODUCTION

Lentils originated in the Indus valley and North-west Alluvial plains of India. *Lens culinaris* is the domesticated species. Wild subspecies known as *Lens culinaris* subspecies *orientalis* is found scattered all over from Turkey in the west to Gangetic plains in the east. It encompasses regions in Iran, Iraq, Lebanon, Israel, Jordon and Afghanistan. Lentils are being grown in the Mediterranean region and Indo-Gangetic region, since early Neolithic period. Excavations of Neolithic sites in Israel have proved that lentils were cultivated 6800 years ago. Lentils have been mentioned in ancient Hebrew literature, in bible and in Ancient Indian literature. Ancient Iranians used lentils regularly in their diets. Ancient Syrians used lentils along with barley and einkorn wheat. It is believed that lentils were domesticated in Northern Syria and Iraq by the inhabitants residing on the banks of river Euphrates [1]. Archeological studies indicate that lentils were introduced into Greece and Egypt during Bronze Age. Evidences suggest that lentils were used by native communities living around Lake Biel in Switzerland during ancient era [2].

10.2 SOILS, AGROCLIMATE, WATER REQUIREMENTS AND CROPPING SYSTEMS

Lentil is an edible crop grown for grains and forage. It is a bushy plant reaching up to 1.6 m in height. Pods contain 2–3 seeds. Based on grain color, we can identify yellow, green, red and brown types. Major types recognized by farmers and market experts are Brown/Spanish, French Green, Green, Yellow, Red chief, Eston Green (small cotyledons), Petite golden, Masoor (brown skinned), Petite Crimson red and Macachados (Mexican yellow) (*see* Table 1; Plate 1).

TABLE 1 Lentil: Botany, Classification, Nomenclature, Uses and Nutritional aspects—A summary.

Lentils were domesticated in North-west India. Wild species are found scattered in many regions within Asian minor and Indo-Gangetic Plains. Lentils were used by inhabitants of Fertile Crescent and adjoining areas in Israel, Iran and Iraq since Neolithic period.

Botany and Classification

Classification of lentils is follows: Kingdom-Plantae; Order-Fabales; Family Fabaceae; Subfamily Faboideae; Tribe-Vicieae; Genus-*Lens*; species-*Lens culinaris.*

$2n = 14$

Germ plasm centers for *Lens* species are available at International Centre for Agricultural Research in Dry Areas (ICARDA), Aleppo in Syria; Indian Institute for Pulses, Kanpur, India; Agriculture Canada Experimental Station, Lethbridge, Alberta, Canada.

Nomenclature in different languages:

Bulgarian-Lenze; Czech-Koska; Danish-Linze; Dutch-Linze; French-Lentile; Irish-Lentile; Italian-Lentichio; Greek-Faki; Portuguese-Lentibe; Spanish-Lentija; Swedish-Lin; Hindi-Masoor; Marathi-Masur; Bengali-Masuri; Kannada-Masoor Bele; Tamil-Masoor Paripu; Armenian-Osp.

Uses

Lentil is used in preparing protein rich soups, stews and salads. Lentils are said to be high in fiber, low in fat and cholesterol-free. They are generally higher in proteins, vitamins, but low in Na content. Lentils are useful in managing type-2 Diabetes. In South Asia, Red lentils are extensively used in curries and soups. Lentils are mixed with cereal flour to prepare cakes and baby food. Lentils are used along with rice in many Asian countries. In India and Pakistan, preparation called *Kichidi* is a mix of rice and lentils. *Mujaddara* or *Mejadra* is a culinary item of West Asia that has lentils. Egyptians prepare *Kushari* which contains lentils in large proportion along with other cereals. Lentils that contain more of fibers are used to feed animals. Lentil is also used as green manure and recycled into fields when they are still succulent. Lentil is an N-fixing legume; hence it is a preferred legume in mixed pasture in Europe and Middle Eastern nations. In North African countries, lentils are used to prepare dishes such as soup, stew, tamia (flafla) and stuffings [3].

Nutritional value of 100 g raw dry lentil grains

Energy 1477 kJ (353 kcal), Carbohydrates-60 g; Sugars-2 g; Dietary fibers-31 g; Fat-1 g; Protein-26 g; Water-10.4 g; Thiamine (Vit B12)-0.87 mg; Folate (Vit B9)-479 μg; Magnesium-122 mg (34%); Phosphorus-451 mg (20%); Sodium-6 mg; Zn 4.8 mg. Lentils are deficient in Methionine and Cysteine.

Source: Several.

PLATE 1 Grains of Black and Yellow Lentil.

Lentils adapt to wide range of soil types that occur in different agrarian regions. They prefer sandy or loamy soils kept under arable conditions without stagnation. In the Pacific North-west; they thrive on Inceptisols and Mollisols. CalciXerasols and Alfisols found in West Asia support moderate grain yield of lentils. In Europe, Cambisols, Chernozems and Alluvial soils support a good harvest of lentils for forage and grains. Mollisols, Inceptisols, Alfisols and Vertisols are the major soils types that support lentil production in the Indian subcontinent.

Lentils are grown in wide range of agroclimates. They are grown in North America, Europe and West Asia in regions ranging from cool temperate, temperate and subtropical conditions. Lentil thrives well under to tropical conditions that occur in Asia. Lentils are generally a cool season crop. Optimum temperature for cultivation ranges from 15°C–28°C depending on geography and crop genotype. Yet, lentils can tolerate cool temperature that occurs during winter. Lentils require 550–600 mm water to reach maturity. Maximum absorption of water and dissolved minerals occurs during flowering and reproductive stages. Lentils are grown in arable soils that support oxidative processes and rich soil microflora. Soil N status is augmented to a certain extent by N-fixing Bradyrhizobia.

Lentils grown in Pacific North-west receive about 55–60 kg N/ha. A portion of N is supplemented through biological nitrogen fixation. Lentils may derive sizeable portion of N from soil-N pool that derived from mineralization of organic matter. The rate of decomposition of organic matter is an important factor that determines amount of N absorbed. In North America, small red lentils are preferred as a substitute to uncropped fallows. It may use 35 to 50 cm of precipitation that occurs during the season, just before the start of the major crop in the rotation. Lentils are grown each year instead of a fallow or once in a three-year rotation. Lentils are a good option, since they produce sufficient quantities of N rich organic residue. Generally, organic residue generated by lentils is incorporated a couple of weeks ahead of seeding the main cereal crop. Lentil residues add organic-C, nutrients and also acts as mulch when applied in between rows [4].

10.3 EXPANSE AND PRODUCTIVITY

Lentils are grown world-wide. It is an important legume in North America, Arabia and Asian agrarian regions. The global lentil produce during 2009 was estimated at 3.92 m t. Major lentil cropping zones that contribute grains occur in Canada-1.5 mt (42%), India-0.9 m t (22%), Turkey-0.36 mt (8%) and United States of America-0.25 mt (6%). Several other countries also contribute lentils. They are Australia-0.14 mt, Ethiopia-0.12 mt, Syria-0.10 mt and Iran-0.10 mt. [5]. Canada is the largest exporter of lentil grains [6].

In North America, lentil production zones are conspicuous in cool regions of Washington State, Idaho and Montana in United States of America, and Saskatchewan in Western Canada. Productivity of lentils ranges from 825 kg/ha to 1550 kg grains/ha plus forage, depending on soil fertility, genotype and inputs [7]. Canadian lentil production zones are dominated mostly by green and red seed coat types. Cultivars such as Laird, Glamis, Sovereign and Grandora are important green coat types. Red lentil varieties such as Crimson, Blaze and Redwig are spread all through the lentil cropping zones of western Canada. In Syria, lentil cropping zones are concentrated in Aleppo, Al-hassakeh, Edlib and Hama regions [1]. Lentils are cultivated in most West Asian nations, wherever precipitation and soil fertility permits a decent crop yield of 800–100 kg/ha. India, Pakistan and Afghanistan are important lentil producing zones in South Asia. Lentils cultivation in China is mostly confined to dry areas with precipitation ranging from 400–750 mm annually. Lentil cultivation began in 1980s in the

Australian agrarian zones. South Australia and Victoria are major lentil growing provinces. Lentils are also cultivated in Western Australia. Lentils are best suited to regions with 350–400 mm precipitation per season. Lentil crop grown on sandy or silty loams, yield about 0.8 to 2.5 t/ha. Its productivity depends on fertilizer and water supply.

KEYWORDS

- **Kichidi**
- **Kushari**
- **Lens culinaris**
- **Mujaddara**
- **Orientalis**

REFERENCES

1. Al-Issa, Y. Lentils and Chickpea. National Agricultural Policy Center, Syria. Commodity Brief No 7, **2006,** 1–12.
2. MDidiea, History and Story of Lentils: from ancestors to modern times. MDidiea Exporting Division. http://www.mdidiea.com/products/new/new06004.html **2010,** 1–7 (May, 2012).
3. El-Mubarak Ali, A. Utilization of some important Food Legumes in the Sudan. In: Uses of Tropical Grain legumes. Jambunathan, R. (Ed.). International Crops Research Institute for the Semi-Arid Tropics, Patancheru, India. **1991,** 89–93.
4. Veseth, R. Small Red lentil as a Fallow substitute. PNW Conservation Tillage Handbook Series No. 10, **1989,** 1–7.
5. FAOSTAT, Lentils. Food and Agricultural Organization of the United Nations, Rome, Italy, http://www.faostat.org **2010,** 1–8 (May, 2012).
6. AAFC. Lentils. Agriculture and Agri-Foods, Manitoba, Canada, **2007,** 1–3.
7. Oplinger, E. S.; Harman, L. L.; Kaminski, A. R.; Kelling, K. A.; Doll, J. D. Lentil. Alternative Field Crops Manual. University of Wisconsin-Extension and Minnesota Extension Services. **1990,** 1–8.

EXERCISE

1. Mention names of three top Lentil producing countries in terms of Total area, Production and Productivity.
2. Discuss the global production trends during past 5 years.
3. Discuss the origin and classification of *Lens esculentius* and *Lens culinaris*.
4. Describe soils that support Lentil cropping zones in North America, West Asia and India.
5. Discuss nutritive value of grains and forage of Lentils.

FURTHER READING

1. Erskine, W.; Meulbhaeur, F. J.; Sarkar, A.; Sharma, B. *Lentils.* CAB International Inc.: Oxford, England, **2009,** pp. 457.
2. *Lentil Production in Manitoba.* Agdex Agriculture, **1982,** pp. 89.
3. Yadav, S. S.; McNeil, Y. D.; Stevenson, P. C. *Lentil: An Ancient Crop for Modern Times.* Springer Verlag Inc.: Heidelberg, **2007,** pp. 461.

USEFUL WEBSITES

www.faostat.org;
www.icarda.org
www.pea-lentil.com;
www.napcsyr.org;
www.saskpulse.com/media/pdfs/ppm-lentil.pdf
www.hort.purdue.edu/newcrop/cropfactsheets/lentil.html
www.legumefutures.de/images/Grain_legumes_57_Lentils.pdf

CHAPTER 11

CHICKPEA CROPPING ZONES

CONTENTS

11.1 INTRODUCTION

Chickpeas are worldwide in distribution. Its cropping zones occur in all the continents wherever agricultural crops flourish. Chickpeas are predominant in the temperate regions; however they are also grown in subtropics and tropics. Chickpeas were domesticated in the 'Fertile Crescent' region encompassing Turkey and Syria. Ladizinsky [1] contends that chickpeas originated in south-eastern Turkey and parts of Syria. Van der Maesen [2] suggests that chickpeas originated in Southern Caucasus and Northern Persia. Archaeological studies, mainly radiocarbon dating of chickpea samples found in caves of L'Abeuorador Department, Aude, Southern France indicate that wild chickpeas were used around 2700 B.C. It was domesticated during Neolithic period (3000 B.C.) in West Asia. The cultivated species of chickpeas were introduced into South Asia through human migration, trade and conquests. Geographic region comprising Afghanistan-Pakistan-India is said to harbor maximum genetic diversity for *Cicer* species. At present, 40 wild species of *Cicer* have been traced in this region. *C. reticulatum* is the progenitor of *Cicer arietinum.* So far, archeological samples of seeds found at Hacilar near Burdur in Turkey, seems historically the earliest record dated at 7500 B.C. [3]. Chickpea grains have been traced in the remains of Egyptian Pharoes dating between 1580–1100 B.C. Chickpeas are also traceable as insignia on crowns of monarchs, buildings and other articles belonging to Ancient Greek and Roman empires. The ancient Greeks called it *Erbinthos*, which is mentioned in Illiad of Homer (800 B.C.–1000 B.C.) (Table 1).

TABLE 1 Chickpea: Origin, Distribution, Botany and Uses—A summary.

Chickpeas originated in West Asia. The area of maximum genetic diversity lies in Afghanisthan, Pakistan and North-west India. The Vavilovian center of origin for chickpea is in Mediterranean region. Chickpeas are actually cultivated in all continents, but preferred more in temperate regions. In South Asia, chickpea cultivation is well spread and fairly intense in Gangetic plains.

Botany and Classification

There are 40 wild species of the genus *Cicer*. A few of them, about 13 are perennial and others are annuals

Cicerreticulatum(Progenitor) » *Cicerarietinum* (Domesticated species)

$2n = 16$

The International Crops Research Institute for the Semi-Arid tropics (ICRISAT), near Hyderabad, India is the major germ plasm repository for chickpeas grown in India and other South-east Asian countries. Currently, it holds over 17,000 accessions. The other germplasm center that caters to Middle East, Europe and Americas is at International Centre for Agricultural Research in Dry areas (ICARDA) Aleppo, in Syria. Agricultural Institutions in Southern Europe, West Asia, India and China also possess germplasm collections of *Cicer* species.

Nomenclature of Chickpeas in Different Languages

Middle East and European languages: *Erbinthos, Kryos* in Greek; *Chickpea, Bengal gram* in English; *Garbanzo* in Spanish; *Pois Cheche* in French; *Ceci* in Italian; *Kicherberse* in German; *Nohud, lablabi* in Turkish; *Shimbra* in Ethiopean; *Homes, Hamaz* in Arabic languages;

South Asian languages: *Chana* in Hindi; *Khalva, Khalaya, Chanaka* in Sanskrit; *Kadale bele* in Kannada, *Kadala parippu* in Malayalam; *Sanagalu* in Telugu; *Kadalai* in Tamil; *Harbara* in Marathi; *Chola* in Bengali; *Chana* in Gujarati; *Butmah* in Assamee.

Uses

Chickpea leaf extracts and acid depositions serve as astringent. In rural areas of Europe, chickpea decoction is used in place of coffee. A combination of broken chickpea seeds, honey, onion and spices is said to possess aphrodisiac qualities. A common food item consumed by Indians since *Rigvedic* period is called '*Sattoo.*' It is prepared from chickpea floor and jaggery. The other most common food prepared using chickpea was '*Soopah*' – a soup prepared from cooked dhal. The popular *garbanzo* or *kabuli* types that are large seeded are used in salads (Plate 1). They are also consumed directly after boiling, roasting or canning. In Pakistan, Chickpeas are used to prepare dishes known as Pukora, Missi Roti, Halwa and Chats. The small, dark seeded types found in India are called *desi* types and consumed by preparing a dish called Dhal. Bengal gram flour is used widely in different dishes and sweets prepared in South India. Its flour is used to prepare chutney. Roasted Bengal gram and popped rice is a common snack in South India. *Mirapakayi* or *Menasinakyi Bajji* is a common hot snack in Southern India prepared using chickpea flour. The chickpea stems and leaves are fed to domestic cattle. The crop residue is used to prepare FYM and Composts.

Nutritional Aspects

Chickpea or Bengal gram is a rich protein source. The nutritional value of 100 g of Bengal gram (whole grain) is as follows:

Energy (calories)-360; Protein (g)-17.1; Fat (g)-5.3; Calcium (mg)-202; Phosphorus (mg)-39; Iron (mg)-10.2; Thiamin (mg)-0.3; Riboflavin (mg)-0.15; Niacin (mg)-2.9; Vitamin C (mg)-3; Vitamin A (mg)-189. According to ICRISAT [4], on an average, chickpea grains contain 23% Protein, 47% Starch, 5% Fat, 6% Crude fiber, 6% Soluble sugar, and 3% Ash. Minerals found in higher concentrations are P (340 mg 100/g), Calcium (190 mg 100/g), Mg (140 mg 100/g), Iron (7 mg 100/g; Zinc (3 mg 100/g).

Source: Refs. [5–8]; http://en.wikipedia.org/wiki/chickpea.

11.2 AGROCLIMATE, SOILS, WATER REQUIREMENTS, CROPPING SYSTEMS

Chickpeas are preferred in the temperate climate prevalent in Southern Europe and West Asia. Chickpea is a post-rainy, cool season crop in Afghanistan and Indo-Gangetic plains (Plate 1). It is usually sown in October/November as the monsoon rains end. Chickpea crop thrives on stored soil moisture. Residual moisture in areas receiving 700 mm seems to suffice seedling stages. Chickpeas need at least 400 mm annual rainfall and subtropical cool season to complete cropping season. Whenever, rabi rains are unreliable, supplementary irrigation is needed. In the semiarid regions residual moisture and dew fall in winter seems to help the moisture requirement of the crop partly. Chickpea is a long-day plant; hence it needs 12–14 h sunshine. The average atmospheric temperature during post rainy period ranges from 20°C to 30°C in the Gangetic Plains. Cold tolerant varieties are preferable if temperatures are prone to reach below 5°C during peak winter. In the Canadian plains region chickpea needs 1120 degree days to mature, which is about 65 degree days more than that required by other major legume namely lentils cultivated in that region. Incidentally, Canadian chickpea varieties have been selected for early maturity. They are sown during early May and harvested by the end of October well before winter approaches. Most varieties mature in 110–130 days.

Chickpea cultivation is scattered on Mollisols found in Canada and Northern United States of America. Chickpeas are well adapted to Cambisols, Loess and Podzols of European plains. The West Asian chickpea belt thrives on Calci Xeralfs. In North Africa, it grows on Oxisols that need nutrient enrichment. These soils are also low in organic matter content. In South Asia, Xeralfs, Inceptisols, Mollisols Alfisols and Alluvial soils found in Indo-Gangetic plains support chickpea cultivation. Vertisol plains of Central India too are used to grow chickpea. In South-east Asia, chickpea cultivation is confined to winter season. It is confined to regions with Alfisols and Alluvial soil. Mollisols and Alfisols of North-east China allow high yields of chickpea crop.

Chickpea is a versatile legume (Plate 1) and fits into several combinations of mixed cropping and rotations. In Europe, chickpea-cereal-fallow sequences are common. It is also intercropped with other legumes. In West Asia, it is rotated with cereals such as wheat and barley. In the Indo-Gangetic plains it is a highly preferred post rainy crop and is rotated with wheat. Chickpea is also intercropped with cereals such as maize. In South-east Asia and China, chickpeas are rotated with arable crops like wheat, maize, millets, etc.

PLATE 1 A Chickpea plant with ripened pods. Bottom Left: Ripened pods; Bottom Right: Large seeds of Kabuli type.
Source: Krishna K.R. Bangalore, India.

11.3 EXPANSE AND PRODUCTIVITY

Chickpeas are cultivated worldwide in temperate, subtropical and tropical regions. Chickpea cropping zones are traceable in 50 different countries. Chickpea is an important legume constituting about 20% of global pulse production. India, Pakistan and Mexico are the three major producers of chickpeas. Currently, global chickpea cropping belt spreads into 11 million ha and contributes about 8 m t grains annually. Chickpea is an important cash crop in Northern Great Plains and Southern Canada. Canada contributes about 8–10% of total global chickpea production. Natural factors such as drought, cold temperature and *Ascochyta* blight seems to affect nutrient dynamics and productivity of chickpea growing regions of Canada. As stated earlier, chickpea is an important legume in the Canadian Prairies. The chickpea growing regions in Canada extend into over 1.4 m ha. Chickpea belt expanded rapidly in the Northern Great Plains during late 1990s [9]. The productivity of chickpea has ranged from 550 kg grain/ha to 1450 kg/ha in the Canadian plains, although projected potential is high at 2800 kg grain/ha. Chickpea is also consumed as a vegetable in the Northern Plains.

Chickpea has been cultivated in India since Neolithic period. Chickpea cultivation, like other legumes, got intensified during ancient, medieval and modern period. Today, its expanses are conspicuous in Gangetic Plains, and on Vertisol zones of Karnataka and Andhra Pradesh. Chickpea is an important legume crop in Indo-Gangetic plains. It is a staple protein source for most people in North Indian plains. Its cultivation is relatively intense in North Indian plains. It is mostly grown as an intercrop or rotation crop during rabi season. The chickpea agroecosystem of India accounts for 61% of total global chickpea area and 68% of production. In addition to India, Turkey (8%), Pakistan (7%), Iran (3%) and Mexico (3%) also contribute to global chickpea produce [4, 7]. Actually, chickpea cultivation in India spreads into 8.2 m ha and annual harvest is 5.7 m t. The productivity is low at 792 kg grain/ha. Yet, total chickpea production is insufficient to match annual demand. Hence, it is imported from countries like Turkey and Canada. During recent years, area under chickpea belt has marginally depreciated in Northern Indian plains.

KEYWORDS

- **Chickpea**
- ***Cicer***
- ***Erbinthos***
- **Fertile Crescent**
- ***Sattoo***

REFERENCES

1. Ladizinsky, G. A new *Cicer* from Turkey. Notes of the Royal Botanic Garden, Edinburgh, Scotland, **1975,** *34,* 201–202.

2. Van Der Maesen, L. J. G. *Cicer* L.; Origin, History, Taxonomy, and its Ecology Cultivation. In: The Chickpeas. Saxna, M. C.; Singh, K. B. (Eds.). Commonwealth Agricultural Bureau, London, International Cambrian News, Ltd. Aberystwyth, United Kingdom, **1972,** 11–34.
3. ICARDA. Introduction to Chickpeas. International Center for Agriculture in Dry Areas, (ICARDA), Aleppo, Syria, http://www.icarda.org/publications/cook/chickpea/chickpea.html **2007,**1–3.
4. ICRISAT. Chickpea. http://www.icrisat.org/chickpea/chickpea.htm **2007,** 1–3.
5. ICRISAT. Uses of Tropical Grain Legumes: Proceedings of a Consultants Meeting. International Crops Research Institute for the Semi-arid Tropics. Hyderabad, India. **1991,** 31–113.
6. ICRISAT. Chickpea, Pois Cheche *Cicer arietinum*. http://www.icrisat.org/text/coolstuff/crops/gcrops5.html **2006,** 1–5 (January 12, 2011).
7. Krishna, K. R. Agroecosystems of South India: Nutrients and Productivity. BrownWalker Press Inc.: Boca Raton, Florida, USA, **2010,** 556.
8. Saxna, M. C. Utilization of Chickpea in West Asia and North Africa. In: Uses of Tropical Grain legumes. Jambunathan, R. (Ed.). International Crops Research Institute for the Semi-Arid Tropics, Patancheru, India. **1991,** 211–215.
9. Mackay, K.; Miller,; Jenks, B.; Riesselman, J.; Neill, K.; Buschena, D.; Bussan, A. J. Growing chickpea in the Northern Great Plains. http://www.ag.ndsu.edu/pubs/plantsci/crops/a1236w.htm (July 8 2012) **2002,** 1–13.

EXERCISE

1. Discuss the origin, centers of genetic diversity, wild and domesticated genotypes of chickpea.
2. Delineate the chickpea cropping zones of the world on a map.
3. Mention the area, total production of top five chickpea producing nations.
4. Describe the agronomic procedures involving fertilizers, irrigation and harvest of chickpeas.
5. Describe various uses of chickpeas. Mentions names of at least five dishes prepared in West Asia and India using chickpeas.

FURTHER READINGS

1. Corp, M.; Machado, S.; Bell, D.; Smiley, R.; Petrie, S. Chickpea Production Guide. Oregon State University: Extension Service. Corvallis, USA, **2004,** pp. 14.
2. Singh, R.; Diwakar, B. Chickpea Botany and Production Practices. International Crops Research Institute for the Semi-Arid Tropics (ICRISAT), Patancheru 502–324, Andhra Pradesh, India, **1995,** pp. 57.
3. Yadav, S. S.; Redden, R. J.; Chen, W.; Sharma, B. Chickpea Breeding and Management. CAB International Inc.: Oxford, England. **2006,** pp. 648.

USEFUL WEBSITES

http://www.icarda.org
http://www.icrisat.org

CHAPTER 12

COWPEA FARMING ZONES OF AFRICA AND ASIA

CONTENTS

12.1 INTRODUCTION

Cowpeas are native to humid and semiarid tropics of West Africa. The genetic diversity of cowpea species is maximum in West Africa and parts of Central Africa. Archeological evidences suggest that cowpea was domesticated around Neolithic settlements situated on the banks of river Niger in West Africa, sometime during 3rd millennium B.C. [1, 2]. These events led to a much enlarged cowpea agroecosystem that spreads into different continents. Earliest evidence for cultivation of cowpea in West Africa is derived from *Kintampo* culture [3]. Cowpeas, later spread to southern and eastern Africa and later found its way into Indian subcontinent. The spread of cowpea to North and South America was mediated through slave trade and seafarers. It reached the North American shores, mainly Florida, North Carolina and Texas during 1700s.

Seeds for enlarged cowpea agroecosystem in India were sown during Neolithic period. Southern Indian plains and hills received cowpeas via maritime travelers and merchants from African main land [4]. Cowpea (*Vigna unguiculata*) must have reached South Indian hills and plains around 2nd millennium B.C. through human migration. Later, through intensive human selection two cultigroups, namely, *biflora* and *sesquipedalis* evolved in India and other parts of South Asia. In due course, large genetic variability became available in the Gangetic plains and Southern India. Hence, arguably, we may consider India as yet another center of genetic diversity for cowpea (Table 1). Archaeological excavations at Harappa in North-west India has shown that cowpea was an important grain legume consumed by the local populace around 3200–3000 B.C. Fuller and Harvey [5] reported that archaeological remains, including seeds and other plant material of *V. unguiculata* is traceable in many Neolithic sites situated at *Sanganakallu, Hullur* and *Tekkalkota* in South Indian plains.

TABLE 1 Cowpea: Origin, Distribution, Botany and Uses—A summary.

Cowpeas were domesticated in Tropical West Africa during 3rd millennium B.C. The area comprising West and Central Africa is considered the primary center of origin for cowpea. Tropical West Africa harbors maximum genetic diversity for cowpeas. Cowpea was introduced to South Indian plains around 2nd millennium B.C. Archaeological evidences indicate that Neolithic settlers (2200 B.C.) in South India cultivated cowpeas. Cowpea cultivation is distributed in the entire subcontinent of India and South-east Asian countries. In India, two cultigroups *Biflora* and *Sesquipedalis* evolved from *Unguiculata*. Intensive selection process has created wide genetic diversity in India. Hence, India is considered as yet another center of genetic diversity for cowpeas – perhaps secondary center of origin. The International Institute for Tropical Agriculture at Ibadan in Nigeria holds world's largest germplasm collection of cowpeas. Its collection exceeds 16,000 accessions of cowpea. According to FAO reports, globally, there are over 85,000 accessions of cowpea germplasm held in different nations.

Botany

There are four subspecies of cowpea (*V. unguiculata)*, namely:

V. unguiculata subsp cylindrica-Catjang bean; *V. unguculata subsp dekindtiana* – Black-eyed bean; *V. unguiculata subs psesquipedalis*-Yardlong beans; *V. unguiculata subsp unguiculata* – Southern peas

---

Classification

Kingdom-Plantae; Division-Magnoliophyta; Class-Magnoliopsida; Order-Fabales; Family Fabaceae; subfamily Faboideae; Genus-*Vigna*; Species-*V. unguiculata* (L.)

Synonyms

V. unguiculata; V.coerulea; V. baoulensis; V..dekindtiana; V. unguiculata var dekindtiata; V. unguiculata var mesensis; V. unguiculata var cylindrical; V. catjang; Phaseolus cylindricus

$2n = 22$

Nomenclature in Different Languages

Catjang bean, Indian cowpea, Catjang pea in English; *Dolic asperge, Dolic des vaches, Dolique mongette, Dolique Catjang; Spargelbohne, Ctajanbohne in* German; *Dolique de Chine, Dolique de Vaches, Dolique Mongette, Dolique catjan* in French; *Judia esperago* in Spanish; *Vigna kitaiskaia, korovii goroshek* in Russian; *Fagiolono asparago* in Italian; *Fejio del occhio* in Italian; *Katjang pandjang in* Dutch; *Vignabonne* in Danish; *Fajio chicote* in Portuguese; *Lubya baladi* in Arabic, *Fagiolog del Occhiao*; *Hata sasage, Yakko sasage* in Japanese, *Kachang mera* in Malay; *Thua khaao, thua rai* in Thai, *Dua ca, Dua trang* in Vietnamese; *Kacang tungagak* in Indonesian; *Faun Dou, Fan jian dou, Mei dou, Yang doujiao, Duan Jia Jiang* in Chinese; *Ye jiang dou, Dhaua gok* in Cantanese;

Indian languages: *Rajmash* in Sanskrit; *Lobiya, Chauli* in Hindi; *Chavalya* in Marathi; *Alasande* in Kannada; *Alasandulu, Tella Bobbarlu, Erra Bobbarlu* in Telugu; *Karamani* in Tamil; *Vellapayiru* in Malayalam;

Uses

Succulent pod is used as vegetable. Dried beans are an important source of protein to Southern Indians (Plate 2). Boiled beans are consumed by preparing variety of hot dishes (e.g., Sambar). Cowpea is an important green manure and forage crop. Its haulms are rich source of nutritional factors to domestic animals. In South India, cowpeas are used for ensilage, hay making and for developing grazing land surrounding villages. Cowpea fodder is superior to other sources with regard to protein, minerals and fiber content.

Nutritive Value (per serving of 145 g)

Energy-548 k cal; Carbohydrtes-27.3 g; Dietry fiber-7.3 g; sugars-4.4 g; protein-4.3 g; Vitamin A 1185 IU; Vitamin C-3.6 mg; Folate-244 mg; Ca-183 mg; Mg-74 mg; P-76.8 mg; K-625 mg; Na-5.8 mg Zn-1.1 mg.

Sources: Refs. [1, 2, 4–7]; http.plantnames.unimelb.edu.au/sorting/Vigna.html; http://en.wikipedia.org/wiki/cowpea.

12.2 AGROCLIMATE, SOILS, WATER REQUIREMENTS, CROPPING SYSTEMS

Cowpea is extensively cultivated in subhumid and semiarid zones of Africa and Asia (Plate 1; Fig. 1). With regard to altitude, its cultivation spreads from sea level in coasts to 1200 m above sea level. The optimum temperature for cowpea production in semiarid tropics is 27°C −32°C, but it withstands temperatures from 12°C−15°C during winter to 35°C–38°C in summer. Cowpea is grown throughout the year in humid and semiarid tropics. Cowpea genotypes specifically selected for drought tolerance, allow optimum grain and forage yield even in drought prone areas and arid landscape. It is preferred both as sole and intercrop with cereals. A short duration genotype or forage species is used during summer season.

PLATE 1 A Cowpea crop on Red Sandy Loam of Southern Karnataka, South India.
Note: In the dry lands of Southern Asia, wowpea contributes 20–60 kg N/ha to soil each season, through Biological Nitrogen Fixation. It serves as an excellent short season crop during third season. Cowpea serves both as regular pulse crop and vegetable. In addition, during fallow, it could be cultivated as green manure. Upon incorporation into soil, its succulent stem and leaf tissue release nutrients rapidly
Source: Dr Krishna, K.R. 2006.

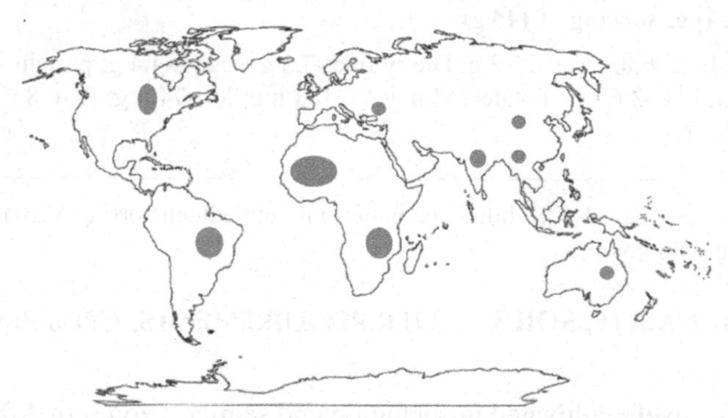

FIGURE 1 Cowpea growing regions of the World. Its cropping is conspicuous in tropical and subtropical regions with moderate to high rainfall. Cowpeas grown in West Africa and Dry Land areas of India are endowed with greater degree of tolerance to scanty precipitation conditions. They are often intercropped with staple cereals of the region. Cowpeas serve as good bet in soils with poor fertility, especially N-deficient soils, since they are able to derive atmospheric-N through symbiosis with Bradyrhizobium.

Cowpea is mostly grown on soils/fields relatively poor with regard to soil fertility. Cowpea is a good bet on sandy Ultisols in Southern USA. Oxisols with maladies such as high Al and Fe content, P-fixation and acidity are used to cultivate cowpea in the Cerrados of Brazil. Cowpea cultivation proceeds on moderately fertile Mollisols, Inceptisols and Alfisols in the Pampas of Argentina. Lateritic soils too support cowpea production. Major soil types that support cowpea production in Africa are Xeralfs, sandy Oxisols, Alfisols, Inceptisols and Coastal sandy soils. In the Indian subcontinent, cowpea is often allocated to fields/regions with soils low in fertility. Cowpea is efficiently grown on Inceptisols and Mollisols in the Indo-Gangetic plains. It is a preferred crop in the dry land expanses. It grows well on Vertisol and Alfisol plains of South India, laterites of West Coast and Sandy soils of Eastern coast of India.

Cowpea is mostly grown as a rain fed crop in the dry lands of different continents. The annual precipitation in cowpea cropping zone ranges from 650–900 mm. However, its cultivation proceeds well in drought prone areas that receive scanty precipitation (<500 mm). Supplemental irrigation from tanks, ponds, open wells, check dams and irrigation channels are required to satisfy water requirements of the crop. As such, water resources are feeble in many locations within cowpea cropping areas. Therefore, farmers tend to supply water at critical stages during crop production. The sensitive stages for irrigation are flowering and pod formation. The productivity of cowpea given full irrigation (560 mm) is 8.7 q/ha, which is four folds higher than rain fed (180 mm) crop (2.29 q/ha). A crop supplemented with water (360 mm), at critical stages (flowering and pod fill) yields 5.96 q grain/ha. The water use efficiency of cowpea given supplemental irrigation at critical stages was 155 kg DM/ha mm, compared with 123 kg DM/ha mm for fully irrigated crop and 109 kg DM/ha mm for rain fed crop [8]. Over all, for optimum productivity of cowpea, be it entirely irrigated or partly rain fed, it requires 360–450 mm water per season. Cowpea cultivated during rabi season depends largely on stored soil moisture. To overcome insufficiency supplemental irrigation is required. We should note that cowpea is often intercropped and rotated with other crops like cereals and oil seeds. In that case, water requirements of inter crops or entire crop sequence should be considered. Residual moisture after each crop plays vital role in sustaining cowpea during seedling stage and later. Short season summer crop of cowpea needs irrigation between 250–360 mm during the season.

Most commonly, cowpea is intercropped with sorghum, maize, finger millet, sunflower or pigeonpea. In the irrigated regions, cowpea is intercropped with different types of cereals, legumes, oil seeds and vegetable. Cowpea is also intercropped with rice during first 6–7 weeks. Cowpea is harvested as a green manure at 6–7 weeks, later rice fields are pounded with water and continued till harvest. Cowpea is also intercropped with oil seeds like groundnut, sunflower or sesamum and pulses like pigeonpea. Cowpea is a good under story or floor crop in fruit tree plantations as well as in perennial agroforestry. As stated earlier, cowpea is grown all through the year. The rain fed crop is sown during rainy season. A second crop is sown between September to December. During summer, cowpea is cultivated on rice fallows as a sole crop or intercrop. In the perennial plantations, cowpea is a good floor crop. Cowpea is a useful intercrop in fields with Tapioca as major crop.

Cowpea is most often rotated with cereals such as rice, maize, sorghum, millets and other legumes depending on soil nutrient and water resources. It is rotated both during rainy and post rainy seasons. During summer a short duration cultivar is preferred. Forage or green manure types grown during summer are usually fast growing. They are harvested early and succulent foliage/haulms are incorporated into soil. It recycles nutrients efficiently. Cowpea is also cultivated for forage purposes. It suits to grow a short duration cowpea along with maize as intercrop during the short gaps that occur in rice-based rotations. Long-term rotations like sorghum/cowpea-maize/cowpea-maize/cowpea are common in semiarid tropics of Asia and Africa. Cowpea acts as a 'catch crop' and recycles nutrients within the farm efficiently.

12.3 COWPEA CROPPING EXPANSE AND PRODUCTIVITY

Recent reports suggest that, globally, cowpea is planted on 11 m ha with an annual production of 5.4 m t. The productivity is low at 0.5–0.8 t/ha. The low productivity is attributable to low fertility soils/fields selected for cowpea production, in preference to highly productive fields that are generally allocated for wheat or rice production. We should note that productivity of many of the legumes grown in developing nations is relatively low. Cowpea cropping zones expanded rapidly into Southern United States of America during the period from 1930s to 1970 s. Production of cowpea is said to have peaked in Texas in 1930 with more than 1.2 million acres. Most of the American cowpea belt was filled with 'Black-eyed beans' (Plate 2). Currently, cowpea belt in USA, spreads into 50,000 acres with a total production of 45,000 tons annually.

PLATE 2 Grains of Black-eyed Beans.
Note: Dried cowpea grains are an important source of dietary protein for people in Tropics of Africa and Asia
Source: Krishna K.R., Bangalore, India.

The cowpea expanses are most conspicuous in the tropics and semiarid zones of Africa. African continent accounts for 92% of cowpea area and productivity. Cowpea cropping zones are predominant in the West African tropics and semiarid regions. Cowpea is mostly grown as a rain fed, subsistence crop in the Sub-Sahara. The sandy soils with low nutrient status, reduces productivity of cowpeas in the Sahelian zone. Cowpea is most often intercropped with sorghum or pearl millet. Nigeria is the main cowpea-producing nation. It accounts for 58% of global cowpea production [9]. Cowpea is an important legume in most of the Southern and Eastern African nations. It is cultivated both under subsistence and medium fertility conditions.

Cowpea is major legume in the Indo-Gangetic plains. Cowpea productivity is moderate at 1.0–1.2 t/ha. Cowpea is often rotated with major cereals like wheat or rice. It is a preferred short season crop. It fits as an excellent fodder crop during short summer season. In the Southern Indian hills and plains cowpea is a protein source. It is gown as intercrop or rotated with cereals. Cowpea production regions are well distributed in the South-east Asia and fareast (Fig 1). Cowpea expanses are traceable in all continents wherever agricultural cropping occurs. Cowpea is easily accepted by Australian farmers due to several reasons. For example, cowpea is multipurpose legume. It helps in improving soil-N status. It is easy to establish cowpea crop. It adapts to wide range of soils. It is drought tolerant [10]. Cowpea is a fast growing short season crop in many places. It reaches harvest stage relatively rapidly and offers good quality pods, seeds and forage.

KEYWORDS

- ***Biflora***
- **Black-eyed beans**
- **Chickpea**
- **Cowpea**
- ***Kintampo***
- **Sahelian**
- ***Sesquipedalis***

REFERENCES

1. Ng, N. Q. Cowpea. *Vigna unguiculata* (Leguminoseae, Papilionideae). In: Evolution of Crop Plants. Smartt, J.; Simmonds, N. W. (Eds.). Longman Scientific and Technical Publishers, London, **1995,** 326–332.
2. Fuller, D. O. African Crops in Prehistoric South Asia; A review In: Food, Fuels and Fields. Neumann, K.; Butler, A. (Eds.). Progress in African Studies. Heinrich-Barth Institute, Germany, **2003,** 239–271.
3. D'Andrea, C.; Casey, J. Pearl Millet and Kintampo subsistence. African Archaeological Review **2002,** *19,* 147–161.
4. Ng, N. Q.; Marechal, R. Cowpea Taxonomy, Origin and Germplasm. In: Cowpea research, Production and Utilization. Singh, S. R.; Rachie, K. O. (Eds.). John Wiley and Sons Chichester, United Kingdom, **1985,** 11–21.
5. Fuller, D. O.; Harvey, E. L. The Archaeobotany of Indian Pulses: Identification, Processing and Evidence for cultivation. Environmental Archaeology **2006,** *11,* 218–246.
6. Krishna, K. R. Agroecosystems of South India: Nutrient Dynamics and Productivity. BrownWalker Press Inc.: Boca Raton, Florida, USA. **2010,** 556.

7. Fuller, D. O.; Korizettar, R.; Venkatasubbaiah, C.; Jones, M. K. Early plant domestication in Southern India: Some preliminary archaeobotanical results. Vegetational History and Archaeobotany, **2004,** *13*,115–129.
8. Selvaraju, R. Effect of Irrigation on Phenology, Growth and Yield of Cowpea (*Vigna unguiculata*). Indian Journal of Agronomy **1999,** *44,* 377–381.
9. IITA, Cowpea. (*Vigna unguiculata*). International Institute for Tropical Agriculture. Ibadan, Nigeria, http://www.iita.org **2011,** 1–3 (March 3, 2012).
10. Cook, B. Cowpea- fact sheet. GRDC Dairy, New South Wales, Australia, **2008,** 1–3.

EXERCISE

1. Discuss Origin and spread of Cowpea in Africa and Asia.
2. Mention the different types of Cowpea and their Scientific names.
3. List top Cowpea producing regions of the World. State the Total Production, Productivity and Area.
4. Mention various cropping systems that include Cowpeas.
5. Mention the various uses of Cowpea.
6. Discuss about Cowpea-Rhizobium symbiosis with regard to N derived from atmosphere in different continents.

FURTHER READINGS

1. Dugje, I. Y.; Omoigui, L. O.; Ekelemea, F.; Kamara, A. Y.; Ajeigbe, H. Farmer's guide to Cowpea Production in West Africa. International Institute for Tropical Agriculture: Ibadan, Nigeria, **2009,** pp. 20.
2. Franklin, W. The History of the Cowpea and its Introduction into America. Ulan Press: New York USA, **2011,** pp. 36.
3. Siddiq, M.; Uebersax, M. A. Dry Bean Pulses: Production, Processing and Nutrition. Willey-Blackwell Publisher: New York, **2012,** pp. 408.

USEFUL WEBSITES

http://www.iita.org
www.ikisan.com/crop%20 specific/eng/links/ap_fc_cowpea.shtml
http://www.sare.org/Learning-Center/Books/Managing-CoverCrops-Profitably 3rd-Edition/Text-Version/Legume-CoverCrops/Cowpeas

CHAPTER 13

PIGEONPEA AGROECOSYSTEM OF ASIA, AFRICA AND CARIBBEAN ISLANDS

CONTENTS

13.1 INTRODUCTION

Pigeonpea expanses are predominant in the plains of Peninsular India, parts of East Africa and Caribbean islands. Pigeonpea is the most important legume species consumed by Southern Indian populace since prehistoric times. Pigeonpea was domesticated in the Western and Eastern Ghats of Southern India. Fuller and Harvey [1] suggest that center of origin for pigeonpea (red gram) is in the region encompassing Northern Andhra Pradesh, parts of Chhattisgarh and fringes of Eastern Ghats in South India. *Cajanus cajanifolia* is the progenitor of cultivated species *C. cajan* [2] (Table 1). Neolithic settlements (3rd and 2nd millennium B.C.) at Sanganakallu, Hullur and Western Ananthapur district contained specimens of *Cajanus* species [3]. Over 17 species of *Cajanus* are traceable in South India. The wild species of *Cajanus* are mostly confined to Western and Eastern Ghats of Southern India and hilly zones of Sri Lanka. *Cajanus* seeds and plant material belonging to late Neolithic and early Chalcolithic age (1300–1400 B.C.) have also been reported from Gopalpur in Coastal Orissa. *Cajanus* species moved north into Gangetic Plains through human migration. This is an example of counter current of crop diffusion from South India to Indo-Gangetic Plains [3]. Human migration, preferences and nutritional qualities of grains must have induced the expansion of pigeonpea agroecosystem into Kenya and other east African countries, as well as to Caribbean Islands.

TABLE 1 Pigeonpea: Origin, Distribution, Botany and Uses—A summary.

Pigeonpea is native to Southern Indian plains and hills. The wild progenitor *Cajanus cajanifolia* has been traced, although rarely in the Eastern Ghats [2]. At least 17 species of Cajanus have been found distributed in the wet and deciduous forests of South India. Domesticated pigeonpea-*Cajanus cajan* was regularly cultivated around the Neolithic settlements (3rd and 2nd millennium B.C.) in South India. Pigeonpea then moved to Deccan and Gangetic plains during late Neolithic phase (1st millennium B.C.). Later, pigeonpea found its way to East Indian archipelago. Pigeonpeas moved to East African coast during late Neolithic and Ancient era. African countries such as Kenya, Mozambique, Zambia and Tanzania must have received pigeonpea from Southern Indian ports through maritime connections. A few *Cajanus* species are also grown in Senegal, Togo, Ghana and Nigeria. Explorers and slave trade seem to have mediated introduction of domestic species of *Cajanus* into West Indies and its surrounding islands. Today, pigeonpea is also cultivated in Florida in main land USA, and even as far as Hawaii, wherever tropical climate suits its production. Presently, pigeonpea is cultivated in over 25 tropical and subtropical countries.

Botany and Classification

Kingdom-Plantae; Division-Magnoliophyta; Class-Magnoliopsidae; Order-Fabales; Family-Fabaceae; Genus-*Cajanus*; Species-*C.cajan.*

Synonyms: *Cajanus bicolor, C.indicus, C.flavus, C. luteous*

Progenitor: *Cajanus cajanifolia* Domesticated Species: *Cajanus cajan*

$2n = 22$

The International Crops Research Institute for the Semi-arid Tropics (ICRISAT) at Patancheru in South India is world's largest germ plasm repository for Pigeonpea and its Wild relatives.

Nomenclature in Different Languages

Indian languages: *Adhaki, Udaaraka, Kakshi, Tuvaraka, Tuvara in* Samskrit, *Tur, Tuvar, Arhar in* Hindi; *Togari bele in* Kannada, *Kandhi pappu in* Telugu, *Paruppu* in *Tamil; Tuvara parippu, Vanpayiru in* Malayalam.

Others Languages: *Red gram, Pigeonpea, Gungo pea, Congo pea in* English; *Ambervade, Pois d'angole* in French; *Arvejo de Angola, Guandul, Cachito* in Spanish; *Pizello d'Angola* in Italian; *Straucherbse, Taubenerbse* in German; *Feijoa-guandu, guandu, guizente de Angola* in Portuguese; *Kachang* in Malaysian.

Uses

Pigeonpea is a staple protein source to South Indians. Fresh beans are used as vegetable. The dried peas are cooked directly. Sprouting enhances its digestibility. In West Indies, canned pigeonpeas are popular. In South India, split pigeonpea, known, as Tur dhal is the most relished product. Several types of soups, (e.g., *Sambar, Rasam*) are prepared using the tur dhal. Fresh leaves and their decoctions are used to overcome tooth ache and sore gums. Decoctions from its flowers are used to treat common coughs, bronchitis and pneumonia. The tender leaves of pigeonpea serve as forage. Pigeonpea is a good green manure source. In some countries fresh leaves are fed to silk insects. Pigeonpea stalks are an important source of firewood in rural South India. Woody stems are also used to prepare fences and baskets. Pigeonpea supports lac or sticklac growth on its stem and branches.

Nutritional Aspects

Dried Seeds 100 g contain: 345 Calories, 9.9% Moisture, 19.5–22 g Protein, 1.3 g fat, 65.5 g Carbohydrate, 1.3 g Fiber, 3.8 g Ash, 161 mg Ca, 285 mg P, 15 mg Fe, 55 μg B-Carotene, 0.72 mg Thiamine, 0.14 mg Riboflavin, and 2.9 mg Niacin.

Immature seeds100 g contain: 117 calories, 69.5% moisture, 7.2 g protein, 0.6 g Fat, 21.3 g carbohydrate, 3.3 g fiber, 1.4 g ash, 29 mg Ca, 135 mg P, 1.3 mg Fe, 5 mg Na, 563 mg K, 145 μg B-Carotene, 0.40 mg Thiamine, 0.25 mg Riboflavin, 2.4 mg Niacin and 26 mg Ascorbic acid. Fresh Green Forage contains 70.4% Moisture, 7.1% Crude protein, 10.7 Crude fiber, 7.9% N-free extract, 1.6% Fat and 2.3% Ash.

Source: Refs. [1, 3–7];
http://www.hort.pudue.edu/newcrop/duke_energy/cajanus_cajan.html;
http://en.wikipedia.org/wiki/pigeonpea; http://www.fas.fed.us/global/iitf/pdf/shrubs/Cajanus%20cajan.pdf

During ancient and medieval times (1st to tenth century A.D.) pigeonpea agroecosystem was confined to fertile riverine zones of North Karnataka, Andhra Pradesh, and parts of Tamil Nadu in South India. It was a preferred legume in Central India. Pigeonpea cultivation got intensified in Black Cotton Soil belt of Andhra Pradesh and Karnataka during Vijayanagara period (Medieval period) [8, 9]. Slowly, but surely, pigeonpea agroecosystem became prominent and stabilized in area and productivity all over South India [10]. Since 1950s, pigeonpea agroecosystem spread rapidly into many areas of South India. Its production got intensified, since it was easily accepted as intercrop with dry land cereals. The expansion of pigeonpea belt was actually attributable to factors such as large irrigation projects, advent of inorganic fertilizers, improved tillage and agronomic procedures, fiscal support and governmental legislations. Human preference and palatability has played its role in inducing pigeonpea cultivation in India, Africa and the Caribbean.

13.2 AGROCLIMATE, SOILS, WATER REQUIREMENTS, CROPPING SYSTEMS

Pigeonpeas are adapted to tropical climatic conditions (Plate 1; Fig. 1). Most congenial temperature during crop season is 18–30°C, but it tolerates higher ranges of temperatures prevalent in dry lands of Peninsular India and Indo-Gangetic belt. The precipitation levels acceptable for pigeonpea production ranges widely from as low as 400 mm in dry belts to 4,000 mm in wet tropics found in upper reaches of Western Ghats. Regarding altitudes, it is grown at sea level in Coastal Andhra Pradesh and Tamil Nadu and at 1,200 m above sea level in Western Ghats. Pigeonpea is highly vulnerable to stagnating water. Root activity and plant growth are inhibited, if fields get inundated.

PLATE 1 A Pigeonpea crop thriving on Red Alfisol plains of South India.
Note: Pigeonpea genotypes that mature early (above) are preferred because they are amenable to variety of crop rotation practices. Farmers are also prone to grow medium and late types depending on cropping systems.
Source: [7].

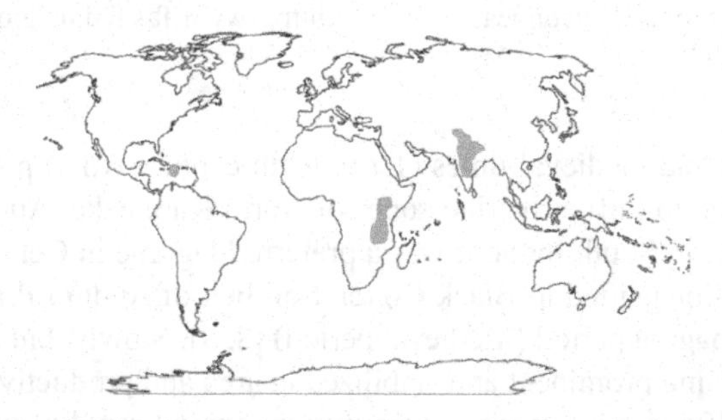

FIGURE 1 Pigeonpea Cropping zones of the World
Note: Pigeonpea is predominantly grown on Vertisol plains of Central India that encompasses parts of Maharashtra, Madhya Pradesh and Gujarat. In North Karnataka and Andhra Pradesh it forms a vast dry land agroecosystem. Pigeonpea fields are scattered in the entire Indo-Gangetic plains.Pigeonpea production zones are dispersed well in East Africa and Caribbean islands.

Pigeonpea is cultivated on Alfisols, Vertisols, Entisols and Inceptisols with pH 5.5 to 8.2. It is sown both during kharif and rabi seasons. Short duration genotypes that mature early within 105 to 155 days are preferred during rainy season. Pigeonpea is sown during June/July and harvested by late October. The medium duration genotypes that require 150 to 170 days may exceed rainy period. Therefore, crop stays in the field until December/January well into rabi season. Medium duration genotypes are sown in June-July but harvested late during January/February.

Pigeonpea is mostly confined to semiarid zones of India. Here, precipitation ranges from 700–1,000 mm annually. Rainfall pattern could be erratic. About 60–80% of annual precipitation occurs during monsoon period in 4–5 months from June to October. The water requirements of sole crop of pigeonpea or that grown as intercrop depends on several factors related to soil, cropping system, genotype, irrigation method and productivity level envisaged. As stated earlier, weed flora and its intensity can be a prime factor in reducing WUE by crops. The consumptive use of a sole crop of pigeonpea ranges from 512 to 583 mm. The water-use rate ranges from 3.06 to 3.23 mm/day and WUE from 2.1–2.6 kg grain/ha mm depending on location and genotype [7].

Pigoenpea is often grown as an intercrop with arable cereals like sorghum and maize. Actually, intercropping of pigeonpea with other legumes improves WUE. For example, pigeonpea/cowpea intercrop needs 515 to 590 mm water per season. Its water-use rate fluctuates between 3.1 to 3.8 mm/day and WUE from 2.2 to 3.6 kg grain/ha-mm. Similarly, pigeonpea-green gram intercrops are efficient and they consume 530–610 mm in a season at a rate of 3.2–3.5 mm/day. The WUE fluctuates between 2.2–2.9 kg grain/ha-mm [7, 11].

13.2.1 SOILS

Pigeonpea is cultivated on Oxisols in the Caribbean and adjoining areas. Farmers in Kenya, Tanzania and Eritrean regions grow it on Alfisols and coastal sandy soils. It is a preferable crop on Alfisols and Inceptisols in the Gangetic plains. Major soil types used to produce pigeonpea in South India are Vertisols, also called the black cotton soils. The Vertisol region is confined to North Karnataka (Gulbarga, Bidar, Bijapur and Raichur) and adjacent districts of Andhra Pradesh. In Southern Karnataka, Tamil Nadu and parts of Kerala pigeonpea is cultivated on Red Alfisols and lateritic soils. Most of the regions are low in fertility. They are prone to soil moisture stress and paucity of nutrients. Actually, incessant cultivation without nutrient replenishment schedules has led to dearth for major nutrients such as N, P and K. During recent years, dearths for S and micronutrients too have appeared in the legume belt in general. In addition, soil related maladies such as salinity and stagnation too affects pigeonpea cultivation in South India. Salinity causes impairment in nutrient availability and reduces nutrient recovery by pigeonpea roots

13.2.2 CROPPING SYSTEMS

Pigeonpea cropping zones occur as both large stretches of monocrops and as mixed crop. Pigeonpea mono-cropping expanses are conspicuous in Central Indian plains and North Karnataka region (Plate 1). Pigeonpea is grown both during kharif and rabi season. It is a versatile crop. Hence, several combinations of intercrops and rotations are possible with pigeonpea (Plate 2). Major rotations are paddy-paddy-pigeonpea, groundnut/pigeonpea (intercrop)-sorghum, groundnut–pigeonpea, soybean-pigeonpea, black gram-pigeonpea, green gram-pigeonpea and pigeonpea-cotton.). Most commonly preferred intercrops are pigeonpea/mungbean or urdbean or cowpea (1:1 rows). Pigeonpea/sorghum intercrops are frequent in North Karnataka and Andhra Pradesh. The pigeonpea/groundnut mixturc (4:2 rows) is preferred on Alfisol zones of Tamil Nadu and Southern Karnataka. Pigeonpea/groundnut is preferred in drought prone areas [12]. Sorghum/pigeonpea and Finger millet /pigeonpea are stable intercrop systems practiced since past 3 decades in Southern India (Plate 2). Pigeonpea is also intercropped with vegetable species. A few examples are pigeonpea/bhendi, pigeonpea/cowpea, pigeonpea/chilies, pigeonpea/ dolichos, etc. In high crop diversity farms, pigeonpea is rotated with several other crops such as jowar, green gram, hibiscus, field bean, cowpea, black gram. In Andhra Pradesh, such rotations involve different genotypes of pigeonpea such as *Erra Thogari, Tella Thogari, Nalla Thogari, Burka Thogari*. As a method to enhance crop diversity, farmers tend to mix pigeonpea with diverse crop species having widely different root traits and growth pattern. It enhances nutrient recovery and crop produce. In addition, it acts as an insurance against crop failures. For example, pigeonpea/groundnut, pigeonpea/black gram, pigeonpea/pearl millet and pigeonpea/sorghum are crop mixtures that enhance nutrient recovery and productivity ranging from 11 to 198 % more in grain equivalence than sole crops [13, 14]. These are proven crop mixtures in South India.

PLATE 2 A Pigeonpea-based mixed crop.
Note: Crop mixtures involving Pigeonpea (foreground), Castor and Finger Millet are common in south Indian Plains. They are efficient in terms of land use and productivity.
Source: Krishna, K.R., 2011.

13.3 EXPANSE AND PRODUCTIVITY

Globally, pigeonpea is grown in 5.25 m ha. Nearly, 65% of global pigeonpea originates in India. During 2006, pigeonpea agroecosystem in entire India occupied 3.7 m/ ha. It contributed 2.4 m t grains at an average productivity of 685 kg grain/ha [15].

Pigeonpea is predominant in the agrarian zones of Central Indian plains. It thrives exceedingly well on the Vertisol plains of Madhya Pradesh and Maharashtra. In Southern India, Andhra Pradesh, Karnataka and Tamil Nadu are major pigeon pea producing states. The pigeonpea cultivation is relatively intense in the Vertisol zones of North Karnataka and Andhra Pradesh. Its cultivation is spread all across Tamil Nadu, wherever dry land conditions prevail and soils are loamy.

Pigeonpea belt in Andhra Pradesh was relatively small at 0.16 m/ha in 1955. It expanded to occupy 0.4 m ha in 1999 then to 0.5 m/ha in 2006. The production improved from 155,000 t in 1999 to 230,000 t in 2006 [15]. The grain productivity improved from 358 kg/ha in 1999 to 435 kg/ha in 2006. Improvement in pigeonpea agroecosystem within Andhra Pradesh could be partly attributed to introduction of short duration varieties that fit as sole crop during kharif season. In Tamil Nadu, pigeonpea belt expanded steadily since 1970 when it was 63,000 ha with a productivity of 461 kg grain/ha. In 1990, it was 112,000 ha with total production at 69,000 t and productivity of 612 kg/ha. In 2005 area planted with pigeonpea was 0.14 m ha. Total production was 0.12 m t annually with a productivity of 864 kg/ha. Productivity has improved partly due to short duration varieties and modifications in nutrient dynamics. In Karnataka, pigeonpea cropping area extends into 0.42 m/ha and productivity is 499 kg grain/ha. Specifically, hybrids of pigeonpea (e.g., COPH2) that are spreading at a faster pace in the South Indian pigeonpea belt yield 1050 to 1350 kg grain/ha. In due course, they are expected to intensify pigeonpea belt, enhance nutrient turnover rates and grain production.

KEYWORDS

- ***Cajanus cajanifolia***
- **kharif**
- **Pigeonpea**
- **rabi**

REFERENCES

1. Fuller, D. O.; Harvey, E. L. The Archaeobotany of Indian Pulses: Identification, Processing and Evidence for cultivation. Environmental Archaeology **2006,** *11,* 218–246.
2. Van Der Maesen, L. J. G. Pigeon Pea: Origin, History, Evolution and Taxonomy. In Pigeonpea. Nene, Y. L, Hall, S. D.; Sheila, K. (Eds.) CAB International. United Kingdom, **1990,** 15–46.
3. Fuller, D. O.; Korizettar, R.; Venkatasubbaiah, C.; Jones, M. K. Early plant domestication in Southern India: Some preliminary archaeobotanical results. Vegetational History and Archaeobotany, **2004,** *13,*115–129.
4. ICRISAT. Uses of Tropical Grain Legumes: Proceedings of a Consultants Meeting. International Crops Research Institute for the Semi-arid Tropics. Hyderabad, India **1991,** 31–113.
5. ICRISAT, Pigeonpea hybrids. http://www.icrisat.org/New&Events/hybrid_pigeonpea_trigger _pulse_revolution.htm **2007** (August 10, 2012).

6. Nene, Y. L. Indian Pulse through Millennia. http://wwwagri-history.org/pdf/ Indian_pulses.pdf **2006,** 1–22 (September 10, 2012).
7. Krishna K. R. Agroecosystems of South India: Nutrient Dynamics and Productivity. Brown walker Press Inc.: Boca Raton, Florida, USA, **2010,** 313–417.
8. Morrison, K. D. Fields of Victory. Manoharlal Munshiram Publishers, New Delhi, **2000,** 201.
9. Kotraiah, C. T. M. Irrigation systems and Vijayanagar Empire. Directorate of Archaeology and Museums. Mysore, Karnataka, **1995,** 1–191.
10. Buchanan, F. A Journey from Madras through the countries of Mysore, Canara and Malabar. W.; Bulmer & Company, Cleveland row, St James, London, **1807,** *1,* 1–370; *2,* 1–510; *3,* 1–440. (in three volumes).
11. Rana, K. S.; Pal, M. Productivity and water use in pigeonpea-based intercropping systems as affected by weed control in rain fed conditions. Indian Journal of Agronomy **1997,** *42,* 576–580.
12. ICAR, Pulse Crops. http://www.icar.org.in/drought/drought1.htm **2006,** 1–2 (August, 10, 2012).
13. Anandraj, D. *Cajanus cajan*. Agriculture Man and Ecology. Bangalore, India. http://ecoport.org/ep.htm **2000,** 1–3 (May 1, 2012).
14. Sharma, S. K.; Singh, K.; Jat, M. L.; Shukla, A. K. Crop diversification-Problems and Prospects. Fertilizer News **2003,** *48,* 63–96.
15. IKISAN Redgram. http://www.IKISAN.com/links/ap_redgramhistory.shtml. **2006,** 1–3 (August 10, 2012).

EXERCISE

1. Discuss the origin, domestication, and area of maximum of genetic diversity of *Cajanus cajan.*
2. Mention top five pigeonpea producing countries and mark its Agroecosystem on a map.
3. Discuss Pigeonpea production trends in Caribbean Islands and Kenya.
4. Discus Pigeonpea-based cropping systems followed in South India.
5. How much Nitrogen do soils gain due to Pigeonpea-Rhizobium symbiosis?

FURTHER READING

1. Faujdar Singh; Oswalt, D. L. Pigeonpea Botany and Production Practices. International Crops Research Institute for the Semi-arid Tropics, Patancheru, Andhra Pradesh, India, **1992**, pp. 35.
2. Nene, Y. L.; Hall, S. D. The Pigeonpea. International Crops Research Institute for the Semi-Arid Tropics (ICRISAT). Patancheru, AP, India, **1990,** pp. 490.
3. Remanandan, P.; Singh, L. Pigeonpea: Conservation and Use of Plant Genetic Resources in CGIAR Centers. In: Biodiversity in Trust. Cambridge University Press, Cambridge, United Kingdom, **1997,** 156–167.

USEFUL WEBSITES

http://www.icrisat.org
http://www.fs.fed.us/global/iitf/pdf/shrubs/Cajanus%20cajan.pdf
www.eolss.net/Sample-Chapters/C10/E1–05A-44–00.pdf

CHAPTER 14

BLACK GRAM AND GREEN GRAM BELTS

CONTENTS

BLACK GRAM (*PHASEOLUS MUNGO*)

14.1 INTRODUCTION

Black gram is native to Southern India. Its origin and area of domestication seems to be located in Southern Indian savannahs and fringes of Ghats. Archaeobotanical analysis suggests that Neolithic settlers in South India domesticated *Vigna mungo*. The Vavilovian center of origin for black gram too lies in Southern India. Fuller and Harvey [1] reported occurrence of seeds and other remnants of black gram in and around Neolithic settlements in North-eastern Andhra Pradesh, North Karnataka (*Hullur, Budhihal, Hatibelagalu*) and Deccan region. Carbon dating of these archeological samples of black gram seeds and other plant remnants derived from large number of Neolithic sites indicated a period between 2300 B.C. and 1700 B.C. *Vigna mungo* and *V. radiata* (green gram) seem to share a common progenitor *V.sublobatus*. According to Fuller and Harvey [1] it is common to trace both *V. mungo* and *V.radiata* in Neolithic settlements of South India. Archaeological remains of *V. mungo* have also been traced in several Neolithic sites between rivers Godavari and Krishna (Table 1). Black gram seems to have diffused into Deccan region of Maharashtra, Gujarat and Mid-Gangetic plains, during Neolithic period (2nd millennium B.C.) through human migration. It subsequently traveled into Eastern plains, North-east India and Burma. Fuller and Harvey [1] report that *V.mungo* was traced in Eastern Harappan and upper Ganges locations belonging to 3rd millennium B.C. According to them it indicates several independent domestication events for *Vigna* species, for example in Mid-or North-west Gangetic plains, hills of Rajasthan (Mt Abu) and foothills of Himalayas. *Vigna mungo* might have been domesticated independently by Neolithic northern Indians, in seclusion from related events occurring in Southern Plains (Table 1).

TABLE 1 Black Gram: Origin, Distribution, Botany and Uses—A summary.

The origin and primary center of genetic diversity of Black gram is in South India. Areas comprising plains between rivers Krishna and Godavari, and Eastern Ghats harbors large number of genetic variants for *V. mungo*, which is also traced in Western Ghats. Archaeological evidences for its domestication and regular use by human consumption are found among the Neolithic settlements in South India (2200–1500 BC.) especially in locations such as Sanganakallu, Hullur and others in North Karnataka. Human migration during Neolithic period aided spread of black gram into most parts of South India, Deccan and Gujarat. Occurrence of several variants of black gram has been reported from sites in Harappa and Western Gangetic plains. According to Fuller et al. [2], these evidences support independent domestication events for *V. mungo* in Northern plains. Black gram spread into other Asian countries such as Burma, Indonesia and China. Burma is a major producer of Black gram. It is also grown in Africa, Australia and America.

Evolution: Black gram (*Vigna mungo*) and green gram (*V. radiata*) share common progenitor (*V. sublobata*).

Classification

Kingdom-Plantae; Division-Magnoliophyta; Class-Magnoliopsida; Order-Fabales; Family: Fabaceae; Genus-*Vigna*; Species-*V. mungo* (L.) Hepper

Synonyms: *Phaseolus mungo (Roxb)*; *Phaseolus mungo var chlorospermus*; *Phaseolus viridissimus*, V. *mungo var viridis*

$2n = 22$

Nomenclature in Different Languages

Black seeded mung bean, Urd, Black mapte, Wooly Pyrut, Black gram, Black Lentil in English; *Haricot Vel, Amberique vert* in French, *Fagiolo Urd* in Italian; *Linesenbohne, Urdbohne* in German; *Feijao da, Hei lu dou, Xiao hei dou* in Chinese, *Feijao urida* in Portuguese; *Fasol mungo, Fasol Vidov* in Russian; Frijol *Mungo* in Spanish; *Mat pe* in Burmese; *Hime tsuru Azuli, Ke tsuru azuki* in Japanese; *Mchooko mweusi* in Swahili;

Indian Languages: *Urad Dhal* in Hindi; *Masha* in Sanskrit; *Uddina Bele* in Kannada; *Minumulu, Udhu pappu* in Telugu; *Ullundu* in Tamil; *Uzhunnu, Uzhunnu parippu* in Malayalam.

Uses

Black gram is a highly prized pulse in South India. It is used along with rice grits to prepare puddings known as *Idlis*, pancakes (*Doshas*) and *Vada*. Thin wafers called *Papad* are crisp items. *Barian* are spiced balls prepared from powdered black gram.

Nutritive value (100/g edible grains)

Protein (%)-20 to 25; Fat (%)-1.3; Ash (%)-3.4; Crude Fiber (%)-4.2; Starch (%)-40 to 47; Vitamin A 300 IU; Vitamin B1 (mg 100/g)-0.52 to 0.66; Vitamin B2 (mg 100/g)-0.29–0.22; Niacin (mg 100/g)-2.0; Vitamin C (mg 100/g)-5.0; Iron (mg 100/g)-7.8; Calcium-145.

Sources: Refs. [1–4].

It seems cultivation of *Vigna mungo* (also *V.radiata*) spread into most areas in South India, Orissa and Gangetic plains during Iron Age. It was aided by invention of iron implements and ability to achieve deep tilth through iron plows [1]. Evidence for mass cultivation of black gram is also available in ancient Indian literature. *Masha* or *Masha Kalaya* has been mentioned as food legume in famous ancient Sanskrit treatizes like *Brihadharanyaka* (5500 B.C.), *Mahabaratha* (2000 B.C.) *Krishi Parashara* (400 B.C). Samskrit treatizes belonging to Ancient period (100 B.C. to 500 A.D.) refer black gram as *Masha* or *Mashaparini* [3, 5]. In Kashayapa's *Krishisukti* (800 A.D.) it has been mentioned that a particular variety of *masha* known as *mugi* was usually intercropped with pearl millet.

14.2 AGROCLIMATE, SOILS, WATER REQUIREMENTS, CROPPING SYSTEMS

Black gram is adapted to warm climatic conditions prevalent in South India. It is grown mostly as kharif crop but during recent decades, it is grown equally well during winter. Since, it is a short duration crop; winter season gap that may occur during rotations could be used to raise a crop. Moreover, winter is not very cool in most parts of South India. Black gram tolerates drought and warm temperatures. It is susceptible to water logging. The optimum temperature during crop season is 27°C–35°C. The rainy season crop is sown in June/July. Early planting allows better precipitation use efficiency. Summer crops are sown in late February and March. Black gram is grown in regions receiving at least 800 mm precipitation in a year. However, a crop could be grown to maturity in semiarid zones receiving 600 mm rainfall. Long dry spells may be detrimental to root proliferation and nutrient recovery. Hence, quantity of grain production depreciates. Black gram is mostly grown as rain fed crop in kharif, otherwise it is planted on rice fallows, wherein it thrives on residual moisture. However, irrigations could be applied on summer crop. Usually it is done in 12–15 day intervals. The critical stages for irrigation are flowering and pod development. Weeds should be cleared timely, if not, they divert a large fraction of irrigation. Long term studies at Coimbatore in South India have shown that consistent weeding improves WUE of rice/black gram intercrops [6].

Regarding soils, black gram is grown on sandy soils in coastal plains of Andhra Pradesh and Tamil Nadu. Red Alfisols and Vertisols found in Karnataka support good crops. Similarly, lateritic soils found in Kerala are also suitable for black gram production. Sandy loam with a pH range of 4.7 to 7.5 is ideal. Saline and Alkaline conditions are detrimental to root growth and nutrient absorption.

Black gram is intercropped with cereals like sorghum, maize, finger millet and pearl millet. In the Vertisol regions of North Karnataka it is intercropped with cotton. On black soils in Tamil Nadu, cotton/black gram intercrop is rotated with sunflower [7]. Black gram is also intercropped with other pulses like pigeonpea. It is grown as floor crop in horticultural orchards that support tamarind or mango production. Important rotations followed in South Indian plains and Hills are black gram-chickpea; sesamum-black gram; pigeonpea/green gram-black gram; black gram-fodder sorghum; rice-black gram; maize/sorghum-black gram. Rotating black gram on rice fallows is popular trend among rice farmers in intensive rice producing districts of Tamil Nadu and Karnataka and Coastal Andhra Pradesh.

14.3 EXPANSE AND PRODUCTIVITY

The black gram-cropping belt is spread all across Indian subcontinent and other South East Asian countries. India, Sri Lanka, Pakistan, Burma and Indonesia support relatively larger areas of black gram cultivation. It is extensively grown in Gangetic belt as rotation crop during rice-wheat sequence. Three South Indian states, namely Andhra Pradesh, Karnataka and Tamil Nadu are major producers of this pulse crop. It is grown as intercrop with cotton or sorghum on the Vertisols of North Karnataka and Andhra Pradesh. It is a preferred rotation crop in the rice fallows of Tamil Nadu. The black gram cropping zones have steadily improved in expanse in South India. The gain in area is mostly due its acceptance as an intercrop with other major cereals and oil seeds. At present, black gram crop in South India yields 850–1200 kg grain/ha and 2–3 t haulms/ha.

In Andhra Pradesh, black gram cultivation zone has gained steadily from a mere 0.1 m ha in 1955 to 0.23 m ha in 1970, then to 0.49 m ha in 1990 and 0.63 m ha in

2005. Total production increased from 25,000 t in 1955 to 35,000 t in 1970, then to 40,000 t in 2005. There was a spurt in both area and productivity during 1985–1988. The average productivity of black gram in Andhra Pradesh was 242 kg/ha in 1955. It increased to 425 kg/ha in early 1970s and stabilized at 450 kg/ha for a while until 1980. During early and mid-1980s again productivity improved significantly to 680–720 kg grain/ha [4].

The black gram belt in Tamil Nadu occupied 97,000 ha in 1970. It has expanded steadily to 180,000 ha in 1980, then to 293,000 ha in 1990 and 336,000 ha in 2005. The expansion of black growing regions is attributable its acceptance as a rotation crop in the rice fallows. The production improved from 31,000 t annually in 1970 to 61,000 t in 1980, 135,000 t in 1990–117,000 t in 2000 and 210,000 t in 2005. The productivity of black gram has increased from 315 kg/ha in 1970 to 518 kg/ha in 2005.

Black gram is among important legumes in Sri Lanka. It is mainly a rain fed crop grown in regions medium range precipitation. During past decade average productivity of black gram cropping zones was 0.8 t grain/ha, although potential grain yield attainable is high at 1.5–2.0 t grain/ha. The black gram belt currently extends into 6500 ha and contributes 5400 t grains annually [8].

KEYWORDS

- **kharif**
- ***mugi***
- ***V. radiata***
- **Vertisol**
- ***Vigna mungo***

REFERENCES

1. Fuller, D. O.; Harvey, E. L. The Archaeobotany of Indian Pulses: Identification, Processing and Evidence for cultivation. Environmental Archaeology **2006,** *11,* 218–246.
2. Fuller, D. O.; Korizettar, R.; Venkatasubbaiah, C.; Jones, M. K. Early plant domestication in Southern India: Some preliminary archaeobotanical results. Vegetational History and Archaeobotany **2004,** *13,* 115–129.
3. Nene, Y. L. Indian Pulse through Millennia. http://wwwagri-history.org/pdf/ Indian_pulses.pdf **2006,** 1–22 (June 12, 2012).
4. Krishna, K. R. Agroecozystesms of South India: Nutrient Dynamics and Productivity. BrownWalker Press Inc.: Boca Raton, Florida, USA, **2010,** 556.
5. Ayachit, S. M. K*ashyapiya-krishisukthi* – A treatize on Agriculture by Kashyapa. Agri-History Bulletin No. 4 Asian Agri-History Foundation. Secunderabad India, **2002,** 158.
6. Ramamoorthy, K.; Arokiaraj, A.; Balasubramaniam. A. Effect of Soil moisture and continuous use of Herbicides on Weed dynamics and their control in Upland Direct-seeded Rice-Black gram intercropping. Indian Journal of Agronomy **1997,** *42,* 564–569.
7. Subbaiah, K., Research Report-Kovilpatti, Tamil Nadu Agricultural University, Coimbatore, Tamil Nadu. http://www.tanau.ac.in/kovil.html **2005,** 1–6 (June 12, 2–12).
8. Department of Agriculture. Crop Recommendations: Black gram. Department of Agriculture, Government of Sri Lanka, Colombo **2006,** 1–4.

9. Sadhale, N. *Krishi Parashara* (Agriculture by *Parashara*) Agri-History Bulletin No 2. Asian Agri-history Foundation, Secunderabad, India **1999**, 94.

EXERCISE

1. Discuss origin and domestication of Black gram in Asia. Mark the green gram regions on a map.
2. Describe trends in area and production of black gram in India and Burma.
3. What are the major reasons for low potential yield of black gram?

FURTHER READINGS

1. INSEDA, Black gram. Booklet No 164 Pulse Crops PCS 5 www.inseda.org/Additional%20 material/…/Black%20 gram-164.doc, **2012**.
2. Siddiq, M.; Uebersax, M. A.; Nasir, M.; Sidhu, J. C. Common Pulses: chickpea, Lentil, Mungbean, Blackgram, Pigeonpea and Indian Vetch. John Willey and Sons: New York, **2013,** pp. 396.

USEFUL WEBSITES

www.ikisan.com/Crop%20Specific/…/Blackgram/Blackgram.htm
www.eolss.net/Sample-Chapters/C10/E1–05A-44–00.pdf
en.wikipedia.org/wiki/Vigna_mungo

GREEN GRAM (*VIGNA RADIATA*)

14.1 INTRODUCTION

The botanical origin, area of maximum genetic diversity and location of domestication for Mung bean or Green gram is said to be in South India. De Candolle [1] believes that mung bean originated in the plains of Peninsular India. Vavilov [2] too considers mung bean as native to Southern India, of course in addition to Central Asia. *Vigna sublobata* is considered as its wild progenitor, which is well distributed in India and Indonesia. The cultivation of green gram spread to various other nations in South Asia through human migration during early Neolithic age. According to Fuller and Harvey [3] probable center of origin for green gram (*V. radiata*) is in Southern Indian savannah and Western Ghats. Examination of seed specimens from Neolithic settlements at Sanganakallu in North Karnataka indicate, that *Vigna radiata* was cultivated and consumed during 2nd millennium B.C. [4, 5]. Archaeological specimens of *V. radiata* from settlements on the banks of rivers such as *Krishna* and *Godavari* suggest that Neolithic people around 2nd millennium used green gram regularly. Fuller and Harvey [3] suggest that green gram was also found in Eastern Ghats during Neolithic period. Overall, a wide range of variations in seed size, seed coat color, shapes and forms have been reported from the Neolithic sites of southern India [4–6]. There are innumerable evidences to show that green gram cultivation spread rapidly during Neolithic period into western and northern Indian plains. Archeological analyzes indicate that green gram was cultivated during 2nd millennium B.C. in Maharashtra and Mid-Gangetic Plains [3]. Examination of green gram specimens from Harrapan culture and upper reaches of river Ganges (3rd millennium B.C.) indicates independent domestication of green gram in addition to its diffusion from Southern Indian plains. Occurrence of wild species of *Vigna* around the western foothills of Himalayas is indicative of independent domestication of mung in Western Gangetic zone (Table 1).

TABLE 1 Green Gram: Origin, Distribution, Botany and Uses—A summary.

The area comprising Southern Indian plains and Western Ghats is considered the center of origin for Green gram. This region holds maximum genetic diversity for green gram. It ranges from bush types to vines, annuals and perennial types. Southern Indian natives domesticated green gram during Neolithic period (2nd millennium B.C.). At present, green gram genotypes are well distributed in Africa, Australia, Australasia, Caribbean, Central America, Europe, Indian Subcontinent, Middle East, Russia, Ukraine and Central Asian republics of Tadzik, Uzbek, Kyrgyz, and Turkmenistan.

Botanical Aspects

Three Vigna species in the subgenus Ceratotropis are important agriculturally in South India. They are derived from wild populations. Both *Vigna radiata* (green gram) and *Vigna mungo* (black gram) are derived from a common progenitor V *sublobata*. According to Fuller and Harvey [3] several wild populations of *Vigna* were grouped under a common species *V.* sublobata. *V. trilobata* is a wild species with smaller sized seeds.

Classification

Kingdom: Plantae; Division: Magnoliophyta; Class: Magnoliopsida; Order; Fabales; Family: Fabaceae; Genus: *Vigna*; Species: V. *radiata* (L.) Wilzeck

Synonyms: *Phaseolus aureus, Phaseolus radiatus, Phaseolus radiatus var aurea, Phaseolus hirtus*

Progenitors: *Vigna sublobata or Vradiata var sublobata* Domesticated species: *Vigna radiata*

$2n = 22$

Nomenclature in Different Languages

Indian Languages: *Mudgaparini* in Sanskrit; *Mung* or *Mung dhal* in Hindi; *Hesaru bele* in Kannada, *Chirupatturu, Cheru payuru, Pasipayir* in Tamil; *Pesarapappu* in Telugu; *Passipayir, cheru payir, Pachapayir* in Malayalam.

Other Languages: *Green gram* or *Mung bean* in English; *Chicksaw pea, Golden gram, Cho Suey bean* in USA; *Pois de Leruzalem, Amberique* in French; *Judia de Mungo* in Spanish; *Fagiolo Semi Verde, Fagiolo Mungo* in Italian; *Mungobohne* in German; *Fasol Zolotistaya, Vigna luchistaya* in Russian; *Fasola Zlota* in Polish; Feijao-mungo verde; *Atly Lobia* in Azerbaijanis; *Sarimahaly, Voango, Antandro, Amberique* in Madagascar; *hei lu Dou* in Chinese; *lukh dou* in Cantonese; *Munggo* in Philippines; *Hime tsuru azuki* in Japanese; *Pe di, Peid Sien, Pe-nauk, To-pi-si* in Burmese; *Thua Khieo, Thua thong* in Thai; *Exiu Xanh* in Vietnamese; *Mchooko Mchoroko* in Swahili.

Uses

Green gram is used in several ways in different parts of the world. It is cooked whole and consumed. Broken cotyledons are made into porridges. Green flour is an important ingredient in pancakes (*Doshas*). Preparations made of mung dhal are consumed along with rice or rotis. *Kichdi* is a mixture of well-cooked rice and green gram. Pongal is a common food item prepared using green gram and rice in Tamil Nadu. In China, green gram is eaten whole in sprouted form or made into soup. The whole mung bean is used to make a very popular soup known as *Tong sui*. The starch from green gram is also used to make noodles and jellies. In Indonesia, green gram is used to make porridge like snack called *Kacang hijau*. Mung bean sprouts are consumed in sizeable quantities in China and Far East. They are called *Dou Ya* (bean sprout) in Chinese, *Ya cai* (sprouted vegetable) or *Yin ya* (sprouts). In Malaysia, green gram is called *Tougeh, Moyashi* in Japanese and Thau-*ngok* in Thailand (see Plate 1).

Nutritional Aspects

Mung bean is an important source of dietary protein to human populace in South India. It also supplies fiber, potassium, calcium and B vitamins. It is low in cholesterol and fat content. Mung beans provide a range of mineral elements to human diet. However, phytates are known to complex with minerals like Ca, Mg and Fe and reduce their bioavailability. The bioavailability of Fe found in mung bean is said to be extremely low at just 2.6%. Soaking, fermentation, cooking and other types of processing may improve bioavailability of minerals by digesting and reducing phytate content. Mung bean sprouts are rich in protein (21–28%), calcium, phosphorus and certain vitamins. Sprouts contain high quality proteins that are easily digestible. It is relatively richer in amino acids such as lysine, leucine, phenylalanine, valine and isoleucine. Green gram is preferred because it is almost free of flatulence. Further, germination and de-hulling of green is supposed to improve nutritional properties of flour.

Chemical Composition of Green Gram Grains:
Crude protein (%) =24; Fat (%) =1.3; Carbohydrate (%) =56.6; Calorific value (Cal 100/g) =334.
Minerals (mg 100/g) Ca = 140; Fe = 8.4; P = 28 Vitamin B = 0.5.

Source: http://en.wikipedia.org/wiki/mung_bean; Refs. [3, 4, 11–14].

During medieval period, green gram cultivation had spread all over South India. It was planted along with staple cereals on dry lands. Archeological excavations around Vijayanagara and other locations in North Karnataka indicate regular use of green gram and other legumes [15]. During 18 and 19th centuries, green gram cultivation was in vogue among farmers in interior southern Karnataka and Tamil Nadu. It was intercropped and/or rotated with cereals. Green gram was a preferred short duration legume in the plains of North Karnataka and Andhra Pradesh. Green gram cultivation spread to other locations in South-east Asia like Burma, Vietnam, China and Philippines. Mung bean cultivation was in vogue in USA, by 1830s. It was grown for human consumption, mostly as nutritious sprouts. At present, green gram cultivation is in vogue both as mono-crop and intercrop in South-east Asia and many other agrarian regions.

14.2 SOILS, AGROELIMATE AND CROPPING SYSTEMS

In South India, it is grown mostly on Alfisols, Entisols or Vertisols (black cotton soils) with pH 5.5 to 8.2. Lateritic soils are also suited for cultivation of green gram. It prefers lighter soils with good drainage. Planting density varies depending on season, soil fertility status, fertilizer inputs and yield goal. Traditional planting methods allow 200,000 seedlings/ha. Whereas, 300,000 seedlings/ha are maintained for an intensively grown summer crop that receives irrigation and fertilizer based nutrients. Summer crop is usually thickly planted with more than 250,000 seedlings/ha.

Mung beans are short duration legumes maturing within 60 to 90 days from planting (Plate 1). Green gram is usually sown in June/July at the beginning of the rainy season (Kharif). Rabi crop is sown at the end of the rainy season around October/November. Summer crop that begins around February/March is almost always irrigated. Green gram is cultivated from sea level to 2000 m.a.s.l. The temperature during cropping season ranges from 27°C–35°C in kharif and 22°C–30°C in rabi. Summer temperatures are higher at 32°C–38°C in the semiarid tropics. Green gram is mostly grown as rain fed crop in the dry lands of South India. It tolerates drought to a certain extent, but is susceptible to water logging and frosty conditions. Irrigating green gram at 0.7 IW/CPE ratio induced rapid nutrient recovery; as a consequence grain yield reached 1.1 t/ha and stover 1.9 t/ha. On Vertisols of Western India, irrigation along with P inputs interacted positively and productivity improved further [16].

14.2.1 CROPPING SYSTEMS

Green gram is grown both as sole crop in rotation with cereals and as intercrop. As intercrop, it combines well with wide range of cereals and oil seed crops. Rice-green gram and maize-green gram are important rotations adopted in South Indian States. Mungbean/maize, mungbean/sorghum and mungbean/pigeonpea intercrops are common in most parts of Southern India. The grain yield equivalence improves by 330–720 kg/ha. Mung bean is grown in sequence with other crops in several combinations. Since it is a short duration legume, it fits well with most rotations. Cereals such as rice, maize, sorghum and finger millet are frequently rotated with green gram, since it adds to soil N status. It also affects nutrient dynamics in field by reducing weed intensity and loss of nutrients to weeds. It reduces soil erosion and nutrient loss. At the same time, green gram adds to soil fertility whenever its haulms are recycled. In the deep black soil areas of Tanjavur in Tamil Nadu, green gram is consistently rotated with rice and cotton. Similarly, in Vertisol belt of North Karnataka, cotton-green gram rotation is practiced, since it improves soil N status.

Green gram is mostly cultivated as rain fed crop during kharif season. It does not require extraneous irrigation, if the precipitation pattern is well distributed. The consumptive use ranges from 380–510 mm per season. The rabi crop thrives on stored moisture and supplemental irrigations. Summer crop requires irrigation throughout the season. Regarding irrigation to a sole crop of green gram, it is said that branching, preflowering and pod development are critical stages. Nutrient recovery was high *whenever* irrigation was timed to coincide with these three critical stages. As a consequence, productivity was also higher in fields provided with three irrigations at critical stages. Over all, a summer crop is given at least 3 irrigations, first at 20–25 DAS to coincide with preflowering stage; second one at flowering 25–40 DAS and third at grain filling stage.

PLATE 1 Green gram seedlings (top) and grains (bottom).
Source: Krishna, K.R. Bangalore.

14.3 EXPANSE AND PRODUCTIVITY

Mung bean or green gram is a small bush or vine-like herb that has adapted itself to different agroclimatic conditions. It is predominantly cultivated in South-east Asia, especially in India, Sri Lanka, Bangladesh, Burma and Thailand [11]. In Burma, it is an important component of the rice-based cropping systems. It supposedly covers 8.5% of the total cultivated area of Burma. The productivity of green gram in Burma averages at 880 kg/ha in farmer's fields, although potential grain harvests in experimental farms is as high as 3000 kg/ha [17]. It is cropped in other Asian countries such as Iran, Pakistan, China, Cambodia, Indonesia, Malaysia and Vietnam. It is also grown in parts of Africa, Australia and USA. Mung bean is an important legume grown in tropics and subtropics of Southern Africa. Green gram is cultivated, although in smaller patches in West Indies. Almost all of the green gram cultivation in USA, is confined to Oklahoma [18]. It is planted sparsely as an intercrop on droughty soils of Missouri [19]. Green gram is widely cultivated in Australasian countries.

In India, mung bean was cultivated on over 3 million ha during the year 2006 with annual grain production of about 1.0 million ton. The six major green gram-producing states in India are Maharashtra, Rajasthan, Bihar, Andhra Pradesh, Karnataka and Tamil Nadu. Together, these 6 states contribute 87% of country's green gram. The three southern states Andhra Pradesh (17.35%), Karnataka (17.4%) and Tamil Nadu (7.25%) account for 40–45 % of total area under green gram and contribute 38–40% of annual grain harvests. The productivity of green gram is highest in the southern state of Kerala (856 kg/ha). The yield potential of green gram genotypes in research plots is 10–12 q/ha but on farmer's fields grain harvest ranges 8–9 q/ha. The national average is still low at 4–5 q/ha. Therefore, a yield gap of 4–5 q/ha needs to be covered by improving nutrient schedules, water supply and high yielding genotypes (Table 1) [20].

KEYWORDS

- **Harrapan culture**
- ***Mudgaparini***
- ***V. radiata***
- ***Vigna sublobata***

REFERENCES

1. De Candolle Origin of Cultivated Species. Hafner, New York, USA, **1886,** 385.
2. Vavilov. Studies on the origin of Cultivated Plant Species. Bulletin of Applied Botany **1926,** *26,* 248.
3. Fuller, D. O.; Harvey, E. L. The Archaeobotany of Indian Pulses: Identification, Processing and Evidence for cultivation. Environmental Archaeology **2006,** *11,* 218–246.
4. Fuller, D. O.; Korizettar, R.; Venkatasubbaiah, C.; Jones, M. K. Early plant domestication in Southern India: Some preliminary archaeobotanical results. Vegetational History and Archaeobotany **2004,** *13,*115–129.
5. Ayachit, S. M. K*ashyapiya-krishisukthi* – A treatize on Agriculture by Kashyapa. Agri-History Bulletin No.4 Asian Agri-History Foundation. Secunderabad India, **2002,** 158.

6. Nene, Y. L. Indian Pulse through Millennia. http://wwwagri-history.org/pdf/ Indian_pulses.pdf **2006,** 1–22 (May 22, 2012).
7. Boivin, N.; Fuller, D.; Korizettar, R Petraglia, M. First farmers in South India: The role of internal processes and external influences in the emergence and transformation of South India's earliest settled societies. In: First Farmers in Global perspective. R. K Tiwari (Ed.) Uttar Pradesh State Archaeology Department. Lucknow, India, **2007,** 1–26.
8. Vishnu-Mittre Plant Economy in Ancient Navdotali-Maheswar. In: Technical report on archaeological remains. Sankalia, H. D. (Ed.). Department of Archaeology and Ancient Indian History. Deccan College and University of Poona Publications 2, Poona, **1961,** 13–52.
9. Kajale, M. D. Ancient Plant Economy in Sathavahana and Indo-Roman Period. Bulletin of the Deccan College Postgraduate and Research Institute, Poona, **1977,** *36,* 48–61.
10. Weber, S. Plants and Harappan subsistence, stability and change from Rojdi. Oxford and IBH Publishers, New Delhi **1991,** 278.
11. Poehlman, J. M. The Mungbean. Oxford and IBH Publishers Pvt Ltd. New Delhi, **1991,** 348.
12. IKISAN, Green Gram: History. www.IKISAN.com/links/ ap_greengramHistory.shtml **2006b,** 1–3 (May 22, 2012).
13. Krishna, K. R. Agroecosystems of South India: Nutrient Dynamics, Ecology and Productivity. BrownWalker Press Inc.: Florida, USA, **2010,** 556.
14. Del Rosario, R. R. Processing and Utilization of Legumes with particular reference to Mungbean in the Philippines. In: Uses of Tropical Grain legumes. Jambunathan, R. (Ed.). International Crops Research Institute for the Semi-Arid Tropics, Patancheru, India. **1991,** 211–215.
15. Morrison, K. D. Fields of Victory. Manoharlal Munshiram Publishers, New Delhi, **2000,** 201.
16. Parmar,; Thanki, J. D. Effect of irrigation, phosphorus and biofertilizer on growth and yield of rabi green gram (*Vigna radiata*) under Gujarat conditions. Crop Research **2007,** *34,* 100–102.
17. Kywe, M.; Buerkert, A.; Finch, M. Phosphorus-use efficiency and amino acid composition of different green gram (*Vigna radiata*) from Myanmar. http://www.tropentag.de/2005/ abstracts/posters/365.pdf. **2005,** 1–3.
18. Oplinger, E. S.; Hardman, L. L.; Kaminski, A. R.; Combs, S. M.; Doll, J. D. Mungbean. http://www.corn.agronomy. wisc.edu/alternativecrops/Mungbean.htm **1990,** 1–8 (June 15, 2012).
19. TJAI, Mungbean. Thomas Jefferson Agricultural Institute, Columbia, Missouri, United States of America, http://www. jefferesoninsitutet.org/mungbean.php. **2007,** 1–4 (May 22, 2012).
20. Masood Ali, Singh, A. A.; Saad, A. A. Balanced Fertilization for Nutritional quality in Pulses. Fertilizer **2004,** *49,* 43–56.
21. Achaya, K. T. A historical dictionary of Indian food. Oxford University Press, New Delhi, India, **1998,** 348 .
22. Krishnamurthy, K. S. The wealth of *Susrutha.* International Institute of Ayurveda, Coimbatore, Tamil Nadu, India, **1991,** 582.

EXERCISE

1. Discuss origin and domestication of Green gram. Give the names of Wild types and at least five recently developed cultivars.
2. Delineate the Green gram growing regions of world on a map.

3. What are the major uses of Green gram?

FURTHER READINGS

1. Erulan, S.; Kumran, S.; Nallasamy, T. System of Crop Intensification in Green Gram: An Innovative Approach. LAP Lambert Academic Press Publishing Company: New York, **2012,** pp. 92.

USEFUL WEBSITES

www.ikisan.com/…/eng/…/ap_greengramfiater%20 Management.sht
www.ikisan.com/…/ap_greengramClimate%20And%20Soils.shtml
www.ikisan.com/crop%20 specific/…/ap_greengramMorphology.sht…
www.ikisan.com/Crop%20Specific/Eng/…/up_greengramffarieties.sht…
www.ikisan.com/…/ap_greengramNutrient%20 Management.shtml

CHAPTER 15

HORSE GRAM FARMING ZONES ASIA AND AFRICA

CONTENTS

15.1 INTRODUCTION

Horse gram agroecosystem confines itself into parts of Africa, Indian Subcontinent, China and Philippines. However, its cropping zones are also traceable in Australia. Horse gram often thrives as intercrop with major cereals and other species. Horse gram is a preferred crop in dry land regions of the world. It is the best bet when precipitation patterns are not congenial and droughts are frequent. Horse gram is an important legume grown extensively in the drier regions of Peninsular India. It has been used as a food legume in South India, since 5 millennia. In almost all archeological sites of South India, such as Sanganakallu, Hullur and others on the banks of Thungabadhra, seed and other remnants of horse gram were noticed. In fact, *kulthi* or horse gram seems ubiquitous among Neolithic human dwellings found in South India [1]. Boivin et al. [2] suggest that southern Indians used horse gram as an important source of protein during Neolithic period. There is no knowledge about progenitors of horse gram (*Macrotyloma uniflorum)*. The domesticated species of horse gram is currently known as *Macrotyloma uniflorum.* Horse gram was domesticated in the plains and hills of South India.

Southern Indian plains and hills are considered as the area of domestication for this important pulse crop of South India. According to Mehra [3], Southern Indian plains and Eastern Ghats seem to be the geographical limits for its domestication. Horse gram cultivation diffused into western and northern parts of Indian subcontinent during Neolithic period through the counter migration of human beings (Table 1; [3–5]). Fuller and Harvey [1] opine that horse gram was widely cultivated across the Gangetic belt during Neolithic age (2500–2000 B.C.). The trapezoidal flattish grains and cotyledons of horse gram could be traced in Neolithic settlements selected all over Indo-Gangetic belt. Several wild species of *Macrotyloma* are traceable in African woods and savannahs, but none seems to have contributed its gene pool to the evolution of domesticated southern Indian horse gram [1]. Ancient Indian Sanskrit literature indicates that horsegram was grown and consumed by the populace. It also served the protein requirements of medieval Indian civilization. At present, horse gram is cultivated for proteineceous grains and forage.

TABLE 1 Horse gram: Origin, Distribution, Botany and Uses—A summary.

Horse gram is widely distributed in tropical Africa encompassing countries in East and Northeast Africa such as Sudan, Ethiopia, Zaire, Kenya, Tanzania, Zimbabwe, South Africa, Mozambique and Angola. In Asia, India is a major producer of horse gram. Its cultivation is spread all over the country from Uttharakhand in North to Southern tip of Tamil Nadu, Gujarat in west to Bengal in east. China, Philippines, Bhutan, Pakistan, Srilanka and few other countries in Asia also produce large amounts of horse gram. Horse gram is cultivated in small areas within Queensland in Australia. Horse gram thrives from 0 to 1500 m above sea level, tolerates drought, low soil fertility and a few other soil maladies. Optimum temperature is 20–30°C during growing season but it withstands 35–40°C in semiarid belt in South India.

Botany and Classification

Kingdom-Plantae; Division-Magnoliophyta; Class-Magnoliopsida; Order-Fabales; Family-Fabaceae; Subfamily Faboidea; Tribe-Phaseolaceae; Genus-*Macrotyloma*; Species-*M. uniflorum* (L) Verdc.

Synonyms:

Macrotyloma uniflorum var benadirianum (formerly Dolichos benadirianus Chiov); Macrotyloma uniflorum var stenocarpum; Macrotyloma uniflorum var uniflorum (formerly Dolichos biflorus auct.); Macrotyloma uniflorum var verrucosum

The genus *Macrotyloma* has 24 species native to tropical Africa and Asia. Wild Species; Unknown as yet

$2n = 20$

Southern Indian Agricultural Universities hold a large collection of germplasm of horse gram (e.g., University of Agricultural Sciences, GKVK Campus, Bangalore, India).

Nomenclature

Horse gram, madras gram in English; *dolic biflore* in French; *Pferdebohne* in German; *Faveira* in Portuguese; *Frijol Verde* in Spanish.

Indian Languages: *Kulatha* in Sanskrit; *Gahat* in Hindi; *Huruli* in Kannada; *Kollu* in Tamil; *Vulavu* in Telugu; *Parippu* in Malayalam.

Uses

Southern Indians relish soups and boiled bean based preparations. Soups are rich in protein and minerals. *Huruli saru* in Karnataka; *Vulavu Charu* in Andhra Pradesh, *Kollu Rasam* in Tamil Nadu and Kerala are most common preparations made from decoctions of boiled horse gram. Horse gram sprouts are also common food items in South India. Roasted horse grams smeared with hot chili powder forms good snack. In North India, horse gram is used to prepare *Khadai*—a kind of soup. *Ras* prepared from horse gram is favorite among Kumaonis in Utharakhand. Horse gram decoctions are known to possess important medicinal values. Folklore medicines based on horse gram are popular in southern Indian villages. Horse gram, husks, crude powders of pods and grains, haulms and residues are churned with water to prepare one the most nutritious animal feed. Horse gram based pastures and forages are common in South India.

Nutrient Value

Horse gram is an excellent source of minerals, especially Iron and Molybdenum. Horse gram seeds possess relatively higher amounts of polyphenols and haeme-agglutinins. They are deficient in amino acids like methionine and tryptophan. Proximate composition of horse gram grain is as follows:

Whole Grains (%): Moisture11.5; Crude Protein-22–24; Crude Fat – 1.5; Crude Fiber-5.07, Carbohydrate-58; Ash-3.3; Ca-238 mg 100/g

Dehulled Grains (%): Moisture9.7; Crude Protein-22; Crude Fat −1.6; Crude Fiber-1.9; Carbohydrate-61; Ash-2.9; Ca-223 mg 100/g

Horse gram has poor cooking quality. However, sprouted seeds are easily cooked and consumed. Seeds contain certain antinutritional factors. For example, Oxalic acid 11.4 mg 100/gin whole grain and 33.2 in de-hulled ones. Tannins in whole seed ranges from 117–289 mg 100/g.

Source: Refs. [4, 6–9];
http://www.ars-grin.gov/cgi-bin/npgs/html/taxon.pl?23076; http://en.wikipedia.org/wiki/horse_gram; http://www.ildis.org/legumeweb/6.00/taxa/1804.shtml; Krishna, 2012

15.2 SOILS, AGROCLIMATE, WATER REQUIREMENTS AND CROPPING SYSTEMS

Horsegram is grown on Inceptisols, Alfisols and Coastal sandy soils of East Africa. It cultivation is scattered and sparsely traced on the Oxisols found in Central Africa. Horse gram is often grown on virgin soils. It is a hardy drought tolerant crop and withstands vagaries in soil fertility. Hence it is grown to reclaim soils poor in soil fertility, those prone to drought and other maladies. In Asia, horse gram is grown as a cover crop on Alfisols, Inceptisols, laterites and sandy soils (Plate 1). Horse gram is the best bet on soils with low-N and SOM, since it is derives advantages from BNF. In Australia, psamments, red earths and sandy soils are used to grow horse gram.

Horse gram is grown in areas that experience subhumid and semiarid climate. Its cropping zones are traceable from sea level in Coastal South India to 1500 m..a.s.l. in Western Ghats. Horse gram withstands drought, poor soil fertility, as well as pests and still provides farmers with proteineceus grains at subsistence level. The optimum temperatures are 25–32°C, but in drought prone areas of Andhra Pradesh and Karnataka it tolerates up to 40°C. It is a short day plant and requires 12 h sunlight in a day to flower.

Horse gram is intercropped with various cereals such as sorghum, maize, pearl millet and little millet. The ratio of intercrops depends on season, soil fertility and genotype. Horse gram serves as an excellent cover crop during fallows, since it adds to soil-N content. Horse gram is rotated with a variety of other crops. It is relatively a short duration crop. Therefore, it easily fits crop sequences that involve two main crops. Horse gram suits well as a short duration summer legume. Horse gram is also favored as under story crop within plantations, since it adds to soil-N and organic matter. of the plantation in addition to SOM.

15.3 EXPANSE AND PRODUCTIVITY

Horse gram are scattered in tropics and subtropical regions of Africa. It is grown as a mixed crop or a subsistence crop during drought spells. The productivity is naturally low since inputs are feeble. In North India, horse gram is a grown during summer. During the short season, it serves as a catch crop and provides both grains and forage. It thrives on residual fertility and moisture still available after the rabi crop. The productivity is relatively low at 300–600 kg grain/ha. Horse gram productions are also traceable in China and Australia.

Horse gram is an important dry land crop in South Asia. It is grown in drought prone areas. When most other pulse crops fail it withstands intermittent drought spells common to South Indian dry lands and yields marginally. Hence, it is a preferred legume during subsistence farming. Horse gram cropping belt flourishes mostly in areas with marginal productivity (Plate 1). During recent decades, horse gram has faced stiff competition from other legumes, even in dry lands. Therefore, its cropping zones have shrunk since past 40 years. At present, in 2006–2007, grain production is around 40,000–50,000 tons. The reduction in grain production is mainly due to sharp decline in cropping area. It is interesting to note that irrespective of fluctuations in preference for horse gram and decline in its cropping area, productivity improved from 1.7 q/ha in 1955 to 2.2 q/ha in 1960s and later picked up steadily to 2.8 q/ha in 1980s and to

3.5–4.1 q/ha in during 2000–2005. Enhanced productivity of horse gram has offset decline in cropping area of horse gram.

PLATE 1 Top: A Horse gram field on coarse Alfisol (Red Sandy soils) found near Doddaballapur, Bangalore South India. Bottom: Close-up view of Horse gram plant.
Source: Krishna K.R., 2011.

KEYWORDS

- **horsegram**
- ***kulthi***
- ***Macrotyloma uniflorum***
- **rabi crop**

REFERENCES

1. Fuller, D. O.; Harvey, E. L. The Archaeobotany of Indian Pulses: Identification, Processing and Evidence for cultivation. Environmental Archaeology **2006,** *11,* 218–246.
2. Boivin, N.; Fuller, D.; Korizettar, R Petraglia, M. First farmers in South India: The role of internal processes and external influences in the emergence and transformation of South India's earliest settled societies. In: First Farmers in Global perspective. R. K Tiwari (Ed.) Uttar Pradesh stateArchaeology Department. Lucknow, India, **2007,** 1–26.
3. Mehra, K. L. History of Crop cultivation in Prehistoric India. In: Ancient and Medieval History of India; Agriculture and its relevance to Sustainable Agriculture in twenty-first century. Choudhary, S. L.; Nene, Y. L. (Eds.). Rajasthan College of Agriculture, Udaipur, Rajasthan, India **2000,** 11–16.

4. Fuller, D. O.; Korizettar, R.; Venkatasubbaiah, C.; Jones, M. K. Early plant domestication in Southern India: Some preliminary archaeobotanical results. Vegetational History and Archaeobotany **2004,** *13,*115–129.
5. Vishnu-Mittre Plant Economy in Ancient Navdotali-Maheswar. In: Technical report on archaeological remains. Sankalia, H. D. (Ed.). Department of Archaeology and Ancient Indian History. Deccan College and University of Poona Publications 2, Poona, **1961,** 13–52.
6. Krishna, K. R. Agroecosystems of South India: Nutrient Dynamics and Productivity. Brown Walker Press Inc.: Boca Raton, Florida, USA, **2010,** 372–382.
7. Subba Rao, A.; Sampath, S. R. **1979,** Chemical composition and nutritive value of Horse gram (*Dolichos biflorus*). Mysore Journal of Agriculture *13,* 198–205.
8. Sudha, N.; Begum, M.; Shambulingappa, K. G.; Babu, C. K. Nutrients and some antinutritional factors in Horse gram (*Macrotyloma uniflorum*) Verdc. Food and Nutrition Bulletin **1995,** *16,* 11–14.
9. Verdcourt, B. A Revision of *Macrotyloma* (Leguminosae). Planta **1982,** *38,* 37–39.

EXERCISE

1. Discuss botany and classification of Horse gram. Mark it cropping zones in India, Africa and other locations on a map.
2. Mention the advantages of growing horse gram as catch crop.
3. Mention various uses and nutritive value of Horse gram.

FURTHER READING

1. Kumar, D. Horse gram in India. Scientific Publishers Ltd.: Jodhpur, Rajastan, India, **2006,** pp. 144.

USEFUL WEBSITES

en.wikipedia.org/wiki/Horse_gram

CHAPTER 16

CANOLA CROPPING ZONES

CONTENTS

16.1 INTRODUCTION

Seeds for a large expanse of Canola were sown some 4,000 years ago in the plains of Central Asia. Archaeological reports indicate that canola plants were cultivated as early as 2000 B.C. The Euro-Siberian region is considered primary center of origin and region of maximum genetic diversity. China and North Indian plains too hold large genetic variation for *Brassica* species (Table 1). Brassica species have been cultivated in Europe, Mediterranean, China and Indo-Gangetic plains since ancient period. It was introduced to North America through human migration. Canola was first grown commercially in Canada in 1942 to produce a lubricant for war ships and for its use as oil for lamps. Canola was first grown commercially in Australia in 1969. Canola is currently grown across wheat belt areas of temperate Australia.

16.1.1 SOILS, AGROCLIMATE, WATER REQUIREMENTS AND CROPPING SYSTEMS

Canola or oilseed rape adapts to several different types of soil. Its cropping belts thrive on fertile Chernozems (Mollisols) found in United States of America. Black and Grey soils found in the Canadian Plains too support good crop of Brassica. Moderately fertile Inceptisols, Spodosols, Histosols, Inceptisols, Alfisols and Alluvial soils found in different continents also support canola production. Canola thrives well on soils with poor fertility, but produces low seed yield. It adapts to Xerasols and sandy Oxisols of West Asia and Central Asian dry tracts.

The Canolas or Brassicas are worldwide in distribution. Oil seed brassicas are versatile and adapt to a wide range of agroclimatic conditions (Plate 1). Canola is traced in the wet boreal region of Canada and other parts of northern hemisphere and temperate regions, subtropical plains, tropical regions and semiarid regions (Fig. 1; [1]). In the Canadian plains, canola is best suited to be grown on black and gray soils. Canola plants are cut at early pod stage, so that it preserves quality as good fodder. Water requirement of canola varieties depends on geographic region, genotype and yield goals. Canola thrives in regions that receive 500 mm to 2300 mm annually. It tolerates temperature fluctuations ably, but grows best in regions with 6°C–27°C. In the tropics, Canola grows at temperatures ranging from 18°C–35°C. For example, in the Indo-Gangetic plains mustard or rye or lahi that grows during rainy season experiences tropical conditions and high temperatures ranging from 23°C–36°C.

TABLE 1 Brassica species: Botany, Classification and Uses—A summary.

There are over 160 species of Brassica identified. They belong to family Brassicaceae. The origin of cultivated mustard seems to lie in the Eurasian plains. The area of origin includes, North-west India, China and Japan. The secondary origin identified for canola occurs in the area comprising present day Turkey and Iran.

Botany and Classification

Canola is said to be a perennial herb, yet grown mostly as annual or biennial field crop in most agrarian regions. It may reach 1.0 m in height. Branches bear leaves that are variously lobed,

---

toothed or frilled. The crop roots to 1.2 m depth in loose soils. Pods are small and bear seeds that are dark, white or yellow colored and small.

Kingdom-Plantae; Division-Magnoliophyta; Clas-Magnoliopsida; Order-Capparales, Family Brassicaceae; Genus-Brassica, Species-*Brassica napus, Brassica rapa, Brassica campestris; Brassica junceae* [2].

$2n = 18$

Germ plasm centers for canola are maintained in several agricultural Institutions of Europe and Canada. Directorate of Oilseeds at Hyderabad in India has a collection of canola. The Chinese Agricultural University at Beijing also has a germ plasm collection.

Nomenclature

Brassicas or Canola belongs to mustard family and is known by several other names depending on region and language. English: *Rape seed, canola, Black mustard*; German-*Braunsenf or Shwarzer senf*; Russian-Gorchistsa chernaya; French-*Moutarde noire*; Italian-*Senape nera*; Polish-*Kapusta czarna;* Portuguese-*Mostard;* Bulgarian-*Sinap cheran*; Arabic-*Khardal*; Hause, (West African dialect)—*Mastad*; Hebrew-*Hardol shahr*; Chinese-*Goi choi, Jei cai*; Japanese-*Kuro-garashi*; Korean-*Hukkyoja*

Indian languages: Hindi-*Raayi, Lahi, Sarson*, Marathi-*Mohari*; Telugu-*Avalu;* Bengali-*Sorsa;* Kannada-*Sasve*, Tamil-*Kadughu*; Malyalam-*Kadu*

Uses

Canola is used as edible oil in most regions of the world. The leaves, seeds and succulent stem of certain varieties of canola are edible as vegetable. The extracts are nutritious and possess medicinal value. Russians use canola primarily as vegetable oil and margarine during baking. In the Indian subcontinent, mustard oil is an important cooking medium. Green mustard is a relished vegetable in hilly tracts of Himalayas. The leaf extract (decoction) of rayi is popular soup in the Indo-Gangetic belt and Western India. Chinese use it as cooking medium and to prepare variety of pickles.

Canola oil was an important fuel for lamps during middle ages until the advent rock oil and electricity. Rapeseed oil is used as fuel for traditional lamps in India. Canola extracts are used to prepare soaps. It is used in cosmetics. Mustard cake and oil pressings are used for preparing greasy adjuvant. Rapeseed oil is used extensively in preparing high tenacious lubricating oil, plastics and biofuel. Canola is also processed to feed animals. The canola meal is used to feed cattle, swine and poultry. Canola oil may cause disturbance in digestion and metabolism. The quality and composition of rapeseed meal is important. It generally contains about 36 to 38% protein. High level of sulfur found in canola oil or cake or forage may create problems for farm animals. Biodiesel is a useful derivative from rapeseeds. Biodiesel is made through transesterification of rapeseed oil.

Nutritional Value

The rapeseed oil or canola oil is rich 6-omega fatty acids and 3-omega fatty acids. They occur in 2:1 ratio. Its consumption is said to reduce cholesterol accumulation in blood.

Sources: http://www.wikipedia.org/wiki/Brassica_juncea; http://www.rirdc.gov.au/programs/established-ruralindustries /pollination/canola.cfm

16.2 EXPANSE AND PRODUCTIVITY

There was a perceptible decrease in canola cropping zones during 1940s due to large-scale use of petroleum. Yet, brassica or rape cropping zones have expanded steadily during past 50 years, in all the continents. The canola production was mere 3.5 million t/yr during 1950. It increased to 8.8 million t/yr in 1975. However, during the next 25 years, canola production improved rather steeply, both due to expansion and improved productivity. Current global rapeseed or canola production is about 47 m t annually. Major canola producing countries are China (12.2 m t/yr), Canada (9.1 m t/yr), India (6 mt/yr), Germany (5.3 mt/yr), France (4.1 m t/yr), United Kingdom (1.9 mt/yr), Poland (1.6 mt/yr) and Australia 90.5 mt/yr) [3]. Globally, rapeseed oil is third most commonly used edible oil [4].

In North America, canola is being promoted to occupy greater area. The oil seed rape and turnip rape are new introductions to North America. In Canada, rapeseed is a cool season crop. Canola is well adapted to Canadian climatic conditions (Fig. 1). Canada is the largest exporter of canola oil. Hence, fluctuation in canola cropping in this region is largely influenced by demand for canola products in other parts of the world. Mustard cultivation is spread all over middle-east from Turkey to Afghanistan. It is grown during cool season and grain yield is moderate at 1.0–1.2 t/ha.

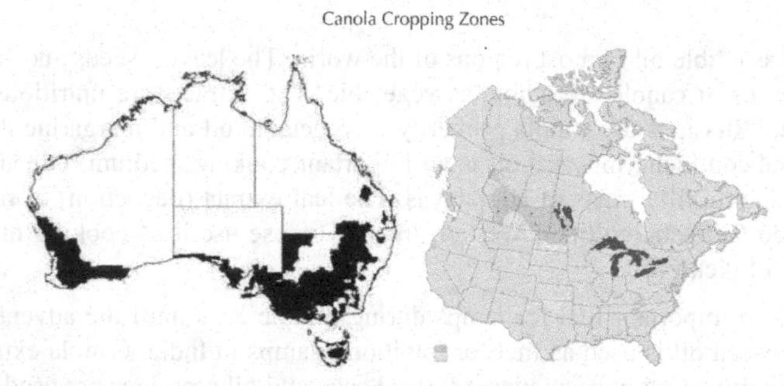

FIGURE 1 Canola production regions within Australia (dark regions) and Canada (dark gray area)

Source: Canola Council of Canada. http://www.canolacouncil.org/images/gallery/CanadaUS%20Growing%20Regions %20 Map.JPG; http://www.rirdc.gov.au/programs/established-ruralindustries/pollination/canola.cfm

Brassica or rapeseed or mustard is cultivated in the entire Indo-Gangetic plains. It is grown during rainy season and is often rotated with wheat during winter. Mustard is also grown in the peninsular India, although at a lower intensity. Mustard intercrops may yield up to 1.2 t grains ha.

In China, total oilseed production is 56.5 m t annually that is actually derived from 27 m ha. Canola or rape seed oil constitutes 13 million t annually, derived from 7.5 m ha. Major canola producing regions in China occur in Yangtze region and South-west provinces [5]. Canola fields in Hunan are said to be high yielding and profitable.

Australia is the world's second largest exporter of canola seed. Australia's exports consistently exceed 1 million tons yearly. Canola is widely grown across south-east Australia and in Western Australia (Fig. 1). Canola adapts to climate and growing conditions of Australia. Canola is largely confined to Western Australia (437,000 ha), New South Wales (194,000 ha), Victoria (190,000 ha), South Australia (143,000 ha). Canola zones in Queensland and Tasmania are negligible. Total canola production in Australia was 1.4 million t during 2010. Western Australia, New South Wales Victoria and South Australia together contribute 80% of Australian canola harvest. Rifkin [6] states that Australian farmers traditionally grew Canola types with Spring maturity. Long season winter types have been less promising. Among several introductions of European Canola, Cultivar CB188802, CB 1206 or CB1206 produced grain yields >7 t/ha. Greater thermal time and longer grain fill period seems to generate higher grain yield (Plate 1).

PLATE 1 Canola field in Central Indian Plains.
Source: Krishna, K.R., Bangalore, India

KEYWORDS

- ***Brassica***
- **Canolas**
- **Chernozems**
- **Rapeseed**

REFERENCES

1. Duke, J. A. Hand Book of Energy Crops. http://www.rirdc.gov.au/programs/established-rural-industries/pollination/ canola.cfm **1983** (September 23, 2012).
2. Renard, M.; Louter, J. H.; Duke, L. H. Oilseed Rape. http://www.OECD.org/datao-ecd/57/49/1946212.pdf **1987** (July 10, 2012).

3. FAO, Rape Seed statistics. Food and Agricultural Organizations of the United Nations. Rome, Italy, FAOSTAT.org **2011,** (November 26, 2012).
4. Downey, R. K. **1990,** Canola: A quality brassica oilseed. In Janick, J.; Simon, J. E (eds.) Advances in New crop. Timber press, Portland, Oregon. http://www.rirdc.gov.au/programs/established-rural-industries/pollination/canola.cfm 1–12 (May *20,*2012).
5. Bioenergy site, China Oilseeds and Products annual report 2011, http://www.thebioenergysite.com/articles/891/china-oilseeds-and-products-annual-report-2011. **2011,** 1–8 (December 3, 2012).
6. Rifkin, P. High yields and quality are achieved from European Canola types grown in the high rainfall zone of South-western Victoria. http://www.australianoilseeds.com/_data/assets/pdf_file/0011/8300/S8_P-Rifkin.pdf **2001,** 1–6 (July 11, 2012).

EXERCISE

1. Mark Canola producing regions of the World on a Map.
2. Mention the various soil types used to produce Canola.
3. Mention major Canola producing nations with reference to production, area and productivity.
4. Discuss origin, area of domestication, centers of genetic diversity and spread of canola into different agrarian regions of the world.

FURTHER READING

Boyles, M.; Bushong, J.; Sandez, H.; Stamm, A. Great Plains Canola Production Hand Book. Kansas State University Agricultural Experiment Station and Cooperative Extension Services, Manhattan, Ks, USA, **2012,** pp. 45.

Dixon, G. R. Vegetable Brassicas and Related Crucifers. CAB International Inc, Oxford, England, **2007,** pp. 327.

Fereidoon Shaidi, Canola and Rape Seed: Production, Chemistry, Nutrition and Processing Technology. Von Nostrand Reinhold, Holland, **1991,** pp. 255.

Gunstone, F. D. Rapeseed and Canola Oil: Production, Processing, Properties and Uses. John Wiley and Sons, New York, **2004,** pp. 223.

Schmidt, R.; Bancroft, I. Genetics and Genomis of the Brassicaceae. Springer Verlag, Heidelberg, **2010,** pp. 677.

USEFUL WEBSITES

http://extension.umass.edu/cdle/sites/extension.umass.edu.cdle/files/research/2009–03-Nitrogen-Use-Efficiency-in-Canola.pdf (Jul10, 2012).

http://www.bourgault.com/ProductsEquipment/Agronomy/RecommendedMaximumFertilizerRates/tabid/550/language/en-AU/Default.aspx (June 10, 2012).

http://www.northerncanola.com (June 10, 2012).

http://www.canola.okstate.edu/productionguides (June 10, 2012).

CHAPTER 17

GROUNDNUT AGROECOSYSTEM

CONTENTS

17.1 INTRODUCTION

South America is the center of origin for almost all species of *Arachis*. *Arachis* species are traceable in the region encompassing Bolivia, Southern Peru, North-west Argentina and Paraguay. Primary center of origin of peanuts encompasses humid zones of Southern Bolivia and North-west Argentina. Archeological analysis of tombs, prehistoric dwellings and plant remains; especially pod/kernels indicate that, both wild and domesticated species of *Arachis* were spread all across Andes, its slopes and adjoining tropical zones (Fig. 1; [1]). Greatest diversity of *Arachis* species, both wild and domesticated types is available in Eastern and Central Bolivia as well as Guarani region of Paraguay. *Matto Grosso* region around Cuiba and Campo Grande in Brazil as well as Serra Gerral in North-east are among the hot spots for diversity of genus *Arachis*. The genus *Arachis* contains 69 described species. There might be several more, at least 15 to 20 species yet undiscovered [2, 3]. Based on archeological reports, it is believed that prehistorically, peanuts were cultivated in the Peruvian oases, beginning around 2500 B.C. [4–6].

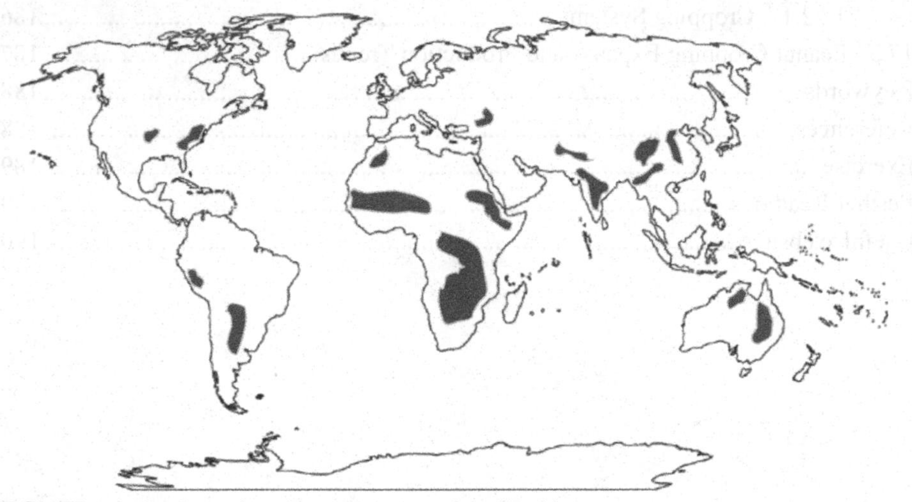

FIGURE 1 The Groundnut Agroecosystem of the World.
Note: Dark areas represent Groundnut Cropping zones in different Continents.
Source: Ref. [7].

Groundnuts were actually brought from their home in Peru to European plains around 16th century. It seems peanut seeds were sent to Nicolas Monaredes in Seville, Spain in 1574. Earliest of the evidences regarding introduction of peanut cultivation in different parts of Americas comes from narrations by Bartolome las Casas, who arrived in Hispaniola (Haiti) in 1502. In his work titled 'Apologetic History' dated 1527, he mentions availability of groundnuts and calls it *mani*. Descriptions about peanuts are also available in other literary works such as, 'Sumario Histario' and 'Historia General de las Indias.' Spaniards introduced peanuts from Mexico into other locations in Pacific Coasts.

Groundnut was transported into Philippines through ship liners. Earliest one, 'AcapulcoManila Galleon' a ship liner was led by the Spanish merchants. It seems Peruvian variety of groundnuts were first introduced into Philippines by Spanish galleons. Groundnut then spread into Chinese mainland, where it was termed as 'Foreign Beans.' During the same period, groundnut moved into Japanese islands and came to be called 'Chinese beans.' It is believed that Chinese travelers, later introduced groundnut to Indonesians and Malaysians. The Dutch conquerors and travelers introduced groundnuts into Indonesian archipelago, picking it most probably from Brazil.

Portuguese travelers carried peanuts from Brazil to African continent, especially to North and West African countries. Initially, Portuguese and Spanish sailors trading in merchandize and slaves brought groundnut to Africa, mainly West Africa. Groundnuts were actually reintroduced into American continent, especially into USA, through slaves and business travelers connected with West African countries. Groundnuts were also introduced into main land North America via Caribbean islands.

Asians received groundnuts from Africa around 15th century. Portuguese sailors shipped peanuts in to southwestern India. Groundnuts were actually introduced into South India by Jesuit travelers and followers of Vasco da Gama, during mid-15th century. Groundnuts were probably received from African locations, because during initial stages of its spread in the Indian plateau it was called 'Mozambique Beans.' Around 1600 s, groundnut were regularly traded and received by local Indians from Portuguese merchants arriving into West Coast of India from African ports. During 18th century, peanuts spread rapidly into several parts of South Indian cropping zones. However, perceptible increase in peanut culture and productivity occurred during twentieth century. During the past 400 years, groundnut agroecosystem has spread into all major continents, where ever tropical, semiarid or humid climate persists, soils are loamy and fertile enough to support a cropping season [7] (Table 1).

TABLE 1 Groundnut: Botany, Nomenclature and Uses—A summary.

Groundnuts are native to Southern American tropics that encompass Bolivia, Peru and Uruguay. The primary center of genetic of diversity occurs in Gran Chaco, southern Peru and South-west Brazil. Groundnut cropping zones expanded during medieval period through Spanish explores. Currently, groundnuts are grown in most of agrarian regions that experience subtropical and tropical climate.

Botany and Classification

Kingdom-Plantae; Sub Kingdom-Angiosperms; Order-Fabales, Family Fabaceae; Genus-*Arachis*; Species-*A. hypogaea.*

The genus *Arachis* has been classified into four botanical types, namely 'Virginia,' Peruvian Runners, Valencia and Spanish. Kokalis-Burelle et al. [8] state that Virginia types originated around Amazonia, where maximum diversity of Virginia types is available. The Peruvian runners are most commonly found in archeological sites around Peru and must have originated and developed in Peru. The 'Spanish' variety was domesticated and cultivated by Brazilians. The Valencias originated in the area encompassing Peru, southern Brazil and western Paraguay. Guarani people residing around Paraguay and Parana rivers domesticated Valencia type [9].

Subspecies Botanical Type Center of Origin

hypogaea Virginia Southern Bolivia and Northern Argentina

hirsuta Peruvian runners Peru

fastigata Valencia Peru, Brazil and Paraguay

vulgaris Spanish Paraguay, Uruguay and North-east Brazil

$2n = 40$

Groundnut germplasm is available in many parts of the world. International Crops Research Institute for the Semi-arid Tropics at Hyderabad in India is the largest repository of peanut germplasm. United States Department of Agriculture maintains extensive germplasm at the Universities in North Carolina and Georgia. National Research Center on Groundnut at Junaghad, Gujarat, India also maintains groundnut germplasm.

Nomenclature

North and South America: Argentina-*Mani*; Aztecs/Mexico: *Thalcacahuetl or Ground Cacao*; Bolivia-*Ynchic*; Brazil-*Mandi, Mandubi, Mundubim, Jingu*; Paraguay-*Ynchic*; Peru-Incas called it *Ynchic*; Portuguese-*Amendoim*; USA, and Canada-*Peanut, Earthnut, Groundn Ground peas, Goobers, Goober nuts, Pinders Monkey nut, Hawks nut, Kipper nut.*

European Languages: Czeck-*Podzenice Oleja, Arasidy;* Danish-*Jordnod, Jordnodder*; Dutch-*Aardnoot, Grondnoot;* Eastonian-*Maapahkel*; Esparanto: *La Arachid;* Finnish-*Maapakhina;* French-*Arachide, Cacahuete (ka-kawet*), *Pistachi de tere*; German-*Erdnusse, Arachinusse*; Greek-*Fystiki*; Italian-*Mandorla di terra, Nocciolina, Pistachio di terra;* Polish-*Orzech Ziemny, Orzacha Podziemna;* Russian-*Zemlyanoy Grek* (zem-ya noyarek), *Fistitsche;* Spanish-*Mani, Avellana Americana, Alacgueses*; Svenska-*Jordnut*

Africa: *Azi* in Fond; *Suma* in Mossi; *Ngooba in* Congolese, *Tiga* in Bambara; *Kolanche* in Housa; *Damsi* in Zarma; *Gridje* in Fulani; *Kolji* in Beri Beri; Xhosa: *Amandongoma; Guerte* in Wolof language; *Ful Sudani* in Arabic; *Njugu,* in Zulu.

South and South-east Asia: Cambodia: *Sannadaek dei;* Chinese-*Chang-sheng-gua* (long-life nut), *Xiangdou* (fragrant nut), *Wuhuaguo* (Flowerless nut), *Luo hua sheng* (falling flower-born nut), *Hua Sheng;Didou* (Underground nut); Hindi-*Mungphali.* Kannada-*Kadalekayi*, Telugu-*Verushenega,* Tamil-*Nilakadalai,* Marathi-*Phuimug,* Bengali-*Cheenabadam*; Oriya-*Cheenabadam,* Punjabi-*Mung phalli*, Gujarati-, *Mungphali. Bhoizing*; Indonesia: *Kacang Tanah, Kacang jawa,;* Japanese: *Nankinmame, Piinatsu, Rakkasei*; Korean: *Tiang kong*; Laos: *Twax din, Thwax ho*; Malaysia: *Kacang China, KacngJawa, Kacang goring.*; Napalese: *Mungphalli*; Pakistan: *Mungphalii* in Urdu

Uses

Groundnuts are also used as whole seeds (kernels). Most commonly, kernels are roasted or boiled, if not sometimes eaten fresh. Groundnut flour and concentrates are used to replace wheat or corn flour and to make breads. Groundnuts are useful in preparing imitation milk to extend cow milk. Groundnut protein and butter is employed to prepare cheese. Groundnut butter is extensively used in North America. Groundnuts were well received by West Africans. Groundnut crop rapidly gained in importance as a protein source. West Africans prepare a variety of dishes using peanuts. Groundnut stew is a favorite dish among people in Sahelian zone. In Ghana, it is served with *'fufu.'* In Mali and Senegal, groundnut stew called *'mafi'* is used with chicken and sweet potatoes. Early in 1700s, Nigerians had started preparing groundnut paste and butter. It was used as a spread with bread. Spicy peanut sauce is a traditional Indonesian culinary item.

It is consumed along with an appetizer called 'satay.' In Java, groundnut mixed with rice is a delicacy. In Szechwan region of China, groundnut and chilies paste is used to prepare wide variety of culinary items. Indians use peanuts extensively in their daily meal items such as curries and soups. 'Groundnut chutney' is common dish prepared using paste of kernels with chilies. Groundnut is a staple fat source and cooking medium for southern Indians. At this juncture, we should realize that demand for groundnut and groundnut products are crucial factors that induce farmers to cultivate peanuts. Therefore, expanse and cropping intensity of peanut is dependent partly on various uses of groundnuts. The ability of groundnut to compete with other oil seeds and protein sources is equally important [10].

Nutritive Value (100/g seeds)

Energy-567 k cal; Carbohydrates-16.1 g; Protein-25.8 g; Fat-49 g; Dietry fiber-8.5 g; Cholesterol-nil; Niacin-12 mg; Pantothenic acid-1.8 mg; Pyridoxine-0.3 mg; Riboflavin-0.13 mg; Thiamine-0.64 mg; Vitamina A, C, B9-negligible; Vitamin E-8.3 mg; Ca-92 mg; K-705 mg; P-76 mg; Zn-3.5 mg; Fe-4.8 mg; Mg-1.7 mg; Mn-1.9 mg

Sources: Refs. [6, 7, 10–17]; http://en.wikipedia.org/wiki/peanut; http://forums.egullet.org/index.php?showtopic=1854; http:// www.plantnames.unimelb.edu.au/sorting/Arachis.html (language shown in italics);
http://www.aboutpeanuts.com/reci.html; http:// www.peanut-institute.org/recipes.html;
http://www.aggie-horticulture.tamu.edu/plantanswers/recipes/peanutrecipes.html

17.2 SOILS, AGROCLIMATE AND CROPPING SYSTEMS

Groundnut cropping zones in South and south-east of United States of America thrives on Mollisols, Ultisols, in Texas and Alabama, Piedmonts in Carolinas and Ultisols and Spodosols in Florida and Georgia. The South American groundnut cultivation zones thrive on sandy Oxisols in Brazil. In the Pampas, low fertility Mollisols, Alfisols and Udalfs support groundnut cropping zones, that mostly thrive as a rotation crop with maize or wheat. The groundnut belt in West Africa occurs predominantly on sandy Oxisiols, Xerasols and silty soils found. It grows on low fertility soils hence productivity is relatively low. Groundnut is grown on Xerosols in West Asia and on sandy Alfisols, Entisols or Aridisols in Southern Africa. In India, groundnuts are grown on Inceptisols in the Gangetic plains. Groundnut belt thrives well on gravely Alfisols and Vertisols found in the West and Southern Indian plains. It is an excellent crop on sandy coastal soils. Groundnut belt extends mainly into Aridisols, Entisols and Inceptisols in China.

Groundnut cultivation on the earth extends from 35°S to 40°N, but in Central Asia and North America it spreads as far as 45°N. In terms of altitude, peanuts are found grown up to 1250 m above sea level, although stray patches even above this altitude are possible [5]. Groundnut is a preferred crop in semiarid tropics, humid tropics and humid temperate climates with sufficiently long summers. Groundnut is predominantly grown in tropical regions that experience warm temperatures ranging from 18–32 C. Groundnut crop matures in 120–140 days depending on the genotype and geographic location. Regarding thermal time, it is said groundnut needs 56 for seedling stage, 103 for branching, 538 for flowering, 670 for pegging, and 720 for pod initiation [7]. Groundnut thrives in the semiarid belt that receives about 500–750 mm precipitation annually. The crop needs 550–700 mm per season to mature. However, groundnuts are growing successfully with subsistent or moderate pod yield (0.8–1.2 t pod/ha) using just 400–450 mm water. Groundnuts grown during summer with irrigation supplements and fertilizer supply produce 2.5–4 t pods/ha, using more than 111 mm water (Plates 1 and 2).

17.2.1 CROPPING SYSTEMS

Groundnut is a versatile crop that thrives well under different cropping systems and environments. Groundnut cropping patterns and extent of their spread or popularity with farmers has kept changing during the past three decades. Biotic, environmental and economic considerations have been the major causes for such shifts. In fact, development of most appropriate cropping system that includes groundnut has been a major preoccupation of farmers and researchers concerned with this agroecosystem ([18]; Plates 1 and 2). There are indeed innumerable reports dealing with Groundnut rotations/intercrops with major cereals of the world like rice, maize, wheat, sorghum or millets. Groundnuts are preferred as rotation or intercrop due to compatible growth habit, short duration and most importantly their ability to improve soil N fertility status. Peanut leaves significant amounts of residual nutrients in soil after harvest. The following cereal or other crop species could efficiently harness residual nutrient. Peanut is also intercropped with perennial and deep-rooted species like Cassava, Leucana, Citrus, etc. In such crop combinations, groundnut improves nutrient recovery from shallow layers of soil while deep-rooted crops like Cassava or Leucana explores deeper horizons of soil. In China and India, peanuts are sown as relay crop just a few days ahead of cereal harvest. Groundnut also serves as green manure or catch crop in intensively cropped zones of India and China. A typical rice-rice rotation allows a short duration peanuts to follow as green manure or catch crop that lessens nutrient loss from field ecosystem. In southern United States od America, perennial groundnuts are cultivated as cover crop/catch crop in citrus groves. It helps in conserving nutrient resources and avoids ground water contamination to a certain extent by holding nutrients in the upper horizons of soil.

PLATE 1 Groundnut Production in South India.
Note: Above picture depicts Groundnut Fertilizer trial at GKVK Experimental Station, Bangalore, India. Peanut is an oilseed cum protein source. It is often rotated with cereals like Sorghum (background), Maize or Cotton. Such rotations replenish soil-N content plus offer basic food requirements to farmers, especially so in subsistence farming belts.
Source: Ref. [7].

PLATE 2 Groundnut field with Castor in borders at a location near Bangalore, South India. *Source:* Krishna, K.R., Bangalore, India.

17.3 PEANUT CROPPING EXPANSE AND PRODUCTION TRENDS

Groundnut is an important oil seed crop of the world. It is a popular oil and protein source in many countries. It is cultivated and consumed in over 100 countries of the world. Among oilseeds, groundnut contributed about 10% of global oilseed production of 290 million tons in 2005. We may note that 90% of groundnut production is accounted for developing countries. Groundnut encounters stiff competition in terms of economics and consumer preference from other major oil seed crops. Soybean with 53% global share of oil seeds, rapeseed with 15% and cottonseed oil with 12% score over peanuts [19, 20].

Groundnut cropping is localized predominantly among developing countries of Asia, Africa and Latin America. More than half of peanut cultivating zones on the globe are situated in Asia. About 31% global peanut cultivating area is in India and 19% in China. In Africa, Nigeria with 11% of global peanut area, Sudan with 7% and Senegal with 3% form the major expanses of peanut agroecosystem.

World peanut production has increased steadily from 14.42 million t pods annually in 1972 to 37.21 million t pods in 2005. Regarding peanut production by different countries, annually, China harvests 42% of global peanut crop, India 20%, Nigeria 7%, USA, 5%, Senegal and Sudan 3% each. These countries are major peanut contributors. In terms of yield per unit area, Israel at 5400 kg/ha and USA, at 3450 kg/ha top the list. In terms of area, Asia recorded highest increases in groundnut cropping zones. However, to a great extent, increase in total groundnut harvest in Asian countries was attributable to higher yield per unit area. During 1990s, China recorded significant jump in yield per unit area, almost by 4–6 times over previous yield. In these portions of groundnut agroecosystem, intensive agricultural practices exemplified by high nutrient inputs, irrigation and plant protection are being practiced. Therefore, nutrient turn-over rates within the cropping belt is relatively higher.

Regarding trade, since 1970 s, nearly 5% of global groundnut produce is traded and exported from place of production. Total exports increased marginally in 1980 s. During 1970s and early 80 s, African countries especially, Nigeria, Senegal and Mali were important groundnut exporting countries. However, with shift in purpose from vegetable oil to groundnut for food purposes, as well as presence of aflatoxin, export from African countries has reduced.

Additionally, production in countries, such as Nigeria has declined due to rosette virus and soil fertility related factors. Currently, China, USA, and Argentina are the major peanut exporting countries. During the past decade, these three countries together have account for 61–65% world groundnut exports. Annually, China and USA, export 250 and 180 million US dollar worth groundnuts, respectively. We ought to know that groundnut trade practices too affects the area and intensity of groundnut cropping, not just in the vicinity, but even in countries remotely situated from point of its consumption. Regarding imports, composition of groundnut importing countries has not changed greatly since 1970s. The Netherlands, United Kingdom, Japan, Canada and Germany are major groundnut importers.

KEYWORDS

- **Acapulco Manila Galleon**
- **aflatoxin**
- **Arachis**
- **mani**

REFERENCES

1. Banks, D. J. Genetic significance and implications of Peanut artifacts from Royal Tombs, Sipan, in Peru. Proceedings of American Peanut Research and Education Society, Stillwater, Oklahoma, USA. **1994,** *26,* 34.
2. CGIAR, Scientists fear irreparable loss of peanut crop biodiversity for world food supply. Future Harvest. **2002.**
3. Bertioli, D. J.; Favero, A.; Simpson, C, Seijo, G.; Stougard, J.; Valls, J.; Proite, K.; Leoi, L.; Moretzshon, M.; Gimenes, M.; Guimaraes, P.; Bertioli, S. Genetic mapping and resysnthesis of *Arachis hypogaea* for improvement of the groundnut crop. Plant and Animal Genetics Conference: Legumes, Soybean, Common Beans. San Diego, California, USA. http://www.intl-pag.org/13/abstracts/pag13_p463.html. **2005,** 463 (July 22, 2012).
4. Hammons, R. The origin and history of the Groundnut. In: the Groundnut Crop: A Scientific Basis for Improvement. Smartt, G. (Ed.) Chapman and Hall, New York, **1994,** 22–35.
5. Weiss, E. A. Oil Seed Crops. Blackwell Scientific Co, London, UK. **2000,** 487.
6. Smith, A. F. Peanuts: The Illustrious History of Goober Peas. University of Illinois Press, Chicago, USA, **2002,** 285.
7. Krishna, K. R. Peanut Agroecosystem: Nutrient Dynamics, Ecology and Productivity. Alpha Science International ltd. Oxford, United Kingdom **2008,** 304.
8. Kokalis-Burrelle, N. Compendium of Peanut Diseases. American Phytopathological Society, St Paul, Minnesota, USA, **1997,** 284.
9. Sauer, J. D. Historical Geography of Crop Plants: A select roster. CRC Press, Boca Raton, Florida, USA, **1993,** 248.
10. ICRISAT, Uses of Tropical Grain legumes. International Crops Research Institute for the Semi-Arid Tropics, Patancheru, India. **1991,** 259–323.

11. Shorter, R.; Patanathoi, A. *Arachis hypogaea* L. PROSEA Hand Book on CD-ROM, Jan Kops House, Thailand, **1997,** 1–9.
12. Serre, M. All about peanut. The Worldwide gourmet. http://www.theworldwide gourmet.com/nuts/peanuts/peanut.htm. **2003,** 1–3 (May 10, 2012).
13. Putnam, D. H.; Oplinger, E. S.; Teynor, T. M.; Oelke, E. A.; Doll, J. D. Peanut. Alternative Field Crops Manual. Universities of Minnesota and Wisconsin-Extension Service, Wisconsin, USA, **2004,** 1–11.
14. Yao, G. Peanut Production and Utilization in the People's Republic of China. Peanut in Local and Global Food Systems Series Report No.4. University of Georgia, Georgia, USA, **2004,** 1–26.
15. Tabo, R. Agroecosystems group. International Crops Research Institute for the Semi Arid Tropics-Sahelian Center. Sadore, Niger, (Personal communications). **2005.**
16. Krapovickas, L. The Origin, Variability and spread of Groundnut (*Arachis hypogaea*). In: Domestication and Exploitation of Plants and Animals. Ucko, J.; Dimblebey, J. W. (Eds.). Aldine Press, Chicago, USA, **1969,** 427–441.
17. NRCG. Production Technology. National Research Center for Groundnuts. Junagad, India. http://www.nrcg.res.in /pages/productiotech.htm **2008,** 1–9 (July 22, 2012).
18. AICORPO. Research Highlights-25 years 1967–1992. Directorate of Oilseeds Research, Rajendra Nagar, Hyderabad, India **1992,** 1–10.
19. Revoredo, C. L.; Fletcher, S. M. World Peanut Market: An Overview of the past 30 years. Georgia Experimental Station, The University of Georgia Research Bulletin No. 437, **2002,** 1–16.
20. Talawar, S.; Rhoades, R. E.; Nazarea, V. World Geography of Groundnut. Distribution, Production, Use and Trade. Peanut Collaborative Research Support Program-Report. University of Georgia, Athens, USA, **2005,** 1–6.
21. Centers, Consultative Group on International Agricultural Research, Rome, Italy. http://www.futreharvest.org/news /wildpeanuts.shtml. 1–5 (August 12, 2012).
22. Krishna, K. R. Phosphorus efficiency of tropical ad semidry land crops. In; Accomplishments and Future challenges in Dryland soil fertility research in the Mediterranean area. Ryan, J. (Ed.). International Center for Agricultural Research in Dry Areas. Aleppo, Syria, **1997,** 343–363.
23. Krishna, K. R. Crop Improvement Towards Resistance to Soil Fertility constraints. In: Soil fertility and Crop production. Krishna, K. R. (Ed). Science Publishers Inc.: Enfield, New Hampshire, USA, **2002,** 337–369.
24. Krishna, K. R. Agrosphere: Nutrient Dynamics, Ecology and productivity. Science Publishers Inc.: Enfield, New Hampshire, USA, **2003,** 336.
25. Talawar, S. Peanut in India: History, Production and Utilization. Peanut in local and Global Food Systems Series Report No. 5. University of Georgia, Georgia, USA, **2004,** 1–35.

EXERCISE

1. Discuss origin, domestication and centers of genetic diversity of groundnuts.
2. Mark the global Groundnut Agroecosystem on a Map.
3. Describe groundnut production trends in USA, Sahelian West Africa and India.
4. Collect data on Nitrogen contribution of Bradyrhizobium to soils in the Groundnut belt.

FURTHER READINGS

1. Krishna, K. R. Peanut Agroecosystem: Nutrient Dynamics and Productivity. Alpha Science International Inc. Oxford, England, **2008** pp. 292.

2. PCA. Peanut Production Guide. Peanut Company of Australia, Kingaroy, Queensland 4610, Australia, **2010,** pp. 31.
3. Talawar, S. Peanut in India: History, Production and Utilization. Peanut in local and Global Food Systems Series Report No. 5 University of Georgia, Georgia, USA, **2004,** 1–35.

USEFUL WEBSITES

http/www.icrisat.org
http//www.apresinc.com
www.peanutscience.com
http://www.soyatech.com/peanut_facts.htm

CHAPTER 18

SUNFLOWER AGROECOSYSTEM

CONTENTS

18.1 INTRODUCTION

Sunflower is a versatile crop with world-wide distribution. The primary center of origin and area of maximum genetic diversity for sunflower (*Helianthus annuus*) is said to be in the region comprising Tennessee in Eastern United States of America. Historically, earliest known specimens of sunflower were derived from sites at Hayes in Tennessee that date back to 2300 B.C. The Native American tribes grew sunflower regularly by 1000 B.C. Recent evidences based on Niche model and ^{14}C assays of seeds from Mexico suggest that sunflower (*Helianthus spp*) may have multiple sites of origin. [1] (Table 1). The Incas grew sunflower in southern America during medieval period. Sunflower reached European plains through Spanish conquerors and traders who plied too and fro between South America and European nations during 16th century. Sunflower derived from North America were grown in England by 16th century [2]. It was particularly popular with Russian Orthodox Christian monasteries because oil derived from it was not prohibited for consumption.

TABLE 1 Sunflower: Origin, Distribution, Botany and Uses—A summary.

The primary center of origin and region of maximum genetic diversity for sunflower (*Helianthus annuus*) and its ancestors are in Eastern United States of America in the area comprising Tennessee and Mississippi states. However, multiple sites of origin are possible. Northern region of Mexico is said to be yet another Center of Origin for *Helianthus* [1]. The commercial hybrids of sunflower are grown and used worldwide. Many of these hybrids are supposedly derived from wild species-*Helianthus petiolaris* [3]. Sunflower is relatively a recent introduction into India. It was introduced to South Indian farm belts in 1969 to improve oilseed production. During 1970s, sunflower belt in India was intensified using high yielding varieties from Russia. Earliest experimental hybrids were developed at Hebbal, Bangalore in 1974–75 using cytoplasmic male sterile line (CMS 2, CMS 124, CCMS 204 and CMS 234) and restorer lines (RHA 266 and Rha 274) from Tennessee in USA. The commercial hybrid (Bangalore Sunflower Hybrid BSH1) was grown in South India in 1980. Currently, innumerable hybrids with high grain yield and wide range of other characteristics such as resistance to disease and tolerance to drought are available [4, 12].

Botany and Classification

Kingdom-Plantae; Class-Magnoliopsida; Order-Asterales; Family Asteraceae; Genus *Helianthus*; Species *H. annuus*

Wild Species; H. annuus sp lenticularis (weedy species),

H. petiolaris is the probable ancestor of cultivated species,

Cultivated species: H. annuus ssp macrocarpus.

Hybrids: H. annuus x H. petiolaris

$2n = 34$

Sunflower germ plasm is available several nations such as USA, Brazil, Argentina, Russia, China etc. Sunflower germplasm totaling over 1000 accessions derived from different continents are available at Indian Council Agricultural Research Institutes, Directorate of Oilseeds Research, at Hyderabad, India and Agricultural Universities in South India.

Nomenclature

English-*Sunflower*; Italian-Eliotropia; Spanish-*Girasol*; French-*Tournesol*; Czech-Slunecnice; Portuguese-Girasol; Russian-Kallang; Hungarian-Naproforgo; Assamese-*Beliphool*; Bengali-*Surajmukhi*; Oriya-*Surajmukhi*; Hindi-*Surajmukhi*; Punjabi-*Surajmukhi*; Marathi-*Surajphool*; Tamil-*Suryakanthi*; Telugu-*Proddutirugudu puvvu*; Kannada-*Suryakanthi*; Malayalam-*Suryakanthi*

Uses

Sunflower seed is roasted and consumed as a snack. Sunflower can be processed to form butter, just like peanut butter. Sunflower oil is a highly popular cooking medium in South India. Several types of oils are extracted and prepared from sunflower. Some are rich in Oleic acid. Most importantly, sunflower contains mono-unsaturated fatty acids that are relatively healthy. Sunflower butter is common in China, Middle East, Russia and USA, although not a preferred culinary item in South India. However, hydrogenated fat (*Vanaspathi ghee*) prepared from sunflower oil is common in South India.

Sunflower oil is semidrying, therefore finds use in paint and varnish industry. Sulfonated sunflower is used as a disinfectant. Sunflower oil is extensively used in soap and cosmetic industry. Sunflower oil is also used in treating woolen fabrics.

Sunflower seeds and husks available after processing are used to feed farm animals. Sunflower cake is rich in protein of good digestibility. It may contain up to 40% that could serve as feed for cattle and poultry. Sunflower seeds are rich in amino acids, like Tryptophan and Methionine but limiting in Lysine. Sunflower oil is also used as carrier for paints and as bio-diesel. The sunflower cake that remains after extraction of oil from seeds is used to feed live-stock. Sunflower leaves could be fed green or used to prepare silage. Sunflower stem has long bast fibers and could be used to obtain rough fibers. Sunflower is also good ornamental species.

Nutritive Value (100/g seeds)

Energy 570 k cal; Carbohydrates-18.8 g; Dietry fiber-10.5 g; Fat-49.5 g (saturated fatty acids-5.2, monounsaturated fatty acids-9.5 g and polyunsaturated fatty acids-32 g); Protein-22.7 g; Thiamine-2.3 mg; Riboflavin-0.25 mg; Niacin-4.5 mg; Pantothenic acid-6.75 mg; Vitamin B6–0.77 mg; Vitamin B9–227 μg; Vitamin C-1.4 mg; Vitamin E-34.5 mg; Ca-116 mg; Fe-6.7 mg; Mg-354 mg; Mn-2.0 mg; P 705 mg; K-689 mg; Na-3 mg.

Source: http://en.wikepedia.org/wiki/sunflower; [3, 5, 6].

18.2 SOILS, AGROCLIMATE AND WATER REQUIREMENTS OF SUNFLOWER CROPPING ZONES

Sunflower is well adapted to seasonal variations. It is grown during all three seasons namely rainy, post rainy and summer depending on water resources and cropping sequences envisaged by farmers. The rainy season crop is sown early at the onset of rains and harvested after 3–4 months. Post rainy crop sown immediately after previous crop, thrives using stored moisture and/or irrigation. Summer crop is almost always irrigated.

The optimum temperature range for sunflower is 20–25°C. Yet, sunflower tolerates drought and relatively higher temperatures that occur in the semiarid tropics. High humidity and/or rainfall during seed set are detrimental. High humidity, warm temperature coupled with precipitation has unfavorable effect on composition of fats in seeds. Thus, it leads to higher biomass and grain yield. An early sown, post rainy crop recovers greater quantities of soil nutrients still held in sub soil. Hybrids such as KBSH1 produced 43% more grains (1016 kg/ha) compared to late sown crop ([7–10]; Plates 1 and 2).

18.2.1 SOILS

Sunflower is cultivated on Mollisols and Ultisols in the Central and Eastern plains of USA. Oxisols rich in Al and Fe, with P fixation problem and low in organic matter are used for sunflower production in the Cerrados of Brazil. Mollisols, Inceptisols, and Alfisols found in Pampas are excellent to raise a moderately high yielding sunflower crop. In the Peninsular India, sunflower is an important rotation crop on Black Cotton Soils. It can also be raised on slightly alkaline/saline soils. The optimum soil pH for sunflower is 6.5 to 8.0 [7]. Sunflower is actually classifiable as low salt tolerant crop species, based on its ability to grow well alkaline soils. It withstands 2–4 ds/m. Sunflower is rather not suited for acid soils with pH < 5.5. Sunflower is among the best bets in sandy/gravely Red Alfisols of dry lands of South India (Plate 1).

18.2.2 WATER REQUIREMENTS

Water requirement of sunflower crop varies with geographic location and prevailing agroclimate. Major factors like season, soil type, soil moisture status, crop genotype and yield goal influences water requirement. The grain yield will be moderate or even below average if precipitation is about 300 mm. The productivity is said to increase linearly between 200 to 500 mm. A good crop of sunflower is possible with as little as 500 mm water supplied through irrigation and/or precipitation. The crop stand, biomass, nutrient recovery and productivity are excellent if water supply ranges from 500–750 mm per season [7]. If water supply is exclusively through surface irrigation, about 600–750 mm is considered minimum.

Regarding irrigation frequencies, rain fed crops may receive only 2 or 3 irrigations to stave of critical stages. Irrigated crop receives water in regular intervals of 15–20 days. It is matched with crop's needs during critical periods like – seedling, early bud development (30–40 d), flowering (50–65 d) and seed filling (65–95 d). On heavy soils (black cotton) an irrigated crop of sunflower is provided with 4–6 irrigations. Whereas, on light textured Alfisols 10–12 irrigations are scheduled. Intermittent droughts are common in many sunflower belts. Moisture stress that ensues also influences nutrient dynamics in agroecosystem.

18.2.3 CROPPING SYSTEMS

Sunflower is an important oilseed crop in the Great Plains region. It is rotated with wheat, maize or cotton. Intercrop of sunflower with maize is also common in this region. Sunflower is intercropped with cereals such as sorghum, maize and millets depending on location, soil type, precipitation pattern and environmental conditions. Sunflower/sorghum intercrops are most frequent on Mollisols of Pampas. Sunflower based cropping systems encounter competition from soybeans and cotton in the Cerrados. Sunflower-fallow, Sunflower-legume rotations are traced in semiarid regions of Brazil. Sunflower monocrops and intercrops with wheat are preferred in the West Asian dry lands. Rice-sunflower is a preferred rotation on Vertisols and Alfisols plains of South India.

PLATE 1 Sunflower seedlings on Red Alfisols of South Karnataka. Nutrient recovery rates increase from seedling to head formation and until Grain-fill stages.
Note: Fertilizers are surface banded along the rows of seedlings. It improves nutrient recovery and fertilizer-use efficiency.
Source: Krishna, K. R., Bangalore, India.

PLATE 2 A Sunflower crop in South India.
Note: Sunflower is well adapted to Dry lands of North Karnataka, in India. It is preferred as intercrop with Sorghum or Pigeonpea. It is grown in all three seasons hence it gets rotated with other crops in several combinations. Relatively large amount of nutrients are recovered from soil and partitioned to support grain-fill. Sunflower plants absorb higher amounts of P ands that are required for oil formation and accumulation in seeds. Sunflower stover and heads may recycle about 15 to 20% of nutrients held in the crop.
Source: Krishna, K.R., Bangalore, India.

18.3 EXPANSE AND PRODUCTIVITY

Sunflower is a versatile crop. It adapts well to geographic and climatic variations (Fig. 1). The sunflower cropping zones occur from 40°s to 55° N, but the greatest fraction of production occurs between latitudes 20° and 50° N as well as 20° and 40° S. Globally, during past three years, sunflower production improved from 26.3 mt to 29.8 mt annually [11]. The average productivity of sunflower is 1219 **kg**/ha globally, but it is significantly low at 549 **kg**/ha in India. Globally, major expanses of sunflower cropping belts occur in USA, China, India, Russian Federation, Ukraine, Argentina, and Brazil. India harbors a large sunflower agroecosystem. In India, sunflower occupies 2.4 m ha and contributes 1.8 mt grains annually. It ranks fourth in expanse among oil seed crops. The area under sunflower is increasing consistently every year. It is attributed to crop's day neutrality, wide adaptability to soil types and environment, short duration, high yield potential, good quality oil and better economic returns. Sunflower cropping zones in Australia is relatively feeble. It is confined to Queensland and New South Wales. It is gaining ground in the Ord river region of Western Australia. Canola and cottonseed are major competitors for the spread of sunflower in Australia. Yet, sunflower belt extends into 77,500 ha and total production 82,200 t annually.

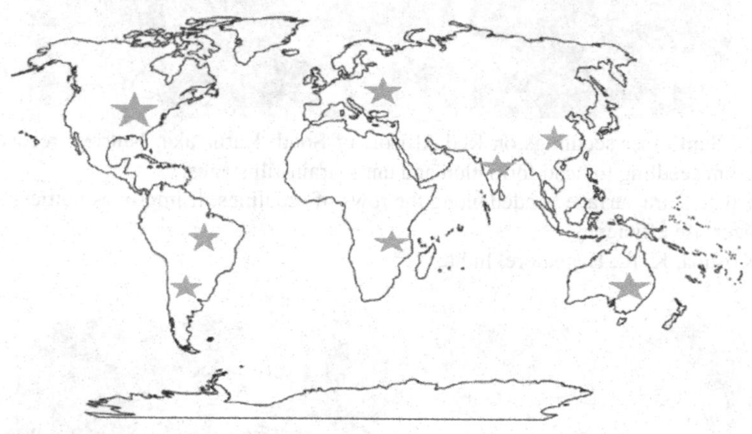

FIGURE 1 Sunflower growing regions of the World.

Note: Green stars indicate regions that support relatively larger cropping zones of Sunflower.

KEYWORDS

- **Alfisols**
- ***Helianthus annuus***
- ***Helianthus petiolaris***
- **Pigeonpea**
- **Vertisols**

REFERENCES

1. Lentz, D.; Robert, B.; and Victor, S. New evidence for sunflower (*Helianthus annuus* L.) origins, Ecological niche model for distribution in Mexico. Botanical Society of America, USA, http://www.2006botanyconference.org/ engine/search/index.php?func=detail&aid=881 **2006,** 1–2 (April 12, 2012).
2. Cockrel, T. D. A. Specific and Varietal characteristics of annual Sunflowers. The American Naturalist **1915,** *49,* 609–622.
3. Riesberg, L. H.; Sunflower Botany. Department of Botany, University of British Columbia. Canada. http://www.botany. ubs.ca/people/riesberg.htm **2008,** 1–4 (May 10, 2012).
4. IKISAN, Sunflower: Seed Varieties. http://www.IKISAN.com/links/ap_sunflower-Seed%20 Varieties.shtml **2008b,** 1–4 (March 20, 2012).
5. Maiti, R. K.; Vidyasagar, P.; Shahapur, S. C.; Singh, Research Trends on food value and chemical composition of sunflower (*Helianthus annuus*) seeds – a review. Research on Crops **2005,** *6,* 179–189.
6. Maiti, R. K.; Singh, V. P.; Purohit, S. S.; Vidyasagar, P. Research Advances in Sunflower (*Helianthus annuus*). Agrobios Publishers, Jodhpur, **2007,** 512.
7. IKISAN, Sunflower: Soils and Climate. http://www.IKISAN.com/links/ ap_sunflowerSoils %20 And%20Climate.shtml **2008d,** 1–5.
8. IKISAN, Sunflower: Crop establishment. http://www.IKISAN.com/links/ap_sunflower-Crop %20Establishment.shtml **2008e,** 7.
9. IKISAN, Sunflower: Nutrient Management. http://www.IKISAN.com/links/ap_sunflower Nutrient%20 Management.shtml **2008f,** 1–8 (August 20, 2012).
10. IKISAN, Sunflower: Hybrid Sunflower. http://www.IKISAN.com/links /ap_sunflowerHy-brid%20Sunflower.shtml **2008g,** 1–6 (May 20, 2012).
11. FAO, Oilseeds, Oils and Oil meals. Food and Agricultural Organization of the United Nations. Rome, Italy, http://www.fao.org/docrep/009/j8126e/j8126e05.htm1–8 (March 10, 2012).
12. IKISAN, Sunflower: History. **2008a** http://www.IKISAN.com 1–7 (March 20, 2012).

EXERCISE

1. Describe origin and spread of Sunflower Agroecosystem.
2. List total production, area and productivity statistics for past five years for Sunflower in Asia.
3. Mention soil types that support Sunflower Production in different continents.

FURTHER READING

Maiti, R. K.; Singh, **V**. P.; Purohit, S.; Vidyasagar, P. Advances in Sunflower. Riddhi International Inc. New Delhi, **2007,** pp. 512.

Schneiter, A. A.; Seiler, G. J.; Bartels, J. H. Sunflower Technology and Production. American Society of Agronomy, Madision, Wisconsin, USA, **1997,** pp. 834.

USEFUL WEBSITES

www.sunflowernsa.com/growers/production-resource-books
http://www.icarda.org
www.sunflowernsa.com/growers/production-resource-books
http://www.jeffersoninstitute.org/pubs/sunflower.shtml

CHAPTER 19

MINOR OILSEEDS

CONTENTS

OLIVES (*OLEA EUROPAEA*)

19.1 INTRODUCTION

Archaeological evidences suggest that earliest of the domesticated olive trees were in vogue in the Eastern Mediterranean region during 3rd millennium B.C. The oldest Olive orchards that date to 3000 B.C. occur in Palestine, Turkey and Coastal Syria [3]. Olive cultivation practices moved to areas in Western Spanish coast during 1st millennium B.C. Olive species spread to Italy and several parts of GrecoRoman farming zones during ancient period around 1000 B.C. [2]. Olives, spread to Arabia, Pakistan and India through trade roots. Chinese farming zones received Olives via silk route. Olive orchards were established south-west of United Sates of America by European settlers during 1870s. Earliest of the orchards were planted in the San Diego and Sonoma region of California. Olives were later established in the Sierra Nevada region [3]. Olive orchards were established during early 1700s by the Franciscan monks who brought olive from Spain. In Argentina, olive cropping belt got initiated during 18th century with the advent of Spanish settlers.

Olive is among most frequently quoted tree in ancient literature. Theophrastus (4th century B.C.), the 'Father of Botany' mentions it as a tree species that lasts for over 200 years. Greeks have used olives as symbols on their coinage dating to 170 A.D. Greeks used olive oil to burn lamps and eternal flames. Israelites used olives during first century A.D. to prepare variety of cuisines, to smear their bodies, during religious functions and as ointment to cure surface ailments. It seems Olive trees and oil has been mentioned at 30 different places in Bible. The 'Olive Plantations' that occurred in the Eastern region of Jerusalem has been most conspicuously mentioned in Ancient Christian literature. It seems Olives have been mentioned seven times in Quran. The Islamic clergy used olive during Ramadan. Its leaves were used to prepare decoctions.

Olive trees last for longer durations. Their longevity has been mentioned in literature and trees with very long active life and fruit bearing traits occur in many places in the Middle East and Southern Europe. Olive trees that are 1600 year old are found in Greece, Montenegro, Istria in Croatia, Bshaale in Lebanon, and in Arraba, Deir Hanna and Gelilee in Israel [1, 2] (Table 1).

TABLE 1 Olives: Origin, Classification, Nomenclature and Uses—A summary.

Olive agroecosystem is predominant in West Asia and Southern European farming zones. Olives were domesticated in the Mediterranean region during 3rd millennium B.C. It spread to Spain and different regions in Africa during 2nd millennium B.C. Olive cultivation spread to regions in Pakistan, India and China during Ancient period. Commercial orchards in California were raised by European settlers during 1870s. Spanish settlers in Argentina initiated olive orchards during 1850s.

Botany and Classification

Kingdom-Plantae; Sub-kingdom-Angiosperms; Order-Lamiales; Family-Oleaceae; Genus-*Olea*; Species-*O. europaea*.

The genus *Olea* has over 35 species that are well distributed into different farming zones of African continent, Mediterranean, Arabia, India, Malaysia and China. *Olea europaea* is the cultivated variety that is most common. *O. chrysophylla* is a wild type and a possible ancestor of *O. europaea.* It is also traced in Central and Eastern Africa, Arabian Peninsula, Iran, Pakistan and even in China. In these regions, it is referred as *O.ferrugina. O.laperrrinei* is found in Southern Algeria.

There are six natural subspecies of *O. europaea* that occur in different olive cropping zones. They are: *O. europaea* sub-sp. *europaea* found in Middle East Asia; *O.europaea* sub-sp. *cuspidate* found in South Africa, Arabia and South-west China; *O.europaea subsp guanchica* found in canaries islands; *O.europaea* sub-sp. *cerasiformis* (tetraploid) found in Madeira, *O.europaea* sub-sp. *maroccama* (hexaploid) found in Morocco; and *O. europaea* sub-sp. *laperrinae* found in Algeria, Sudan and Niger). The global olive cropping zones are actually an assortment of several subspecies of *O.europaea.*

$2n = 46$

Major germ plasm center for *Olea species* is available at World Germ Plasm Bank of the Centre de Investigacion y Formacion Agraria (CIFA) 'Alameda del Obispo' in Cordoba, Spain. There is another World Olive Germplasm collection Centre at Marrakech in Morocco. National Clonal Germ Plasm Repository for Tree fruit and nuts at One Shields Avenue, University of California also holds Olive (*Oleasp*) Germplasm.

Nomenclature

English-Olive; Spanish-Aceituna; Russian-Muslina, Olivka; Swedish-Oliv; French-Olive; German-Olbaum; Greek Elia; Italian-Oliva; Polish-Oliwka; Slovanain-Oljka; Dutch-Olijf; Albanian-Ullir; Latin-Oliva; Portuguese-Azeitona; Georgian-Zetis; Yiddish-Mazline; Bulgarian-Muslina; Romanian-muslin; Arabic-Zaith; Quranic-Al-Zaitun; Turkish-Zeytin; Hebrew-Zaith, Aramic-Zayta; Maltese-Zebbug; Kazakh-Zaytun; Armenian-Jiteni; Kurdish-Zaitun; Sumerian-Zirdum; Parsi-Zetitun; Swahili-Zeituni; Olyf-Afrikans; Indonesian-Zaitun; Malay-Buah Zaitun; Hindi-Jaitoon; Bengali-Jalpai; Assamesse-Jolfai; Saidun-Tamil; Punjabi-Jaitun; Mangolian-Chidun; Chinese-Minchung ganlan.

Uses

There are two main types of Olives based on their commercial use. They are table types (e.g., California Black Olives). There are cultivars used for both oil and table purposes. Olives contain relatively high proportion of cholesterol-free oil. There are three classes: low oil types contain <18% oil; medium 18–22% oil and high 23–33% oil. In West Asia, Europe and some regions of California, olive oil is used as cooking medium. Olive oil is used in extracting perfumes. It is also used as solvent in industries. Olives are used in preparing soaps. Olive oil is used as base for many medicines. Olive wood is used to preparing furniture. Olive leaf residues are used prepare farm yard manures.

Nutritional Aspects

A 100 g of fresh green olive fruits contain:

Calories-145, Fat-15.3 g, Carbohydrates 3.4 g, Fiber-3.3 g, Protein-1.0 g, Cholesterol-nil.

Source: Refs. [3–5]; en.wikepedia.org/wiki/Olive; http://ndb.nal.usda.gov/ndb/ foods / show/2384; FAO Primary Crops Statistical Database. FAO, Rome, Italy.

19.2 SOILS AND AGROCLIMATE

Olive agroecosystem thrives on several different types of soils found in West Asia, Africa, southern USA, Europe and China. In general, olive trees grow well on sandy to clayey soils. They prefer light textured soils that are deep and well drained. Clayey soils are also used to raise olive orchards, but clay content has to be <20%. High water retention in clayey soils could be detrimental to root growth. Olives do not withstand long duration of water logging. In the arid region, dry sandy textured soils may hold little water, but farmers select deep soils so that water held in lower horizons support tree growth. In fact, olive belt in Tunisia thrives on deep, sandy soils that are able to supply enough moisture to trees. Farmers avoid clayey water logging zones. We should note that olive trees produce deep roots that reach 3–5 m into soil. Olive orchards in the hilly tracts of West Asia are grown on soils classified as Eutric Leptosols, and Lithic Xerorthents. These are strongly calcareous, brown to dark brown and light textured. They are slightly alkaline. In Turkey and Syria, olive orchards occur on Luvic Calcisols, Calcic Gypsisols, Calcorthids, and Calcic Xerorthents [3, 6]. In the European region, olive orchards occur on light textured Mollisols, Cambisols, and Alfisols. In California, olive thrives on San Joaquin soils (brown loams) of wide range of nutrient status. It is grown more soils with poor fertility. High soil fertility and excessive N inputs keep the orchards in vegetative phase for longer period. High N can affect fruit quality.

Fertilizer and organic manure amendments are required to achieve better fruit yield. In Southern Italy and Mediterranean region, trees are provided with 1.5–3.0 kg N as urea per tree. Blanket recommendation of fertilizer for olive in most regions is 1,000 g N, 300–600 g P and 500–1000 g K/tree. Fertilizer inputs also depend on inherent soil fertility, precipitation pattern and yield goals. Olive trees that yield 30 kg fruits/tree are given 1.5 kg N. In some orchards, 2.5–3.9 kg P and 0.8 kg K is supplied as basal dose at the time of planting the young saplings. Annual input of P and K together does not exceeded beyond1.0 kg fertilizer/tree.

19.2.1 AGROCLIMATE

Regarding distribution of Olives at different altitudes, it is said that olives grow from sea level up to 700–800 m.a.s.l. on north facing slopes and 900–1000 m.a.s.l. on south-facing slopes. Olives do grow up to 1200 m.a.sl. The productivity depreciates [7].Olive trees tolerate a certain degree of intermittent drought stress in the Mediterranean region. Olives require 200 mm precipitation in the dry regions of Syrian deserts. Olive plantations found in humid areas with normal rainfall pattern need 1300 mm annually. Drought stress depreciates shoot growth and formation of leaves. Olives are known to use moisture from atmosphere through morning dew [3]. Drought spells affect tree growth, foliage formation and fruiting. It affects nutrient acquisition pattern and fruit productivity. Water requirement of olives may vary depending on stage of the tree, its age, soil type, fertilizer-based nutrients supplied and yield goals set for each orchard. Total evapotranspiration ranges from 1700–1935 mm and extra water supplied through irrigation ranges from nil to 44 mm [3]. Olive farmers adopt agronomic measures that enhance water use efficiency. Water harvesting in the hill slopes is a good storage measure. Soil bunds of 15–30 cm high around tree helps in storing water in soil. Reduction in surface flow and infiltration and allowing soils to accumulate water in the lower horizon is beneficial. Stone–mulches reduce water loss via evaporation. Olive mill waste is again a good mulch. Recycling olive leaf and fruit residues help enhancing soil organic matter and water holding capacity of soil. Farmers also adopt deficit irrigation principle. Farmers in drought prone areas and those with low irrigation facilities, find it difficult to meet water requirements of olives. They try to improve water use efficiency using variety of methods including deficit irrigation systems. For example, with certain 'Table types of Olive' grown on clayey soils, supply of 275 mm or 474 mm through irrigation did not alter fruit yield [3]. Surface irrigation is common under traditional production systems. Porous pipelines are also used to supply irrigation water efficiently. Drip irrigation is practiced to suppress loss of water via evaporation, surface runoff and percolation. Olive orchards are situated more closely to sea and coastline in the Mediterranean region. In the arid and semiarid regions of West Asia, water quality especially its salt content and alkalinity are important. Excessive salinity may affect root growth and regeneration during each cycle. Accumulation of Na and B in soil and irrigation water should be monitored carefully. It may affect absorption of major nutrients by trees.

Olive agroecosystem negotiates different temperature regimes depending on geographic location and its characteristics. It adapts to low temperature and frost to a certain extent. Olive tree has a physiological requirement for low temperature (vernalization). Olive orchards in West Asia are usually exposed to 150 to 300 h of 9.0°C that occurs between November and January. In European belts and California, olive may negotiate 1000 h of 7°C to produce flowers. In general, Olive orchards found in Italy are cold tolerant compared to those in other regions. Olive orchards found in Morocco and other regions of north-west Africa are more tolerant to high temperatures say 30–33°C. Ambient temperature of 35°C affects flowering and at 37°C fruit set and maturation are affected [3]. Olive trees withstand low temperature of −7° to −8°C for short duration.

19.3 PRODUCTION SYSTEMS

19.3.1 OLIVE TREE: PHYSIOLOGY

The Olive agroecosystems are permanent orchards of trees of different growth and physiological stages. Olives are slow growing tree species. In West Asia it takes 3–4 years to bear first few fruits. Economically feasible and profitable fruit bearing begins after 8–10 years of growth in the field. High fruit bearing seasons occur usually 18–20 years after planting saplings. The olive trees have long survival and growth period ranging from to 120–150 years. There are stories that suggest that, in Palestine, Syria and Israel a few of the Olive orchards belong to Roman era. They are still in fruit bearing cycles. Hence, those Olive orchards are nicknamed 'Roman.' Olive plantations experience growth cycles as follows:

Fall: Fruit harvest and vegetative growth;

Winter: Dormancy, chilling that results in vernalization and flower bud primordial formation;

Spring: Leaf bud formation, flowering and flush of rapid foliage;
Summer: Early stages of fruit development (June), growth and ripening (August to early fall)

Olives show up excessive flower drop. Fruits are formed out of only 1–5% of flowers that develop and get pollinated. Hormonal imbalance and environmental vagaries cause the flower drop [3]. The productivity of olive tree is mostly governed by alternate-bearing habit and fruit abscission during post -bloom period. Fruit weight ranges from 2–6 g also affects oil extraction.

19.3.2 AGRONOMIC PROCEDURES

The Olive agroecosystem actually differs markedly in terms of production systems adopted. Orchards differ with respect to planting density, soil management, fertilizer and irrigation supply, soil amendments, tree maintenance, harvest practices and residue recycling trends. Obviously, each genotype has its impact on nutrient recovery, translocation and accumulation. The olive cultivar does influence several of the agronomic practices and nutrient recycling trends in the orchards. Vossen [8] identifies at least three different types of Agronomic procedures based on planting density, fertilizer and water inputs. They are, Traditional Production Systems, High Density Production Systems and Super High Density Production Systems.

Traditional Procedures are supposedly inefficient. Planting density is low at 170–180 trees/ha. Currently it is not popular. Chemical fertilizer supply and FYM inputs are relatively low. Traditional system produced very low fruit yield at 1.1 to 4.5 t/ha. The trees attained fruit bearing after a long duration of 15–40 years. Fruits are harvested by hand. The residue recycling methods are inefficient.

High Density Production Procedures are preferred in most modern olive orchards. It involves closer planting. Usually 540–830 trees are planted/ha. Orchards start bearing fruits early between 7–10 years after planting. Mechanical harvesting is practiced.

Super High Density Production Procedures are common in many Southern European regions and Middle East. For example, in Catalonia (Spain) high density planting accommodates 1,700–3,000 trees/ha. Cultivars such as Arbequina, Arbosana and Koroneiki are preferred more in orchards adopting super high-density production systems. Fertilizer and irrigation supply is relatively more under super high-density production systems. The productivity is superlative at 3–17.5 t fruit/ha [8]. Over all, we should appreciate that olive belts scattered in Europe, Mediterranean and Arabia are affected by variations in production systems. The nutrient supply, its recovery by trees, accumulation patterns and recycling within orchards or entire agroecosystem is invariably influenced by production systems.

Some important Olive (*Olea eurpaea*) cultivars and regions where they flourish: Arbequina-Catalonia in Spain; Empeltre found more in Aragon and Balearic island of Spain; Hojiblanca-grown widely in Cordoba, Spain; Manzanilla-common in Seville region of Spain; Picual-a most common and dominant variety in Spain. It is responsible for over 20% of total olive oil produced in the world. Clearly, such cultivars bestow maximum influence on the Olive orchards and affect the agroecosystem most with regard to soil fertility, nutrient dynamics and productivity. Kalamata-common in

Greece; Koroneike-dominant in Peloponese in Greece; Amfissa-Central Greece; Amfissa-Central Greece; Patrinia-common in Aigiareia region of Greece; Frantoio-grown in Tuscany, Italy, Lucques-common in France; Picholine-common in Southern France; Gemlik-adapted to Northern Turkey; Maalot-is traced in Eastern Mediterranean, North Africa and parts of Israel; Nabali-common most regions of Israel; Souri-found more in Lebanon; Barri Zaitoon-common to Indus plains in Pakistan; Mission-Originated in California as black table types, common to South-west USA, [4, 9, 10] (Plate 1).

PLATE 1 Olive Orchards in California.
Source: Olive Growers and Producers, California, USA;
http://www.oliveoilsourcc.com/page/crop-management;
http://www.olives101.com/2007/10/30/iran-olive-groves-have-increased.

19.4 EXPANSE AND PRODUCTIVITY

Worldwide, the expanse of Olive agroecosystem, its productivity and total harvest were as follows:

Global Olive growing regions extended into 9.4 m ha, contributed about 20.6 m ton olive fruits at an average productivity of 2.2 t fruits/ha [11]. Olive orchards are prominent in Spain. The Spanish olive belt extends into 2.01 m ha and contributes 8 m ton fruits annually. Italian olive belt is 1.1 m ha in expanse and offers 3 m t fruits annually at 2.7 t fruits/ha productivity. Countries such as Tunisia (1.64 m/ha), Greece (0.83 m ha), Morocco (0.73 m ha), Turkey (0.83 m ha), Syria (0.63 m ha), Egypt (0.65 m ha), Algeria (0.55 m ha) and Argentina (0.2 m ha) are regions that possess fairly large olive based agroecosystem. Olive orchards extend into other nations such as Jordan, Israel, Libya, Palestine and Lebanon, where orchards spread into less than 0.1 m ha. For example, in Lebanon olive belt spreads into 57,000 ha with over 13 million trees. Nearly 40% of olive orchards occur in northern part of the nation, 40% in south, 15% in Mount Lebanon region and 8% in Bekaa valley [12]. Overall, it is clear that Olive agroecosystem is conspicuous and concentrated in southern Europe, Mediterranean and Arabia region. More than35% of olive orchards found in Lebanon are over 50 years old.

The olive belt in California was only 2000 ha in 1885 and meant mainly to produce ‘Black Table types.’ Currently, the olive belt is over 31,000 ha large [8]. In California, Olive orchards are more conspicuous in counties such as Tulare, King, Fresno, Glenn,

Tehama and Butte. Olive orchards also occur in New Mexico, Arizona and Texas. The recent surge in interest to produce olive oil was supported by its need in perfume industry. Preference for fresh fruit (black table type), oil, and use in perfumes and medicine drive the expansion and productivity of olives in USA. Argentina is a major olive producing nation. The cropping zone extends into 200,000 ha. It exports most of the olive fruits and extracted oil to European nations. The Olive agroecosystem also extends into Peru and Chile.

The European Olive cropping zones are most conspicuous in Spain, but it is also cultivated in Southern France, Portugal, Italy and Greece. Spain has the world's largest Olive Agroecosystem. The cropping zone supports over 215 million trees covering 2.2 m ha [14]. Olive belt in Spain accounts for 30% of total olive growing regions found in the world. Greece has one of the oldest olive cropping regions. Some of the regions are known to have grown olives for over 3,000 years. In Italy, several types of olives are grown. They are grown both for table and oil purposes. Olive orchards are also found in Australia.

The productivity of olive agroecosystem varies depending on several factors related to geographic region, topography, soil type, its fertility, irrigation, olive species and genotype, and agronomic measures. Olive belts in Egypt are best in terms of productivity and offer the farmer 4.7 t fruits/ha. It is followed by olive orchards in Spain at 3.8 t fruits/ha, Jordan at 2.8 t fruits/ha, Italy at 2.7 t fruits/ha, Morocco, Greece each at 2 t fruits/ha. Olives in countries such as Turkey, Syria, Algeria, Lebanon and Portugal offer between 1–2 t fruits/ha. Orchards in Palestine, Libya, and Tunisia are low yielding at 0.5–1.2 t fruits/ha. Over all, Olive cultivation is intense in Southern Europe and most regions of Mediterranean. Here, it is supplied with fertilizers, irrigation and other amendments. Hence they harvest between 2–4 t fruits/ha. The Olive orchards are intensely nurtured in nations such a Syria, Palestine, Lebanon, Tunisia and Libya.

KEYWORDS

- **Eutric Leptosols**
- **GrecoRoman**
- **Lithic Xerorthents**
- ***Oleasp***
- **Theophrastus**

REFERENCES

1. Drinkwater, C. The Olive Route. Weidenfeld and Nicholson, New York, USA, **2006,** 145.
2. Lewington, A.; Parker, E. Ancient Trees. Collins and Brown Ltd. London, **1999,** 110–113.
3. Tubeilah, A.; Brugsewen, A.; Turkelboom, F. Growing Olives and other tree species in Marginal Dry Environment with examples: from the Khanessar Valley, in Syria. International Centre for Agricultural Research in Dry Areas, Aleppo, Syria, **2004,** 1–58.
4. Dell Carter Foods, Lindsay Olives: Varieties. http://www.lindsayolives.com/olives-101/olive-varieties.html **2012,** 1–2 (September 23, 2012).
5. FAOSTAT. FAO Primary Crops Statistical Database. FAO, Rome, Italy. **2003.**

6. FAO/UNESCO, FAO/UNESCO Soil map of the World 1, 5,000,000). FAO, Rome, Italy. **1994.**
7. Pansiot, F.; Rebour, H. Improvement in Olive Cultivation. FAO Agricultural Studies, FAO, Rome, Italy. **1961,** 249 p.
8. Vossen, Olive oil: History, Production and Characteristics of the World's Classic Oils. HortScience **2007,** *42,* 1093–1100.
9. Fotiadi, E. Unusual olives. Epikoura magazine. Quoted In: Olive. En.wikipedia.org.wild/olive.htm **2006,** 1–15 (September 24, 2012).
10. Belaj, Z.; Satovic, I.; Rallo, L.; Trujillo, L. Genetic diversity and relationships in Olive. Germplasm collections as determined by RFLP. Theoretical and Applied Genetics **2002,** *105,* 638–645.
11. FAOSTAT, Oilseed Statistics. Food and Agricultural Organization of the United Nations. Rome, Italy. FAOSTAT.org **2010** (November 12, 2012).
12. Ministry of Agriculture of Lebanon, Overview of Lebanese Olive Oil Industry. http://www.lebaneseoliveoil.com/ Overveiw.htm **2012,** 1–3 (September 22, 2012).
13. De Candolle, A. Origine des Plantes Cultivées. Laffitte Reprints, Paris, France. **1883** [In French.].

EXERCISE

1. Mention names of Domesticated and Wild type of Olives. Mark the Olive growing regions of the World.
2. List the current trends in Global Olive growing area, production and productivity.
3. Discuss the three different types of Olive production systems-with reference to planting density, number of trees, production per tree and productivity per ha.
4. Mention the various use of Olive Oil.

FURTHER READING

Fairbanks, D.; Hess, W. M.; Welch, J. W.; Driggs, J. K. Botanical Aspects of Olive culture Relevant to Jacob 5. Maxwell Institute for Religious Scholarship. Brigham Young University, Utah, **2012,** pp. 108.

Ferguson, L.; Sibbett, G. S.; Martin, G. C. Olive Production Manual. University of California, Division of Agriculture and Natural Resources, Oakland, CA, USA. **1994,** 3353.

Gupta, S. K. Technological Innovations in Major World Oil Crops. Volume-1. Springer, Heidelberg, Germany, **2012,** pp. 267.

USEFUL WEBSITES

www.internationaloliveoil.org
www.icarda.org
http://www.fao.org/scripts/olivo/query/olcoll2.idc
http://www.ncbi.nlm.nih.gov/pmc/art
http://www.olive-bar.com/site/1263206/page/612827
http://www.oleadb.eu/

NIGER (*GUIZOTIA ABYSSINICA*)

19.1 INTRODUCTION

Niger is an oil seed crop native to Ethiopia and adjoining North-east African agrarian area. It seems Niger was domesticated in the Ethiopian highlands. The domesticated species is supposedly derived from wild relative known as *Guizotia scabra*. The primary center for genetic diversity for *Guizotia* species lies in the region covered by Ethiopian Highlands and Rift Valley in Kenya. Cultivation of Niger spread to Indian plains, Bangladesh, Burma and other South-east Asian agrarian regions during 2nd millennium B.C. [1, 2]. Niger was also cultivated by farmers in Central and Western Africa during ancient period, but its acceptability has subsided. Currently, Niger cultivation is confined to Ethiopia and Indian Subcontinent (Table 1).

TABLE 1 Niger: Origin, Classification, Nomenclature and Uses—A summary.

Niger was domesticated by native tribes of Ethiopian Highlands during 2nd millennium B.C. The area of maximum genetic diversity occurs in North-east Africa. Its cultivation zone spread to other parts of African continent and to Indian subcontinent during ancient era. At present its cultivation is conspicuous in Ethiopia and some regions of India.

Classification

Kingdom-Plantae, Sub-Kingdom-Angiosperms, Order-Asterales, Family Asteraceae, Genus-*Guizotia*, Species *G.abyssinica*

Synonyms: *Guizotia oleifera, Heliopsis platyglossa, Jaegarai abyssynica, Parthenium lutcum, Polymnia abysinica, Polymnia frondosa, Ramtilla oleifera, Verbesina sativa,* and *Veslingia scabra* [1].

Wild Species: *Guizotia scabra* subsp *schimperi*

$2n = 30$

Nomenclature

English-Niger; Spanish-Negrillo; French-Noug, Grains du Niger; German-Nigersaat; Swedish-Nigerfro; Hindi-Ramtil; Kannada-Hucchellu; Marathi-Karde; Tamil-Payellu; Bengali-Sarguza;

Uses

Niger oil is used as cooking medium. The keeping quality of Niger oil is poor due to high content of unsaturated fatty acids. Niger oil is used to prepare soaps and lubricants. Niger oil is used in perfume industry, mainly to extract fragrance of flowers. Niger cakes derived from seed chaff after extraction of oil is used to feed farm animals. Niger meal has 30% protein and 17% crude fiber. Niger residues are also used as green manure to enhance organic matter content of soil. Niger oil is used to cure cough. Niger sprouts mixed with garlic and tej is a concoction that is supposedly good to suppress cough.

Nutritive Value

Niger seed contains 40.06%-oil, 23.3% crude protein, 10.3 % crude fiber, 22.4% carbohydrates, 4.3% ash, 0.4% calcium and 0.4% phosphate. Biochemical composition of Niger oil is fallows: Oleic acid 30–40%, 48–55% Linoleic acid, 5–8% Palmitic acid, 3–5% Stearic acid, 0.3–0.5 Arachidic acid and 1.0%; Llinolenic acid [3].

Source: Refs. [2, 4, 5].

19.2 SOILS, AGROCLIMATE AND CROPPING SYSTEMS

Niger is cultivated on wide range of soil types. It tolerates low soil fertility and moisture paucity to a certain extent, but yields low. Niger belt in Ethiopia thrives on clayey loams. Here, Niger (noug) is grown extensively in the Highlands on dark-brown clays that occur around Begemdir region. In the Gojam region, red-brown clay is used. Loamy soils are used for Niger production in the areas surrounding Addis Ababa [1, 4]. Soil pH preferred is 5.2–7.3. Niger needs deep soils with good drainage, yet it tolerates water logging. Niger is tolerant to low oxygen tensions commonly encountered in poorly drained water logged soils. Niger also tolerates salinity, but flowering and crop duration gets extended. Niger tolerates drought stress better than other oil seed crops. In India and adjoining regions, Niger is cultivated on clayey soils with poor water resources. In Madhya Pradesh and Southern Plains it is cultivated on black clays loams (Vertisols) and loamy or clayey red soil (Red Alfisol). In Bangladesh and Burma, Niger is cultivated on fine textured loamy soils and silty Alluvial soils. Soils or fields selected for production of Niger may often be poor in fertility. Hence, fertilizer-based nutrient replenishments are required. Nutrient supply depends on soil tests for major nutrients, organic matter and yield goals. In India, farmers apply 50 kg N, 10–40 kg P/ha. High fertilizer-N input depreciates oil content of seeds.

Niger belt in Ethiopia is confined to cooler regions. The Ethiopian genotypes flower best at an average temperature of 18°C. Temperatures above 23°C may delay flowering. Niger is a day neutral plant and needs 12 h day light during growth and till maturity. The Indian genotypes prefer relatively short day pattern. Over all, Niger puts forth good growth and seeds at temperatures ranging from 18°C–23°C. In the Indian plains, Niger crops are sown during 2nd or 3rd week of August when the soil is moist. Niger crop stays in the field for relatively longer duration. It is harvested after 130–150 days. The Ethiopian belt is usually sown immediately after the onset of rains during 1st or 2nd July. Early sowing allows better precipitation use efficiency. Sowing date affects seed yield [6–8].

Niger growing regions in Ethiopia thrives at relatively higher altitudes compared with other crops. It grows at altitudes ranging from 1600 to 2980 m.a.sl. In India, Bangladesh and Burma, it grows on plains, undulated topography and in delta region. It thrives from sea level to 1500 m.a.s.l. altitude. Niger is well adapted to semiarid regions with 700–900 mm precipitation annually. Productivity is higher, when it is cultivated in regions with 1200–1500 mm precipitation annually. In regions with precipitation higher than 2200 mm, flowering and seed formation may be affected. Vegetative phase is not affected much by high precipitation.

Niger adapts to wide range of cropping systems. It is cultivated as a mono crop or in combination with other species. In the Gangetic belt and Madhya Pradesh in India, it is intercropped with legumes such as pigeonpea, soybean, groundnut or horsegram. It is also intercropped with finger millet, sorghum or maize. The productivity of Niger is 150–200 kg/ha, when it is intercropped and 300–450 kg seed/ha, if grown as mono crop (Plate 1).

PLATE 2 A Mono-crop of Niger grown on Red Alfisol near Bangalore, India.
Source: GKVK Agricultural Experiment Station, Bangalore, India.

19.3 EXPANSE AND PRODUCTIVITY

As stated earlier, Ethiopian Niger belt is confined to highlands and adjoining zones. Niger is an important oilseed species in Ethiopia. It constitutes about 50% of total oilseed region in that nation. Annually, Niger agroecosystem in Ethiopia extends into 105,000/ha and contributes 200,000 ton at an average productivity of 520 kg seeds/ha [9].

At present, Niger agroecosystem spreads into areas in Ethiopia, India, Bangladesh, Burma and to a small extent in West Indies and United States of America [4]. Niger cropping belt is relatively large in India and Ethiopia. In India, it extends into 437,000 ha, contributes 111,000 ton seeds at an average productivity of 253 kg seed/ha [9]. The productivity of Niger depends on inherent soil fertility, irrigation resources, and season and yield goals. It ranges from 150 kg/ha to 645 kg/ha in the Indian plains. The Indian Niger agroecosystem is conspicuous in states such as Orissa, Madhya Pradesh, Chhattisgarh, and Jharkhand. It is also grown in Andhra Pradesh, Assam, West Bengal and Karnataka. Niger cropping belt in India is dominated by cultivars such as Birsa Niger, KRN-1, Sahyadri, and Ootacamund.

Niger cropping zones in Bangladesh appear mostly during the post rainy (rabi) season. It is grown both as a monocrop or mixed with legumes as intercrops. Niger cropping zones in wet tropics of Bangladesh are conspicuous in Brahmambaria, Magura, Jamalpur and Gopalgunj. The productivity of the crop is relatively high at 1050 kg seeds/ha. Seeds are rich in oil continent at 30–50% [5].

KEYWORDS

- **groundnut**
- ***Guizotia scabra***
- **horsegram**
- **pigeonpea**
- **rabi**
- **soybean**

REFERENCES

1. Getinet, A.; Sharma, A. *Guizotia abysinica.*ecoport.org/ep **2003,** 1–3 (August 25, 2012).
2. Hiremath, S. C.; Murthy, H. N. Domestication of Niger (Guizotia abyssinica). Euphytica **1988,** *37,* 225–228.
3. Weiss, E. A. Oilseed Crops. Longman PublishersInc.; Essex, England, **1983,** 428.
4. Duke, J. M. *Guizotia abyssinica.*In: Hand Book of Energy Crops.http://www.hort.purdue.edu/newcrop/duke_energy/Guizotia_abyssinia.html **1983,** (August 25, 2012).
5. Banglapedia, Niger. http://www.banglapedia.org/httpdocs/HT?N_0175.htm **2006,** (August 25, 2006).
6. Ahlawat, I. S. Niger (*Guizotia abyssinica*). Division of Agronomy, Indian Agricultural Research Institute, New Delhi. Internal Report **2010,** 1–34.
7. Mohan Kumar, B. N.; Basavegowda, B. S.; Vyakarnahal, K.; Kenchangoudar, Influence of sowing dates on production of seed yield in Niger (*Guizotia abyssinica*). Karnataka Journal of Agricultural Sciences **2011,** *24,* 232–235.
8. Patil, H. S.; Dhadge, S. M.; Bodake, S. Effect of Macro and Micronutrients and organic manure on yield attributes, seed yield of Niger (*Guizotia abyssinica*) under rainfed conditions in Western Ghats zone of Maharashtra State. Agricultural Science Digest **2006,** *36,* 233–234.
9. FAOSTAT, Niger statistics. Food and Agricultural Organization of the United Nations. Rome, Italy http://www. faostat.org **2006,** (August, 23, 2012).

EXERCISE

1. Give the names of wild type and cultivated Niger.
2. Mention the top three Niger producing Nations with regard to Total production, Area and Productivity.
3. Mention different cropping systems that include Niger.
4. What is the potential yield of Niger? Mentions soil types that support Niger production in different continents.
5. Discuss nutritive value of Niger.

FURTHER READING

1. Getinet, A.; Sharma, S. M. Niger: *Guizotia abyssinica.* Biodiversity International, New York, **1996,** pp. 69.
2. Surhone, L. M.; Tennoe, M. T.; Hensonnow, S. F. Guizotia abyssinica. Betascript Publishing Co.: Great Britain, **2010,** pp. 327.

USEFUL WEBSITES

http://www.ars-grin.gov/cgi-bin/npgs/html/taxon.pl?18068 (December 2, 2012)
http://www.pfaf.org/user/Plant.aspx?LatinName=Guizotia+abyssinica (December 10, 2012)
www.hort.purdue.edu/newcrop/…/Guizotia_abyssinica.html (December 10, 2012)
http://www3.botany.ubc.ca/noug/ (December 12, 2012)

SESAMUM (*SESAMUM INDICUM*)

19.1 INTRODUCTION

Sesamum is an important oil seed crop in parts of Africa, Arabia and Indian subcontinent. It's cropping zones extend into South America. Botanical studies and archaeological analysis of prehistoric sites in different nations indicate that Sesamum originated in Africa. The primary center of origin and region of domestication is in North Africa. Seeds for sesamum agroeocsystem were actually sown sometime 4000 years ago in the African continent. It seems Sesamum spread in to Mesopotamian region during 2nd millennium B.C. During Neolithic age, sesamum crop was distributed to different parts of Mediterranean from Iraq. Ancient Egyptians grew sesamum as early as 1300 B.C. It seems sesamum seeds were one of the items traced in Tutankhamen's tomb [1]. Sesamum became an important oilseed crop in Babylon and Assyria [2]. Later, during second century B.C., sesamum became popular in many Asian regions including vast regions of China.

TABLE 1 Sesamum: Origin, Classification, Nomenclature and Uses—A summary.

Sesamum originated in Africa. It was domesticated during 2nd millennium B.C. Sesamum was cultivated as an oilseed crop by Ancient Babylonians and Assyrians. There are a few centers of genetic diversity from sesamum. They are Africa, Middle East, Indian subcontinent and China. The wild and closely related species occur in several cropping zones Africa and Indian Subcontinent [2, 3]. Currently, sesamum agroecosystem is relatively conspicuous in South America, Africa, Arabia and South Asia.

Classification

Kingdom–Plantae; Order– Lamiales; Family– Pedaliaceae; subfamily Asternae; Genus– *Sesamum*; Species– *S. indicum; S.alatum, S. radiatum*

Synonym: *Sesamum orientale*; related genus – *Ceratotheca.*

There are over 35 species of Sesamum traced in Tropical and Subtropical Africa.Sesamum germplasm is available at ICAR Research Centre, Jabalpur in Madhya Pradesh, India. It holds over 9300 accessions from different parts of the world. *S. indicum* is the most commonly cultivated species worldwide. There are six other species cultivated in different agrarian regions. They are:*S. radiatum* is grown in India, Africa and Sri Lanka; *S. angustifolium* in Congo, Mozambique and Uganda; *S.occidentale* in Africa, Sri Lanka and India; *S.calycinum* in Angola and Mozambique; *S.bauymi* in Angola. The agroecosystem is affected according to the species that dominates and its influence on soils fertility and productivity. There are two major types of sesamum namely, black and white seeded.

$2n = 26$ for *S. indicum, S. alatum, S.mulayunum*

$2n = 32$ for *S. prostratum, S. lacianiatum, S.angustifolium*

$2n = 64$ for *S. radiatum, S. occidentale*

Nomenclature

Sesame, Gingelly-English; Susam-Bulgarian; Sesam-Catalan; Sezam-Czech; Sesam-Dutch; Kunzut-Estonian; Sesam-German; Szezemfu-Hungarian; Sesemo-Italian; Kunjit-Kazakh;

Gergelim-Portuguese; Kuzut-Russian; Ajonjoli-Spanish; Julijilan, Simsim-Arabic; Shooshma-Armenian; Kuncut-Azerrbaijan; Shumshum-Hebrew; Bijlan-Malayan; Goma-Japanese; Wijen-Indonesian; Dee la-Thai; Cay Vung-Vietnamese; Cham Kkae-Korean; Til-Hindi; Yellu-Kannada;Telugu-Nuvulu-Telugu; Ellu-Tamil; Chitellu-Malayalam; Rashi-Oriyan; Til-Bengali; Ashaditil-Marathi.

Uses

Sesamum is used as an oilseed crop. Its seeds contain between 32–50% fat and 22% protein. Seeds of some cultivars contain 50–63% fat. Sesamum oil contains 47% oleic acid and 39% linoleic acid. Sesamum oil is used as cooking medium. Sesamum oil has relatively longer shelf life among different vegetable oils. Sesamum oil is used to prepare some of the best quality margarines. Sesamum is used by Arabians to prepare 'Tahini' a kind of butter. Sesamum based products serve as appetizers in Arabic households. It is usually mixed with chickpeas, garlic and lemon and consumed. In USA, sesamum is used in burgers, pies, bun and bread. 'Open Sesame Pie' is a popular preparation in North America. Sesamum oil is an important cooking medium in the Indian subcontinent. It is also used in many types of dishes to enhance fat content and taste. Sesamum oil is rich in antioxidants. Sesamol is an important antioxidant. Its oil is used to prepare soaps, lubricant oils and a range of pharmaceuticals. Sesamum residue available after oil extraction is used to prepare sesamum meal or cakes that are of excellent value as animal feed. Sesamum oil is used in many types of medicines. It is used to prepare Iodinal and Brominol that are employed to treat surface skin conditions. Sesamum oil is used in preparing several types of ointments and plasters.

Nutritive Value (100/g seeds)

Energy-567 k cal; carbohydrates-26.04 g; Fat-48 g; Protein 17 g; Tryptophan-0.4 g; Threonine-0.7 g; Isoleucine-0.7 g; Leucine-1.2 g; Lysine-0.5 g; Methionine-0.6 g; Cystine-0.3 g; Phenylalanine-0.9 g; Tyrosine-0.7 g; Valine-0.9 g; Arginie-2.5 g; Histidine-0.5 g; Alanine-0.9 g; Aspartic acid-1.6 g; Glutamic acid-3.7 g; Glycine-1.2 g; Proline-0.7 g; Serine-0.9 g; Water 5 g; Vitamin C nil; Ca-131 mg; Fe-7.78 mg; P-774 mg; K-406 mg.

Source: Refs. [2, 4, 5, 6]; http://www.ecoplanet.com/Herbsandplants/Sesamum%20Indicum.htm (August 31, 2012); http://en.wikipedia.org/wiki/Sesamum (September 23, 2012); http://www.cultivator.in/sesame.htm (September 23, 2012).

19.2 SOILS, AGROCLIMATE AND CROPPING SYSTEMS

Sesamum is a versatile crop and adapts to soil types encountered in different continents. It is grown on Calcareous Xerasols in Turkey and Syria, on Inceptisols, Alfisols and Vertisols in the Indian subcontinent and on Entisols in China. Sesamum produces a well-branched extensive root system to absorb moisture and nutrients. It needs deep soils with good tilth and drainage. Fertile soils with medium texture are best suited. Sesamum adapts to neutral pH, but withstands a certain degree of acidity. It is susceptible to alkaline soils and flooding. Sesamum needs fertile soil for optimum pod yield. Fertilizer based nutrient supply is in vogue in most sesamum growing belts. The nutrient supply is dependent on soil tests and grain yield expectations. Fertilizer inputs also depend on soil organic matter continent. Soils with 2–5% SOM are best suited for crops intended to yield high. On an average, 50 kg N, 25 kg P and 25 K/ha is supplied crops in grown on medium textured soils found in dry land regions of Asia. Micronutrient such as Zn is supplied once in 3 years @ 25 kg Zn/ha. Liming is essential, if the acidity greater than 5.6 [4]. Nutrient supply could be split into starter-N and splits could be applied at seedling and flowering stage. In the South-west region of Nigeria, sesamum is provided with 75 kg N, 45 kg P and 22.5 kg K/ha.

Sesamum adapts to different cropping regions of the world. Sesamum cultivation proceeds in the plains at seas level, in the undulated regions and hills at 1,200 m.a.s.l. It needs a minimum of 90–120 days to mature and offer seed yield. In India and China, sesamum is grown in rainy, post rainy and even during summer provided irrigation source is assured. The rainy season crop is sown early at the onset of rains during June/July. In Nigeria, sesame is sown during early July. The sesame belt is confined to guinea-savannah regions with 1000 mm annual precipitation and lies between 6° and 10° N latitude. It is confined to tropical regions of South-west Nigeria [7]. The optimum growing temperature for sesamum is about 22–28°C. At temperatures below 15°C its growth slows down. Sesamum is a photoperiod sensitive crop. Long day conditions are known enhance seed oil content. Sesamum is a drought tolerant species. It tolerates intermittent drought spells in the dry regions of Middle East and Asia. The precipitation pattern and quantity decides the expanse and productivity of sesamum belts. Sesamum needs 500–650 mm water per growing season, although crop thrives and produces seeds with just 400 mm water per season. Drought spells and paucity of water reduce its yields. Hence, assured irrigation during critical stages such as flowering and pod fill is necessary. Rainfall pattern affects growth, duration and pod formation. Late season rains are detrimental, since seed loss due to shattering can be severe.

The major expanses of sesamum found in India and China are monocrops. It is also grown as intercrop with cereals and legumes during rainy season. In the Indian subcontinent, common crop mixtures that include sesamum are sesamum/red gram, sesamum/green gram, sesamum/pearl millet, sesamum/groundnut, sesamum/cowpea. Sesamum relay crops with short season cereals such as finger millet or sorghum is common in Southern India plains. In China, sesamum monocrops and sesamum with cereals such wheat or maize is common. In Southern USA, sesamum is mostly a monocrop. However, in parts of Texas plains, sesamum/sorghum intercrops are possible. Sesamum is intercropped with cereals such as oats or barley in West Asia. It is also grown as monocrop depending rainfall pattern.

19.3 EXPANSE AND PRODUCTIVITY

Currently, sesamum agroecosystem extends into Mexico, Texas and Venezuela in Americas; Turkey, Iran, Egypt, Ethiopia and Syria in Middle East; India, China, Burma, Pakistan and Sri Lanka in Asia; and Nigeria, Uganda, Sudan and Ethiopia in Africa. Sesamum is also cultivated in small areas in Turkmenistan, Uzbekistan, Azerbaijan, and Krasnodar region of Russia. Globally, sesamum cropping belt extended into 7.43 m ha and produced 3.28 m t seeds during2005 [8]. The average productivity of sesamum agroecosystem in the world is 441 kg seeds/ha. Sesamum seeds were originally shattering type and fruits (capsule) would open showing seeds and spilling them on ground. However, since mid-1900s, cultivars with indehiscent fruits are grown. They possess high seed yield potential. Currently, sesamum belts that occur are filled with either shattering or non-shattering genotypes. Both types do occur in different regions in a cropping zone, depending on farmers' preferences.

Sesamum seed yield in South Asian dry lands range from 250 to 550 kg/ha [4]. Post rainy season crop grown with assured irrigation and high fertilizer inputs may off 750 to 900 kg seed yield/ha. The Asian sesamum cropping belt is mostly filled with genotypes selected for high seed yield, early maturity and resistance to *Phytophthora* disease. The genotypes that dominate the cropping belt such as Improved-Sel, Madhavi, Gujarat-till, TMV-3, Patan 64, Pratap, and TMV-4 yield between 3–400 kg seeds/ha under dry land conditions prevalent in India. Sesamum agroecosytem is largest in India and spreads into 1.85 m ha. It contributes 0.7 m t seeds at an average productivity of 368 kg/ha. The Chinese sesamum belt is relatively small at 0.65 m ha, but contributes over 0.8 m t seeds annually. The high production is easily attributed to better productivity at 1100 kg seed/ha. The sesamum belt in India is relatively intense in the Western, Central and Northern regions of the country. Major sesamum producing zones occur in Gujarat, Rajastan, Uttar Pradesh, Madhya Pradesh, and Maharashtra. In Punjab, sesamum belts are preferred less due to dominance of rice-wheat cropping system [9]. Sesamum cropping zones are also well spread out in the Southern Indian Plains, mainly in the states Andhra Pradesh, Tamil Nadu and Karnataka. West Bengal in the Eastern Gangetic plain has a large belt of 163,000 ha [10]

The Nigerian sesamum belt extends into 0.16 m ha and contributes 0.08 m t seeds annually at an average productivity of 450 kg seeds/ha [11]. In Ghana, sesamum seed yield ranges from 383–688 kg seeds/ha for Asian varieties. However, grain yield potential is high and it ranges from 1100 to1170 kg seeds/ha in some locations within Ghana [12] (Plate 1).

PLATE 3 Left: Sesamum crop in South India; Right: Small pods that mature usually dry and shatter to show up seeds.

The commercial production of sesamum began during mid-1950s in the Southern plains of USA. Sesamum culture occurs in small area in different states of USA. In southern USA, especially in Texas, sesamum yield ranges from 700–1700 kg seed/ha for shattering types and 860–1900 kg seeds/ha for nonshattering types. Both shattering and nonshattering genotypes occur in the cropping belt. Most of these varieties are medium in crop duration and mature in 120 days. Prominent shattering cultivars are Blanco and Margo; nonshattering types are Baco and Paloma [2].

At present total production and productivity of top sesame producing countries are:

	Burma	India	China	Ethiopia	Sudan
Total Production (m t/yr):	0.72	0.62	0.59	0.31	0.25
Productivity (t/ha):	0.46	0.34	1.22	0.99	0.19

Source: en.wikipedia.org/wik/Sesame

KEYWORDS

- **Open Sesame Pie**
- ***Phytophthora***
- **sesamum**
- **Tahini**
- **Tutankhamen's tomb**

REFERENCES

1. CNP, Sesamum (*Sesamum indicum*). Cultivator Natural Products. Jodhpur, India, **2011,** 1–3 http://www.cultivator.in/sesame.htm (September 23, 2012).
2. Oplinger, E. S.; Putnam, D. H.; Kaminski, A. R.; Hanson, C. V.; Oelke, E. A.; Schulte, E. E.; Doll, J. D. Sesame. Alternative Field Crops Manual. University of Wisconsin-Extension Services. http://www.hort.purdue.edu/ newcrop/afcm/sesame.html **1997,** 1–6 (September 23, 2012).
3. Josh, A. B. Sesamum. Indian Central Oilseed Committee. Hyderabad, India Internal report. **1961,** 109.
4. AICORPO, All India Co-ordinated Project on Oilseed crops, Directorate of Oilseed Research, Hyderabad, India Research Highlights-25 years. **1992,** 39–43.
5. Aurora, R. K.; Riley, K. W. Sesame Biodiversity in Asia. Conservation, Evaluation and Use. International Plant Genetic Resources Institute-South Asia Region. New Delhi, **1994,** 1–88.
6. Gangaiah, B. Sesame-Kharif Crops. Internal Report of Division of Agronomy, Indian Agricultural Research Institute, New Delhi, **2010,** 1–142.
7. Olowe, I. O. Optimum planting date for sesame in the transition zone of South-west Nigeria. Agricultura Tropica et Subtropica **2007,** *40,* 156–164.
8. FAO, Oil Seed Statistics. Food and Agricultural Organization of the United Nations, Rome, Italy. FAOSTAT.org **2006,** (October 12, 2012).
9. Grover, D. K.; Singh, J. M. Sesamum cultivation in Punjab: Status, Potential and Constraints. Agricultural Economics Research Review **2007,** *20,* 299–313.
10. Damodaran, T.; Hegde, D. M. Oilseeds situation. A statistical compendium. Directorate of Oilseeds Research, Hyderabad, **2005,** 65–75.

11. FAOSTAT, Sesame Production Statistics. Food Agricultural Organization of the United Nations. Rome, Italy. **2005.**
12. Ofosuhene, H.; Yeboah-Badu, I. I. Evaluation of Sesame (Sesamum indicum) production in Ghana. Journal of Animal and Plant Science **2010,** *6,* 653–662.

EXERCISE

1. Write a short note on domestication and spread of Sesamum.
2. Give the Nutritive value of Sesamum oil.
3. Write about current status Sesamum expanses and productivity in Asia.

FURTHER READING

1. Bedigian, D. Sesame: The Genus Sesamum. CRC Press, Boca Raton, Florida, USA, **2010,** pp. 432.
2. Joshi, A. B. Sesamum. Indian Central Oilseeds Committee, Hyderabad, India, **1961,** pp. 109.
3. Sesame and Safflower New Letter, Institute of Sustainable Agriculture, CSIC, Apartado 4084, Cordoba, Spain.
4. Shehu, H. E.; Kwari, J. D.; Sandabe, M. K. Nitrogen, Phosphorus and Potassium in Mubi, *Nigeria. Res. J. Agronomy*, **2009**, *3*, 32–36.
5. Weiss, E. H. Sesame. In: *Oil seed crops*. Longman: New York, **1983,** 282–340.
6. Weiss, E. A.; De la Cruz, Q. D. In: *Sesamum orientale*. Van Der Vossen, H. A. M.; Umali, B. E. (Eds.) Plant Resources of South-East Asia. Backhuys Publishers Inc.: Leiden, Netherlands, **2001,** 123–128.

USEFUL WEBSITES

http://www.hort.purdue.edu/newcrop/afcm/sesame.html pp. 1–6 (September 23, 2012).
http://en.wikipedia.org/wiki/Sesamum (September 23, 2012).
http://agricoop.nic.in/tmop&m/RVO_DOR121010.pdf (September 23, 2012).
http://www.preservearticles.com/2012020322591/complete-information-on-area-andproduction.htm pp. 1–3 (August 31, 2012).
http://www.jeffersoninstitute.org/pubs/sesame.shtml (August 31, 2012).

CHAPTER 20

COTTON CROPPING ZONES

CONTENTS

20.1 INTRODUCTION

Cotton cultivation began in the Indus valley around 3000 B.C. Cotton crop and its products have been mentioned in the *Rig-Veda* and other ancient Indian texts of 1500 B.C. It seems cotton cultivation spread to many regions within the Indian subcontinent during the period between 2000–1000 B.C. [1]. Egyptians cultivated cotton in the Nile valley during 3rd millennium B.C. Cotton was independently domesticated and grown by natives in the New World. Cotton fabrics discovered in the caves in Tehuacan in Mexico offer evidence indicating that cotton crop was cultivated in this region during 5800 B.C. Several archeological evidences suggest that cotton cultivation occurred in Mexico during 3rd millennium B.C. Ancient Peruvians cultivated cotton in the Coastal regions, North Chaco and Nazca. It was mostly constituted by *G. barbedense*. Persian region supported cotton cultivation during fifth century B.C. It seems Greeks and Arabs initiated cotton production during the period of Alexander the Great (323 B.C.). Cotton cropping zones spread to Asia Minor and Europe from Gangetic belt.

Cotton species were also transshipped by Spaniards during their voyages to new World. The seeds for a cotton belt in Florida were sown by Spaniards around 1550 A.D. During early 1600s, cotton cultivation had been initiated by English colonists. They had sown cotton along the James River in Virginia. Cotton production in Alabama was well entrenched by late 1700s. One of the earliest to plant cotton, perhaps initiate a cotton belt in Alabama was a gentleman named Joseph Collins, a surveyor for the Spaniards during 1772 [2]. Cotton was imported from West Indies and grown on sprawling farms of Georgia around 1785 [3]. English workers migrated to USA, and initiated cotton mills. Inventions of machines such as cotton gin by Eli Whitney in 1793, stripper machines and cotton picker vastly improved efficiency of cotton production. It induced expansion of cotton cropping belts in the Southern Plains. This is an example for an agroecosystem being expanded due to development of machinery and devices useful during cultivation and processing of a crop. The cotton belts in Africa were created during early 1900s. Commercial cotton farming in Uganda, Tanzania, Kenya and other regions were initiated by British and German settlers during early 1900s [4]. A cotton agroecosystem took shape rather permanently in Southern Africa by early 1900 s. Cotton was transshipped into Australia with the arrival of British colonists during later part of 17th century. Since, then the cotton belt in Australia has grown into a flourishing enterprise.

The cotton cultivation zones expanded rapidly during medieval period owing to its introduction into many regions of the world. Europeans began importing cotton from the Indian subcontinent during medieval times. Gradually, with the British conquest of India and adjoining areas, rules were framed to encourage only new cotton. Cotton was grown in the plains of Southern India but exported to British Isles for processing. The demand for cotton and textiles increased markedly during mid-1800s. Hence, New World Cotton species namely *G. hirsutum* and *G. barbedense* were encouraged (Table 1). King Cotton became an important export item for Americans. They traded it with Europeans. Cotton cultivation in Southern Plains of USA, decreased perceptibly during civil war. During this period, cotton cultivation in Egypt became highly profitable, and hence it was expanded. During post civil war period, emancipation induced cotton cultivation to spread into northern states. During recent years, mechanization and electronic controls on cotton production and processing has induced its spread into many agrarian regions in Asia and Africa, in addition to America. The genetically modified BT cotton has been replacing the erstwhile high yielders rapidly, in most regions of the world. It is said that BT cotton requires almost half the quantity of pesticide compared to conventional genotypes. Cultivation of BT cotton does not seem to affect regular need for soil fertility, fertilizer based nutrients and farmyard manure. The productivity is well stabilized. During recent years, many farmers in America and Asia have changed over to organic farming. This step allows them to use organic matter in larger quantity and avoids inorganic nutrient inputs. It helps in improving soil quality and sustains productivity in a long run.

TABLE 1 Cotton: Origin, Classification, Nomenclature and Uses—A summary.

Wild cotton genotypes occur in Australia, Africa, Central America (Mexico), South America (Brazil), and South-east Asia (India, Pakistan and China). Currently, the cotton agroecosystem encompasses five different types namely, Egyptian, Sea island, American, Pima, Asiatic and Upland. The Vavilonian centers of origin for different types of cotton occur in Mesoamerica (Upland Cotton), South America (Egyptian cotton); Indian Centre (Oriental cotton and tree cotton).

There are four domesticated cotton species that fill the cotton belts found across different regions of the globe. They are:

Gossypium hirsutum (2n = 26; tetraploid)— It is found in the Cotton cropping zones of Central America, including Mexico, Caribbean islands and Florida.

Gossypium barbedense (2n = 26; teraploid) – It is native to Tropical regions of Brazil and adjacent regions. It is known as extralong staple cotton.

Gossypium arboretum (2n = 13; diploid) – This species is known as Old world cotton that is native to India and Pakistan.

Gossypium herbaceum (2n = 13; diploid) – It is also known as Levant cotton. It is native to tropical Africa and Arab regions.

New world species are dominant in most regions. The global cotton agroecosystem is predominantly constituted by *G.hirsutum* (90%), *G. barbedense* accounts for < 2% area, *G.arboreum* < 2% and *G.herbeceum* < 2%.

Classification

Kingdom-Plantae; Sub-Kingdom-Angiosperms; Order-Malvales; Family Malvaceae; Genus-*Gossypium*; Species-Herbaceum

Nomenclature

Cotton-English; Bavouna-Byelorussian; Bomuld-Danish; Katoen-Dutch; Puuvill-Estonian; Puvillar-Finnish; Baumwolle-German; Kapas-Hindi; Pamut-Hungarian; Cadas-Irish; Kapas-Indonesian; Hathi-Kannada; Kokvilnas-Latvian; Kapas-Malay; Bawlina-Polish; Bamuk-Romanian; Zlopok-Russian; Algodon-Spanish; Pamba-Swahili; Bomull-Swedish; Pamuk-Turkish; Bong-Vietnamese; Parutti-Tamil; Pratti-Telugu.

Uses

Cotton is mostly used to prepare variety of textiles. Fibers of different strengths, nets, filters, explosives (nitrocellulose), paper and boards are also prepared using cotton. Cotton seed oil is prepared from seeds that remain after ginning. It is refined and consumed as vegetable oil or as fuel. Cotton husk or meal is useful as animal feed. However, gossypol could be toxic to animals.

Nutritive Value

Cotton seed cooking oil: Free fatty acid-0.05%; Peroxide value-1.0 Meq/kg; Lovibond color-2.0–6.0; Iodine value-103–106; Fatty acids-Saturated fatty acid-27%; Monounsaturated fatty acids-18% and Polyunsaturated fatty acids-55%. Major fatty groups found are Linoleic acid (18:2), Palmitic acid (16:2), Oleic acid (18:1), Stearic acid (18:0) and Myristic acid.

Sources: http://www.cottonjourney.com/storyofcotton/ pp. 1–12 (August 3, 2012); en.wikipedia.org/wiki/Cotton pp. 1–8 (July 26, 2012); en.organisasi.org/translation/cotton-in-other-languages (August 14, 2012); [3].

20.2 SOILS, AGROCLIMATE AND CROPPING SYSTEMS

20.2.1 SOILS AND TILLAGE

Cotton requires deep tillage and seed bed formation. In areas prone to soil erosion, fields are kept under conservation tillage and crop residues are placed as mulches to reduce loss of nutrients from the field. Cotton farmers are advised to follow certain procedures before deciding on cotton. Soil evaluation for physical and chemical conditions is necessary. Cotton is a deep-rooted crop with ability to adapt to clayey Vertisols and loamy soils that occur in different regions. Hence, evaluation of soil characteristics such as soil texture, depth of horizons, occurrence of hard pans or plow pans, soil fertility status, pH and salinity/alkalinity is necessary [5]. Deep tillage to remove hard pans is common in Vertisols belts where cotton predominates. Soil chemical evaluation for major and micronutrients is mandatory in most regions. This allows farmers to ascertain fertilizer requirements based on yield goals. Soil management plans vary significantly with cotton belts. Precision farming techniques based on GPS/GIS, computer models and variable dispensers are being used in some areas of American cotton belt. They may affect the cropping zone by improving fertilizer efficiency and avoiding soil deterioration.

The cotton belt in Southern Plains of USA, flourishes on few different soil types. Most frequently encountered soil types are Mollisols, Sandy to loamy Psamments in Kansas, Oklahoma, North Texas, Alabama and Mississippi; Ultisols in Georgia and Florida; Piedmonts in Carolinas and Virginia. Periodic soil testing, fertilizer and FYM input to match yield goals keep the cotton belt fertile and productive. Cotton crops grow better in soils with slightly acidic pH at 5.5–6.5. The cotton producing regions in Brazil is built on acidic Oxisols that are traced all over the Cerrados plains, in the south around Sao Paulo and North-east. The sandy Oxisols are moderately fertile, high in Al/Mn resulting in toxicity. Gypsum application to correct soil reaction and Al toxicity is almost compulsory in most regions of cotton belt.

The Egyptian cotton growing regions occur in Giza, Nile valley and Coastal regions. The sandy Oxisols and Xeric soils are usually rich in Ca, slightly alkaline and moderately fertile with regard to major nutrients. Periodic input of major nutrients and organic manure maintains soil quality and productivity. Small areas of cotton occur on Ferralsols and Nitisols in the Southern African coastal and inland regions. Cotton belt in the Indian subcontinent thrives on different types of soils. Larger patches of cotton occur on Vertisols plains found in Central and Southern India. They are popularly termed 'Black Cotton Soils.' Black Cotton Soils (Vertisols) are deep and rich in clay content. They are mostly endowed with Montomorillonite and Illite clays. These black cotton soils exhibit swell-shrink character, also possess high buffering capacity for water and nutrients. Alfisols and Inceptisols are other soil types preferred by cotton farmers in India. Black Cotton Soils preferred in South India possess at least average water holding capacity of 100–500 mm. They are moderately well drained with soil depth ranging from 0.6–1.2 m and pH of 7.5–8.2. In the Indus plains, cotton belt thrives on sandy soils in Sindh region and Mollisols of Punjab. The Chinese cotton belt occurs on Inceptisols and alluvial soils found in the valleys of Yellow river and Shandong region. The cotton belt in North-west China thrives on sandy Alfisols, Inceptisols and Alluvial sands [6].

Majority of cotton agroecosystem in Australia thrives on black earths or its variants and brown clayey soils. These are clayey cracking soils commonly referred as Vertisols. In Queensland, red-brown earths are used. They are sodic or saline. In the Dawson and Callide valley, cotton belt occurs on alluvial soils. The clayey Vertisols are deep (1.0 m) to very deep. Surface is hard, but cracked in many places. The brown and gray soils are slightly alkaline pH >7.0 and calcareous. Salinity is pronounced in deeper horizons. Redish-brown earths used for cotton cropping are sandy, silty or loamy. Sometimes clay loams are also encountered. The deeper horizons are yellowish or olive-brown. Such soils have subsoil horizons that are sandy. Most soil types that support cotton farming need to be replenished with organic manures. Supply of organic matter supposedly increases soil microbial component, rooting and nutrient recovery by the crop [7].

20.2.2 AGROCLIMATE

Cotton cultivation anywhere requires about 160 frost free days at the minimum. Hence, much of the cotton belt is confined to area between 45°N and 30°S. Ample sunshine, water and fertile soil are minimum requirements. Cotton belts occur in both rain fed and irrigated zones. Irrigation is usually achieved using furrows, sprinklers or drip tape, depending on availability of water and yield goals (Plate 1).

PLATE 1 Top left: A Cotton crop in Vegetative Phase. Top Right and Bottom: A crop with ripe bolls in Southern India.
Source: Krishna, K.R., Bangalore, India.

In the Southern plains of USA, and adjoining area, cotton is sown during March/April immediately after winter ends and it is harvested by end of August, when rains stop and a dry sunny season lasts for couple of weeks.

In the North-west India and Pakistan, cotton is grown predominantly during rainy season. The crop is sown with the onset of rains in May/June and harvested by October. Most cotton genotypes grown need 180–210 days to maturity. It is a warm climate crop in Pakistan. The temperature during crop season fluctuates between 11°C to 25°C. It does not tolerate freezing temperatures. In western and southern Indian plains, crop is sown with monsoon rains in June and harvested by October. The average temperature during crop growth ranges from 22°C to 30°C. Factors such as precipitation pattern, nutrient supply, genotype and sowing time, all interact and affect the crop yield to different extents. In most parts of India, cotton is sown in May/June, by which time fields should have received at least 30–40 mm precipitation. Delayed sowing beyond July 15th affects crop growth, nutrient recovery pattern and boll formation.

20.2.3 WATER REQUIREMENTS

Cotton crop grown in vast plains of North America need at least 700–1300 mm water. Fertilizer inputs are based on irrigation resources, inherent soil fertility and yield goals set by farmers. Several methods of irrigation are practiced by the farmers in this belt as well as those in Brazilian Cerrados. Most of them aim at improving water use efficiency and minimizing on economic costs. In the North-west India and Pakistan, cotton belt thrives on moderate levels of precipitation ranging from 750 mm to 1100 mm. The drier region with only 650 mm precipitation also supports a cotton belt, but it needs assured irrigation at least, during critical stages of boll formation and maturity.

Cotton agroecosystem in Peninsular India occurs in semiarid region that receives 700–1100 mm rainfall annually. The drier tracts of Vertisol supports cotton despite slightly lower levels of precipitation because the black cotton soils are known for their ability to buffer nutrients and soil moisture better. The clayey soils store more water than sandy regions. The Vertisol belt also supports a slightly larger area of irrigated cotton. Irrigated cotton is grown during rainy and post rainy or even during summer.

20.2.4 SOIL FERTILITY ASPECTS

Fields meant for cotton are usually given deep plowing once 2–3 years and two shallow plowing during summer. Crop residue incorporated is an important source of nutrients. Farm Yard Manure (15–20 t/ha) application is mandatory. About 2 kg/ha Phosphate Solubilizing Bacteria and Vermin-Compost are other inputs. Fertilizer supply is usually based on soil test data. It may range from 120–180 kg N: 40–60 kg P and 60–80 kg K/ha. Green manuring with neem, *Glyricidia* or *Sesbania* is common. Cotton fields are primed with small amounts of fertilizers to induce rapid rooting and plant establishment.

20.2.5 CROPPING SYSTEMS

In the Great Plains regions, cotton is intercropped with maize or soybean. It is rotated with crops such as wheat, maize, legumes and vegetables. In Motto Grasso, cotton is a mono crop or cultivated as rotation crop with soybean. Cotton is also rotated with maize, soybean and pastures.

Cotton crop is often rotated with cereals such as sorghum, maize or finger millet in the southern Indian plains. Cotton is intercropped with maize or pearl millet in Maharashtra and Gujarat. Cotton is also intercropped with legumes such as black gram or cowpea. Mono-crops of cotton are also frequently encountered in India and Pakistan.

20.3 EXPANSE AND PRODUCTIVITY OF COTTON AGROECOSYSTEM

The global cotton agroecosystem that is scattered all over in different continents is currently large and extends in to 31 m ha. India has largest cotton belt in terms of area (Plate 1). During 2010, world cotton production was 102.7 million bales. However, it was lower than previous years. There was 14% decrease in cotton production relative to 2008 [8]. The cotton agroecosystem is largely contributed by agrarian regions of following 10 nations, namely Peoples Republic of China (33 m bales); India (27 m bales); United States of America (18 m bales); Pakistan (10.3 m bales); Brazil (9.3 m bales); Uzbekistan (4.6 m bales); Australia (4.2 m bales); Turkey (2.8 m bales); Turkmenistan (1.6 m bales) and Greece (1.4 m bales) [9]. There are over 80 nations that support cotton expanses. Three major cotton-farming nations namely China, USA, and India together account for 60% of global cotton produce [10].

During past two decades, fluctuations in cotton cultivating area and total production were feeble. The annual growth rate was marginal at 0.4–0.6% during 1990s and later years. Forecasts suggest that cotton production would increase to about 25 m t by 2012. About 16–17 m t of cotton would be contributed by developing nations in Asia and Latin America. Cotton belt in South America increased in area by 2.1% annually. Brazil has large cotton belt. There are reports that cotton belt in Latin America may expand. However, in Argentina, it may experience loss in area that gets offset by marginal expansion in Paraguay. Cotton belt in the African continent is growing rapidly. Production of cotton has increased by 3.2% annually, to reach 1.7 m t by 2010. African cotton belt exports about 4.4% of its produce to other nations.

Cotton growing regions in the Middle East has experienced a decline in area and production. Cotton farming zones also occur in Turkey and Syria. The forecasts suggest that export of cotton from Middle-east may not shrink, because it has to meet a large local demand for cotton textiles and other products.

Asia supports a large cotton belt. It contributes about 44% of global cotton equivalent to 11–12 m t/yr. China has a very large cotton agroecosystem that provides about 6.2 m t annually. India produces 3 mt/yr and Pakistan 2 m t/yr. During past two decades, cotton producing regions in India, China and Pakistan did not perceive fluctuation in area. The development of transgenic cotton and improvised agronomic methods seem to affect cotton agroeocsystem [11]. Legislations and pricing have also affected cotton growing area in all continents.

The cotton agroecosystem of Great Plains of North America is historically important. It is a vast stretch that has provided textiles, seed oil and lively hood to a large populace. Currently, USA, has the second largest cotton-cropping zone in the world. Fluctuation in cotton cropping area in Southern plains is governed by demand for it from nations in other continents. Annually, USA, contributes about 4.2 m t cotton. About 40% of cotton produce is exported to different regions. Cotton agroecosystem dominates the landscape in at least seven southern states of USA, namely Texas, Georgia, Mississippi, California, Arkansas, North Carolina and Louisiana. These regions in USA, account for 60% of total cotton produced in USA, [10]. The cotton belt in the Southern plains has fluctuated based on several natural factors such as soil fertility, rainfall pattern, disease and insect pressure. Man-made factors like Civil war, World Wars 1 and 2, economic recession, lack of labor and subsidies have affected area of cotton cropping belts. Cotton belt has generally declined during twentieth century. Actually, between 1952 and 1985 cotton agroecosystem has decreased by 1.2 m ha [2].

The cotton belt in Texas is accentuated in the Northern Highlands. Here, it is grown intensely despite the area being classified as arid with unreliable rainfall pattern. Sandy tract found along Gulf of Mexico is other region that supports cotton crop in Texas. The cotton agroecosystem in Texas has fluctuated in area from 1.8 to 2.4 m ha during the years 1986 to 2008. The productivity of cotton has ranged from 440 kg/ha in 1985 to 880 kg/ha in 2008 (Plate 2; [12]).

PLATE 2 Left: A close-up view of Ripe Cotton Boll. Right: Storage of Cotton.
Source: USDA, Texas, Alabama, Mississippi, Arkansas and Tennessee.

The cotton belt in Alabama is predominantly covered by Upland cotton (*G. hirsutum*). It is a relatively long duration crop of 180–200 days that suits farms in Alabama. If a crop is sown in March/April, then bolls ripen during August/September when days are sunny. Cotton farming in Alabama is pronounced in the Mississippi river valley. The cotton belt in Georgia is large and concentrated in the Coastal plains. The cotton belt in California occurs around the San Joaquin valley, Sacramento region, Palo Verde and the Imperial. The cotton belt is filled predominantly with genotypes that are characterized as long and strong fibered cotton [13]. Much of the cotton produced in California is exported. Hence, the extent of cotton belt fluctuates based on demand from other regions. The productivity is moderate at 600–900 kg/ha. Acala and Pima are the two dominant cultivars that fill the California cotton regions.

The Brazilian cotton cropping zone is composed of large mechanized farms with an average size of 2000 ha each. In the Brazilian savannas, Matto Grasso region has over 1 m ha equivalent to 90% of the nation's cotton belt (Plate 3). The cotton cropping belt is highly mechanized in the Savannas, but less so in the North-east. The Brazilian cotton is predominantly rain fed. About 3–4% of region is irrigated. The Brazilian cotton agroecosystem encompasses several different genotypes of cotton. Currently, popular genotypes in Matto Grasso are FMT 701 and Delta Opal. Whereas, in the North-east, BRS 187, CNPA 8H a and BRS Serido are dominant. Obviously each genotype has its influence on nutrient dynamics within the cotton belt. Fluctuations in cotton area are largely due to internal consumption of cotton products and a moderate share of exports. The annual production of cotton in Brazil exceeds 1.6 m t, at an average productivity of 1487 kg/ha. The cotton belt supports a large human population with jobs and textiles. The Argentine cotton thrives on the Mollisols and Inceptisols of Pampas region. Cotton is mostly rotated with maize or soybean, if not it is intercropped with major cereals wheat and maize. The Argentine cotton belt contributed about 220,000 tons during 2011 [14].

PLATE 3 A Cotton expanse in the Matto Grasso region of Brazil.
Note: Cotton is rotated with Maize and Soybean on moderately fertile Oxisols that are generally prone to Phosphorus deficiency. It is customary to advice higher levels fertilizer-P to overcome soil P fixation problem caused due to high Al and Fe salts. Gypsum treatment to adjust soil pH is mandatory in many locations within Cerrados.
Source: EMBRAPA, Campe Verde, Brazil.

The cotton belt in Russia and Eastern European Nations is large and extends into many agrarian regions. The expansion of cotton belt occurred markedly during early years of Socialist Revolution. Major cotton producing regions occur in Transcaucasia, Central Asian Republics, Uzbekistan, Kazakhstan and Kirghizstan. Cotton production technology in Russia and neighboring countries was improved during mid-1900s. Cotton farming is intense in Uzbek region. About 60% of cotton in erstwhile Soviet Union was contributed by Uzbekistan. During recent years, Uzbekistan has produced over 1.2 m t cotton annually [14]. Tajik and Armenian region are also important cotton producing regions [8].

African cotton agroecosystem is large and well distributed into tropical and semi-arid zones. Uganda in Central Africa possesses a cotton belt that has fluctuated in area due to several reasons. The natural factors like soil fertility, precipitation, disease/insect pressure and man-made reasons like insecure governance, fiscal policies, market demand have affected the size of cotton belt in Uganda. Cotton belt that got initiated in early 1900s was well stabilized in the central province of Uganda. However, in due course, cotton belt took shift to west due to competition from cereals and coarse grains that were preferred in that region. The annual production of cotton in Uganda has experienced upheavals due to World War II, Independence movement, Idi Amin regime and policy reforms during recent decades. The annual cotton production in 1930s till 1970 averaged 60,000–65,000 tons. Cotton belt experienced rapid decline during Idi Amin regime from 1971–1979. It dipped to <15,000 tons during revolution. During recent years from 2004, the cotton belt has grown into larger area, reaching an annual production of 38,000 t [4].

Cotton is an important cash crop of Tanzania. The commercial cotton belt was initiated in 1904 by German settlers. Cotton belt became conspicuous around Lake Victoria in Mwanza, Shinyanga, Mara, Tabora, Kigoma and Singida regions. The productivity of cotton farms improved perceptibly during 1930s. Local varieties dominated the cropping zone. During mid-1960s, cotton production in Tanzania reached over 80,000 tons.

The cotton belt in Middle East extends into regions in Egypt, Israel, Syria, Iraq and Iran. Egypt supports relatively larger cotton belt, which is mostly irrigated and supplied with fertilizer and organic manures. During 2011, Egypt produced 0.3 m t and Syria 0.32 m t cotton. These are major cotton producing nations.

The cotton crop was domesticated and cultivated in North-west Indian plains about 3000 years ago. It is a center for genetic diversity for cotton. India has the largest cotton cropping area and contributes 29 m bales annually. The cotton agroecosystem in India is mended and sustained by 33 million people, who are employed either directly or indirectly by cotton farming and industrial enterprises. Indian cotton agroecosystem encompasses all four cotton species and hybrids. Cotton belt is spread into different regions of India, but it intense in the Vertisols regions of Peninsular. The Indian cotton belt can be classified into North zone comprising Punjab, Haryana and Rajasthan where *G. hirsutum* and *G.arboreum* dominate. Central zone is filled with *G. hirsutum, G.arboreum, G.herbaceum* and hybrids. It encompasses major states like Gujarat, Maharashtra and Madhya Pradesh. The southern zone supports cultivation of *G. hirsutum, G.arboreum, G.heraceum* and *G barbedense* plus hybrids [15]. Cotton cultivation is intense in Gujarat (10 million bales/yr), Maharashtra (6.7 million bales/yr), Andhra Pradesh (4 m bales/yr) and Karnataka (1.8 million bales/yr). The average productivity of cotton fields in India ranges from 590 kg/ha to 930 kg/ha depending on inherent soil fertility, input and yield goals. Together, these states account for 75% of nation's cotton zone. The cotton agroecosystem in India is largely supported and sustained by large internal demand from spinning mills.

The cotton agroecosystem in Pakistan flourishes predominantly in the Punjab and Sindh regions. It is grown as a rain fed or sometimes as irrigated crop. Cotton is relatively intense in Multan. Cotton cropping belt also extends in Bahawalpur, Dera ghazi Khan, Faisalabad, Sargodha, Lahore and Rawalpindi divisions. The New world cotton predominates but desi or Old world species also coexist in the cropping belt. Pakistan supports a large cotton belt extending into 7.85 m ha. Major varieties that occupy cotton belt in Sindh are Simnast, Qalindri and Desi varieties. In Punjab important varieties are Delta pine, Nayab 86, MHH 93 and Bt-cotton [16]. Cultivars that dominate dictate the nutrient dynamics of the cotton belt.

The cotton cropping zones are wide spread into almost all provinces of China. About 24 of 31 provinces in the mainland China support cotton farming. Cotton production during the past decade has fluctuated between 4.0–5.7 m tons.

The cotton agroecosystem in Australia is predominant in the Central Border region and it is traceable between 23° 30′ S and 32° 32′ mainly in New South Wales and Queensland. Majority of cotton belt is irrigated (76%). Rain fed cotton occupies only 25% of total area. It is found in Dawson, Callide valleys, Darling downs and Emerald areas. In New South Wales, cotton belt occurs east of Moree and west of Warialds [7]. The cotton belt in Australia extends into both irrigated and rain fed regions. The irrigated belt is large and spreads into 269,000 ha, but rain fed cotton belt is very small at 15,000 ha. The total lint production is 345,000 tons annually, whereas seed cotton amounts to 916,000 tons/yr [17, 18].

KEYWORDS

- **Black Cotton Soils**
- ***G. arboreum***
- ***G. barbedense***
- ***G. heraceum***
- ***G. hirsutum***
- **Rig-Veda**

REFERENCES

1. Fuller, D. Q. The spread of textile production and textile crops in India beyond the Harappan zone: An aspect of the emergence of craft specialization and systematic trade. In: Linguistics, Archeology and the Human past. Indus project Occasional papers 3 series. Osada, T.; Uesugi, A. (Eds.). http://www.ucl.ac.uk/archeology/ people/staff/fuller/ usercontent_profile/Textilesbeyondindus.pdf **2008,** 1–18 (August 1, 2012).
2. Phillips, K. E.; Janet, R. Cotton. Encyclopedia of Alabama. http://www.encyclopediaofalabama.org/face /Article.jsp?id=h-1491 **2011,** 1–8 (August 3, 2012).
3. GHC, Cotton. Georgia Humanities Council. The New Georgia Encyclopedia. http://www.georgiaencyclopedia.org /nge/Article.jsp?d=h-2087. **2012,** 1–3 (August, 4, 2012).
4. Baffe, J. The cotton sector of Uganda. African Region Working paper series No 123 http://www. worldbank.org/ afr/wps/index.htm **2009,** 1–15 (August 8, 2012).
5. Silvertooth, J. C. Soil Management and Soil Testing for Irrigated Cotton Production. University of Arizona Extension Services, Arizona, USA, http://cals.arizona.edu/crops/cotton/soilmgt/soilmanagement.html **2001,** 1–8 (April 4, 2012).
6. Brady, N. C. Nature and Properties of Soil. Prentice Hall of India. New Delhi **1975,** 575.
7. NDPI, 'Soilpak-Cotton-Growers-Readers' note. Third Edition. NSW Department of Primary Industries http://www.dpi.nsw.gov.au/agriculture/resources/soils/guides/soilpak/cotton **2011,** 1–32 (August 4, 2012).
8. FAOSTAT, Cotton statistics. Food and Agricultural Organization of the United Nations. Rome, Italy. http://www.FAOstat.org **2011** (August 3, 2012).
9. NCCA, Cotton Rankings. http://www.cotton.org/econ/cropinfo/cropdata//rankings.cfm. **2011,**1–8 (August 3, 2012).
10. USDA, Facts and figures: the Cotton Trade. http://www.pbs.org/now/shows/310/cotton-trade.html **2007,** 1–2 (August 3, 2012).
11. FAO, Cotton. Food and Agricultural Organization of the United Nations. Rome, Italy. http://www.fao.org/docrep/006/ y5143e/yr5143ele.htm **2002,** 1–6 (August 3, 2012).
12. Robinson, C.; McCorkle, D. A. Trends and Prospects for Texas cotton. Cotton Outlook http://agecon2.tamu.edu/ people/faculty/robinson-john/Cotlookarticle.pdf **2006, 1**–4 (August 4, 2012).
13. Scheuring, A. F. A guidebook to California Agriculture. University of California Press. Berkeley, USA, **1993,** http:// www.ccgga.org/cotton_information/cotton.html (August 3, 2012).
14. USDA, **2011,** Cotton-World supply and Demand table. USDA Economic and Statistics system. http://www. spectrumcommodities.com/education/commodity/statistics/cottontable.html (August, 12, 2012).
15. Pal, S. Indian Cotton Production: Current Scenario. Indian Textile Journal **2010,** *65,* 42–56.

16. Cotton Research Centre. Cotton in Pakistan. http://www.madebypakistan.com/2011/02/cotton-in-pakistan/ 1–9 (August, 10, 2012).
17. ABS **2008,** *Agricultural Commodities Small Area Data, Australia 2005–06.* RIRDC Pub. No. 10/081. **2011.**
18. ANRA, Agriculture Cotton Industry-Australia. Australian Natural Resources Atlas, Canberra http://www.anra.au/ topics/agriculture/cotton/index.htm **2011,** 1–11 (August 11, 2012).

EXERCISE

1. Demarcate the Cotton Agroecosystem in different continents.
2. Write an essay on Cotton producing zones of India, Nutrient Management and Productivity trends.
3. Describe the geographic locations in Australia that support Cotton cultivation.
4. Collect data on Area and Productivity of cotton in Australia.
5. Write about major constraints to Cotton culture in Australia.

FURTHER READING

1. Clayton, B. D. In: *King Cotton: A Cultural, Political and Economic History Since 1945.* University of Mississippi Press: MS, USA, **2011,** pp. 440.
2. Wayne, S. C.; Cothern, J. T. In: *Cotton: Origin, History, Technology and Production.* **1999,** pp. 850.

USEFUL WEBSITES

www.cotton.org.news (July 23, 2012).
http://www.georgiaencyclopedia.org/nge/Article.jsp?id=h-2087 (August 4, 2012).
http://www.ica-ltd.org/ http://www.plantcultures.org/plants/cotton_landing.html (August 4, 2012).
http://www.cotton.missouri.edu/InOurLives-Producersfiorld.html (August 4, 2012).
http://wwwbrazilintl.com/agsectors/cotton/map_cotton_brazil.html (July 28, 2012).
http://www.mahacot.com/cotton.html (August 3, 2012).
http://www.cotcorp.cogov.in/organization.asp (August 1, 2012).
http://www.cnep.ch/etu/publications/synth_china.pdf 57–122 (July 30, 2012).
http://www.anra.gov.au/topics/agriculture/cotton/index.html 1–11 (August 11 2012).

CHAPTER 21

SUGARCANE PRODUCING ZONES

CONTENTS

21.1 INTRODUCTION

Sugarcane is native to tropical swampy regions of Yuanan province in South China and Assam in North-east India. This region is supposedly primary center of genetic diversity for *Saccharum* species. Sugarcane cultivation was in vogue in Persia and Greece during 4–6 century B.C [1]. Arab traders aided the spread of sugarcane into Mediterranean region. Regions such as Iraq, Egypt, North Africa and Andalusia received sugarcane between 8 and tenth century A.D. [2]. Arabs had set up several sugar refining and production factories by 1200 A.D. Sugar refining techniques were introduced to China through voyages of Buddhist monks. Chinese travelers/envoys visiting India during sixth century introduced sugarcane native to Indian plains and sugar crystallization technology into the Chinese mainland [1]. Crusaders from European kingdoms transshipped sugar from Arabian region to European mainland during 12th century [3].

Sugarcane was introduced into Americas by Portuguese from Madiera island. Sugarcane reached Hispaniola, i.e., Caribbean region through the voyages of Christopher Columbus in mid-1500s. Some of the earliest sugar factories were erected in Hispaniola (1501), Jamaica and Cuba (1520s) [4]. Currently, sugarcane is an important wetland crop in Meso-America, especially Jamaica, Dominican Republic, Guyana, Belize and Haiti. Sugarcane and sugar production was mechanized during late 1700s in the Caribbean Islands. Portuguese introduced sugarcane cultivation techniques into Brazilian plains during 1540. This effort was followed by erection of large number of sugar mills in Saint Catarina. Sugarcane and processed sugar was transshipped into New England and Europe regularly during 17th century. The spread of sugarcane cultivation in different parts of Americas, it seems, induced human migration rather conspicuously. By late 18th century, sugar consumption had become routine in many parts of the world. At present, sugarcane-farming zones extend into all major agrarian zones of the world, excepting cold temperate regions. Sugarcane was introduced into Australia in 1788. It was initially grown in New South Wales and around Brisbane in Queensland. Capt. Louis Hope and John Buhot established earliest of the sugar plantations in Queensland. Sugarcane production, harvesting and processing was mechanized by early 1900s in Australia (Canegrowers Australia, 2010; Table 1; Fig. 1).

TABLE 1 Sugarcane: Origin, Botany, Uses, Nutritional Aspects—A summary.

Sugarcane was domesticated 6,000–8,000 years ago in the area comprising Southern China, Indo-China and North-east India. It is believed that among different species of sugarcane, *Saccharum barberi* orginated in North-east India, Whereas, *S.edule* and *S.officinarum* were antive to New Guinea. Sugarcane is predominantly grown in the tropical and warmer regions of the world. It is native to South and South Asian tropical farming zones. It is grown in Southern USA, and Meso-American countries. Brazil has a large area under sugarcane cultivation. In West Asia, sugar cane producing regions are situated in regions with facility to irrigate fields. Sugarcane production zones are well spread in Africa. The Indo-Gangetic Plains and Chinese agrarian zones are other major sugar cane producing centers. Globaly, 80% of sugar is derived from sugarcane and rest 20% from beet that is grown predominantly in temperate regions.

Classification

Kingdom-Plantae; Order-Poales; Family Poaceae; Subfamily Panicoideae; Tribe-Andropogonae; Genus-*Saccharum*; Species-*Saccharum arundinaceum, S.bengalense, S.edule, S.munja, S. officinarum, S. procerum, S. ravennae, S.robustum, S. sinense, S. spontaneum*

$2n = 107–115$ (Aneuploid)

Nomenclature

Sugar-United States of America, Britain, Australia; SucreFrance, Canada; Suikar-Belgium, South Africa, Netherlands; Sukker-Denmark, Norway, Estonia; Sucao-Ireland; Sukori-Finland; Socker-Sweden; Azucar-Spain; Acucar-Portuguese; Caxar-Russian; Cukr-Czech Republic; Cukrus-latvania; Zucchero-Italy; Cukier-Poland; Sharker-Uzbekistan; Shequer-Albania; Shiker-Tataristan; Kant-Kazakisthan, Kirgisthan; Shaqari-Georgia; Wekep Macedonia; Asukal-Philippines; Gula-Indonesia and Malyasia; Satou-Japan; Chini or Shakar-India, Bangladesh, Pakistan; Tang or Teng-China.

Uses

Sugarcane was consumed by Neolithic people in the Indus Valley in crystalline form some 5000 years ago. Indians living in Gangetic plains discovered the procedure to crystallize and use it freely in larger amounts with other culinary items, sometime during 350 A.D. Sugarcane is also used to produce ethanol, alcoholic beverages, molasses and bagasse. Sugarcane reeds are used to make mats, screens and roofing. Sugarcane molasses is converted into variety of fermented products like liquors (rum), vinegar, cosmetics, pharmaceutical, solvents, coatings, base material for paints. Molasses is also used to produce organic acids like citric acid, lactic acid, butyric acid. Etc. Molasses is a good substrate to produce yeasts for industrial use. Sugarcane industry byproducts are also used to manufacture wax, polishes and insulation material.

Sugar rich juice is consumed directly in many regions. In most countries sugar crystals are produced by boiling sugar rich juice and adding sugar dust for crystals to grown on them. Molasses is a byproduct derived during sugar crystal production. Sugarcane Bagasse is an important starter material for biofuel. It is useful in preparing material that replaces plastics and polystyrene. It is used to manufacture paper. Bagasse's mixed with charcoal is an important source of energy in many industries.

Nutritional Aspects

Nutritional value of one serving of sugarcane juice equivalent to 28.5 g is follows: Energy-111 kJ; Carbohydrates-27.5 g; Sugars-27 g; Protein-0.27 g; Ca-11 mg; Fe-0.37 mg; K-42 mg and Na-17 mg.

Source: Refs. [5–7]; http://www.floridacrysals.com/content/125/production.aspx; http://www.Wikipedia.org/wiki/Sugarcane; http://www.sugarcane.res.in/index.php/knowledge-bank/genetic-studies? start=3

21.2 CROP GROWTH PATTERN, AGROCLIMATE AND SOILS

Sugarcane is a tropical, perennial grass that grows to 3–4 meters height (Plate 1). There are cultivars that reach to only 2 m and those reaching 6 m in height. The stem is 4–5 cm thick and cylindrical with tissue rich fructose. A mature stalk has 11–16% fiber, 12–17% sugar, 2–3% nonsugar and 63–70% water. The average productivity of sugarcane is about 6–70 t/ha/yr. Sugarcane is propagated using setts, settlings or budchips. Setts are most commonly used planting material in Asia, Africa and Caribbean. Sugarcane crop stays in the field for 3–6 years at a stretch. The first crop that arises from setts is termed 'mother crop' or 'plant crop.' It reaches maturity in 9–24 months depending on cultivar and geographic conditions. The succeeding 4 or 5 ratoons allow the crop to survive and produce for next 4–5 years [8]. Therefore, once established, a sugarcane field/zone lasts for longer duration in the area affecting soil and nutrient dynamics in ecosystem rather conspicuously, compared to annual cereal crop of 3–4 months e.g., wheat.

21.2.1 AGROCLIMATE

In India, sugarcane is grown under tropical conditions that exist in the plains. Sugarcane needs moderate temperature that ranges from 26°C–32°C. In the African continent, especially in Southern parts, sugarcane farming is confined to humid tropics that receive 1200–1500 mm precipitation per year [9]. Precipitation quantity and pattern affects cane production. Sugarcane does not thrive in areas with possibility of temperatures >50°C. It grows well in areas with rainfall between 750–1200 mm. Sugarcane is considered a wet land crop in many regions, since water requirement is high. This crop needs wet soil throughout the profile, although not flooded fields like rice.

21.2.2 SOILS

Sugarcane production occurs on several types of soils. It requires deep, fertile and well-drained soil. It grows luxuriantly on Mollisols and Ultisols in North America and Meso-America. Sugarcane thrives on fertile Spodosols and Ultisols in Southern Florida. Sugarcane belt in Cerrados of Brazil thrives on acid Oxisols. Sugarcane agroeocsystem thrives on low fertility Alfisols and Inceptisols in Eastern and Southern Africa. In Egypt, sugarcane farming zones occur on sandy soils that are supplied with manures. Soil amelioration and planting legumes during fallow seems important to preserve soil fertility in sugarcane farming regions. In Asia, sugarcane cultivation proceeds on Mollisols and Inceptisols of Indo-Gagnetic Plains. Sugarcane prefers well-drained loamy soil. Lighter soils need to be irrigated frequently. It is grown on moderately fertile Alfisol and Vertisols in the Southern Indian plains. Sugarcane growing region in China and Fareast is supported by Inceptisols, Alfisols and sandy Oxisols [10]. Sugarcane cropping zones also thrive on Histosols and Andisols. It is grown on low fertility podzolic soils in the Southeast Asian farming belts. Soil tillage prior to planting setts is crucial to establishment and sustenance of crop. Sugarcane stays in the field for longer duration of up to 5 years. Hence, deep tillage that disturbs and mixes the upper horizon is essential. During ratooning, crop allows only intercultural operations. Hence, soil is prone to compaction and weed growth.

21.3 EXPANSE AND PRODUCTIVITY OF SUGARCANE AGROECOSYSTEM

Sugarcane agroecosystem extends into agrarian regions that occur between 22° N and 22° S (Fig. 1). Its cultivation is in vogue from sea level to 1600 m.a.s.l. For example, sugarcane farming zones in the Andes occur at 1000–1200 m.a.s.l. In western ghats of

India, sugar farming occurs at 900–1000 m.a.s.l. Whereas, in the Coastal Plains of India or in Cuba or in Florida, sugarcane ecosystem thrives at sea level. Globally, sugarcane cultivation occurs mostly in warm tropics. Sugarcane agroecosystem spreads into 90 different countries, totaling an area of 23.8 m ha. Annual cane harvest during 2010 was 1.7 billion tons [11]. Sugarcane growing regions are found in Brazil (672 mt), India (285 mt), China (116 mt), Thailand (66 mt), Pakistan (50 mt) and Mexico (49 mt).

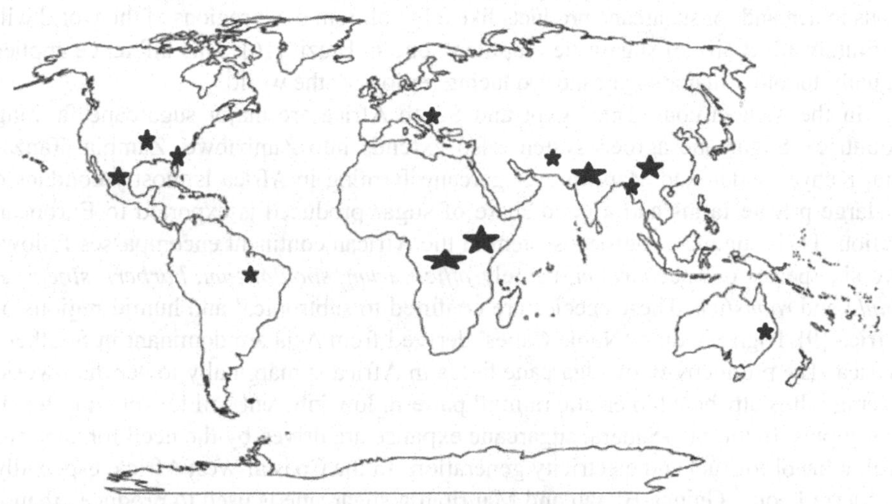

FIGURE 1 Sugarcane Farming zones of the World.
Note: Dark stars indicate regions wherein sugarcane cultivation zone is intense or conspicuously large.

Sugarcane agroecosystem is mostly confined to wetlands of Florida and Louisiana in USA. The wetlands of South Florida including regions around lake Okeechobee and Beleglade support large sugarcane farms. Sugarcane fields in Florida are provided with green manure species to reduce soil erosion and improve soil organic matter content. Leguminous green manures help further in adding soil-N via BNF [12, 13]. Sugarcane farmers in Everglades of Florida adopt soil tests and plant analysis while dispensing fertilizers and FYM into fields. They follow critical limits for various nutrients. This allows balanced nutrition and helps in regulating yield goals [14]. Perennial peanuts are popular as cover crops between sugarcane rows. Sugar cane cropping extends well into tropical Meso-America. Mexico is a major sugarcane-producing nation. Wet regions in Guatemala and Honduras also support sugarcane agroecosystem. The Sugarcane agroecosystem is intense in the Caribbean region. Agrarian regions in Cuba [15], Haiti, Hispaniola, Bahamas and Puerto Rico support large-scale sugarcane production.

At present, Brazil has a vast sugarcane-farming zone. Major sugarcane belts occur around Sao Paulo in Southern Brazil. Here, the terrain is flat with temperate climate and fertile soil. North-east Brazil encompassing regions in the states of Algolas and Pernambuco also support large scale sugarcane cultivation. Here, the terrain is slightly gravely and climate is tropical. The cropping belts in Brazil extend into over 60 m ha, out of which 6 m ha is covered by sugarcane [16]. Brazilian sugarcane agroecosystem encompasses over 60 different cultivars with wide range of adaptability and sugar content [6]. They also adapt to soil fertility variations and dearth's for water. Sugarcane agroecosystem in Brazil is dependent on demand for cane from sugar processing industry and export. Brazilian sugar is first processed into VHP sugar and then refined to ICUMSA 45 for export. In Brazil, expanse, inputs and productivity of sugarcane agroecosystem is partly driven by the demand for ethanol for vehicles. Nearly 80% of the transport vehicles are flexible to use either petroleum or gasohol (ethanol mixed petrol). Fluctuations in demand for sugarcane products like ethanol from other regions of the world will definitely affect size of sugarcane cropping zone in Brazil [16]. This inference applies equally for other major sugarcane producing regions of the world.

In the African continent, Egypt and South Africa are major sugarcane farming countries. Sugarcane agroecosystem also extends into Zimbabwe, Zambia, Tanzania, Kenya, Sudan and Mauritius. Sugarcane farming in Africa is mostly conducted in large private farms and a good share of sugar produced is exported to European nations [17]. Sugarcane agroecosystem in the African continent encompasses following six species of *Saccharaum*, namely *officinarum*, *spontaneum*, *barberi*, *sinensee*, *edule* and *robustum*. These species are confined to subtropical and humid regions of Africa [9]. High yielding 'Noble Canes' derived from Asia are dominant in Southern Africa. The productivity of sugarcane fields in Africa is marginally lower than world average. It is attributed to erratic rainfall pattern, low inherent soil fertility and fertilizer inputs. In the Sub-Sahara, sugarcane expanse are driven by the need for sugar as well ethanol for fuel and electricity generation. In the Coastal West Africa, especially in Sierra Leone, Guinea-Bissau and Mauritania sugarcane is used to produce ethanol that generates electricity [18]. This trend of developing sugarcane cultivation zones to support generation of electricity via ethanol seems is rapidly spreading in Southern African nations [19].

The sugarcane agroecosystem in India, especially its area, inputs and yield goals are under the influence of a stable demand by over 1.2 billion people and industries that depend on byproducts. Sugarcane agroecosystem extends into several states within India. Its cultivation is intense in states such as Uttar Pradesh and Maharashtra. Its cropping zones are moderately intense in Andhra Pradesh, Karnataka, Tamil Nadu and Gujarat. In most other states it is grown with low inputs. Sugarcane is a cash crop grown intensely with optimum supply of irrigation and fertilizers. Sugar cane is a long-term crop, removes large quantities of nutrients from soil into above ground ecosystem. In most parts of India sugarcane is a high input cash crop. Farmers supply 275 kg N, 62 kg P and 112 kg K/ha. Nitrogen fixing microbes such as *Azospirillum* is applied at the beginning of first crop. Balanced nutrition is achieved by using commercial mixtures that contain micronutrients. Ratooning up to 5 crops is possible. Each ratoon crop is supplied with nutrients fresh [20]. Sugarcane belt in South India encompasses both high input and moderate input farms. Soil fertility testing and Nutrient management techniques holds the key for productivity of farms situated in river valleys and fertile plains of south-eastern coast of India [21].

The fertile plains of Southern Pakistan support sugarcane farming belt. The sugarcane agroecosystem in Pakistan encompasses several landraces, high yielding varieties and poly crosses produced at National Sugar Crops Research Institute, Thatta

(Sindh), Pakistan and elsewhere in other regions of the world [22]. During recent years, several varieties obtained from China have been introduced into fertile plains of Sindh in Pakistan.

Philippines supports sugarcane farming in relatively large area. Sugarcane monocultures of Philippines are important contributors to sugar industry in the Fareast. Sugarcane agroecosystem in Philippines is predominant in Negros, Luzon, Panay and Mindanao. The island of Negros is known as the sugar bowl of Philippines. More than half of agrarian zone in Negros is used to produce sugarcane. During recent years sugarcane farmers in Philippines have resorted to ecological farming, which involves optimum levels of nutrient recycling via residue incorporation and avoiding burning of sugarcane trash. Sugarcane fields are provided with inorganic fertilizers to refurbish lost fertility. Usually, 210 kg N, 55 kg P and 74 kg K is supplied/ha/yr. Decomposing sugarcane trash adds to soil fertility immensely [23]. Sugarcane agroecosystem may expand in future due to demand for biofuel. The productivity of sugarcane in Philippines, considered in terms of ethanol is 4550 lts/ha/yr. It is attributed to low cane yield of 70 t/ha/yr compared to other countries like Brazil that harvest over 100 t/ha cane/yr.

Sugarcane expanses in Australia are concentrated in the river valleys and coastal plains. It is prominent in Queensland and New South Wales. Queensland accounts for 95% of sugarcane cropping zone in Australia. There are over 4,000 sugarcane farms in this region, each ranging between 100 and 1,000 ha in area. Annually, Australia produces about 32–35 m t of cane, equating to 4.5–5.0 m t sugar. Australian sugarcane agroecosystem depends excessively on demand for its export to nations in Asia and Europe. Fluctuations in demand for sugarcane export has an effect on expanse, inputs and yield goals. About 80% of sugarcane is processed and exported. Domestic consumption is satisfied with just 20% of sugar production. Australian sugarcane farmers adopt efficient residue recycling. They usually spread sugar trash on fields, so that its acts as a mulch and reduces soil erosion, provides optimum soil temperature and avoids buildup of weeds. Australian sugarcane farmers are experimenting with Precision farming techniques [24, 25] (Plate 1).

PLATE 1 A Sugarcane field near Tungabhadra Project in Bellary district of Karnataka in South India.
Source: Krishna, K.R., Bangalore, India.

KEYWORDS

- ***Azospirillum***
- **budchips**
- **mother crop**
- ***S. edule***
- ***S. officinarum***
- ***Saccharum barberi***

REFERENCES

1. Sharpe, P. Sugarcane: Past and Present. Ethnobotanical Leaflets 1–5 http://www.ethnoleaflets.com/leaflets/ sugar.htm **1998,** (July 26, 2012).
2. Watson, A. Agricultural Innovation in the early Islamic World. Cambridge University Press. Cambridge, Great Britain. **1997,** 26–27.
3. Ponting, C. World History: A New Perspective. London: Chatto and Windus. **2000.**
4. Benitez-Rojo, W. The Repeating Island. Duke University Press, Durham, USA, **1996,** 233.
5. ANPL. Crops Information: Sugarcane. Agrocommerce Network Private Ltd. http://www.agroecommerce.com/ Agroecom/BriefInfo/ Crops/SugarcaneInfo.asp **2002** (July 25, 2012).
6. Brabus, P. Sugar from Brazil. http://www.peterbrabus.com/sugar-cane-in-brazil/ **2008,** 1–5 (July 26, 2012).
7. Soloman, S. Sugarcane byproduct based Industries in India. Sugar Technology **2011,** *13,* 408–416.
8. Purseglove, J. W. Tropical crops: Monocotyledons. Longman Group Ltd London, United Kingdom, **1979,** 607.
9. Tarimo, A. J.; Takamura, Y. T. Sugarcane Production, Processing and Marketing in Tanzania. African Study Monographs **1998,** *19,* 1–11.
10. Brady, N. C. Nature and Properties of Soils. Prentice Hall of India, New Delhi, **1975,** 575.
11. FAO, Sugarcane Production statistics. Food and Agricultural Organization of the United Nations, Rome, Italy, http://faostat.fao.org **2011,** (July 26, 2012).
12. Baucum, L. E.; Rice, R. W. An overview of Florida sugarcane. http://edis.ifas.ufl.edu/sc032. **2009,** 1–9 (July 25, 2012).
13. Muchovej, R. M. Rotational Crops for sugarcane grown on Mineral soils. http://edis.ifas.ufl.edu/sco53. **2003,** 1–7 (July 25, 2012).
14. Rice, R. W.; Ezenwa, I. V.; Lang, T. A.; Baucum, L. Sugarcane Plant Nutrient Diagnosis. http://edis.ifas.ufl.edu/sc75. **2010,** 1–13 (July 25, 2012).
15. Peters, Cuba downsizes its Sugar Industry: Cutting Losses. Lexington Institute, Arlingtron, USA, **2003,** 1–12.
16. Lagercrantz, J. Ethanol Production from Sugarcane in Brazil: Review of Potential for Social and Environmental labeling of ethanol production from sugarcane. Grona Bilizer, Stockholm, Sweden, http://www.gronabilister.se **2006,** 1–10 (July 25, 2012).
17. CTA, Regional sugar trade and Production in Eastern and Southern Africa growing. Technical Centre for Agricultural and Rural cooperation. Agritrade http://www.agritrade.cta.int.htm **2011,** 1–3.
18. Smith, R. O. Sub-Sahara sugar industry diversifying into fuel and Power. ESI-Africa.com http://www.esi-africa.com/Sub-Sahara/sugar/industry/divrsifying/info/fuel/power.htm **2012,** 1–2 (July 26 2012).

19. Johnson, F. Sugarcane resources in Southern Africa. http://ww.tiempocyberrclimate.org/portal/archive/issue35/t35a1.htm **2012,** 1–5 (July 26, 2012).
20. TNAU. Nutrient Management: Sugarcane (*Saccharaum officnarum*). Http://agritech.tnau.ac.in/agriculture /agri_nutrientmgt_sugarcane.html. **2008,** 1–4 (July 25 2012).
21. IKISAN. Nutrient Management in Sugarcane. http://www.IKISAN.com/crop%20Specific/Eng/links/tn-Sugarcane% 20NutrientManagement. **2012,** 1–5 (July 26, 2012).
22. PARC, National Sugar Crops Research Institute. http://www.parc.gov.pk/1SubDivisions/SARCKHY?ncsr.html **2011,** 1–11 (July 26, 2011).
23. R. E. A. P. Ecological Sugarcane Farming: From Sugarcane Monoculture to Agro-Ecological Village. http://www.reap-Canada.com/international_dev_4_3_3.htm **2012,** 1–4 (July 26, 2012).
24. Cane-growers Australia. About the Australian sugarcane Industry. http://www.cangrowers.com.au/page/ industry_Centre/about-the-australian-sugarcane-industry.htm **2010,** 1–14 (July 25, 2012).
25. Cane-growers Australia. The story of Sugarcane: Paddock to Plate. www.canegrowers.com.au/page/case-studies/Pest_diseasee_management/ **2012,** 1–5 (July 26, 2012).

EXERCISE

1. Mark the sugar cane producing regions in Asia.
2. Mention different types of sugar canes in South Asia and Meso-America.
3. Describe various agronomic procedures involved during sugar cane production.
4. Mention average global sugarcane production, area and productivity during past five years.

FURTHER READING

1. Alvarez, J.; Castellanos, L. P. Cubas Sugar Industry. University of Florida Press, Gainesville, Florida USA, **2001,** pp. 326.
2. Balasundaram, N.; Thagarajan, R.; Chandran, R. Advances in Sugarcane Production Technology. Sugarcane Breeding Station (Indian Council of Agricultural Research), Coimbatore, Tamil Nadu, India, **2003,** pp. 194.
3. Hagleberg, G. B. Sugar in the Caribbean:Turing Sunshine into money. The Woodrow Wilson International Centre of Scholars, Washington, D.C., **1985,** pp. 42.
4. Gill, P. S.; Srivastava, S. C.; Johri, D. P. Manual of Sugarcane Production in India. National Institute of Science. Council of Scientific and Industrial Research, New Delhi, **1998,** pp. 78.
5. Johnson, F. Sustainable Energy Program. Stockholm Environment Institute, Stockholm, Sweden http://www.sie.se. **2012.**
6. Rice, R. W. Florida Sugarcane Hand Book. University of Florida Extension Services, Gainesville, Florida, USA, **2009,** pp. 148.
7. Solomon, S.; Singh, G. B. Sugarcane diversification: Recent developments and future prospects. In *Sugarcane: Agro-industrial alternatives*, Singh, G. B.; Solomon, S. Ed.; Oxford IBH Co. New Delhi, India, **2005,** 523–541.
8. Verma, R. S. Sugarcane Production Technology in India. International Book Distributors. New Delhi, **2004,** pp. 628.

USEFUL WEBSITES

http://english.unica.com.br/
http://www/nfsp-philippines.org/
http://www.srdc.gov.au

http://www.pinoy-entrepenuer.com/2011/06/02/sugarcane-production/
http://www.parc.gov.pk/1Subdivisions/SARCKHY/nscr.html
http://www.netafimindia.com/agriculturesugarcane.html pp. 1–4 (July 26, 2012)
www.iisr.nic.in (October 12, 2012)

CHAPTER 22

FORAGES AND PASTURES

CONTENTS

22.1 INTRODUCTION

Forage and Pasture development is an ancient practice. Occurrence of domesticated farm animals meant need for regular supply of forage grasses/legumes. Historical data suggests that forages such as Alfalfa was in vogue in Turkey and Asia Minor region during 1300 B.C. Evidences for forage crops could be derived from painting in Egypt dating 2000–1300 B.C. Archeologists suggest that silos were regularly used by Carthegians during 1200 B.C. [1]. Forages were popular during Roman Era. Writings by Varro (116–27 B.C.) and Cato (234–149 B.C.) indicate culture of forages such as alfalfa, cowpea, turnip, lupines and vetches.

Hirata [10] opines that several forage species were brought into cultivation in European Plains and Mediterranean region during Early Christian era. In the Mediterranean region, forage species such as *Trifolium* and *Alfalfa* dominated the pastures during tenth century. However, a kind of plateau occurred until Middle Ages regarding spread of forages. During dark ages, forage culture was given primacy. The invasions and fall of Roman Empire reduced interest in rearing of farm animals and forage cultivation. It is said even forages like Alfalfa and grasses vanished from expanses in Northern Italy. Revival of forages occurred in Spanish Pyrenees during 8th and 9th century. Cultivation of turnip, red clover and vetches became popular in Northern Europe and British isles during 14th century. Australian farmers received forage species during late 18th century, but standard rotations that included forage production began during early 1900s [2].

During 20th century, forage was valued more by the farmers. Several forage species were identified and their cultivation was intensified in different parts of the world (Table 1). Expansion and intensification of forage and pasture was actually driven by Farm animal population.

TABLE 1 Forage Grasses and Legumes: Origin, Nomenclature, Uses—A summary.

Forage production is an ancient system. Archeological evidences suggest that pasture grass/legumes were grown during 2nd millennium B.C. in Turkey and other regions of Middle-east Asia. Development of pastures and forage zones around citadels became a popular practice during medieval period. Presently, pastures and forage cropping belts are widely spread. They occupy one third of agrarian zones on earth. Reports suggest that about 10,000 crop species suit to be cultivated as forage/pastures. Yet, we routinely cultivate only 100–150 species to develop pastures and forage fields [3]. Several grass and legume species have been domesticated to be used as forages to feed farm animals. A few examples are listed below:

Forage Grasses

Avena nuda, Cynodon species, Brachiaria species, Lolium perenne, Pennesitum purpureum, Panicum maximum, Paspalum notatum, Cenchrus ciliaris, Chloris guyana, Avena spp, Hordeum spp, Setaria italic, Sorghum sudanense, Zea mays.

Forage Legumes

Madicago sativa, Mellilotus sp, Trifolium repens, Stylosanthes sp, Dolichos sp, Macrotyloma uniflorum, Trigonella, Calapogonium sp, Crotalaria, Cajanus sp, Cenrosema pubescence, Glycine max, Vicia sp, Vigna unguiculata, Leucana leucocephala.

Nomenclature (Expression for the Word Forage in Different Languages)

Bulgarian-Furazhni Rastenyi; Croatian-Stocna hrana; Czech-Picninarstvi, Danish-Fodrarealer; Ductch-Voer; Finnish-Rehuala; French-Fourareges; German-Futter; Hindi-Khurak, Telugu-Meta, Kaya; Kannada-Mevu; Italian-Foraggio; Norwegian-For; Polish-Pestewne; Portuguese-Forragen; Romanian-Furagere; Persian-Kortha; Spanish-Forraje; Swedish-Grovfoderareal; English-Forage.

Uses

Pastures are used primarily to feed farm animals with nutritious green leafy fodder. Dried forage material is stored and used as 'hay.' Sometimes forage is silaged and processed by inoculating with microbes known as cellulose digesters. Microbial inoculation affects biochemical constitution of forage. It enhances silage quality. Forage grass could be used to mulch crop fields. Forage portion not used as feed, could be effectively recycled to improve soil fertility and quality. Recycling forage improves C-sequestration in the agroecosystem.

A few of cereal forages are used a bird feed. Cereal/legume stems could be retted and used in preparing baskets and other handicraft material. A few of cereal/grain powder are used in preparing medicines. Leaf decoctions from legumes are used to cure medical ailments.

Nutritive Value

Nutritive value of cereal/legume forage species varies markedly based on species and its genotype. The geographic location, season, soil fertility status and agroclimate may all affect the mineral and biochemical constitution of forage species.

Source: Refs. [4–6].

Note: List of forage grass and legumes includes only few examples of grasses and legumes. It is not exhaustive.

22.2 SOILS AND AGROCLIMATE

Forage productivity is actually influenced by interaction of factors such as physiography, topography, soil and climate. Intensity of forage culture and productivity are actually affected by complex interactions of several natural resources. Forage belt in North America thrives on variety of soil types that vary with regard to parent material, profile characteristics, texture, structure, nutrient content, availability, organic matter content, microbial load and productivity. Parent material of soils that support forage grasses and alfalfa in USA, could be a combination of limestone, sandstone, shale, metamorphic rocks, igneous rocks and loess material. Natural grasslands and forages occur on soil types such as Mollisols, Ultisols, Oxisols, Histosols, Andisols, Spodosols, Entisols and Inceptisols in North America. Mollisols found in the Central plains supports intensive production of forage grasses and legumes. Aridisols found in New Mexico and Arizona are usually associated with dry shrubs, dry savanna grasses, but they are also used to cultivate forage grasses. Aridisols may exhibit accumulation of Ca in the profile. Forages are also grown on Gelisols found close to permafrost in Alaska [7].

PLATE 1 Field with Pasture Grass near Corvallis in Oregon, USA. The Oregon State of USA, is reputed to produce highest quantity of Forage grass/legume seeds.
Source: Krishna, K.R., Bangalore, India.

Pastures and forage belts in USA, experience different agro-climatic patterns (Plate 1). Precipitation is 500 mm annually in Eastern USA, with a maximum of 1700 mm in Mexican Gulf region and 3,000 mm in the Appalachians. The precipitation in western USA, depends on Pacific currents. It ranges from 100 mm in the dry deserts of Arizona and New Mexico to 5000 mm in mountains, for example in Cascades. Pastures endure severe cold temperatures of −25°C in the Northern Great Plains and Canadian Prairies. Pastures in Western USA, experience cold temperatures wherever cool-air pooling occurs. However, in southern region, forage crop production is conducted at an average temperature of 8°C in Alabama and Texas. Maximum temperatures may reach 25°C–31°C in the Southern regions of Florida [8].

Brazil supports one of largest forage cropping belts. It spreads into variety of tropical soils and agroclimatic zones. Forage belt in Brazil flourishes on low fertility Oxisols in most part of Cerrados and Sao Paolo region. It encounters loamy soils in the Amazonia and lateritic belts in the Hilly regions. Alfisols and low fertility Ultisols are traced in pasture belts of Matto Grasso and fringes of Gran Chaco region. In Cerrados, about 32 pasture species are cultivated on Oxisols and Ultisols. The agroclimate is classifiable as semiarid. The average annual temperature may range from 10°C–18°C depending on the location and topography. The pastures may experience drought stress and encounter acid soils with Al and Mn toxicity. It is common to spread or sprinkle gypsum to correct soil pH and improve P absorption by the pastures.

Pampas in Argentina is a great repository of forage species. Pastures and forage fields are large and include both cereals and legumes. A few crop species such as sorghum, maize and barley are also grown to serve as forages to feed farm animals. Soil types that support forage farming are classifiable as Mollisols, Inceptisols and Alfisols. Soils are moderately fertile. They show dearth for N and P. Major forage farming zones occur in Cordoba, Pampas and Santa Fe regions. The general agroclimate could be categorized as subtropical, Mediterranean or temperate. Temperature is low in the Patagonian region where only cold tolerant forage species thrive. Rainfall pattern, length of the growing season and inherent soil fertility dictate forage yield. Large scale legume pastures have been useful in enhancing soil-N status. Forages are grown as part of rotations that involve a major cereal like wheat or maize. The productivity of forage is moderate at 2–3 t/ha. Currently, pastures are vast and are amenable for assessment through remote sensing. Influence of factors such as forage species, seasonal variations, fertilizer sprays if any and precipitation could be easily estimated using Normalized Difference Vegetative index [9].

Vast stretches of forage and natural pastures occur in the European plains. They thrive on fertile Chernozems, Cambisols, Podzols and Peaty soils. Productivity of European forage zones is high at 4–5 t/ha. The European pastures are composed a large number of different grass/legume species. Forage production occurs mostly on soils that are less fertile. Forage belt negotiates cool winters or mediterranian climate depending on geographic location. Precipitation pattern influences forage species and intensity cropping. Legume pastures are usually inoculated with *Rhizobium* to improve soil-N status.

The dry regions of West Asia, North Africa and even Sub-Saharan Africa support vast stretches of forage farms. Soils in West Asia are deep, sandy, Xerasols. They are generally rich in Ca content but low in major nutrients such as N, P and K. Moisture holding capacity is moderate. Precipitation pattern holds the key to forage productivity. Rainfall is often scanty ranging from 300–500 mm. The relative humidity is low in the dry lands and in the fringes of desert. Cropping systems often involves a major cereal such as wheat sequenced with fallow or mixed pastures. Pastures endure cool temperatures ranging from −3 to 10°C during winter and 18°C–32°C during hot summer period. In West Africa, small farmers adopt slash and burn methods to develop pastures. These pastures are kept growing for 3–4 years. The climate is usually dry, precipitation is low at 300–550 mm. Average temperature ay range from 8°C–28°C. Soils are sandy and low nutrient buffering capacity. Residual moisture plays a key role during winter and early summer months. Pastures regenerate with advent of rainy season [10].

In East Africa, mixed pastures are popular in scanty rainy fall zones. They are developed on Alfisols and sandy soils. Pastures in Kenya, Uganda, Tanzania and Malawi receive relatively low rainfall at 500 mm/yr [11]. However, there are regions in Ethiopia and Kenya where pastures occur in regions that receive 1500 mm precipitation. Pastures thrive in tropical climate with temperature ranging from 24°C–36°C. Hilly zones may encounter low temperatures of 10°C–15°C.

In South Asia, pastures are widespread and traceable in temperate Himalayan regions, subtropical regions, humid tropics and dry lands. The productivity depends on soil type, agroclimate encountered and inputs if any. Pastures are traced at high altitudes of 3000 m.a.s.l. in the Sivaliks and Himalayan hill ranges. They experience temperate climate. Temperature often reaches subzero. Weather could be frosty or snowy during winter, Precipitation is not a major factor. Cold tolerance traits in forage species is of utmost importance. Forage farming in the Gangetic plains is conducted during off-season or in fields marked as low fertility. Forage crops usually thrive on stored moisture and residual fertility available after the harvest of main cereal/legume crop. Major soil types used for forage farming are Inceptisols, Mollisols, Alluvial soils, dry sandy soils. The relative humidity is high during monsoon, reaching 75–80%. During summer, relative humidity is low around 40–50% and fields may experience droughty conditions until the onset of monsoon. In the Southern Indian Plains, forage and mixed pastures are included in normally practiced rotations. Productivity depends on soil fertility and inputs. Major soil types used are Vertisols, Alfisols, lateritic soils and coastal alluvial soils. Precipitation ranges from 500–900 mm annually in the plains. However, pastures are allowed to flourish during post rainy season, mainly to use residual moisture in soil and feeble showers that occur during winter and early summer. Forages grass/legumes are cultivated from sea level to hilly regions with 1800 m.a.s.l. in the Western Ghats of South India (Plate 2).

PLATE 2 A field with forage grass at a Subtropical location near Bangalore, India.
Source: Krishna, K.R., Bangalore, India.

The forage production zones in China are traced mostly in regions with poor soil fertility and harsher climatic conditions, namely arid and semiarid plains. Pastures and forage growing regions are spread into thermal zones classifiable as frigid (1300°C), temperate zone (1300–5300°C), subtropical zone (5300–8000°C) and Tropical zone (>8000°C) [12]. Farmers usually match forage species with temperature regimes and precipitation pattern. Annual precipitation ranges from 250–400 mm in arid zones, 500–900 mm in the semiarid belt and from 1,000–2,500 mm in humid tropics. Grass/legumes species well adapted to drought and low rainfall pattern are obviously dominant in the arid zones of North-west China. Their productivity is low at 1–2 t forage/ha. Chinese farmers adopt both perennial and short duration pastures. Usually wheat, maize or soybean is rotated with annual forage species.

22.3 EXPANSE AND PRODUCTIVITY

Globally, land area is 134 million km^2 out of which about 27% is occupied pastures/forage cropping zones. Total agricultural area is 49.6 million km^2. At 35 million km^2 world pastures and forage belts occupy about 70% of agricultural area [13]. Continent-wise, following is the expanse of pastures and forage cropping zones: Asia-1106 m ha; Africa-869 m ha, America-808 m ha, Oceania-419 m ha and Europe-182 m ha [3, 13].

Globally, forage and legume seed production is dominated by few regions namely United States of America (418 m t), European Union (149 m t), Canada (48 m t), Czech (5.7 m t), New Zealand (29.8 m t), Australia (9.8 m t) and Argentina 7.6 m t). Seed production area, quantity and forage species are partially indicative of forage/pasture area and intensity of cropping that may occur in the next season/yr. Globally, forage grasses that hold primacy with regard to annual seed production are Perennial Ryegrass (185 m t), Annual Rye grass (171 m t), Tall Fescue (123 m t), Red Fescue (80 m t), Hybrid Rye grass 7.3 mt), Bromegrass (6.1 m t) (Wong, 2005). Forage agroecosystems support several legume species. Among the legume species following are species that dominate. They are Alfalfa (947 m t), Common Vetch (16 m t), Red Clover (10.2 m t), White Clover (15.1 m t), Egyptian Clover (2.5 m t), and Crimson clover (1.6 m t).

Canada is a premier forage/hay producer in North America. This forage belt supplies products such as forage grass, hay and straw to farmers in North America, as well as to those in Europe and Asia. The Canadian Forage belt is composed of several grass species and two major legumes, namely Alfalfa and Trifolium [14]. The Canadian forage belt is 36 m ha in expanse. About 26 m ha is actually under native grasses/legumes, 4 million ha is occupied by cultivated pastures and 6 m ha is under forage grasses [15]. Forage and pastures are predominant in following 5 states, namely Saskatchewam (64 m ha), Alberta (52 m ha), Manitoba (19 m ha), Ontario (8 m ha). States such as Nova Scotia and New Brunswick region support about 1.0 m ha forage grass/legume mixtures. Together, above states contribute 57% of total forage/hay produced by Canada.

Agroecosystems classifiable as grassland, rangeland, pasture land and crop land are major sources of forage to farm animals and other species. In USA, about 55% of land surface (405 million ha) is used up by forage producing systems. Forage and hay production occurs on natural expanses and planted regions from the Canadian Prairies in the North, extending into Great Plains, North-west Pacific cropping zones, Semi-arid regions, North-east and Piedmont regions. Forages encompassing grasses and crops occur in Southern plains and Florida. Overall, wide variations of soil types and agroclimate are endured by forage crops [16]. Pastures could also be traced very close to Taiga and Tundra regions in Alaska and North Pacific region.

Forage agroecosystem in USA, is large expanse occupying entire region at different intensities. We can classify the belt into silage, hay, alfalfa and green chop. The cropping meant for silage extends into 5 m ha. It offers 10.9 m t silage annually at an average productivity of 2.0 t/ha. Hay production zone in US extends into several states covering 36 m ha. It produces 65 m t ha at a productivity of 2.3 t/ha. Alfalfa producing zones are predominant in Wisconsin, California, New York State, Minnesota, Idaho and South Dakota. It occupies aout 19 m ha and contributes65 m t annually. Forage cropping for green chop is also important in USA. It spreads into 4.3 m ha and produces 31 m t annually at an average productivity of 7.2 t/ha. (USDA Statistics, 2011).

Silage production in USA, involves crops such as Corn and Sorghum. The corn belt meant for silage production extends into vast area. Total corn silage production is about 108 m t annually. Major corn silage zones occur in Wisconsin (15.7 m t/yr), California (12.3 m t/yr), New York State (7.5 m t/yr), Pennsylvania (6.5 m t), Massachusetts (6.2 m t/yr), Michigan (5.4 m t/yr) and Iowa (3.8 m t/yr). Sorghum silage cropping zones are prominent in Texas (112,000 t/yr), Arizona (396,000 t/yr), Colorado (260,000 t/yr), South Dakota (275,000 t/yr), and Georgia (272,000 t/yr). Totally, sorghum belt in USA, contributed 2,300,000 t silage [17].

Alabama in Southern USA, is among major forage/hay producers and users. The Forage cropping zones here extend into 292,000 ha and produce 1.97 m t ha, at an average productivity of 1.2 t dry hay/ha [18]. Forage and pastures are concentrated mostly in Northern Alabama.

Forage seed production is an important enterprise in the Pacific North-west of USA. About 700,000 ha of forages and natural pastures are used for seed production in USA. Oregon has a large area of forage marked exclusively for seed production. Seed farms extend into 220,000 ha, which is equivalent to 38% of forage cropping zone in USA. Missouri, Idaho, Washington, California, Kansas and Minnesota are other forage seed producing region in USA. The agroclimatic requirements are slightly different for seed farms. Fertilizer supply and other agronomic procedures are commensurate with higher nutrient needs of seed farms. Alfalfa is a major forage legume seed produced in California and Washington States. Legumes such as Red clover and vetch are raised for seeds in states such as Oregon, Missouri and Kansa. Vetch seeds are predominately produced in Oregon and Minnesota [19].

In the Mid-west region of USA, crop production is a major preoccupation. Crop production involves high input technology and is relatively intense. Forage production on fertile Mollisols too proceeds with high input. Most important forage grasses and legumes that fill the agroecosystem are as follows:

Grasses: KentuckyBluegrass, Perennial Rye Grass, Timothy grass, Smooth Brome Grass, Sudan grass, Canary grass, Tall fescue.

Legumes: Alsike Clover, Annual Lespedza, Birdsfoot Trefoil, Crownvetch, Ladino Clover, Red Clover, Serica Lespiza, Sweet Clover, Dutch Clover. Other forage species: Big blue stem, Indian grass, Pearl millet, Switch grass, Triticale [20].

Oklahoma and adjoining states grow a wide range of pasture grasses and legumes. Major forage grasses are Bermuda grass, Crabgrass, Bluestems, Bahiagrass and native species. Legumes such as *Medicago* and *Trifolium* are predominant in the Southern Plains [21].

Forages and pastures are key to sustenance of farm animals in most parts of Africa. In Sub-Sahara, factors such as dry weather conditions, scanty and erratic precipitation patterns support forage/pastures rather feebly compared to wet tropics. Yet, crop residue recycling and forage farming are essential aspects of nutrient dynamics in sub-Saharan Africa. Maintaining pastures has several advantages to farmers, in addition to supplying fresh forage to farm animals. They improve land and water use efficiency, enhance biomass production and C –sequestration in the farm, reduce soil erosion and nutrient loss from fields. Sahelian farmers use several species of grasses and legumes to initiate pastures. Mixed pastures offer greater advantages in terms of soil-N fertility, because legumes add N through BNF. The entire stretch of agrarian zones in Sahelian West Africa supports mixed pastures/forages. The productivity is low at 2–3 t fresh forage. However, mixed pastures in Northern Nigeria and adjoining areas contribute 20–45 kg N/ha through BNF. Recently, International Livestock Research Centre evaluated forage mixtures involving *Panicum, Brachiaria, Centrosema, Macuna, Stylosanthes* and cowpea in Western Nigeria. Olanite et al. [22] found that grass/legume mixtures were relatively more productive and offered 9–10 t forage/ha. Solc cereals such as *Brachiaria* gave mere 5.5–6 t forage/ha. Mixed pastures also add to nutritional quality of fodder. They improve mineralization processes in soil upon incorporation.

Forage and pastures are common to Dry regions of North Africa and West Asia. Forage grasses are cultivated during fallow period. The productivity of pastures is low at 1–2 t forage/ha. Forage species such as oats, rye, vetch, alfalfa, lentils are wide spread in West Asia and North Africa.

Pasture and forage farming is an important preoccupation in large tracts of Southern Africa. Forage grass/legumes are grown on soils with low fertility. Forages are prominent in South Africa, Mozambique, Botswana, Angola and Malawi. During recent decades production of pastures have drawn greater attention because of higher priority to milch cattle. During 1970 s, many of the countries in this region went for systematic evaluation of large number germplasm lines of pasture grasses and legumes. Pasture grass from different regions of the world were tested for survival and productivity in South Africa and Swaziland. In case of Swaziland, about 100 grass/legume species were tested for performance as forage. About 17 species were used to expand the pasture zones in Swaziland. The legume/grass mixtures produced 4–5 t forage/ha. A few high yielding *Stylosanthes* species produce about 7.1–7.5 t fresh forage/ha [23]. It is said that legume forages can contribute sizeable amounts of N via BNF, but it may not suffice to support entire season. Therefore, certain combinations of grass/legume pastures are supplied with a 80–100 kg N/ha.

Forage cultivation is in vogue in most parts of China. They are well distributed in semiarid and humid regions of Central and Eastern China. Grass/legume pastures can be classified into grazing pastures (264 m ha), warm season pastures (117 m ha), cold season patures (64 m ha), dual purpose pastures (67 m ha) and Year round pastures (82 m ha) [12]. Among crop species, *Medicago sativa* (1.8 m ha), Astragalus sinicus (1.7 m ha), Caragana koshinski (1.1 m ha) and Vicia villosa [24] are dominant.

The Australian forage agroecosystem is relatively recent. Forage production systems were introduced into Australia during late 18th century. Wheat and forage crop rotations were standardized around 1900 A.D. Cereal-forage sequence became popular in Southern part of Australia. Soon, annual legume forages were developed in different states of Australia. At present, we can identify 4 major farming-grazing systems. They are prominent in farming regions of Australia that experience Mediterranean climate. They are: Ley faming system, Phase Farming, Forage Farming and Permanent pastures [2]. Ley farming is practiced in areas with relatively high precipitation levels. Legumes species such as *Trifolium subterrainium, T. vesiculosum, T. glanduliferum, Medicago polymorpha, M littoralis, M truncatula* and *Trigonella balanse* are cultivated as ley in zones receiving about 350 mm rainfall. Sandy soils with poor water holding capacity fertility are preferred. Phase Farming involves establishment of perennial legumes/grasses. Usually, dry land Leucerne, perennial ryegrass, *Dactylis glomerate* and *Festuca arundinaceae* are grown under Phase farming systems. Phase farming is preferred when seed cost is low and repeated seeding is a clear possibility, seed dormancy is moderate, legumes are able to fix sizeable amounts of atmospheric-N and maturity period is not long [2]. Forage farming is dominant in areas where crop production is given a long break. Major cereal such as wheat or maize is not to be planted for few season, a forage grass/legume is sown. Late maturity legumes with ability improve soil-N status are most preferred. Permanent pastures are also wide spread in Australia. During recent years, main cereal such as wheat or maize is followed by a minor mille such as finger millet. Seeds produced are used to feed birds and forage to nourish farm animals [25].

Let us consider the forage species. The forage belt in Australia encompasses wide range of grass, legume and other species that offer nutritious feed to farm animals. In Australia we encounter both natural pastures and planted forage production zones. The forge is dominated by following grass species that for pastures. They are: African Star grass (*Cynodon lemfluensis*), Angelton grass (*Dichanthium aristum*), Bahia grass (*Paspalum notatum*), Birdwood grass (*Cenchrus setiger*), Bluegrass (*Bothroichloa insculpta*), Buffel grass (*Cenchrus ciliaris*), Blue couch (*Digitaria sp*), Guinea grass (*Panicum maximum*), Rhodes grass (*Chloris guyana*) and Elephant grass (*Pennesitum purpureum*). Legume species that are conspicuous in the forage belts of Australia are *Calapogonium mucunoides, Centrosema pubescese, Trifolium subterranium, Trifolium repens, Vigna unguiculata,* V. *sinensis,* V. *species, Desmodium intortun, D. unicinatum, Dolichos auxilaris, D. uniflorus, D.lablab, Glycine wightii, Luecana leucoscephala,* etc. [27]. The productivity of forage belts is moderate in most parts of Australia. It ranges from 2–2.5 t forage/ha.

KEYWORDS

- **Alfalfa**
- **Carthegians**
- **Rhizobium**
- **Stylosanthes**
- **Trifolium subterrainium**

REFERENCES

1. Hirata, M. Forage Crop Production. The Role of Food, Agriculture and Fisheries in Human Nutrition-Forage Crops. Encyclopedia of Life Support Systems (EOLSS). **2010,** 1–23 http://www.eolss.net/Eolss-sampleAllChapter.aspx (November 10, 2012).
2. Norman, H. C.; Ewing, M. A.; Loi, A.; Nutt, B. J.; Sundral, G. A. The pasture and forage industry in the Mediterranian bioclimates of Australia. In: Proceedings for Mediterranean Forge Crops, Pastures and Alternative Uses. Sasseri, Italy, **2001,** 469–475.
3. Panunzi, E. Are Grasslands under threat. Food and Agricultural Organization of the United States of America. Rome, Italy, **2008,** 1–34.
4. University of Minnesota, Forage Legumes. University of Minnesota Extension /sevice bulletin. http://www.extension.umn.edu/distribution/cropsystems/DC5963a.html. **2012,** (October 20, 2012).
5. Miles, J. W.; Lascano, C. E. **1997,** status of *Stylosanthes* development and utilization in South America. Tropical Grasslands 31, 454–459.
6. Louw-Gaume, A. E.; Rao, I. M.; Frossard, E.; Gaume, A. Adaptive strategies of Tropical Forage Grasses to Low Phosphorus. In: Hand book of Plant and Crop Stress. 3rd Edition. Pessaraleti, M. (Ed.). CRC Press: Boca Raton, Florida, USA, **2010,** 1111–1144.
7. Brown, J. R.; Hannaway, D. B.; Fribourg, H. A.; Pieper, R. E. Soil and Topography. In: Country Pasture and Forage Resources profiles of United States of America. Hannaway, D. B.; Fribourg, H. A. (Eds.) http://www.fao.org/ag/AGP/ AGPC/doc/Counprof/usa/usa.html **2012,** 75 (November 9, 2012).
8. Daly, C.; Taylor, G. H.; Hannaway, D. B.; Fribourg, H. A. Climate and Agro-Ecological zones. In: Country Pasture and Forage Resources profiles of United States of America. Hannaway, D. B.; Fribourg, H. A. (Eds.) http://www.fao.org/ag/AGP/ AGPC/doc/Counprof/usa/usa.html **2011,** 75 (November 9, 2012).
9. Di Bella, C. M.; Negri, I. J.; Posse, G.; Jaimes, F. R.; Jobbagy, E. G.; Garbulsky, M. F.; Deregibus, A. Foragae production of the Argentine Pampa Region based on Land Use and Long-term Normalized Difference Vegetation Index data. Rangeland Ecology and Management **2009,** *62,* 163–170.
10. Adjolohon, S.; Bindelle, J.; Adandedjan, C.; Buldgen, A. Some suitable grasses and legumes for ley pastures in Sudanian Africa: The case of the Borgou region in Benin. Biotechnology, Agronomy, Society and Environment **2008,***12,* 405–409.
11. Wheeler, J. L.; Jones, R. L. The potential for forage legumes in Kenya. Tropical Grass lands **1977,** *11,* 273–285.
12. Zizhi, Hu and Degang, Z. Country Pastures-Forage Resources Profile. http://www.fao.org/ag/AGP/AGPC/ doc/Counprof/china/china1.htm **2006,** 1–22 (December 3, 2012).
13. FAOSTAT, Forage Statistics. Food and Agricultural Organization of he United Nations. Rome, Italy, http://www.Faostat.org **2012,** (November 10, 2012).

14. Small, E. Forage Crops. The Candian Encycylopedia. http:www.thecanadianencyclopedia.com/articles/ forage-crops **2012,**1–2 (November 12, 2–12).
15. Canadian Forage and Grassland Association **2011,** Canada: Premier Forage Supplier. http://www.canadianfga. ca/# 1 (November 10, 2012).
16. Hannaway, D. B.; Fribourg, H. A. Country Pasture and Forage Resources Profiles. http://www.fao.org/ ag/AGP/AGPC/doc/Counprof/usa/usa.html **2012,** 75 (November 9, 2012).
17. USDA Statistics, 2011–2011, US Forage Statistics. http://www.progressivepublish.com/downloads/2012/ general/2012-fg-stats-lowers.pdf 1–4 (November 12, 2012).
18. Jaeger, N. Turning Green Grass into Greenbacks. Alabama Farmers Federation, Montgomery, Alabama, **2011,** 1–3.
19. McLellan, B. Forage, Turf and Legume Seed Production in the United States. Alberta Ag-Info Centre. http://www.Alberta.ca.htm **2012,** (November 11, 2012).
20. Johnson, K. Forage Grasses and Legumes. http:www.purdue.edu/newcrop/cropmap/Indiana/crop/forage.html **1998,** 1–3 (November 7, 2012).
21. Samuel Roberts Noble Foundation Forage Management: Summer Pasture Grass Choices. Samuel Roberts Noble Foundation, Ardmore, Oklahoma, USA, **2012,** http://www.noble.org/ag/pature/horse-forage/summer-pasturegrass/ 1–3 (November 1, 2012).
22. Olanite, J. A.; Tarawali, S. A.; Akenova, M. E. Dry matter yield and botanical composition of three grasses and two legume mixtures grazed by cattle in a derived savanna area of Nigeria. International Journal of Agricultural Sciences, Environment and Technology. **2009,** *9.*
23. Ogwang, B. H. Research on Forage Legumes in Swaziland. http://www.fao.org/wairdocs/ilri/ x5488e/ x5488eOK.htm 1085, 1–9 (November 11, 2012).
24. Baoshu, C. Cultivation of Forage Grasses and Forage crops. China Agricultural Science and Technology Press, Beijing, China, **2001,** 248.
25. Lacey, T. Forage millet growing in Western Australia. Department of Agriculture, Farmnote, **2007,** *93,* 1–4.
26. Humphreys, L. R. A guide to better Pastures for the Tropics and Sub-Tropics. Wright Stephenson and Company. Silver water, New South Wales, Australia, **1980,** 1–108.

EXERCISE

1. Mention botanical names of at least top five legumes and grass forage species in each major agrarian zone.
2. Write a short assay of 20 lines on Alfalfa-the queen of pastures.
3. List ten legume forage species and state the extent of atmospheric-N fixed through Biological Nitrogen fixation.
4. Mention a few rotations that involve forages.
5. Mention the advantages of growing grass pastures.
6. Mention top five forage producing countries. State the area, productivity and production of forage in these nations.

FURTHER READING

1. Burson, B. L.; Young, B. A. In: *Breeding and Improvement of Tropical Grasses.* CRC Press: Boca Raton, Florida, USA, **2001,** pp. 248.
2. Hopkins, A. In: Grass, its Production and Utilization. Blackwell Science Ltd: Oxford, **2000,** pp. 440.

USEFUL WEBSITES

http://plants.usda.gov/100_native_grasses.pdf (November 10, 2012)
http://www.crcnetbase.com/doi/abs/10.1201/9781420038781.ch5 (November 10, 2012)
http://www.ilri.cgiar.org/InfoServ/Webpub/fulldocs/X5488E/X5488E0P.HTM (November 10, 2012)
http://aciar.gov.au/files/node/555/proc115.pdf (November 10, 2012)
http://www.gov.mb.ca/agriculture/crops/forages/bjd29 s01.html. 1–4 (November 9, 2012)

CHAPTER 23

PLANTATION CROPS OF THE WORLD

CONTENTS

APPLE ORCHARDS OF THE WORLD

23.1 INTRODUCTION

Apple is the most popular fruit consumed all over the world. It is an important gift from temperate agrarian zone to humans all over the globe. Apples are perhaps the earliest of the fruit trees domesticated by humans. Apple is a native horticultural species of Central Asia. Apples were domesticated in the region comprising present day Kyrgyzstan, Kazakhstan, and Tajikistan. There are 35 species of the genus *Malus*. Most of these are small deciduous trees and native to Northern Hemisphere. They are traced in Europe, Asia and North America (Plate 1). The cultivated species of apple is called *Malus domesticus*. It was probably derived through selection from wild species known as *Malus sieversii*. Apples reached Mesopotamia during 3800 B.C. and later became an integral part of Greece and Roman food production systems [1]. Apple production has been in vogue in Italy since ancient times. It was concentrated in Apennines. Cultivars such as Gelata, Annurca and Limoncell were cultivated by ancient romans [2]. Apples were introduced into New South Wales and Tasmania by early English settlers (Table 1) [3, 4].

TABLE 1 Apple: Origin, Classification, Nomenclature, Uses and related Salient information—A summary.

Malus domesticus was domesticated in Central Asia by the natives of Kazakhstan, Kyrgyzstan, Tajikistan and Xinjiang in North West China. The center of origin and region of maximum diversity occurs in the mixed forests of Central Asia. Apples have been consumed by the Central Asian tribes since 7,000 years. Currently, Apple orchards are widely distributed in many regions of the world. It is among the top most fruit crops of the world, in terms of expanse and production.

Classification

Kingdom-Plantae; Sub-Kingdom-Angiosperms; Class-Magnoliopsidae; Order-Rosales; Family Rosacea; Subfamily Maloideae; Tribe-Malae; Genus-Malus; Species-*Malus domesticus*
$2n = 17$

Wild Species-*Malus sieversii*; Related species-*Malus sylvestris*

There are 7,500 varieties of apples grown in different parts of the world. Major germ plasm centers are available in Europe and Washington State, USA. Many of the apple varieties grown in North America have ceased because of excessive preference to only few delicious cultivars. Hence, conserving the large number of apple cultivars is a major task for geneticists. It seems in USA, alone; over 6,500 cultivars have not found favor with farmers for cultivation. Many of the apple genotypes have become nostalgia. Only 100 cultivars are traceable in the Apple belt of North America [1, 5].

Nomenclature

Catalan-Pomes; Czech-Jablko, Jablecny; Danish-Aeble; Dutch-Appel; French-Pomme; German-Apfel; Italian-Melas; Polish-Jablko; Portuguese-Maca; Romanian-Mere; Russian-Yavlonya, Zrachok, Yavloko; Spanish-Manzana, Poma; Swedish-Apple; Turkish-Elma; Hindi-Seb; Kannada Seb, Bengali-Seb.

Uses

Apples fruits are consumed raw and cooked much like vegetables and consumed. Ripened, sweet fruits are most common worldwide. There are innumerable delicious fresh fruit varieties consumed in different parts of the world. Apples are consumed in large quantities after extracting juices, as jellies and jams. Apples fruits are used as substrates to produce alcoholic beverages and other fermented products. Apple cider is produced in large scale in temperate regions. Apples are used extract starch for home and industrial purposes. Raw apples are supposedly good for those affected with constipation. Apples are also consumed after baking. They are prescribed for patience with diarrhea and dysentery. Apple milk and curds are consumed in Tibet. Apple toffees are traditional food items in England. Similarly, apple candies are consumed in large quantities in United States of America. Apple and honey mixtures are common in Middle East Asia. Apple syrup is used as a base for many medicines. Apple based sweets are popular in European nations. Apple pulp is used as FYM, pith material and for generating biogas.

Nutritive Value

A 100 g of apple fruit contains:

Energy 52 cal; carbohydrate-13.8 g; Sugars-10.4 g; Dietary fiber-2.4 g; Fat-0.17 g; Proetin-0.26 g; Water-85.6 g; Vitamin-3 μg; Thiamine-0.017 mg; Riboflavin-0.026 mg; Niacin-0.091 mg; Pantothenicacid-0.061 mg, Vitamin B6–0.041 mg; Folate-3 μg, Vitamin C-4 mg; Calcium 6 mg; Iron-0.12 mg, Magnesium-3 mg; Phosphorus-11 mg; and Zinc-0.04 mg.

Source: USDA National Data base http://www.nal.usda.gov/fnic/foodcomp/research/. (October 1, 2012).

www.newworldencyclopedia.org/entry/Apple (August 25, 2012).

PLATE 1 Top Left: Apple twigs that develop into fruits. Top right and Bottom: Apple Fruits. *Source:* http://ikenberryorchards.com/orchard.php (June 10, 2012) and Krishna, K.R. Bangalore, India.

23.2 SOILS AND AGROCLIMATE

Apples are versatile and adapt to wide range of infertile and fertile soils available in temperate regions of the world. Basically, apple trees need deep soils that are sandy or silty and are well drained. In North America, apple orchards occur on gravely soils found in Pacific North-west and Northern Great Plains (Plate 1). In Brazil, apple orchards are raised on soils that are acidic, low in P and optimum for K availability. Soils are usually amended with P and K based on soil tests and gypsum is added to enhance soil pH to 6.5 [6]. In Europe, Cambisols, Chernozems and loess soils support the apple belt. In West Asia and southern Europe, Calcic Xerasols are used to raise apple orchards. In India, apples are gown on sandy or silty loams of the Himalayan region. Soil are not very deep, but are well drained and devoid of stagnation. Soil fertility is low to moderate and needs fertilizer replenishments periodically [7]. In Australia, apples are produced on fertile soils that are deep, well drained and possess optimum levels of available nutrients. The optimum pH range is 5.5–6.5.

Apple trees require a temperate climatic regime for growth and fruit production. The average temperature range in apple belts is 12°C–24°C. the winter temperatures may reach −3°C or −4°C. the apple orchards most often need a chilling period of 1,000–1,500 hr at 7°C or below. This chilling period allows normal fruit bearing. Apple orchards thrive under rain fed conditions in regions with 300–400 mm precipitation. Apple farms are also irrigated using different methods such as sprinklers, drip and flat beds.

23.3 EXPANSE AND PRODUCTIVITY

Globally, about 69.5 m t apples were harvest during 2010 [8]. Peoples Republic of China is the worlds' top most apple producer. It contributed 33.2 m t fruits during 2010. Other major apple producers during 2010 were United States of America-4.2 m t, Turkey-2.6 m t, Italy-2.2 m t, India-2.16 m t, Poland-1.8 mt, France-1.7 m t, Iran-1.6 m t, Brazil-1.27 m t, Chile-1.1 m t. Productivity of apple orchards in different regions varies enormously based on topography, soils, its inherent fertility, a fertilizer, irrigation and other agronomic measures, and yield goals.

Apple orchards in Canada are concentrated in Annapolis valley in Nova Scotia, St John River valley in New Brunswick region, Southern Quebec, St Lawrence valley, Great lakes area in Ontario and Okanagan valley in British Columbia.

The Apple agroecosystem in United States of America supports over 2,500 cultivars. However, only 100 cultivars seem to dominate the apple landscape in Washington and other regions of the nation. Commercial apple nurseries were initiated in United States of America during 1730 s. Reports suggest that earliest family farms produced apples during 1870. Simultaneously, apple orchards were also developed by large corporations [9]. Organic farming is currently popular among apple orchards in Washington State. In fact, supply of fertilizer-N, organic manures and irrigation are important factors that influence apple fruit production [10].

The apple belt in United States of America is large and concentrated in Washington, New York, Michigan, California, Pennsylvania and Virginia. Together these states contribute 83% of nations' apple fruit. Washington State alone produces 60% of the nation's apples. The apple belt in Washington occurs on the fringes of Cascade Mountains. Nearly, 60% of apples are consumed as delicious sweet fruits, 21% get processed to Juice and Cider and rest is converted into other products such as jams, jellies and other food items.

Argentina is the world's second largest apple producer. It contributes 15% of global apple produce. During 2011, Argentina harvest 9.7 m t apple fruits. During 2011, apple belt extended into 30,000 ha in Argentina. About 28,000 ha were harvested rest was filled with nonbearing trees. Apple Agroecosystem is mostly made up of cultivars such as Red Delicious, Gala, Fuji, Telstar, Golden Delicious, Granny Smith. Among Yellow Apples, Golden delicious is most common. Over all, apple is dominated by Red Delicious and Granny Smith. Red Delicious orchards produced 65% of Argentina's total fruit harvest during 2011. Granny Smith is said to occupy 15% of apple agroecosystem in Argentina. Royal Gala is a variety that has been gaining in area in Argentina. Nearly 15% of fruit harvested during 2011 were Gala [11]. In Argentina, apple orchards are also traceable in Patagonia. The orchards around 20–50 ha large and about a 1,000 ha are used to produce apples.

In Europe, temperate climate and fertile Cambisols allow large-scale apple farming. Majority of apple orchards occur in the regions covering Spain, France, Germany, Italy, Great Britain, Poland, Ukraine and Southern Russia. France (1.7 mt fruits/yr), Germany (1.1), Poland (1.8 m t fruits/yr), Russia (1.47 m t fruits/yr), Ukraine (0.71 mt/yr), Hungary 90.57 mt fruits/yr) and Spain (0.68 m t fruits/yr) are major apple producing nations. Annually, these nations contribute over 1.7 m t fruits. Apple cropping zones in Italy are predominant in the Alps region. However, apple production in Italy occurs throughout the agrarian regions of the nation. The Italian apple orchards occur from seal level to 800 m.a.s.l. Orchards are high input production systems in some areas and low-input organic farming enterprises in other regions. Major cultivars the fill the apple belt in Italy are Gala, Fuji, Braeburn, and Goldrush. Organic farming systems to regulate nutrient dynamics and productivity are being popularized [2]. Pruning and training systems, multilayer cropping and use of semidwarfing root stocks have been encouraged.

Apple plantations are conspicuous in several regions within North Africa, West Asia and Central Asia. Turkey contributes 2.6 m t apple fruits annually. Iran has a large apple belt that produces 1.6 m t fruits annually. Turkey has a large apple agroecosystem extending into Isparta, Nigde, Karaman and TransTaurus. Apple orchards are supplied with fertilizers based on soil tests and irrigated using drip irrigation. The apple productivity in Turkey improved markedly during past 4–5 decades from mere 585 kg fruits to 2186 kg fruits/ha [12]. Apple orchard size and fertilizer supply seems to affect the productivity and economic efficiency of enterprise. There are several other apple producing nations in Middle East and Central Asia. They are Egypt (0.55 mt/yr), Uzbekistan (0.58 m t/yr), Syria (0.36 mt/yr), Azerbaijan (0.2 mt/yr), Tajikistan (0.18 m t/yr), Kyrgyzstan (0.13 mt/yr), Armenia (0.11 m t/yr) and Turkmenistan (0.11 m t/ yr). The productivity of apple orchards in these regions is moderate at 1–3 t fruits/ha. Since many of these orchards are held under low input subsistence systems. Apple orchards are also traceable in the cool highlands of Ethiopia in North-east Africa. Apple farming is confined to Amahara region, where in organic farming is preferred to produce apples.

Apple agroecosystem of India is confined to regions with temperate climate, such as the North-west Himalayan region, Jammu and Kashmir, Himachal Pradesh, Uttar Pradesh, Arunachal Pradesh, Nagaland and Meghalaya [13, 14]. Some parts of Nilgiris in South India also apple orchards. In India temperate climate is available mostly in hilly tracts and locations at 1500–2700 m.a.s.l. Apples are exposed to the 1000–1500 hr. of chilling below 7°C in these regions. Apple orchards in the Himalayan zone mostly experience cool climate at 14°C-22°C during the year. It seems during 1950 and early sixties, apple belt were dominated by cultivars such as green English, Macintosh, Baldwin, Jonathan, Golden Delicious and Black Ben Davis. However, during recent years apple orchards are mostly covered by high coloring, early maturing types such as Starkrimson, Well spur, Red spur, Red chief, Silver Super etc. The red delicious types occupy 83% of apple expanse in North India [13]. Many of the spur mutants produce 50% more yield than traditional varieties. Root stocks with dwarfing or semidwarfing genes and robust rooting pattern are preferred by apple farmers in in India. High density plantings allow 250 trees/ha; Ultra high density planting allows 1000–1250 trees/ha. In contrast some European nations practice super high-density plantings of 20,000 trees/ha. Training and pruning is based on farmers' preferences. In India, apple orchards are provided with 700 g N, 350 g P and 700 g K/tree. Micronutrient sprays are made based on soil test and plant analysis [13].

China harvested nearly half of the apples produced in the world. It seems during 2003, the total exchequer to China from apples crossed 10 billion US $. Apple agroecosystem in Australia extends into New South Wales, Victoria, Western Australia, Tasmania, and South Australia. Annually, 200,000 t of apple fruits are harvested from the apple belt of Australia (Plate 2). The apple belt is filled by varieties such as Pink lady, Granny Smith and Sundowner. In Queensland, a variety known as 'Gala' dominates the plantation zones. Apple genotypes are selected based on fruiting season. Australian apple belt is occupied mostly by orchards with high density plantings. These orchards accommodate over 1000 trees/ha. The orchards are developed on root stocks with dwarfing genes [3, 4, 15].

PLATE 2 Apple growing regions of Australia; Right Apple variety (Red) is popular in Australia.
Source: http://www.rirdc.gov.au/programs/established-rural-industries/pollination/apple.cfm (August 23, 2012).

KEYWORDS

- Gala
- *Malus*
- sprinklers
- Starkrimson

REFERENCES

1. GCDT, Apple: Priority crops. Global Crop Diversity Trust, Rome, Italy, **2012,** 1–2.
2. Neri, D. Low-input Apple Production in Central Italy: Tree and Soil management. Journal of Fruit and Ornamental Plant Research **2004,** *12,* 69–76.
3. Aussie Apples, Growing Regions-Tasmania: Australia's Island Orchard. http://www.aussieapples.com au/growing-regions/Tasmania.aspx. **2012a,** 1–3 (August 12).
4. Aussie Apples Growing regions-Western Australia. http://www.apples.smartviewer.com.au/apple_info/growing regions/western_australia-wa/ **2012b,** 1–3 (August 30, 2012).
5. Evans, R. C.; Campbell, C. S. The origin of the apple subfamily (Maloidae: Rosacea) is classified by DNA sequence data from duplicated GBSSI genes. American Journal of Botany **2002,** *89,* 1478–1484.
6. Wilms, F. W.; Basso, C. **2011,** Soil amelioration for Apple Production in Southern Brazil. http://www.actahort.org 1–2 (August 26, 2012).
7. Agrinfo, Soil requirements for Horticultural crops. My Agricultural Information Bank. http://www.Agrinfo.in. **2012.**
8. FAOSTAT, Apple Statistics. Food and Agricultural Organization of the United nations, Rome, Italy. http://www.FAOSTAT.org **2011,** (December 4, 2012).
9. Washington Apple Country, Washington Apple Country History. http:// www.appleorchardtours.com/ history.com **2012,** 103 (August 25, 2012).
10. TerAvest, D.; Smith, J. L.; Carpenter-Boggs, L, Granatstein, D.; Hoagland, L.; Reganold, J. Soil Carbon Pools, Nitrogen Supply and Tree Performance under several ground covers and compost rates in a newly planted Apple Orchard. HortScience **2011,** *46,* 1687–1694.
11. Balbi, M. J. Argentina-Fresh Deciduous Fruit Semi-Annual. Global Agricultural Information Network, USDA Foreign Agricultural Services. Gain Report. **2011,** 1–25.
12. Gul, M. Technical Efficiency of Apple Farming in turkey: A case study covering Isparata, Karaman and Nigde Provinces. Pakistan Journal of Biological Sciences **2006,** *9,* 601–605.
13. Avasthi, R.; Chauhan, S. Apple (*Malus pumila*). http://www.fruitipedia.com/apple.Malus pumila.htm **2011,** 1–13 (August 25, 2012).
14. Uma, Cash crop farming in the Himalayas: The importance of pollinator management and managed pollination. Food and Agricultural Organization of the United Nations, Rome, Italy, 1–8 http://www. fao.org/ docrep/005/yr4586E/yr586e11.htm **2005,** (August 26, 2012).
15. ICT, Apple size and Irrigation Management. http://www.icinternationan.com.au/appnotes/ICT232.htm **2012,** 1–3 (August 26, 2012).

EXERCISE

1. Which is the area of origin of *Malus species*? Mark the Apple growing regions of the World on a map.
2. Discuss the trends in Apple Production, Area and Productivity in different continents.

3. Mention various uses of Apple.

FURTHER READING

1. Feree, D. C.; Warrington, I. J. Apples: Botany, Production and Uses. CAB International: Oxford, England, **2003,** pp. 660.
2. Tresnik, S.; Perente, S. Apple Production. http://www.pan-europe.info/Resources/Reports/Apple_production_review.pdf, **2007.**

USEFUL WEBSITES

http://web.archive.org/web/20080121045236/http://www.uga.edu/fruit/apple.html
http://www.plosgenetics.org/article/info%3Adoi%2F10.1371%2Fjournal.pgen.1002703
www.indiastat.com/agriculture/2/fruitsandnuts/…/apple/…/stats.aspx
http://www.hortinews.co.ke/article.php?id=345
http://www.hort.purdue.edu/newcrop/pri/chapter.pdf

CITRUS PLANTATIONS OF THE WORLD

23.1 INTRODUCTION

Citrus is a native tree crop of South-east Asia. The region of origin and domestication includes Yunnan Province in Southern China, Northern Burma, Malaysian Peninsula and some parts of North-east India [1]. Citrus trees spread into most parts of mainland China and Indian subcontinent through human migrations that occurred during late Neolithic till Ancient period. Citrus cultivation moved into southern Europe, especially Italy during 300 A.D. at the fall of Roman Empire [2]. The citrus belt in Spain was initiated by Moors who conquered Spain during 900 A.D. Later, these Spaniard explorers, conquerors and merchants were most active in spreading citrus and several other plant species into various parts of the world. Citrus was introduced into Florida by followers of Ponse De Lion who landed at St Augustine during mid-1500s. Citrus culture and production systems were extended into Mexico, California and Brazil during 16th century. Today, grape fruits, mandarins and tangerines grow almost in all Citrus belts of Americas. The citrus agroecosystem in Florida and Brazil developed and spread rapidly during 19th and twentieth century. It was induced by the rapid development of infrastructure, such as rail heads, roads, soil fertility tests and fertilizer availability. Development of storage, freezing and transshipment facility has been the key factors, regulating the expanse and productivity of citrus orchards in North America and Brazil (Table 1).

About Lemon, geographic location for its origin and domestication is still not clear, although it is most often suggested as North-west India. Lemon or acid lime reached Italy and parts of southern Europe during Ancient era around 200 A.D. Arabian home gardens had limes by 700 A.D. It seems lime trees occurred in most parts of China by 1200 A.D. Egyptians used lime regularly during medieval times around 1200 A.D. Lime reached Hispaniola through Spanish explorers around 1500 s. Earliest of lime plantation occurred in Florida around St Augustine. Lemon reached California during mid-1700s. Lemon cultivation became intense and popular in Florida and California during 1850s. Currently, lemon cropping zones occur in California, Arizona and Florida [3]. During medieval times (1500s), lemon cultivation also spread into parts of Meso and South America.

TABLE 1 Citrus: Origin, Classification, Nomenclature and Uses—A summary.

Citrus was domesticated in the region comprising southern China, Northern Burma and Malaysia. Citrus cultivation spread into vast regions in China, India, Middle East, Southern Europe, Spain and Parts of African continent during ancient era. Citrus reached Florida and other parts of North American main land during medieval times through Spanish explorers. Citrus production in South America began with European settlers who established orchards in many areas of Brazil.

Classification

Kingdom-Plantae; Sub-kingdom-Angiospermae; Division-Magnoliophyta; Class-Magnoliopsida; Subclass-Rosidae; Order-Sapindales; Family Rutaceae; Subfamily Aurantoideae; Tribe-Citrae; Genus-*Citrus*;

Species-*C.aurantifolia (Lime); C.maxima (Pomelo); C. media (Citroen); C.reticulata (Mandarin)*

Some important citrus crosses that are popular and cultivated in vast stretches are:

Bitter orange (Citrus × aurantia); Persian lime (Citrus × latifolia); Lemon (Citrus × limon); Rangpur lime (Citrus × limonia); Grape fruit (Citrus × paradize); Sweet Orange (Citrus × sinensis); Tangerine (Citrus × tangerine). Actually, citrus hybrids and cultivars can be grouped as follows: *Citrus maxima* based, *C.medica* based, *C.reticulata* based and others.

$2n = 18$

Nomenclature

Catalan-Citrus; Czech-Citrusove; Danish-Citrus; Dutch-Citruvruchten; English-Orange, Tangerines, Lemon, Grape fruit; Citrus; French-Citron, Citronnier; German-Zitrus; Italian-Citro, Limone, Agrume; Polish-Owoce Cytrusove; Portuguese-Citricos; Romanian-Citrus; Russian-Cytrus; Spanish-Limon; Swedish-Citrus; Hindi-Mosambi, Narangi, Malta; Kannada-Kittale, Nimbe, Mosambi; Telugu-Nimmakayi.

Uses

Citrus fruits and other products derived from orchards are used in several ways. Most popularly, citrus fruits are peeled and consumed fresh for nutritious ingredients. Citrus juice extracted from sweet oranges, grape fruit, tangerines and lemon are used in large quantities in many regions of the world. Citrus juice is consumed with variety of foods to enrich them with vitamin C. Lemonades and popular beverages across different regions of the world. Citrus fruits, their rinds and other portions are used to extract wide range of flavors. These flavors or additives are used in culinary and preparing perfumes. The citrus rind oil is bitter but popular culinary item. Marmalades and jams prepared using citrus fruits are very popular, and are consumed with bread and other wheat products. Citrus peels and inner rind is often used flavor coffee.

Citrus has many medicinal uses. Foremost, citrus fruits or juice avoids Vitamin C deficiency. It is prescribed for patients with scurvy. Citrus juice removes fatigue. Citrus fruits and pastes are used as gelling agent in many types of medicines. Citrus fruits/juices are prescribed to reduce the risk of kidney stone. Citrus is known to dissolve stones. Citrus juice is also to prepare variety of tonic and syrups meant for improving general health.

Nutritive Value

Fresh Fruits (100 g contains): Calories-27; Moisture 90%; Protein-1.1 g; Fat-0.3 g; Carbohydrtes-8.2 g; Fiber-0.4 g; Ash-0.3 g; Calcium-26 mg; Phosphorus-16 mg; Iron-0.6 mg; sodium-2 mg; Potassium-138 mg; Vitamin A 20 I.U.; Thiamine-0.04 mg; Riboflavin-0.02 mg; Niacin-0.1 mg; Ascorbic acid-53 mg.

Fresh Juice (100 g contains): Calories-25; Moisture 91%; Protein-0.5 g; Fat-0.2 g; Carbohydrate-8.0 g; fiber-0.4 g; Ash-0.3 g; Calcium-7 mg; Phosphorus-10 mg;Iron-0.2 mg; sodium-1.0 mg; Potassium-141 mg; Vitamin A-20 I.U.; Thiamine-0.03 mg; Riboflavin-0.02 mg; Niacin-0.01 mg; and Ascorbic acid 46 mg.

Raw peels of citrus fruits is richer with regard to carbohydrates (16.2 g), Potassium content (160 mg); Ascorbic acid (129 mg) and Vitamin A (50 I.U.) per 100 g fruit.

Source: Refs. [3–6]; en.wikipedia.org/wiki/Citrus (August 29th, 2012); (November 23, 2012) http://www.fas.usda.gov/data.asp.

23.2 SOILS, AGROCLIMATE AND PRODUCTION SYSTEMS

Citrus trees need deep and well-drained soils. Soil type is not a major constraint since it adapts to wide range soils derived from different parent material. Citrus thrives on soils with pH 5.5 to 7.5. Trees are affected at pH, 4.0 and > 9.0 [7]. The citrus belt in USA, is confined to loamy soils in California, sandy soils in Arizona, low fertility Ultisols in Texas. Much of the citrus belt in Florida thrives on Hammock soils, Histosols, Spodosols and low fertility Ultisols (Plate 1). In Brazil, citrus belt is predominant on acidic Oxisols, low in P but high in Al and Mn. Soil amendments needed are gypsum and organic manures. In some parts of Sao Paolo and Mexico citrus trees do thrive on slightly heavy and clayey soils [2]. In Europe, especially in Italy and Spain citrus thrives on loamy soils that are deep and well drained. In West Asia, Xerasols rich in Ca are used to develop citrus orchards. In India, citrus orchards occur on Vertisols, Alfisols and lateritic soils.

PLATE 1 Top: Citrus Grooves in Central Florida, USA; Bottom: Harvested Citrus fruits.
Source: IFAS, University of Florida Citrus Center at Lake Alfred, Florida, USA; Citrus Mutual Inc., Lakeland, Florida; USDA Foreign Agricultural Service, 2004.

Citrus orchards grow best in tropics and subtropics with an optimum temperature range of 22°–35°C. Best fruit yields are possible with orchards experiencing 13°C–37°C. Yet, citrus belt extends into regions with cold weather pattern, frost and in some areas it negotiates high temperatures. In fact, in Florida, citrus belt is severely affected by cold front that afflicts each year. The expanse and productivity is affected by frost period. Temperature below −4°C is detrimental to trees. However, some citrus types such as trifoliate orange can survive 10°C-20°C. Ambient temperature above 40°C is again detrimental to citrus orchards. Citrus orchards receive about 300–400 mm precipitation per season.

23.2.1 CITRUS PRODUCTION SYSTEM

Citrus grows into a small tree or shrub in nature. Citrus tree can be trained and dwarfed using specific roots stocks or careful pruning. Citrus trees found in most plantations across different continents grow 5–10 m tall. They possess a bushy canopy, built on multiple branches. Leaves are evergreen, but leaf fall occurs incessantly in the orchards. Flowers are solitary and borne in large numbers. Flower drop can be a major problem, if mineral nutrition is not held at optimum levels. Trees have to put forth 3–4 years growth prior to prolific fruit bearing. The citrus fruit is a hesperidium. It is globose, pear shaped or sometime even long obviate. Citrus fruit is nonclimatric with regard to ripening. Citrus fruits vary widely in their shape size color, ripening pattern, nutritional value, taste, juice content and economic value. Citrus fruits are known for the variety of fragrances that they elicit. To a great extent, yearly formation of branches leaves and fruits decide the extent of fertilizers supplied. Nutrient recycling and removal from orchards depends on tree physiology and growth pattern.

Tree density in the main orchards is an important factor that affects nutrients removed from soil phase of agroecosystem and productivity of fruits/ha. The tree densities maintained in Brazil, USA, and Southern Europe were generally low until 1980s. It ranged from 250–280/ha. Such low density planting allows better orchard management. Agronomic measures can be practiced with greater accuracy and efficiency. However, fruit yield is low. In order to improvise on economic advantages, most orchards in Americas maintain over 375 trees/ha. Some orchards in Brazil practice high-density plantings at 500 trees/ha. In Asia, trees are planted much closely and over crowded. Tree density is high at 750–1000/ha. High-density plantings are advantages in terms of nutrient recovery and fruit yield. Fertilizer efficiency is enhanced.

Citrus orchards in North and South America are predominantly rain fed. Yet, 20% of orchards in Florida and Sao Paolo are irrigated using drip or sprinkler irrigation. In India, most orchards are rain fed. A small fraction of orchards are irrigated. In Australia and South Africa agro-climate in the citrus belt is predominately dry. Hence, most, if not all, orchards are irrigated using drip or sprinkler irrigation. The rate of irrigation, frequency and volume depends on the root stock and yield goals set by farmers. Some tree root stocks like Rangpur lime is highly tolerant to drought. Hence, during recent years, most orchards in Brazil, Florida and in Europe use this cultivar to raise citrus orchards. Irrigation can then be suitably monitored.

Citrus orchards in Brazil and Florida are picked during mid-July till October. A small portion is harvested in November. The productivity of citrus tree is dependent on several factors related to agroclimate, citrus genetic stock, agronomic procedures adopted and yield goals set by planters. Fruit yield of 300/tree is satisfactory in some parts of Florida. In Brazil, 500 fruits/tree is common. In India, high-density planting and fertilizer supply has resulted over 900 fruits/tree. The average productivity of citrus orchards in Brazil, Florida, Southern Europe and China is 30–40 t/ha. However, low input techniques, paucity of fertilizer based nutrients and erratic rain fed conditions that prevail many regions of the world may reduce fruit yield to 1–12 t/ha. Citrus orchards generate 20–30 t leaf litter that could be recycled.

23.3 EXPANSE AND PRODUCTIVITY

Citrus belt is wide spread in all four continents, wherever agricultural cropping is possible. About 140 countries are known to support citrus plantations (Plate 1). Citrus belt is most intense and large in Brazil, China and United States of America. Brazil contributed 20.1 m t citrus fruits, mainly oranges. China contributed 19.6 m t and United States of America 10.2 m t during 2007. Together these three nations produced over half of global citrus at 50.2 m t during 2007. The citrus agroecosystem is large in several other nations. Countries such as Mexico (6.8 m t/yr); India (6.2 m t/yr); Spain (5.7 m t/yr); Iran (3.8 m t/yr); Italy (3.6 m t/yr), Nigeria (3.3 m t/yr) and Turkey (3.1 mt/yr) are other contributors of citrus fruits [8].

Citrus orchards in Florida and California are intensely cultivated. Citrus belts in California, Arizona and Texas are meant for fruits. However, Florida citrus belt is predominantly meant for juice extraction. It is filled with Grape fruit and Oranges. The Brazilian citrus agroecosystem is almost entirely made of orange meant for extracting juice. Therefore, citrus cropping belt in both Florida and entire Brazil (Sao Paola and Minas Gerais) are regulated to a great extent by demand for orange juice from other regions of the World [6]. In Meso-America, Mexico supports a large citrus belt that contributes to orange juice pool traded in the world. Much of the produce is again meant for export. Spain and Italy are major citrus producers in Europe. Oranges and Tangerines are dominant citrus types in this region. Citrus orchards in these two nations produced 2.3 m t orange fruits during 2007. Turkey and Iran are major citrus producers of citrus in West Asia. They contributed 6.8 m t fruits, mainly oranges. China has a large citrus agroecosystem. Citrus belt contributes about 18% of that nation's total fruit produce. It is predominantly made of Tangerine. About 14.6 m t is tangerines out of a total produce of 19.6 m t. The citrus belt in Upper and Middle Yangtze River supports orange trees meant for extracting juice. Citrus orchards found in Southern Jiangxi and Hunan is mostly navel orange trees. Much of citrus belt in Zhejiang, Fujian and Guangdong is occupied Pomelo and Mandarins trees.

Citrus is cultivated in different provinces of South Africa for fruits and to extract juice. The citrus agroecosystem in South Africa extends into 53.7 ha. Eastern Cape region has the largest citrus belt within South Africa. It is 14.3,000 ha large and constitutes 26% of citrus belt of that nation. Limpopo with 13.3,000 ha, Mpumalang with 12.03,000 ha and Western Cape with 9.6,000 ha are other major citrus producing regions. These four provinces of South Africa possess 90% of citrus region within South Africa. Planting density is moderate at 660/ha and this has direct impact on fruit harvest [9].

In India, citrus cropping belts are situated in the Central Indian Vertisol plains and hilly tracts of Western Ghats. Citrus belt in India is predominantly made of oranges and grape fruit. Citrus agroecosystem in India is fairly large and extends into 843,000 ha. Citrus belt is confined to states such as Andhra Pradesh (236,000 ha), Maharashtra (236,000 ha), Punjab (35,000 ha), Gujarat (34,000 ha), Orissa (26,000 ha) and Uttharakhand (27,000 ha). Citrus orchards are well spread out in other states such as Karnataka, Madhya Pradesh and Assam. Clearly, citrus belt in India is intense in dry Vertisol regions of Central India and Andhra Pradesh. Together, these above two states account for 43% of citrus agroecosystem in India. [10]. Annually, the citrus belt in India produces 7.5 m t fruits at an average productivity of 9 t fruits/ha. Planting density ranges from 270–400/ha, which is moderate. Fertilizer supply is low at 10–40 t FYM, 100 kg N, 50 kg P and 25 kg k/ha. In comparison, citrus plantations in North America are provided with over 270 kg N 120 kg P and 180 kg K/ha. The citrus belt in India often supports inter crops or cover crops such as beans or other legumes. These crops add to soil-N upon recycling of residues [7].

Citrus was introduced into Australia during recent history by the British Settlers. Citrus plantations are primarily concentrated on the banks of rivers Murray and Murrumbidgee. Citrus plantations in New South Wales, Victoria, South Australia and Northern Territory are mostly irrigated. Citrus is also popular in the Burnett region of Queensland. The citrus agroecosystem in Australia extends into 12,100 ha. It thrives on variety of soils such as duplex soils, Xerasols, red earths, etc. The citrus agroecosystem of Australia encompasses different types of citrus. For example, largest share of 507,000 ha is under oranges, about 75,000 ha are under mandarins. Lemons occupy 26,000 ha. Major lime cultivars that fill the ecosystem are *Citrus austalasica, C. australis* and *C glauca.* Obviously, each type of citrus has its special influence on ecosystematic functions and total produce. The citrus belt in Australia is highly productive. Farmers harvest about 650,000 t annually. New South Wales and South Australia together account for 70% of citrus production.

As stated earlier, global citrus fruit produce was 116 m t during 2007. Citrus agroecosystem of the world is composed of orchards that produce oranges (63%), tangerines (26%), lemons and lime (13%), grape fruits (5%) and other type (7%). Clearly, global citrus landscape is dominated by oranges and tangerine. Orange orchards are well spread and dominant in Brazil, China, United States of America [6], Mexico, India, Spain and Italy. China has the largest tangerine belt in the world. Grape fruits are most conspicuous in United States of America, mostly in Florida. Lemons and lime orchards are largest in Brazil, Mexico and India.

KEYWORDS

- ***C glauca***
- ***C. australis***
- ***Citrus austalasica***
- ***Citrus maxima***
- **Xerasols**

REFERENCES

1. Gmitter, F.; Hu, X. The possible role of Yunnan, China in the origin of contemporary citrus species (Rutaceae). Economic Botany **1990,** *44,* 267–277.
2. Spreen, T. H. The World Citrus Industry. Encyclopedia of Life support systems (EOLSS) **2011,** *3,* 21 http://www.eolss.net/Eolss-sampleAllChapter.aspx (August, 27, 2012).

3. Morton, J. Lemon. In: Fruits of Warm climates. http://www.hort.purdue.edu/nw crop/morton/lemon.html **1987,** 1–12 (August 29, 2012).
4. Hynniewta, M.; Malik, S. K.; Rama Rao, S. Karyological studies in ten species of Citrus (Linnaeus, 1753) (Rutaceae) of North-east India. Comparative Cytogenetics **2011,** *5,* 277–287.
5. Krishna, K. R. The Citrus Agroecosystem of Florida. In: Agrosphere: Nutrient Dynamics, Ecology and Productivity. Krishna, K. R. (Ed.). Science Publishers Inc.: Enfield, New Hampshire, USA, **2003,** 257–279.
6. USDA Foreign Agricultural Services. United States of America and World situation-Citrus. http://wwwfas.usda.gov/ htp/horticulture/citrus/2004%20Citrus.pdf, **2004,** 1–23 (August 29, 2012).
7. NABARD, Citrus Cultivation. http://www.nabard.org/modelbankprojects/plant_citrus.asp 1–7 (August 29, 2012).
8. FAOSTAT, **2010,** Citrus Fruit Producers. Food and Agricultural Organizations of the United States of America. Rome, Italy, http://faostat.fao.org/site/567/DesktopDefault.aspx?PageID=567#ancor. 1–3 (August 29, 2012).
9. Citrus Business Plan, **2006,** Citrus. http.www.tradeand investcacadu.coza/agri/citrus **2007,**1–3. (August 30, 2012).
10. NHB, Indian Production of Citrus. APEDA Agri Exchange. agriexchange.apeda.gov.in/India.production/India_ Productions.aspx?hscode=08059000 **2011,** (August 30, 2012).

EXERCISE

1. Write an essay on origin and distribution of citrus species in the world.
2. Mention the major citrus types.
3. List different cultivars/hybrids of citrus grown in Asia and identify their parentage
4. Describe the agronomic procedures required to maintain citrus Orchards.
5. Write an essay on expanse, production procedures and fruit yield of Florida Citrus Belt.
6. Mention major citrus exporters and importers of the world.
7. Describe the Anatomy, Physiology and growth pattern of Citrus trees.
8. List different types of irrigation methods adopted in Citrus groves.
9. Write an essay on soil and plant tests adopted and fertilizer supply methods adopted in Citrus groves.

FURTHER READING

1. Dugo, G.; Giacomo, A. D. *Citrus: The Genus Citrus*. CRC Press: Boca Raton, Florida, USA, **2002,** pp. 642.
2. Lazlo, P. *Citrus: A History*. University of Chicago Press: Chicago, USA, **2008,** pp. 282.
3. Lyrenne, P. M. *Citrus: Genetics, Breeding and Biotechnology*. CAB International: Cambridge, MA, USA, **2007,** pp. 370.
4. Hartmiki, A. E. *Soil Fertility decline in the Tropics with case studies in Plantations*. Willingford Publishers: England, **2003,** pp. 360.

USEFUL WEBSITES

http://www.crec.ifas.ufl.edu/index.htm (August 29, 2012)
http:///www.uga.edu/fruit/citrus.html (August 29, 2012)
http://www.homecitrusgrowers.co.uk/poncirustrifoliata/poncirus.html (August 29, 2012)
www.fruit-crops.com/citrus.html (November 23, 2012)

http://www.agnet.org/library.php?func=view&id=20110803104519&type_id=1 (November 23, 2012)
http://www.ultimatecitrus.com/info.html

BANANA PLANTATIONS (*MUSA, SPECIES*)

Bananas are currently world-wide in distribution and are encountered in tropical and subtropical regions of different continents. Bananas (*Musa, species*) originated in tropical Asia, probably in the Indo-Malay region. Banana plantations are wide spread in the Indian subcontinent, China and other South-east Asian nations. Bananas were introduced into African main land during prehistoric times (*see* Table 1). Bananas are grown in almost all regions of Africa, excepting regions that experience extreme dry conditions. Bananas were introduced into Central America through Spanish conquests and expansion. It later spread to southern USA. Banana cropping zones were established in Costa Rica in 1870s. Further expansion of Banana plantations in North America were primarily guided by private farms and companies such as Boston Fruit Company, United Fruit Company, etc. These entrepreneurs aided expansion of banana plantations even in other regions, for example in South America. During 1800s, introduction of bananas and development of plantations in Central America involved large-scale conversion of tropical forest vegetation into plantations. Banana companies used drastic measures. They slashed forests to develop plantations. Soils were amended with large amounts of fertilizer-based nutrients. Of course, large stretches of virgin land were also brought into banana cultivation.

TABLE 1 Bananas: Origin, Classification, Nomenclature and Uses—A summary.
Banana is a subtropical fruit crop. Bananas originated in Indo-Malaysian region. Major centers of genetic diversity are available in India, South-east Asia and West Indies.

Classification: Kingdom-Plantae; Subkingdom-Angiosperms; Class-Commelids; Order-Zingerberales; Genus *Musa*; Species-*M.accuminata*

There are three major groups-*M accuminata* is most widely distributed; *M. balbisiana* including hybrids derived are also distributed into different continents. *M. accuminata* and *M balbisiana* are edible bananas. *M callemusa* and *M. schizocarpa* are less widely spread.

$2n = 22$

Germ Plasm centers are located in India and several other Tropical Nations such as West Indies, Nigeria, Sri Lanka, Vietnam, Indonesia and Thailand.

Nomenclature

Catalan-Plantan; Czech-Banan; Danish-Melbanan; English-Banana, Plantian; French-Banana; German-Wagrich, Bananaenart; Italian-Plantaggine; Polish-Plantain; Spanish-Plantaina, Plantano; Romanian-Banana;

Hindi-Kela, Kannada-Bale Hannu, Telugu-Arti Pandu; Tamil-Valzh Palam

Uses

Banana is an edible fruit. When ripe it is sweet with sugars. It accumulates flavor as it ripens. Raw banana fruits are used as vegetables. Bananas are used to make variety of snacks and dishes in the tropical regions. Banana powder is used base material in preparing syrups. Banana stem is also edible and is used as vegetable. Banana leaves are used extensively in tropical regions for serving food, preserving food items and wrapping different items. Banana residues are also used as packing material.

Nutritive Value (100/g fruit)

Energy-89 k cal; Carbohydrates-23 g; sugars-12.3 g; Dietry fiber-2.6 g; Fat-0.33 g; Protein-1.1 g; Vitamin A 3 μg; Thiamine-0.31 mg; Riboflavin-0.07 mg; Niacin-0.665 mg; Pantothenic acid-0.33 mg; Folate-20 μg; Choline-9.8 mg; Vitamin C-8.7 mg; Ca-5 mg; Fe-0.26 mg; Mg-27 mg; P 22 mg; K-358 mg; Zn-0.15 mg.

Source: en.wikipedia.org/wiki/banana; http://www.nal.usda.gov/fnic/foodcomp/search.

PLATE 1 A Crop mixture of Plantains (*Musa, sp*), Pepper vine and useful timber yielding trees at a location near Karwar, in Western Ghats of India.
Note: **Such** two or three tier cropping is known to fix relatively higher amounts of C and form more biomass. They improve land use efficiency perceptibly. Nutrient recovery and turnover are efficient. Carbon sequestration of mixed crops is higher than that recorded for monocrops cereals.
Source: Krishna, K.R., Bangalore, India

Banana cultivation occurs under specific agroclimatic conditions that prevail in the tropics and subtropics. Soils should be fertile and deep to accommodate a deep and well spread out root system. Banana plantations require slightly higher amounts of water. Soils are kept moist but not inundated for a long duration. Soils should be 1–1.2 m deep, well drained and optimum in mineral and organic matter content. Banana tolerates slightly acidic conditions. It thrives well on soils with pH 5.5 to 7.2. Banana plantations are adapted to tropical and subtropical climatic conditions. Warm temperature regime is required. Bananas prefer 20°**C**–35°C during growth and fruit formation. Banana is a long duration requiring 15 months to grow and form fruits. Bananas require frost period of 12 months to bear fruits. There are actually four stages during banana cultivation and export. They are plant growth; fruit formation requires 6–8 months, harvesting, protection of fruits and packaging. Bananas are versatile with regard to soil types on which they can flourish. Bananas grow excellently on Spodosols and Ultisols found in Florida.

The Xerosols and Alfisols of Caribbean support large-scale production of Sweet Bananas. Loamy soil of Central American region including Mexico, Panama, and Costa Rica are congenial for banana cultivation [1]. Banana plantations in Columbia and adjoining regions thrive on Oxisols affected by high Al and Fe, acidic pH and P fixation. In the African continent, bananas are cultivated on wide range of soil types that vary with regard to soil fertility and productivity. Banana cropping zones of Indo-Gangetic belt are developed mostly on Inceptisols with moderate fertility. Bananas are well adapted to Alfisols, Vertisols and lateritic soils of South India (Plates 4 and 5). The South-east Asian banana growing regions are developed on Alfisols, Inceptisols, Mollisols, Alluvial soils and Coastal sandy soils.

PLATE 2 Bananas in a South Indian market.
Source: Krishna, K.R., Bangalore, India

Globally, Bananas are grown in all continents, wherever warm climate prevails. The banana agroecosystem in any region could be filled up with types such as plantains, dessert bananas and highland cooking banana. Plantains are predominant in Africa and Latin America. The Cavendish plantations are most conspicuous in Latin America followed by Asia. Major producers of Cavendish bananas are India, Equador, China, Columbia and Costa Rica. About 47% of global banana harvest comprises Cavendish type or Sweet Banana. These countries contribute half of world Cavendish banana harvest. Highland cooking types are prevalent in Asia and parts of Africa. Global banana production zones expanded rapidly during mid-1900 s. It crossed 10 m ha in area and 105 m t fruit in production during 2010 [2]. Globally, about 19% of banana fruit productions were derived from India, 12% from Equador, 10% from China, 6% from Columbia, 5% from Costa Rica, 5% from Brazil, 4% from Philippines and 4% from Mexico. Other countries contribute 35% of total global banana produce [3]. Generally, it is difficult to accurately gauge banana fruit production, because a bulk of banana consumed is derived from kitchen garden, small plots and backyards. The per capita consumption of bananas is high in Central African region. For example, in Uganda banana is a major source of carbohydrates to the populace. The average per capita consumption is 243 **kg**. In Rwanda, Gabon and Cameroon per capita consumption is over 220 **kg**. In the Central Africa, bananas may contribute about 12–27% of carbohydrate requirements of the populace. It is important to note that nearly 80% of plantain production in Latin America is exported. Therefore, the area planted to sweet bananas and its expansion is highly dependent on demand for bananas from other regions of the world like Europe and North America.

KEYWORDS

- **edible fruit**
- ***M. accuminata***
- ***MUSA***
- **Spodosols**
- **Ultisols**

REFERENCES

1. McCracken, C. The impact of Banana plantation Development in Central America. http://memebers.tripod.com/ **2005** (September *10,*2012).
2. FAO. Overview of world Banana production and trade. www.fao.org/docrep/007/yr5102e/yr5102e04 1–32 **2010,** (September 22, 2012).
3. FAOSTAT**.** Food and Agricultural Organization of the United Nations. Rome, Italy. www.faostat.org **2010** (September 10, 2012).

EXERCISE

1. Mark the Banana cultivating zones of the world.
2. Discuss the production trends of Banana in India, Sri Lanka and Jamaica.
3. Mention various types of Bananas and how they are used by us.
4. Mention the top 5 Banana countries of the world and list Productivity, Area and Total Production for each.

FURTHER READING

1. Dadzie, S. Guide to Banana Production in Jamaica. Natural Resources Institute, Jamaica, **1999,** pp. 198.
2. Hartmiki, A. E. Soil Fertility decline in Tropics: Case studies in Plantations. Willingford Publishers, England, **2003,** pp. 360.
3. Sealey, T. Jamaica Banana industry. The Association, Jamaica. **1984,** pp. 114.
4. Spreen, T. H. The World Banana Industry. http://www.eolss.net/Sample-Chapters/C10/E1–05A-42–00.pdf **2010,** 1–32.
   http://www.eolss.net/Sample-Chapters/C10/E1–05A-42–00.pdf (November 12, 2012).

USEFUL WEBSITES

http://en.wikipedia.org/wiki/Banana_production_in_the_Caribbean (November 12, 2012).
http://www.icar.org.in/node/2276 (November 12, 2012).
http://www.mapsofindia.com/indiaagriculture/fruits-map/banana-producing-states.html (November 12, 2012).

GRAPE VINEYARDS OF THE WORLD

23.1 INTRODUCTION

Grapes were domesticated some 8,000 years ago in the region comprising Georgia, Armenia, Italy, Turkey and Azerbaijan. Middle East and Southern Europe is accepted as region of origin for grapes ([1, 2]; *see* Table 1). Archaeological surveys and analysis of grape genotypes using Restriction Fragment Length Polymorphism suggest that Georgia in Eastern Europe is the center of origin for grapes. Some of the oldest wine making locations with pottery and fermentation utensils occur in Armenia. They are dated to 4000 B.C. During early medieval period, wineries were dominant in Iran and surrounding regions. During 9th century, city of Shiraz was famous for good grape wines. Shirazi wines are popular even today in Iran and other Arabian nations. Purple grapes were, it seems common to Greece, Rome and Egypt during Ancient era. Grapes were part of diets of North Africans during Ancient era. Several types of grapes (*Vitis species*) are native to North America and were consumed by the native tribes situated in different parts of the North American main land. They were less preferred by the European settlers who introduced grapes from Europe and near East.

A probable time line for domestication of grapes, spread of grape agroecosystem into different parts of the world, and development of wine based industries is as follows:

6000 B.C.: Domestication of grapes in Georgia. *Vitis vinifera* is cultivated in Iran and on shores of Caspian Sea.

3000 B.C.: Cultivation of grapes reaches Egypt, Phoenicia.

2000 B.C.: Grapes are made popular in Greece.

1000 B.C.: Grape cultivation begins in Sicily and North African countries like Libya, Tunisia and Algeria.

500 B.C.: Grape reaches European agrarian zones, mainly in Spain, Portugal and France. Grapes, later spreads to British islands. Grapes become common in Arabia and South Asia. Grapes reach several regions of earth via Spanish explorers, especially into southern American regions.

1830s: Grapes vine yards are initiated in California. Several types of grapes from Eastern Europe are introduced into California.

1860s: William Thomson introduces 'Kishmish' from Mediterranean into Sacramento region.

1900s: Grape cultivation is intensified in California and other parts of North America [3].

Grape cultivation spread into Ozarks and foothills during 1870s *Vitis vinifera* was introduced into agrarian zones of Argentina by Spanish settlers who colonized the Cuyo region during mid-1600s. Subsequently, grape cultivation practices spread to other parts of Argentina, namely La Rioja. Domingo Faustian Sarmiento introduced European grape varieties in 1852. This event brought about changes in the cultivar composition and grape landscape in Argentina. European grape varieties such as Cabernet Franc, Cabernet Sauvignon, Merlot, Marbec and Riesling started to dominate the Argentinean grape belt [4].

Grape cultivation was in vogue in the Indian Subcontinent during *Rig-Veda* period around 2500 B.C. Grapes production and its use as beverage, also in medicines has been mentioned Hindu scriptures and Sanskrit literature (*Charaka Samhita*) of 1326 B.C. Grapes have also been mentioned in Kautilyas 'Artha shasthra' that dates to third century B.C. Native grape species in India resemble *Vitis lanata* and *V. palmate* [5]. Grapes from Arabia were introduced by Persians during 1350s. They started cultivation of grapes in Western and Southern India Plains around Daulatabad and Aurangabad. During mid-1800s, British soldiers and Christian missionaries settling in South India started culture of European grapes around locations in Salem, Madurai and Hyderabad. Grapes cultivation was also introduced by the invading armies of Mohamed Bin Thuglak (Table 1).

TABLE 1 Grapes: Origin, Classification, Nomenclature, Uses—A summary.

Grapes originated in Southern Europe within the region comprising present day Georgia, Azerbaijan and Armenia. Grape cultivation was in vogu in Europe, Asia Minor and South Asia since Neolithic period. Grapes spread further into many regions in the New World during medieval times.

Botany and Classification

Kingdom-Plantae, subkingdom-Angiosperms, Order-Family Genus-*Vitis*; species-V.*vinifra*

$2n = 38$

Vitis vinifera is common to Europe, Central Asia and Indian subcontinent. It is consumed as ripened sweet fruit as well used for extraction of juice.

Vitis labrusca occurs in North America. It is meant for table and juice extraction. V. *labrusca* is most common in Eastern United States of America and Canada.

Vitis rotundifolia is used to prepare jams, jellies and wine. It is frequent in Southern United States of America

Vitis amurensis is a grape species more common to Asian grape belts.

Germplasm centers occur in several countries of Europe, West Asia, Southern Asia, China and North America.

Nomenclature

Bulgarian-Gruzde; Croatian-Grozde; Czeck-Grozny; Danish-Vindruir; Dutch-Druiven; Finnish-Viinirypalet; French-Raisin; German-Trauben; Italian-Uva; Norwegian-Druer; Winograma-Polish; Portuguese-Uva; Romanian-Struguri; Russian-Vinograd; Spanish-Uvas; Swedish-Vindroor; English-grapes; Turkish-Uzum; Vietnamese-Nho; Hindi-Angur; Kannada-Dhrakshi; Tamil-Dhraksha; Telugu-Dhrakshalu.

Uses

Grapes are consumed as ripened fruits. Dried grapes are popular in many regions of the world. Grape fruit and its juice are used in variety of culinary items. Grapes are used to prepare wide range alcoholic beverages. Grape fruit wastes after extraction of juice is used as organic matter. Grapes and juice is used in many medicinal concoctions that are aimed at bestowing health. Grapes are used in syrups and porridges. Grapes are used to prepare several types of organic acids, ethanol, ethylene and several other organic chemicals.

Nutritive Value (100 g grape fruit contains):

Carbohydrates-18 g; Sugars-15.5 g; Dietary fiber-0.9 g; Fat-nil; Proetin-0.72 g; Thiamine-0.069 mg; Riboflavin-0.07 mg; Niacin-0.188 mg; Pantothenic acid-0.05 mg; Vitamin B6–0.086 mg-Folate-2 μg; Vitamin B12-nil; Vitamin C-10.8 mg; Vitamin K-22 μg; Calcium-10 mg; Iron-0.36 mg; Magnesium-7 mg; Manganese-0.071 mg; Phosphorus-20 mg; Potassium-191 mg-Sodium-3.02 mg; Zinc-0.07 mg.

Source: Ref. [3]; http://www.Arkansasgrape Crop Profile.htm 1–2 (October 8, 2012). http://www.nal.usda.gov/fnic/foodcomp/search.

23.2 SOILS, AGROCLIMATE AND PRODUCTION SYSTEMS

California grape belt is a major contributor of grapes in North America. This grape belt extends into San Joaquin valley and Coachella Valley. It thrives mostly on sandy and loamy soils. Grapes actually need at least 3 ft deep soils that are well drained and weed free. In a grape orchard, usually soils are inter-cultured, loosened in between rows and fertilized based on soil tests. Soils found to support grape belt in California are classified as Abruptic Duri-Xeralfs. Soils in San Joaquin Valley are well drained, but suffer from excessive runoff. This natural process leads to loss of alluvium and topsoil. They are also susceptible to compaction and duripan formation, if worked excessively. Grape farmers use terraces if the orchards occur on undulated terrains. (Plate 1; [6]). Acid soils are detrimental if the pH is below 5.5. Usually gypsum amendment is applied to raise the soil pH to required level. In South America, especially in Brazil and in other continents, grape orchards may encounter acid soil with Al and Mn toxicity. Addition of gypsum to raise pH and application of excessive fertilizer-P is necessary [7]. Higher amount of fertilizer-P supply is required, because it has to first satisfy the P fixation trends of the soil and then satisfy grape plant's requirements.

Grape orchards in New York State are exposed to cold temperatures of −2°C to 10°C during peak winter. During growing period, orchards thrive under warmer conditions of 20°C–28°C. The number of frost-free days during spring and fall determine grape productivity.

In South-east Asia, grape production occurs under humid tropical conditions. For example in Thailand, grape orchards are raised on loamy and clay loams. Soils with poor drainage are avoided. The ambient temperature ranges from 25°C–30°C, annual precipitation from 1300 to1450 mm and relative humidity fluctuates between 60 and 90% during the season. Grape orchards growing on acid soils in Northern Thailand are supplied with lime and manure to improve pH and organic matter content. Precipitation could be erratic. In such situations, irrigation from rivers and tanks help in providing water to orchards [8].

PLATE 1 Grape Gardens in Central Illinois.
Note: Above-Training system used is known as Telephone pole type; Bottom: A highly productive vine.
Sources: www.plowcreekfarm.com, Illinois, USA.

23.2.1 PRODUCTION SYSTEMS

Production systems for grapes vary enormously based on geographic region, topography, soil type, soil characteristics relevant to nutrient availability, irrigation sources, agronomic procedures, economic value of the product and farmers' yield goals.

In any agrarian zone, minimum requirements and criteria for establishing grape orchards to obtain optimum produce are as follows: soil with good drainage, acceptable soil physicochemical properties such as pH, moderate water holding capacity, irrigation resources, suitable topography with sufficient photosynthetic radiation, optimum ambient temperature. Facilities like cold storage, shipping, market and demand for grapes are equally important.

Mineral nutrition of grape orchards is an important aspect. Fertilizer-based nutrient supply varies widely with agro-ecoregion, intensity of cropping, soil types their fertility and yield goals. Usually, grape vines are supplied with fertilizers based on standard set of soil and plant analysis. Grape vines possess deep roots. They explore water and dissolved nutrients from deeper layers of soil. Therefore, within an agro-ecosystem, soil sampling should cover deeper layers. Nutrient status of subsoil may actually be of higher consequence to grape crop. Grape petiole and leaf tissue are sampled and analyzed periodically to arrive at optimum rates of fertilizer supply [9]. Often, seasonal pattern of nutrient concentrations in different tissues of grape plant are considered to arrive at optimum nutrient inputs.

Overall, a grape orchard that yields 2.5 t grapes removes 8.2 kg N, 3.2 kg P_2O_5, 3.0 kg K_2O, 13.0 kg Ca, 1.4 kg Mg, 0.003 kg B, 0.0052 kg Cu, 0.004 kg Mn, 0.009 kg Zn and 0.022 kg Fe [10]. About 70% of nutrients are garnered by stem, twigs and leafs that are generally recycled each season. Therefore, it adds to soil fertility status. Careful study of nutrient accumulation and recycling patterns have suggested that each ton of grape fruit produced actually removes 2.2 kg N, 1.12 kg P_2O_5 and 5.3 kg K_2O from the ecosystem via the fruits harvested. A large share of biomass is actually recycled. Hence, fertilizer schedules prepared for each field or even the large grape belt should consider the extent of nutrients recycled via crop tissue/dry matter [11].

Fertilizer recommendations are dependent on recommendations by local agricultural agencies that monitor soil fertility status and production potential of soils. In most parts of USA, grape orchards are supplied with fertilizer-N ranging from 35–120 kg N/ha. Fertilizer-N application is regulated, again based on local recommendations. Grape orchards demand N during bud-break and early development of branches and leaves. Split application of fertilizer-N is also done. It is channeled once at bud break and next at bloom. Grape orchards are also supplied fertilizer-N using foliar spray of liquid NH_3 or NO_3 compounds. Calcium nitrate is supplied in two or three sprays at very low concentration.

Grapes are harvested based on several traits such as sugar content, color, size and uniformity of ripening. In California, grape season starts with the harvest of grapes in Coachella Valley. Harvesting lasts till mid-July. Harvesting season then moves to north into San Joaquin Valley. Actually, grape orchard maintenance and harvest are year round activities in California. During winter, vines are pruned and girdled to induce bud formation and fresh branches. Once harvested, grapes are held in controlled temperatures until they reach the market or destinations (*see* Plate 1).

Let us consider grape production systems followed in Asian region, for example in India. The land is prepared using a couple of deep disking, clod crushing and harrowing with tines. Trenches of 75 cm width are dug in north-south direction. Trenches are filled with organic manures and prescribed fertilizers. Planting season in most regions begins during September/October. Spacing varies with soil fertility. For example, 6×3 m, 4×3 m, 3×3 m, or 3×2 m is common. The grape orchards in India are mostly trained using Telephone system, Bower or Flat roof systems. Pruning is usually done in October in hot tropics and in December in subtropical regions. Fertilizer supply is usually high at 500 kg N, 500 kg P and 1000 kg K in areas with sandy soils and 660 kg N, 880 kg p and 660 kg K in clay loams [5]. The potential fruit yield of grape varieties ranges from 50–90 t/ha in India. There are reports of 100 t fruit/ha from Anab-e-Shahi orchards around Hyderabad in South India.

In China, grape orchards are propagated using cuttings. Grape vines are derived by grafting cold-resistant root stocks. To begin an orchard, soil is tilled deep using disks. Land is trenched at 0.6–1.0 m width and 1.0 m deep. Soil trenches are mixed with organic manure and chemical fertilizers based on local prescriptions. Planting density varies depending on several factors related to grape genotype, soil fertility, yield goal, etc. Planting density may vary from 1000 to 5000 vines ha [12]. The Chinese grape farmers use two types of Trellis namely Pergola and Vertical Trellis. The vine density is maintained between 1800/ha and 2980/ha in the grape belts of Northern China and 1100/ha to 2000/ha in South China. Soils vary enormously with regard to inherent fertility and residual nutrients still available after previous crop. Soils are generally periodically analyzed for nutrients. Major nutrients are applied to vines in 3 or 4 splits along with 30 t of organic matter. The Chinese grape belt thrives in humid regions with optimum precipitation patterns. Yet, it may need irrigation, if rainfall is erratic. Irrigation is avoided about 2–3 weeks prior to fruit harvest. It avoids unnecessary succulence and diseases that may occur on fresh fruits. Most of the cultivars grown in China are harvested around September/October, when fruits contain about 18–23% sugar, acid content reaches 7.1–8.5 % and pH ranges from 3.01–3.20 [13].

23.3 EXPANSE AND PRODUCTIVITY OF GRAPE AGROECOSYSTEM

Grape agroecosystem is well spread into different continents. Global grape belt extends into 75,866 km^2. Grape belts are mostly dominated by a cultivar 'Thomson Seedless' also known as Sultana. There are several other grape varieties that dominate the grape belts across different agrarian regions. Global grape agroecosystem can be classified based on the major types/cultivars that fill the belt. Major grape types are grouped into Table and Wine grapes, Seedless grapes, Raisins, Currants and Sultanas. Table types are usually large without seeds, sweet and with relatively thin skin. Wine grapes are very sweet with over 24% sugar. They are highly juicy and are amenable for fermentation. Seedless types are highly nutritious and possess minerals in relatively larger proportion. Most common seedless types are Thomson seedless, Russian seedless, Black Monukka, Eiet seedless, Benjamin Gunnel's seedless, Reliance and Venus. Raisins are dried grape fruits. Raisins or dried grape varieties are common in Europe. Currants or Zante Black Corinth is common in France and other Northern European cropping zones. Sultanas are seedless grapes common to Southern Europe, Middle East and Arabian regions. They are also common to Southern Indian plains. Sultanas are known as Thomson seedless in North America. They are made into dried grapes (raisins) or eaten fresh. Following are few examples of grape varieties that are common and flourish in European zones: Airen, Cabernet Sauvignon, Sauvignon Blanc, Cabernet Franc, Grenache, Tempranillo, Riesling and Chardonnnay. Majority of grape belts occur in Europe and North America. The grape vineyards are large and occur more conspicuously affecting the agrarian belt in Southern Europe. The total area of grape vineyards in different nations during 2011 is as follows (in thousands):

Spain-11.75 km^2; France-8.6 km^2; Italy-8.2 km^2; United States of America-4.2 km^2; Iran -2.8 km^2; Romania-2.48 km^2; Portugal-2.16 km^2; Argentina-2.08 km^2; Chile-1.84 km^2; Australia-1.64 km^2, Armenia-1.45 km^2 and Lebanon-1.12 km^2.

Global grape production during 2011 was 68.31 mt. The grape production trends for 2011 indicate that nearly 35% of total grape production occurs in Europe. About 12% occurs in United States of America. China contributes 12.5 % of total global produce. Following is a list of countries and grape production (t/yr):

China-8.65; Italy-7.78; United States of America-6.22; Spain-6.10; France-5.84; Turkey-4.25; Chile-2.75; Argentina-2.61; India-2.63 and Iran-2.25 [13].

In USA, San Joaquin Valley in California contributes about 85% of table grapes, and the rest is derived from grape vines situated in Coachella Valley. The productivity is moderate and depends on soil management and fertilizer supply. Grape cultivation occurs in foot hill region of Ozarks. It constitutes a small area of grape vines, filled predominantly with *V. vinifera,* V. *rotundifolia,* V. *labruscana.* Grape production is concentrated in the counties such as Benton and Washington in North-west Arkansas. Muscadines and juice types are more common in the Franklin and White counties of Arkansas. Grape varieties common to Arkansas are Sunbelt, Concord, Carlos, Nobel, Summit and Black beauty. New York State has varied topography and climatic conditions, but all of these variations are still suitable for establishment and perpetuation of grape cropping belt. Grape belt actually occurs in three sub-regions. They are Lake Erie region, which is the largest grape belt in New York. Here the grape orchards are meant for juice and wine making. Concord types are prominent. Next. The Finger Lakes Region has 4000 ha of grape vines that supports diverse types of grapes, such as lubruscas, inter specific hybrids etc. Long island is another grape production center in New York State. It supports *V. vinifera* in about 2000 ha. Grape production zones also occur surrounding Niagara, and Lake Ontario.

The Argentinean grape belt is filled with varieties from Southern Europe and North America. European varieties such as Cabernet, Chardonnay, Sauvignon Blanc, Semillon, Merlot and many other grape types have been cultivated now for over 150 years, since their introduction in 1850s. However, during recent years, National Institute Agricultural Technology (INTA) has enthused farmers to cultivated grape varieties/ hybrids such as Tempranillo, Ugni Blanc, Barbera, Lambrusco, Bonarda and Pedro Gimenez. Such change in the varietal composition of grape belt in Argentina has enhanced competitive ability of Argentinean farmers to sell grape products in International markets.

The grape agroecosystem flourishes throughout the Indian subcontinent. It adapts to variations in agroclimate. Farmers are versatile and select cultivars that are apt and adapt well to local variations in soil, its fertility and climate (Plate 2). Grape belt in India can be categorized based on climate into:

Subtropical Region-It mainly covers Northern Plains. The bud break starts around March with the end of cold period in March/April. The vegetative growth lasts for 90–95 days from June, when the first rains start. Harvesting is done in November to January depending on cultivar.

Hot Tropical Region: This region comprises areas in Western Maharashtra such as Nashik, Sangli, Sholapur, Osmanbad; Hyderabad, Ranga Reddy, Mahboobnagar in Andhra Pradesh and districts such as Bijapur, Kalburgi, Bagalkot, Belgaum, Chickballapur in Karnataka. This belt experiences hot temperature ranging from 23°C–36°C. Maximum temperature during summer may rise to 40°–42°C. It is humid during monsoon, but often prone to drought and erratic precipitation patterns. Thomson seedless and its clones, Sharad seedless and few other cultivars dominate the agroecosystem.

Mild Tropical Region: Grape cropping regions in Bangalore, Kolar, and Chickballapur in Karnataka; Chithoor in Andhra Pradesh and Coimbatore, Madurai and Theni districts in Tamil Nadu are grouped under mild tropics.

The growing season temperature ranges from 12°C–36°C. Principal varieties that are grown in mild tropics are Anab-e Shahi, Sharad seedless, Bangalore Blue, Bokhri etc.

PLATE 2 Grape Plantations situated North of Bangalore in South India.
Note: **Major** varieties preferred in this region are Bangalore Blue and Thomson Seedless. They are consumed fresh and used extensively in wine making. Here, grapes thrive on moderately fertile, gravely Alfisols and under subtropical climatic conditions.

'Thomson seedless' that occupies over 34,000 ha dominates the grape orchards in India. Thomson seedless accounts for 550,000 t of fruits of the total Indian produce of 1.0 million ton. Anab-e Shahi and Bangalore Blue are other conspicuous grape cultivars found in the grape belts. These grape orchards are among the most productive at 30–35 t fruits/ha. Obviously, these three cultivars have greater impact on soil fertility, nutrient input trends and nutrient cycling in the vast grape belt. Cultivars such as Muscat Hamburg, Perlette and Sharad seedless are other dominant varieties found in the Indian grape belt.

Grape production in Thailand was initially confined to Central Plains and around Bangkok. Currently, it has spread into many regions in north, north-west and western regions of the nation [8]. The grape belt actually extends into over 3000 ha and contributes 35 t fruits annually. The average grape yield fluctuates between 15–30 t/ha depending on soil geographical region, soil type fertilizer supply and yield goals. Grape agroecosystem in Thailand is composed of cultivars such as Cardinal, Kyuho, Chenin Blanc, Shiraz, white Malaga, Eary Muscat and Black Rose.

Grape agroecosystem in China is actually over 2000 years old. The grape production zones were relatively small until the revolution. Grape orchards expanded mainly during the aftermath of revolution in Peoples Republic of China [12]. During 1949, grape belt occupied only 3200 ha and contributed 39,000 t fruits. However, it spread rapidly under the new regime. Currently, grape production zone in China is concentrated in the following seven regions—North-west (Xinjiang, Ningxia, and Gansu), North-east (Jilin), North-1 (Shanxi, Hualai and Changli), North-2 (Beijing-Tianjin), East Coast-(Shandong), Henan (Yellow River Valley), south-west (Yunnan) [14]. The grape orchards increased by 55–61 times in area. It was caused mainly by greater demand for wine. At present, China is the top most grape fruit producer. It contributes 8.65 m t fruits annually [13]. About 70% of grape fruits and wine are supplied by five major provinces in China. They are Xinjiang, Hebei, Shandong, Liaoning and Henan. Grape agroecosystem in China is composed of Table grapes (e.g., Kyuho, Muscat Hamburg, Thomson seedless, Longyan, Jiangxi, Zana, Rizmat, Fenghuang, Red globe and Fujimori) and Wine grapes (Chardonnay, Italian Riesling, Ugni Blanc, Chennin blanc, Sauvignon Blanc, Cabernet Sauvignon, Saperavi, Carignan, etc.). Clearly, it is these dominant cultivars that dictate the nutrient dynamics and productivity of grape landscape.

KEYWORDS

- **Charaka Samhita**
- **Kishmish**
- **Shirazi wines**
- **Thomson seedless**
- **Vitis amurensis**
- **Vitis labrusca**
- **Vitis rotundifolia**
- **Vitis vinifera**

REFERENCES

1. McGovern, E. Ancient Wine. The search for Origins of Viniculture. http://penn.musem/sites/ biomoleculararchaeology/wp-content/uploads/2010/04/Georgia%20Wine.pdf **2003a,** 235 (August 29, 2012).
2. McGovern, E. Georgia: Homeland of Winemaking and Viticulture. http://winehistory.com/2i.htm **2003b,** 223 (August 29, 2012).
3. California Grape Commission, History. http://www.tablegrape.com/history.php **2012,** 1–3 (August 26, 2012).
4. Winesur, P. Grape Varieties. http://www.winesur.com/wine-guide/grape-varieties.htm **2012,** 1–8 (August 28, 2012).
5. Shikamany, S. D. Grape Production in India. Food and Agricultural Organization of United Nations, Rome, Italy. **2010,** 1–9.
6. USEPA, California Grapes (Northern and Southern). http://www.epa.gov/oppefed1/models/water/met_ca_grapes.htm **2012,** 1–3 (October 8, 2012).

7. Bates, T. R. Improving Wine Grape Production in Acid soils with Rootstocks and Soil Management. Department of Horticultural Science, Cornell University Vineyard laboratory, Cornell, USA, **2003,** 1–6 (internal project report).
8. Nilnond, S. Grape Production in Thailand. Food and Agricultural Organization of the United Nations. Rome, Italy, **1998,** 1–9.
9. Bates, T. R.; Dunst, R.; Taft, T.; Vercant, M. Seasonal dry matter, starch and Nutrient distribution in'concord' Grape roots. HortScience **2002,** *37,* 890–893.
10. IFA. IFA World Fertilizer manual. International Fertilizer Development Association, Paris, France, **1992,** 142.
11. Spectrum Analytic Inc.; Fertilizing Grapes. Spectrum analytic Inc.: Washington, Ohio, **2011,** 1–14.
12. Shao-Hua, L. Grape Production in China. Food and Agricultural Organization of the United Nations. Rome, Italy. Grape Production in china.htm **1998,** (August 28, 2012).
13. FAOSTAT, Grape Production Statistics. Food and Agricultural Organization of the United Nations. Rome, Italy, http://faostat.fao.org/site/567/DesktopDefault.aspx?PageID=567#ancor. **2012,** 1–13 (August 27, 2012).
14. Demai, L. Seven wine regions in China. http://www.grapewall ofchina.com/2009/02/05/seven-wine-regions-in-china-pros-and-cons.htm **2009,** 1–17 (August 27, 2012).

EXERCISE

1. Discuss origin, centers of genetic diversity and spread of grapes in different continent.
2. Mention at least 10 grape cultivars popular in Europe.
3. Discuss Production methods for grapes in India.
4. Mention the top five grape producing regions of the world and mark them on a map.
5. Mention various uses of grapes.

FURTHER READING

Christensen, L. P.; Kasimatis, A. N.; Jensen, R. L. Grapevine Nutrition and Fertilization in the San Joaquin Valley, University of California, Berkeley, USA, **1978,** pp. 278.

Ohio State University, Fertilizing Fruit Crops. Ohio Cooperative Extension Service, Ohio State University, **1985,** *458,* 78.

Martinson, T. A. Grape Production in New York State. Cornell University Extension Service. http://grapesandwine.cals.cornell.edu/cals/grapesandwine/outreach/viticulture/ny-grape-production.cfm, **2012.**

Winkle, A. J.; Cook, J. A.; Kliewer, W. M.; Lider, L. A. General Viticulture. University of California Press, California, USA, **1974,** pp. 710.

USEFUL WEBSITES

http://www.shttp://www.ces.ncsu.edu/resources/winegrape/ (October 8, 2012).

http://www.spectrumanalytic.com (August 29, 2012).

http://www.grapesfromcalifornia.com/history.php (October 14, 2012).

http://www.landwirtschaft-mlr.baden-wuerttemberg.de/servlet/PB/menu/1043205/index.html (October 14, 2012).

http://nregrapes.mah.nic.in.htm 1–13 (August 27, 2010).

http://www.ngwi.org (October 6, 2012).

MANGO ORCHARDS OF THE WORLD

23.1 INTRODUCTION

Mango is native to Southern Asian regions comprising India, Pakistan, Burma and Andaman Islands. Mango cultivation spread to South-east Asian tropics during 5th century B.C. Mangos were introduced to Chinese main land in 645 A.D., during the reign of Tang Dynasty [1]. Persians and people in Asia Minor region received mango from Eastern Africa, through seafarers and merchants during tenth century A.D. Portuguese travelers introduced mangos to West Africa and Brazil during 16th century. Mango cultivation began in West Indies during mid-1700s. It reached Mexico during mid-1800s [2]. It seems, Wilson Popenoe, a botanist introduced several important mango selections from India into Tegucigalpa, Honduras and other regions of Meso-America. Mangos from Yucatan Peninsula also reached Florida. Mango seeds were imported in large numbers in to Miami, in southern Florida during 1830s. Mango variety such as 'Haden' introduced to Florida 1930 has become popular. Cubans plant first commercial mango trees during 1870.

In the United States of America, the US Department of Agriculture introduced several important commercial varieties and useful germplasm lines of mangos into Florida around Homestead, Fort Lauderdale, during 1930s. Mango cultivation has been in vogue in Puerto Rico, since 1790s. California received mango varieties from Guatemala during 1880. It seems Hawaii islands received mangos from Philippines during 1824. There were also introductions from other regions into Meso-America and United States of America.

Australians received several important mango varieties from South India and Sri Lanka during 1875. Initially, mangos were confined to Northern Queensland. Later, it spread to subtropical regions of Western Australia ([2]; *see* Table 1).

TABLE 1 Mango: Origin, Botany, Nomenclature and Uses—A summary.

Mango is a tree crop native to South Asia. It was domesticated in tropical India. It has been cultivated in the Indian subcontinent since 3,000 years. Mango cultivation spread to East Asia during 5th century B.C. Mango species reached East African coast during Ancient period between 2–10 the century A.D. Mango reached South American regions during 16th century. North American planters in Florida began producing mangos during mid-1800s. Mangoes reached Australian continent during 1800s.

Botany and Classification

Kingdom-Plantae; Order: Family Anacardiacea; Genus-*Mangifera;* Species-*Mangiferra indica.*

$2n = 40$

Germplasm Centers

Germplasm Centre at Central Research Institute for Sub-Tropical Horticulture, Kakori, Lucknow, India; Indian Institute for Horticultural Research, Hessarghatta, Karnataka, India.

Nomenclature

English-Mango; Spanish-Mango; Russian-Mango; Bulgarian-Mango; Croatian-Mango; Czech-Mango Je; Italian-De Mango; Portuguese-Manga-Swedish-Mango ar; Romanian-Mango Lui; Chinese-Manggua; Kannada-Maavina Hannu; Telugu-Mamidi Pandu; Tamil-Mangai; Malayalam-manna; Hindi-Aam; Gujarati-Aam; Marathi-Amba.

Uses

Mango fruits are delicious and sweet. They are mostly ripened and consumed, but green fresh fruit is also eaten.

Mango jelly and marmalades are popular in South India (e.g., Mamidi Thandra, Mambazha vettu). Mango jam (mangada) is a method to preserve the fruit for longer duration. Mangoes are cooked with legume grains (Mamidi pappu, mangai Parippu), Mangos are also used prepare curries and nourish soups. Mangos are consumed fresh even at unripe stge. It is topped with salt and chili powder. Dried and desiccated mangos are used along with salt and pepper (Amchor). Mango pickles are common to most South Asian countries. Large variations in spicy ingredients and mango types are available among pickles.

Mangos are used in preparing juices, ice creams, fruit bars, pies and sauces. Mango mixed with condensed milk is a common dish during festivities. Mangos are used along with other fruits in salads.

Nutritive Value (100 g of raw mango fruit contains following biochemical constituents):

Carbohydrates-15 g; sugars-13.7 g; Dietary fiber-1.6 g; fat-0.38 g; protein-0.82 g; Vitamin A-54 µg; Beta Carotene-640 µg; thiamine-0.03 mg; Riboflavin-0.04 mg; Niacin-0.67 mg; Pantothenic acid-0.2 mg; Vitamin B-6–0.12 mg; Folate-43 µg; Vitamin C-36 mg; Calcium-11 mg; Magnesium-0.16 mg; Phosphorus-14 mg; Zinc-0.09 mg.

Source: Ref. [2]; http://www.hort.purdue.edu/newcrop/morton/manog_ars.html; en.wikipedia.org/wiki/Mango.

23.2 SOILS AND AGROCLIMATE

Mango trees are versatile and adapt to different environmental conditions and soils (Plates 1–3). Soils with poor fertility, mainly low in available nutrients and organic matter are preferred to raise a plantation. Mango is a deep-rooted trees species and explores moisture and dissolved nutrient efficiently from a large volume of soil. In the Indian subcontinent, mango plantations thrive on low fertility Mollisol and Inceptisol areas of Indo-Gangetic belt. Mangos are well adapted to most regions of South India. It is a preferred plantation in low fertility, gravelly and virgin soils. Mangos flourish on clay loams and sandy/gravely Alfisols. Laterites in the coastal and hilly regions are also used to grow mango in large expanses.

PLATE 1 A Mango Orchard with drip irrigation facility, situated near Devanahalli International Airport, Bangalore, India.
Note: Drip irrigation is an efficient system to dispense water in orchards.
Source: Krishna, K. R., Bangalore, India

PLATE 2 Mango inflorecense with mature flowers. Mangoes exhibit massive loss of flowers due 'flower to drop.' Fruit set is proportionately reduced.
Source: Krishna, K. R., Bangalore, India

PLATE 3 A Mango tree with fully grown fruits that are eventually stored in baskets with Cereal hay to ripen them. Location-Doddaballapur near Bangalore in South India.
Source: Krishna, K. R., Bangalore, India

In parts of North Africa and Egypt, mango plantations do occur on calcareous and slightly alkaline Xerasols. Mango orchards found in Israel are known to tolerate certain level of salinity. In Florida, mango plantations are mostly confined to gravely, sandy and calcareous regions not used for production of other crops. Low fertility Spodosols found in Southern Florida supports mango production.

Mango cultivation is spread between 25°N and 25°S of equator. Trees thrive well from sea level to 3,000 ft above sea level. Mango orchards do occur in cooler regions of the world, but they are prone to suffer cold damage. Mangoes are best adapted to warm tropics and sub-tropic of the world. Precipitation requirements range from 80–250 cm during rainy season. In India mango orchards adopt to monsoon climate. The rainy season begins during June/July and lasts until October 2nd weeks. Winter rains occur during December to February. Strong gales, storms and heavy down pours during August/September is often detrimental to fruit bearing. In Florida, mangos thrive on the frequent rains from June to September, and intermittent precipitation pattern that occurs from October to February. In several locations within West Asia and North Africa, mangos are cultivated using regular drip irrigation systems, without relying much on natural precipitation.

The growing season in Australia stretches from mid spring to autumn with flowering from June to August. Mango belt in Australia usually receives about 100 cm precipitation during a year. Peak production of fruits occurs during December and January. The first fruit to reach the markets in late September and October is from the northern growing areas such as Darwin, Kununurra and Katherine, while the last fruit on the market is sourced from around Carnarvon in Western Australia [3].

23.3 PRODUCTION SYSTEMS

23.3.1 MANGO TREE

Mango is an erect, multi-branched evergreen tree characterized by its dome-shaped canopy. It may reach 30 m tall, although most trees are less than half that height, living up to 100 years or more. Mango leaves are evergreen and are borne alternately. Fresh young leaves are pink and highly succulent with fragrance. Flowers are produced on terminal panicles. The root system is extensive, reaching 20–25 ft. in depth. The tree arbuscule or canopy of mango tree varies with cultivar and agronomic procedure adopted by the farmers. Mango cultivars vary enormously with regard to fruit characteristics such a size, shape, color at ripening, flesh thickness, weight and nutritive value. The tree grows in frost-free areas of the world from sea level to about 1,200 m. Heavy rains during flowering will drastically reduce fruit production and mangoes have a tendency to biennial bearing, with many cultivars that produce only one good crop in three to four years (Plate 3).

There are indeed several different production systems adopted to establish, nurture and harvest mangoes. Production systems are mostly governed by the geographic region, mango cultivars predominant and well adapted to region, soil fertility status, irrigation resources, yield goals and economic returns. Based planting density in the orchards we can classify production systems into three groups. They are orchards with 10.5 × 10.5 m spacing that accommodates 86 trees/ha; those with medium spacing (15.2 × 15.2 m) that accommodates 45 trees/ha. There are plantations that hold 100 trees/ha. Fertilizer and irrigation resources need to be prepared according to planting density. Saplings are usually placed previous well-prepared pits, fertilized with inorganic nutrients and FYM. Each tree is supplied with at least 500 g N/tree. In India and many other regions, mango plantations are irrigated only till 3–4 years. In some commercial farms, water is withheld 3 months prior to onset of flowering season. Sprinkler irrigation is practiced in some farms in Florida, Mexico and Brazil [2]. Worldwide, productivity of mango plantations vary depending on production systems adopted. Average productivity varies from 12–23 t/ha in India. A single tree may produce 800–3,000 fruits. In Florida, mango fruit yield ranges from 25–35 t fruits/ha.

At present, mango agroecosystem in different parts of the world is occupied by several different cultivars and hybrids. In India, it seems over 1500 varieties are grown in different mango belts. No doubt, aspects such as nutrient dynamics and productivity are influenced by the cultivar/hybrid that dominates the zone. Mango varieties that fill the cropping belt are grouped based on fruit bearing season [2]:

Early or Mid-season: Bombay yellow, Malda, Pairi (Raspuri, Paheri, goha), swarnarekha, Safeda Pasand;

Early to Mid-Season: Langra, Rajapuri, Mid-season: Alampur Baneshan, Alphonso, Bangalora, Banganapally, Dusheri, Gulab khas, Zardalu, KO-11;

Mid-late Season: Rumani, Samarbehist, Vanraj, Neelum;

Late: Fazli, Safeda Lucknowi, Mulgoba, Neelum, Mallika;

Examples of important *hybrids* that fill the mango belts:

Amrapali (Dashehari × Neelam); arka Aruna (Banganapally × Alphonso); Arka Puneet (Alphonso × Janardhan Pasand); Ratna (Neelam × Alphons); PKM-1 (chinnasuvarnarekha × Neelam); Au Rumani (Rumani × Mulgoa); Sindhu (Ratna × Alphonso);

23.4 EXPANSE AND PRODUCTIVITY

Mango production is concentrated in the Indian subcontinent and China. It is also grown in relatively larger quantities in Mexico and Brazil. India is the largest producer of mango at 16.3 m ton fruits yr^{-1}. It is followed by China 4.35 m t, Thailand

2.55 m t, Pakistan, 1.78 m t, Indonesia 1.31 m t, Mexico 1.63 m t and Brazil 1.2 m t yr^{-1} [9]. The mango agroecosystem is large and concentrated in the Gangetic plains and Southern Indian Plateau. The mango cultivation region in India extends into 1 m ha and is equivalent 70% of global mango growing region. It produces about 65% of global mango fruits annually. Mango belt in Mexico is relatively large at 42,000 ha. The mango agro ecosystem in India is highly variegated in terms of cultivars and hybrids that fill the zones. There are over 100 cultivars and several hybrids that fill the mango belts. However, some like Neelam, Alphonso, Dushehri, and Langra are more common. Some cultivars are highly localized and are grown intensely. For example, Bangalora or Thotaharpuri is predominant around Southern Karnataka (Plates 1–3). Similarly, Banganpally is predominant in parts of Andhra Pradesh, Langra and Dushaheri in Uttar Pradesh. We should expect the dominant mango cultivar to exert maximum influence on vegetative landscape, soil nutrient dynamics and productivity of the mango belt.

Mango orchards are predominant in several states of India. Andhra Pradesh has the largest area under mango plantations. During 2009, mango belt in Andhra Pradesh extended into 0.5 m ha and produced 2.5 m t fruits at a productivity of 5.1 t fruits/ha. Productivity of mango orchards is better in states like Uttar Pradesh at 12.7 t fruits/ha. Major mango producing zones occur in Andhra Pradesh, Uttar Pradesh, Bihar, Karnataka, Tamil Nadu, Maharashtra, West Bengal, Orissa, Kerala and Gujarat. The mango growing regions in these states range from 86,000 ha to 497,000 ha. The productivity ranges from 1.6 t fruits/ha to 12.8 t/ha depending on region. Details on soil type, localization of mango orchard and production are available [5–8].

Mango growing regions are conspicuous in many provinces of Pakistan. Major mango belts occur in Sindh, Southern Punjab, Mirpur Khas, and Multan. Mango orchards do spread into other regions such as Hyderabad (Sindh), Nawabshah, Khairpur, Bhahwakpur and Rahim yar Khana. The mango agroecosystem in Pakistan is filled with several cultivars and hybrids. Sindhri is a popular variety that dominates most of mango growing regions. Other cultivars that predominant are Chaunsa, Dasheri, Langra, Sonara, Anwar kle, Saroli, Fajri, Malda, Gulab Khas, Neelam, etc. During 2011, Pakistan contributed 1.78 m t fruits at an average productivity of 2.1 t fruits/ha. Mango belt in Bangladesh is moderately intense and spreads into most regions other country except water-logged zones. Productivity ranges from 1–3 t fruit/ha.

China has a very large mango agroecosystem. It contributed 4.35 m t fruits during 2011 [9]. The mango belt is distributed across different regions of main land China. It is predominant in Hainan, Guangxi, Taiwan, Yunnan, Guangdong, Sichuan and Fujian. Hainan (47,000 ha) and Guangxi (33,000 ha) together constitute 73% of mango expanse in China. The productivity of mango in these two provinces is high at 7.4–8.3 t/ha. Mango cultivars that dominate the orchards in Hainan and Guangxi are tainoung No-1, Nang Klagnwan, Guifei, Red Ivory, Zihua and Jin Hwang [1].

In Philippines, mango production in orchards is mainly to satisfy its own needs. Farmers tend to supply high levels of nutrients and irrigation in order to achieve yield goals. Fertilizers are actually supplied based on soil tests and tissue analysis. Foliar fertilization is also practiced. Currently, fertilizer technology is being improved to raise productivity of orchards [10].

Average production of mangos in the West African region, during past decade was 1.4 m t annually. The major mango producing nations in West Africa are Benin, Burkina Fuso, Cape Verde, Cote d' Ivori, Gambia, Ghana, Guinea Bissau, Mali, Nigeria, Senegal and Sierra Leonne [9]. The average productivity of mango orchards in most locations is about 4–5.2 t fruits/ha [11]. Mango is popular fruit in many of the West African nations.

South African mango growing regions expanded slightly rapidly during past 15 years. Currently, it contributes 2% of global mango production. Mango belt has also experienced decline during recent years [12]. Mango orchards are traced in many regions of South Africa. Major mango growing regions occur in Northern Province. Letsitelle, lower Letaba and Ofcaloca areas account for 60% of nation's mango production. Actually, Mpumalanga and Malelene regions are most important mango production zones. Mangos in South Africa tolerate high temperatures ranging from 20–35°C and precipitation of 300–350 mm. Mango orchards are found from sea level to 1200 m.a.sl. Mango orchards are productive if raised n soils that are well drained, deep and fertile. Mango trees tolerate semiarid and dry climatic conditions.

Mango orchards occur practically in all the regions of tropics and subtropics South America. Mango production mostly satisfies domestic consumption. Mexico exports a certain share of its mangos to Europe and North American markets. Mango cultivars that are popular are Tommy Atkins, Keitt and Heiden. Brazil has a regular mango-breeding program to improve both roots stocks and grafts [13].

Europe has its share of mango growing regions, although not prominent as in Asia. Mango orchards are found in Canary Islands and Andulasia. Mango cropping belt in Spain is concentrated in Malaga province. Mango trees are also traceable feebly and scattered in Coastal southern Spain. Southern Italy and Sicily has a few mango orchards that contribute fruits to satisfy domestic needs. Mangos are grown in small areas in Coastal Greece, Malta, Cyprus and Riviera region of France.

On the world scale, the Australian mango industry is relatively insignificant. However, production has increased over the past decade and it is forecasted that domestic production will double within ten years, making it one of the major domestic horticultural crops [3]. Mangoes grow best in climates, which have low rainfall, and low relative humidity at flowering, fruit setting and harvesting and that are warm to hot during fruiting. However, mangoes will tolerate a wide range of climates from warm temperate to tropical. With these climate requirements, mangoes are grown predominately in northern Australia. Queensland is the major producing state, accounting for 70% of domestic production, by volume. The Northern Territory is also a major mango producer accounting for 20% of production. The remaining production areas are in Western Australia and northern New South Wales [3].

In Australia a grafted mango tree will take approximately three years to bear fruit and will achieve peak production at six to eight years of age. Seedling trees take a year longer to come into production. Currently around 90% of the trees grown commercially in Australia are Kensington Pride with the other established varieties including Irwin, Nam Dok Mai, R2E2, Glenn, Kent, Tommy Atkins and Palmer. A new variety of mango called Calypso is also starting to be produced and is reaching the market, albeit in small volumes at this stage.

KEYWORDS

- **Haden**
- **Inceptisol**
- **Mamidi Thandra**
- **Mollisol**

REFERENCES

1. Gao, A.; Chen, Y.; Crane, J. H. Status and Analysis on Mango Production in China. International Conference on Agricultural and Biosystems Engineering. Advances in Biomedical Engineering **2011,** *1–2*, 472–476.
2. Morton, J. F. Mango (*Mangiferra indica*). In: Fruits of Warm Climates. Miami, Florida, USA, **1987,** 231–239 http://www.hort.purdue.edu/newcrop/morton/manog_ars.html (September 22, 2012).
3. AAG. Mango Market Overview. Australian Agribusiness Group. Sydney. checkhttp://www.ausagrigroup.com.au/ market/market.php?mode=view_doc&doc_id=123 **2006,** (December 5, 2012).
4. FAOSTAT, Mango Production statistics. Food and Agricultural Organization of the United States of America. Rome, Italy, http://www.faostat.fao.org/site/576/DesktopDefault.asp?PageID=567#ancor **2011,** (September 22, 2012).
5. ICAR, Hand Book of Horticulture. Directorate of Information and Publication of Agriculture, ICAR, New Delhi, India **2002,** 876.
6. NHB. Indian Horticultural Data base. National Horticultural Borad. New Delhi, **2009,** 485.
7. Kumar. V. Indian Horticultural Database **2009,** National Horticultural Board, government of India, Institutional Area, Gurgaon, New Delhi, India **2009,** 185.
8. Biswas, B. C.; Kumar, L. Revolution in Mango Production: Success stories of some Farmers. Fertilizer Marketing New **2011,** *42,* 1–24.
9. FOASTAT, Mango Statistics. Food and Agricultural Organization of the United Nations. Rome, Italy. http://www.faostat.org **2004,** (October 25, 2012).
10. Belarmino, C. N. Mango Production Technology. http://www.neda.gov.ph/knowledge-emporium/details.asp?DataD=93 **2003,** 1–3 (August 28, 2012).
11. Deng, Z.; Janssens, M. Shaping the future through pruning the Mango tree. http://www.tropen.uni-bonn.de/new_website/englische_seiten/Research/Research_%20project. **2004,** 1–6 (August 28, 2012).
12. Fivaz, J. Mango production in South Africa as compared to rest of the world. VIII International mango Symposium. http://www.actahort.org/members/showpd?booknramr=820_1. **2010,** 1–2 (August 28, 2012).
13. Galan Sauco, V. The mango in Latin America. VI International Symposium on Mango. http://www.actahort/ books/509/509_11.htm **2011,** 1–3 (August 28, 2012).

EXERCISE

1. Describe growth physiology of Mango tree from seedling till fruiting stage.
2. Discuss Origin and spread of Mangos worldwide.
3. Mention the names of 10 most popular mango cultivars grown in India.
4. Mark the regions of mango production South America and Australia.
5. Provide data on nutritive value of at least five different Mango varieties.

FURTHER READING

1. Biswas, B. C.; Lalit Kumar. Revolution in Mango Production. Success stories of some farmrs. Fertilizer Association of India. *Fertilizer Marketing News,* **2011,** *20,* 1–24.
2. Litz, R. E. The Mango: Botany, Production and Uses. Commonwealth Agricultural Bureau, England, **2009,** pp. 682.

USEFUL WEBSITES

http://era.deedi.qld.gov.au/1647/ (December 12, 2012)
http://atobelensfarm.com/2008/05/21/mango-production-guide/ (December 12, 2012)
http://www.virtualherbarium.org/tropicalfruit/MangoTreeCare.html
http://www.icar.org.in/node/2303 (December 12, 2012)
http://rd.springer.com/chapter/10.1007/978–3-540–34533–6_16#page-1 (December 12, 2012)

COCONUT PLANTATIONS OF ASIA

23.1 INTRODUCTION

Coconuts were domesticated in South Asia. Archeological studies suggest that coconut has been used by native populations in South Asia since 2nd millennium B.C. Ancient texts of Hindus refer to coconuts as 'Kalpa Vriksha,' which means 'Tree of Heaven.' It has variety of uses to mankind. Some of these relics state that coconuts were actually brought to Indian main land from Srilanka [1, 2]. Coconuts originated in tropics of India and adjoining regions in South Asia. Coconuts are highly dispersed species, due to the drupes that float through the oceans into different continents. Hence, it is sometimes difficult to locate the origin of many of the varieties of coconut palms. The coasts in India and other Southern Asian countries seems to be the center of origin of many varieties that occur in Fareast, Pacific Islands, African coasts, Caribbean and other regions of South America. Coconut cultivation was prominent in ancient Sri Lanka. Edicts and relics of *Mahavamsa* state about coconut groves during the Agrbodhi's reign dating to 589 A.D. Occurrence of established coconut groves in Arabian region has been mentioned Ibn Batata's writings *'Al Rihla.'*

Coconut seems to have dispersed naturally from west coast of India to Eastern coastal regions of Africa, through drupes that float across long distances. The Spaniard conquerors and seafarers introduced coconut to shores of West Indies and other Caribbean islands. Portuguese settlers in South America introduced coconut into Bahia region of Brazil. During medieval period, merchants, traders and monks, who traveled across different parts of West Asia and Fareast, dispersed coconut cultivation methods from India [1].

TABLE 1 Coconut: Origin, Botany, Classification, Nomenclature and Uses—A summary.

Coconuts are native to South Asian coasts and plains that experience tropical climate. Coconut agroecosystem spread into vast stretches of tropical South-east Asia, Fareast and Pacific Islands through natural dispersal and human migration. During medieval period, coconuts were introduced to many regions in Latin America by Spaniards. Currently, coconut belt thrives in over 92 countries on the earth. The coconut belt is most intense in South Asia and its coastal regions.

Botany and Classification

Kingdom-Plantae, Class-Monocots, Order-Arecales, Suborder-Commelids, Family Arecaceae, Sub-Family Arecoidae, Tribe-Cocoeae, Genus-*Cocus*, Species-*C. nucifera*

$2n = 32$

Coconut cultivars traced in different continents can be grouped into two natural subgroups, namely Tall and Dwarfs. Commercial coconut plantations prefer high yielding, long duration, and tall varieties. For example, Ceylon Tall, Indian Tall, Jamaican Tall, Panama Tall etc. Tall cultivars are referred as *Cocus nucifera var typical* and dwarf cultivars are known as *Cocus nucifera var nana.*

Genetic analysis of coconut germplasm from different regions of the world that focused on 10 different microsatellite loci, has shown several centers of genetic diversity and origin. There are large numbers of subpopulations with origin in the nations within South Asia and Indian Ocean region. There is second center of genetic population that originates in the Pacific islands. Yet another set of germplasm belongs to Madagascar and Seychelles as center of origin. Genetic analysis has also shown that ancestral populations of coconuts may have originated in Australasian regions.

Germplasm centers occur at Central Plantation Crops Research Institute, Kasargod, Kerala in South India, Coconut Cultivation Board, Battaramulla, Srilanka. Several other germplasm collections are traced in Coasts of Asia and Africa.

Nomenclature

Azerbaijani-Kokos, Palmasi; Belarussian-Palma; Bulgarian-Kokasa; Catalan-Cocoter; Czech-Kokosvnik; English-Coconut; German-Kokospalme; Danish-Kokos; Spanish-Cocos nucifera; Norwegian-Kokosnet; Slavik-Kokosvnik; Finnish-Kookos PalmeSwahili-Mnazi; Indonesian-Kalap; Malayan-Kalapa; Bengali-Narikela;Hindi-Nariyal, Kannada-Tengu or Tenginakayi, Telugu-Tenkayi, Kobbara Kayi

Uses

Coconuts are used in innumerable ways by the farming community and general human population. It is estimated that 48% of coconuts produced are used for edible purposes. About 31%is used as milled copra, 11% as tender coconut and 1% used for nontraditional purposes [1]. Following are major uses: namely Copra or flesh or meat of coconut is used commonly in cakes. Coconut oil is a common product derived by extracting the dried flesh inside fruit. Coconut oil is a popular cooking medium and fat source for people in South Asian countries. Coconut cakes derived from dried and extracted wastes are used as feed for animals. Coconut coir is an important fiber in Coastal zones. Fiber and husk are used as cushiony packing material. Coconut is used as fuel in most villages of South India. Coconut water (also called milk) is a nutritious drink derived from drupes.

Coconut fruit (meat) or 'Copra' is used in several ways. The flesh is dried to 2.5% moisture. Its powder or flakes are used in delicacies, candies and range of confectionary items. Fresh or raw flesh can be squeezed and the milk could be used in different culinary items. The sap from coconut inflorescence is used prepare to brown jiggery. The fermented coconut water/sap is a good alcoholic drink known commonly in Asia as 'Toddy' or 'Arrack.' In the Coastal regions of Asia and Africa, coconut is used as daily vegetable. The per capita consumption of coconut products is high 0.67 kg/yr.

Coconut meal prepared from extracted fruits are useful as feed to farm animals. The dried leaves, calyx and inflorescence material is used as fire wood. The stem and woody portions are used to prepare furniture, canoes and variety of containers in the coastal regions of India. Coconut leaves are used preparing roofs or thatches. Coconut leaves are used for making shelters for nursery beds, mulches, fences and temporary shades for animals. Coconut leaves are weaved into variety of baskets and other useful items for household use. Coconut water is used in preservation of tissue and micro-propagation of seedlings of different plant species. Medicinal uses of coconuts are wide ranging. Coconut water is sterile and is used for hydration of intravenous fluids. The extract from coconut husk is used to treat inflammatory disorders. Coconut water is used as low cholesterol substitute in many food material and tonics. Young coconut juice is said to possess estrogen-like properties.

Nutritive Value of Coconuts (100/g of edible portion)

Raw coconut: calories-354, Protein-3.3 g, Fat-33.5 g, Saturated fats-29.7 g, Mono-saturated fats-1.43 g, Polyunsaturated fats-0.37 g, Carbohydrate-15.2 g, sugars-6.2 g, Crude fiber-9.0%, Thiamine B1–0.066 mg, Riboflavin B2–0.02 mg, Niacin-0.54 mg, Pantothenic acid-0.3 mg, Vitamin B6–0.05 mg, Folate-26 μg, Vitamin C-3.3 mg, Calcium 14 mg, Phosphorus 113 mg, Iron-2.43 mg, Potassium-356 mg, Zinc-1.1 mg.

Coconut Shell (on dry basis): Cellulose-34%, Lignin-36%, Pentoses-29% and Ash-0.6%.

Coconut Shell Ash (burnt product): K_2O 45%, Na_2O-15.4%, CaO-6.26%, MgO-1.32%, Fe_2O_3 + Al_2O_3 1.39%, P_2O_5 4.64%, SO_3 5.75% and SiO_2 4.64%.

Source: Ref. [3]; USDA Nutrient Database; en.wikipedia.org/wiki/Coconut; http://www.unctad.info/en/Infocomm/ AACP-Products/COMMODITY-PROFILE—Coconut2/

23.1.1 COCONUT TREE

Coconut is a monocot tree, a tall or dwarf palm based on genotype characteristics. Tall varieties may reach 30 m in height. Coconut trees form a whorl of leaves at the crown. Leaves are 4–6 m in length and pinnae are 60-cm. Coconut tree produces about 75 fruits/yr. The coconut inflorescence born at the crown has both male and female flowers. Coconut palm is monoecious. The female flower is larger than male ones. Coconuts are mostly cross pollinated. Coconut fruit is a drupe. The exocarp and mesocarp make up the husk. The mesocarp has fiber, more commonly called as coir. The shell has a single seed. When the seed germinates, radicle and embryo pushes out of one of the eyes of shell. A mature coconut fruit has a thick albuminous endosperm that adheres to inside of the testa. The endosperm or meat or flesh is the white fleshy edible part of the coconut. The endosperm has saturated fat. However, it has relatively lower quantities of sugar. Depending on the size of the drupe, coconut may contain 300–1000 ml of coconut water. Coconut water is sweet in a tender coconut. The coconut tree has a fibrous root system. Coconut palms start yielding fruits at 3–7 years depending on weather it is an early dwarf or late tall variety. Each tree may 70–150 fruits/yr. Coconut palms has an extended period of vegetative phase. They produce fruits for more than 50–100 years. The peak fruit yield occurs on plantations that are 20–25 years old.

23.2 SOILS, AGROCLIMATE AND PRODUCTION SYSTEMS

Coconut agroecosystem thrives best on sandy soils. However, coconut palms can be grown on loamy and heavier soils with greater proportion of clays. Coconuts need well-drained deep soils. They can tolerate infertile and saline soils better than other tree species. Coconut trees grow normally even in soils with pH 8.0. However, optimum soil pH is 5.5–7.0. In Philippines, where coconut plantations are intensely cultivated, soils are mostly loamy Alfisols or laterites. In India, coconuts thrive on coastal sands, Alfisols and light textured clay loams in the Southern India plains. It thrives well on the laterites found in South-west coast. Soil pH ranges from 5.5 to 7.8. The fertility of soil found in Indian coconut belt could be classified as moderate. Nutrient deficiencies are common. Hence, routine supply of major nutrients and supply micronutrients periodically is necessary.

Coconut belts thrive from sea level (coast) to altitudes of 500 ft above sea level. They are found mostly at altitudes <1000 ft above sea level. Coconuts tolerate drought spells but not extended periods of water scarcity. Water logging affects its root respiration and nutrient acquisition functions. Yet, the tree can tolerate water-logged conditions for 3–4 weeks. The coconut belt thrives in regions with humid or sub-humid conditions, with mean temperature ranging from 18°C–32°C. It prefers relative humidity of 60–70 % during the year. Coconut plantation needs 1500–2500 mm precipitation during the year.

23.2.1 PRODUCTION SYSTEMS

Coconut trees are deep rooted monocots. Plantations are therefore developed on sites with deep soils of at least 1.5 m and above. Sandy or loamy deep soils are preferred. Shallow soil with hard rocks and low lands prone stagnation of water avoided. Production systems are dependent on factors such as geographic location, topography, soil type, its fertility, water resources and a range of environmental parameters. Coconut plantations in South and South Asia are often situated close to or along with paddy fields. To avoid ill effects of water stagnation, contour bunds or mounding is practiced. Coconut seedlings that are raised in nursery are placed in individual pits and nurtured. The size of the pit is dependent on location, nature of soil and its fertility. In sandy soils, pit sizes could be 0.75 × 0.75 m. In lateritic soils with hard layers, pit sizes need to be larger at 1.2 m × 1.2 m. Mixtures of coconut husks, FYM and fertilizers are applied to each pit. Spacing of coconut trees depends on the genotype, soil fertility status and economic goals. In South India farms prefer 250 trees/ha. Coconut seedlings are transplanted to main field during late May or early June in order to exploit precipitation pattern effectively. During initial stages of tree growth, shading is necessary. Fertilizers are supplied 2 or 3 times in split dosages, depending on SOM content. Inter-culture with cover crops or weeding periodically is necessary to maintain proper nutrient dynamics in the plantation. Lateritic or loamy soils are plowed once or twice during a year. Farmers adopt different types of irrigation, such as spot irrigation at each pit, flat bed flooding and drip irrigation. It is said water requirements of an adult coconut palm is about 40–50 liters/day. Therefore, water holding capacity of soil is an important trait.

Regarding nutrient supply to coconut palms, all three major nutrients N, P and K are supplied to each tree. Nitrogen is supplied in an increasing order from 150 g/plant during first year of sapling to 500 g/plant by 8th year and kept constantly at that level thereafter. Phosphorus supply ranges from 200–400 g/plant and K supply ranges from 300 to 1,000 g/plant. Farm yard manure is applied at 40 kg/plant. In Sri Lanka, coconut plantations are maintained using optimum amounts of fertilizers and irrigation as required based on soil tests. General recommendations by Coconut Cultivation Board [4] suggest that each tree should be provided with 765 g urea, 140 g Eppawa Rock Phosphate and 883 g of Muriate of Potash. Fertilizer rates are revised upwards, if plantations possess hybrids. In Philippines, most plantation owners opt for integrated nutrient management procedures that include chemical fertilizers such as N, P, K and organic manures [5].

Coconut is a tall tree, even the dwarf varieties are tall enough to allow enough photosynthetic radiation and ambient space for intercrops to flourish as understory crops. Coconuts are intercropped with wide range of cereals, legumes and even perennial fruit crops. Coconuts are grown raised terrain in low land paddy cultivation zones of South India (Plates 1 and 2). Coconuts plantations are sown with cover crops in order to restrict loss of soil and nutrients from the plantation.

PLATE 1 Coconut-Paddy intercrops in South India. Location is near Bangalore, South India.
Source: Krishna, K.R., Bangalore, India

PLATE 2 Coconut Palm plantations in the Coastal tracts of Peninsular India.
Note: Location is near Panjim, Dona Poula in Goa. Coconut trees thrive very close to salt water and rivulets joining the sea without being affected by salty water. The lateritic soils, although leached heavily, moderately fertile and rich in Al and Fe salts, support optimum growth and fruit production.
Source: Krishna, K.R., Bangalore, India

23.3 EXPANSE AND PRODUCTIVITY

The coconut agro-ecosystem is wide spread in the subtropical and tropical coasts and interior land. Coconut drupe is light in weight and floats. Hence, its distribution via ocean waves has induced its spread all over in the coasts of topics. Coconut agro-ecosystem dominates the Coasts of South and South-east Asia. It also thrives well in the tropical plains and hills in the tropical Asia. Coconut plantations are common all through the African Coasts that experience tropical climate. It is more pronounced in the East Coast covering regions in Kenya, Tanzania and Mozambique. Coconut fruits are traceable in Hawaii, Norway, Polynesian islands, Caribbean and South American. In USA, coconut palms are grown mostly as ornamental species. They are traced in Hawaii and Southern Florida. United States of America's territories such as Guam, Puerto Rico, Virgin islands and Northern Mariana islands also support coconut palms.

More than 92 countries in the world possess coconut plantations. Globally, coconut agroecosystem stretches into a total area of 11.8 m ha. It produces 61.78 m ton fruits at an average productivity of 5.2 t drupes/ha [6]. The global coconut agroecosystem is most intense in Asian countries. About 78% of global coconut area and produce arise from Asian tropics. In Asia, Philippines has the largest coconut belt with 3.2 m ha, followed by Indonesia 2.66 m ha, India 1.90 m ha, Srilanka 0.45 m ha, Thailand 0.34 m ha, Vietnam 0.14 m ha, and Myanmar 41,000 ha. China has relatively small coconut region with 28.2,000 ha. In Africa, coconut agroecosystem is conspicuous in Tanzania with 310,000 ha, Mozambique 70,000 ha and Ghana 55,000 ha. Coconut belt in Dominican Republic extends into 38,000 ha. Brazil has coconut belt that stretches into 2200 ha.

During 2009, Philippines contributed 16.6 mt, Indonesia 21.5 mt, India 10.1 mt, Srilanka 2.09 mt, Thailand 1.3 mt, Vietnam 1.3 mt and Malaysia 0.5 mt [6]. The productivity of coconut plantations vary with geographic location. It ranges from 4.2–6.7

t drupes/ha in most Asian nations. It is 7.0 t drupes/ha in Brazil [7]. China, USA, and United Arab Emirates are the major coconut importing nations. Any fluctuation in demand for coconut by these nations may impart a certain degree of influence on coconut belts in other regions, regarding input, yield goals and expansion plans. At present, coconut belts in Vietnam, Indonesia, Sri Lanka and Thailand export most coconuts. They export about 264,000 tons of coconut out of a total of 360,000 ton exported globally [6].

Coconut agroecosystem of India is confined to Southern states such as Andhra Pradesh, Karnataka, Kerala and Tamil Nadu. Together, the four southern states account for 92% of total coconut belt in the country. The coconut belt also extends into Orissa, West Bengal, Tripura and Arunachal Pradesh in the east and Maharashtra and parts of Gujarat in the West. It also extends into Lakshadweep and Andaman Islands [2]. The Coconut belt has fluctuated in area from 1,513,000 ha to 1,890,000/ha during the past two decades from 1990 to 2010 [1]. Total production during this same period of 2 decades has ranged from 9700 million nuts to 12500 nuts per year. The productivity of coconut belt has fluctuated from 6300 nut/ha to 7700 nuts/ha during the past 2 decades (Ikisan, 2012). The coconut belt in India is occupied by several tall and dwarf cultivars/hybrids. For example, Chowghat Dwarf Orange, Chowghat Dwarf Yellow, Chowghat Dwarf Green, Malayan Dwarfs are prominent dwarfs. Gangabondam is a semitall type that is popular in Andhra Pradesh. Lacadive Tall, Tiptur Tall, Komadan and Andaman Ordinary are common tall varieties found in India and other regions of the world. Most common indigenous genotypes are West coast Tall, East Coast Tall, Banavalli Green Round and Kappadam. Exotic genotypes that are conspicuous in South India are Fiji Tall, Philippines Ordinary, Lakshganga, Anandaganga and Keraganga. Most of these genotypes yield 80–105 fruits/tree. The proportion of each coconut genotype and its influence on soil nutrient dynamics and productivity needs attention.

Coconut palms extend into coasts of Red sea, Arabian Sea and Persian Gulf. Their cultivation is prominent in Yemen's Al Maharah and Hadramaut region. Sultanate of Oman has coconut palms meant for fruits and oil. They are popularly grown for tourist attraction in the oasis and coasts. Coconut groves in Middle East and Arabia are mostly made of West Coast Tall derived from India, since these are highly drought tolerant. Coconut belt extends into Sri Lanka since it enjoys tropical climate throughout the country. Coconut is among the main dietary source of fat for the people.

KEYWORDS

- ***Al Rihla***
- ***Cocus nucifera var nana***
- ***Cocus nucifera var typical***
- **Copra**
- **Kalpa Vriksha**
- **Toddy**

REFERENCES

1. IKISAN, Coconut. http://www.IKISAN.com/cropspecific/eng/links/ap_coconutHistory.shtml **2012,**1–8 (October 12, 2012).
2. Markrose, T.; Coconuts in India. http://www.bgci.org/education/1685/. 1–2 (August 26, 2012).
3. Woodroof, J. G. Coconuts: Production, Processing, Processing, Products. AVI Publishing, New York, **1979,** 278.
4. Coconut Cultivation Board, Fertilizer Application. Coconut Board of Sri Lanka, Battaramulla, Srilanka. http://www.coconut.gov.lk/index.php **2011,** 1–3 (August 26, 2012).
5. Mantiquilla, J. A.; Canja, L. H.; Margate, R. Z.; Magat, S. S. The use of Organic Fertilizer in Coconut. **1994.**
6. FAOSTAT, Coconut Statitics. Food and Agricultural Organization of the United Nations. Rome, Italy, http://www.faostat.org **2012,** (October 24, 2012).
7. UNCTAD. Coconut-Infocomm-Commodity Profile. Unctad.org **2012,** (August 26, 2012).

EXERCISE

1. Mark Coconut growing regions of the world.
2. Describe various soil types and agroclimatic conditions that support coconut plantations.
3. Mention top five coconut growing regions and productivity levels.
4. Mention various uses of coconuts.
5. Mention names of five coconut genotypes and their Productivity.

FURTHER READING

Liyange, D. C. Increasing Coconut Production: The Challenge. Vijitha Yapa Publications, Colombo, Srilanka, **2005,** pp. 113.

Mahindapala, R.; Pinto, J. L. J. G. Coconut Cultivation. Coconut Research Institute, Srilanka, **1991,** pp. 183.

Mandal, R. C. Coconut (Production and Protection Technology). Riddhi International Inc. New Delhi, **2010,** pp. 168.

Mussig, I. J. Coir: Coconut cultivation, Extraction and Processing of coir. In: Natural Fibers. Coconut Research Institute, Lumunula, Srilanka, **2010,** pp. 85.

Ohler, J. G. Modern Coconut Management: Palm cultivation and Production. Practical Action Inc. Netherlands, **1999,** pp. 458.

USEFUL WEBSITE

http://www.coconutindia.blogspot.com/2009/06/coconut-cultivation-practices-for-india.htmlpp.1–8. (October 12, 2012)

http://www.mapsofindia.com/andiaagricultutre/oil-seeds/coconut-growing-states.htmlpp.1–2 (October 12, 2012)

http://www.ikisan.com/crop.specific/eng/links/ap-coconutHistory.shtml pp. 1–8 (October 12, 2012)

http://www.fao.orgg/docrep/004/ac126e/ac12604.htm pp. 1–4 (August 24, 2012)

http://www.coconutboard.nic.in/publi1.htm (August 24, 2012)

COFFEE PLANTATIONS OF THE WORLD

23.1 INTRODUCTION

Archaebotanical studies suggest that region of maximum genetic diversity and domestication for coffee occurs around the hilly terrains of Ethiopia and Sudan. Coffee (Arabica types) has been cultivated in the plateaus of Central Ethiopia since ages. Coffee plants were domesticated in the Kaficho and Shakicho areas of Ethiopia. Archaeological analyzes of these locations indicate that coffee was consumed along with pepper and ginger to prepare a decoction known as *Chomo*. It traversed to Yemen during sixth century A.D. During the same period, it was also introduced to Egypt and Saudi Arabia [1]. Coffee reached Istanbul in Turkey and Damascus in Syria during medieval period around 1517. Coffee production was well established in southern Europe during medieval times. Coffee houses were found in great number in Venice and other cities of Italy. Coffee consumption increased significantly even in northern European nations during mid-17th century. Coffee was introduced into Brazil by Francisco de Mello Palheta in 1727 [2]. Coffee seeds were transshipped to Srilanka during early 1500s via maritime trade. Coffee was introduced into South-west Indian hill tracts by a pilgrim to Mecca named Baba Budan, during 17th century. The Dutch travelers aided transfer of coffee seeds and establishment of plantations in South-east Asian nations. It is said the Dutch Governor at Malabar in South-west India took active interest in introducing coffee seedlings into the vicinities of Jakarta during 17th century. Coffee seeds were also introduced into Indonesia by colonizers from other European regions. Clearly, like most other crops, coffee plantations too spread across the world with seafarers, traders and conquests. Its spread out of Arabian region and North-east Africa was slow during ancient period, but was rapid during medieval time. Most striking is the rapid spread of coffee agroecosystem in Brazil that occurred during 18th and 19th century.

There are two important types of coffee plantations. They are Arabica and Robusta types. The *Coffea arabica* known as Arabica type covers about 75–80% of global coffee based agroecosystem. *Coffea canephora* known as Robusta accounts for the remaining 20% of coffee belts. Coffee plants grow up to 10 m, if left un-pruned. However, coffee plantations are generally well pruned and trained, so that it allows maximum photosynthetic efficiency. Flowers appear about 3 to 4 years after planting. Root systems are deep and reach 30–45 cm depth. Total root length of a coffee tree may reach 20 to 25 km. Clearly, coffee plantations produce extensive and efficient root system that induces rapid absorption and translocation of nutrients from subsurface and surface soils to above-ground portions. The elliptical thick deep green leaves are waxy in appearance. Coffee plantations are excellent providers of O_2 to the atmosphere. Each hectare of coffee plantation is known to emanate 86 lbs of O_2 into atmosphere ([3]; Plates 1 and 2; Table 1).

PLATE 1 Left: Top Coffee Plantation in bloom. Top Right, Above Left and Right: Young and Ripe Coffee berries.
Source: **Krishna,** K.R., Bangalore, India

PLATE 2 A coffee plantation on hillock near a location called Malinal, south-west of Tepic which is the capital city of Narayit in Mexico.
Source: http://www.coffeeresearch.org/agriculture/varietals.htm

TABLE 1 Coffee: Origin, Nomenclature, Uses—A summary.

Coffee was domesticated in the North-east Africa in the region comprising Ethiopian High lands and Northern Kenya. It later spread to other regions in West Asia, Arabia, India, Vietnam, Mexico and Brazil

Important coffee species grown worldwide are *Coffee arabica var typica*; *Coffee cenephora variety robusta* and *Coffee liberica*

Coffee arabica $2n = 44$

Coffee robusta $2n = 22$

Germ plasm centers are available in several nations such as Brazil, Mexico, Ethiopia, Kenya, Vietnam, India, etc.

Nomenclature

Catalan-Café; Czech-Kava; Danish-Kaffe; French-Café; German-Kaffee; Italian-Caffe, Chico de Caffe; Hungarian-Kave; Russian-Kofe, Spanish-Cafeina, Café; Swedish-Kopp Kafe.

Uses

Coffee is a popular beverage worldwide. Coffee is also used to extract caffeine that has medicinal value.

Coffee residue available after curing is used as organic matter.

Nutritive Value (per cup of 237 g)

Cafiene content: Average-80–135 mg, drip coffee-115–175 mg, Espresso-100 mg, Instant-65–110 mg; decaffeinated coffee-3 to 4 mg.

Major Amino acids: Alanine-7.1 mg; Aspartic acid-11.9 mg; Glutamic acid-47.4 mg; Leucine-11.9 mg; Phenylalanine-7.1 mg; Proline-9.5 mg; Valine-7.1 mg.

Vitamins: Folate 4.7 µg; Niacin 0.5 mg; Pantothenic acid 0.6 mg; Riboflavin-0.2 mg; Ca-4.7 mg; Mg-7.1 mg; Mn-0.1 mg; P-7.1 mg; K-116 mg; Na-4.7 mg.

Source: Several; http://www.coffee-makers-et-cetra.com/nutrition-facts-for-coffee.html

23.2 AGROCLIMATE, SOILS, WATER REQUIREMENTS

Coffee plantations need a tropical climate with warm temperature and moderate relative humidity. They are mostly grown on well-drained soils or on slopes that do not allow stagnation and hold only low amounts of water in the profile. In Brazil, coffee plantations confine to sandy soils occurring at about 700 m.a.s.l. and that retain only low quantities of moisture. The coffee belt in Vietnam and Indonesia thrives on Alfisols and Inceptisols. In the South India hills, coffee thrives well on Lateritic soils rich in Al and Fe with acidic reaction. They are mostly grown on hill slopes, so that moisture stagnation is avoided. Coffee is grown under rain-fed conditions whenever rainfall pattern is congenial. Farmers adopt different kinds of irrigation such as drip, furrows and sprinklers to irrigate coffee gardens. Yet, coffee crop grown in Brazil or elsewhere in Vietnam and India suffers from drought spells. In Brazil (Western Bahia), *veronicas* reduce coffee pod yield, significantly. For example, a sprinkler irrigated coffee garden produces about 4–5 ton pods/ha in first 18 months and about 6 t pods/ha in the next 12 months.

23.3 COFFEE GROWING REGIONS AND PRODUCTIVITY

According to forecasts by the International Coffee Organization, annual production of coffee seeds may reach 131.4 m bags (60 kg per bag) during 2012. It is about 2% less than previous best harvest. Annual production of coffee seeds in different continents during 2010–2011 is fallows:

Major coffee producing countries in Africa are Ethiopia, Uganda, Cote De Ivoire and Tanzania. Production of Arabicas was 7.38 million (m) bags and that of Robustas was 11.1 m bags, with each bag containing 60 kg green seeds. During 2011, total coffee seed production by African continent was 18.5 m bags [4]. In Asia and Oceania, countries such as Vietnam, Indonesia, India and Papua New Guinea are major coffee producers. Total coffee seed production in the Asia and Oceania region during 2011 was 35.4 m bags. Robustas dominated with 29.5 m bags compared to 6.3 m bags of Arabicas. Mexico and Central American coffee regions produced about 18.6 m bags during 2011. Almost all of coffee production in Central America was Arabicas at 18.4 m bags. Major countries that contributed Arabicas are Mexico, Honduras, Guatemala and Costa Rica. The South America is the worlds' most important coffee producer. Major coffee growing regions occur in Brazil, Columbia, Peru and Equador. Together, they contribute 45% global coffee produce during 2011, which are about 58.8 m bags. Arabicas dominate with 47 m bags and Robustas constitute 11.7 million bags. Global coffee seed harvest during 2011 was estimated at 131.382 m bags. Globally, major coffee consumers are Brazil, Ethiopia, Indonesia, Mexico, Philippines, India, Venezuela and Vietnam. Demand created in these regions literally determines spread or shrinkage of the global coffee agroecosystem. Farmers are also sensitive to several other factors like agroclimate, economic advantages. Market price and demand are among most stated factors that induce or repress farmers' interest in coffee production.

Brazil has the largest coffee belt that is filled with different types such as Bourbon, Typica, Caturra and Mundo Novo. The coffee belt is predominant in the states of Parana, Espirito Santos, Sao Paulo, Minas Gerais and Bahia. Most coffee plantations in Brazil are less than 10 ha in size. Actually, about 71% of them are <10 ha, 25% < 50 ha and only 4% > 50 ha [2]. Brazil contributes about 25% of global coffee produce and 80% of global Arabica coffee. Mexican region in Central America supports large-scale cultivation of coffee. Coffee production is conspicuous in Veracruz, Catoopec, Oaxaca and Chiapas. Mexican coffee is generally priced low compared to Colombian or Brazilian types. They do grow several types of coffee. Again, Mexican farmers seem to prefer organic coffee compared to those cultivated using inorganic fertilizers. Coffee plantations also occur in African agrarian regions. They are traced in tropics of Tanzania, Kenya and Ethiopian hills.

In India, states such as Karnataka, Tamil Nadu and Kerala possess major coffee cropping zones. Coffee plantations are localized in the hilly terrain of Western and Eastern Ghats. The evergreen tropical conditions allow luxuriant growth of coffee plants. Coffee production occurs both under exclusively rain fed and irrigated conditions. Productivity of coffee plantations in South India depends mostly on precipitation pattern and fertilizer inputs.

Vietnam is an important coffee-producing region in the world. Coffee plantations are mostly Robusta types (97%). Other coffee types such as Arabica, Chari (Excelsa) and Catimor are also grown, but in very small proportion. Annual coffee seed production during 2011 was 1.17 m t. Coffee production is predominant in the Central Highlands (Dak lak, Gia lai, Kontum, Lam Dong, Buan Me Thout); South-east (Dong nai, Ba Ria, Vung Tou and Bin Phuoc) and Coastal Plains. Indonesia is a major coffee exporting country from South-east Asia. Indonesia coffee producers grow over 20 different varieties of coffee, mainly to serve the needs of importing countries. Coffee types such as Hibrido de Timor, Typica, Ethiopian lines, Caturra from Brazil and Catimors that are crosses between Arabicas and Robusta. More than 90% Indonesian farmers are small holders with <1.0 ha. Productivity is moderate and farmers currently prefer to cultivate coffee with least amount of inorganic fertilizers. Therefore, organic coffee seems to dominate the scene.

KEYWORDS

- **Cafiene**
- **Chomo**
- **Coffea Arabica**
- **Coffea canephora**
- **veronicas**

REFERENCES

1. Coffee Research Institute, Coffee History. Coffeeresearch.org **2006a,** 1–2.
2. Coffee Research Institute, Brazilian Coffee Beans. Coffeeresearch.org **2006b,** 1–2.
3. Coffee Research Institute, Arabica and Robusta plants. Coffeeresearch.org **2006c,** 1–2.
4. International Coffee Organization Monthly Coffee Market report. **2012,** 1–22.

EXERCISE

1. Mention the various types of coffee grown in different continents.
2. What are the salient features of Arabica and Robust coffee types?
3. Describe various agronomic practices adopted in coffee nursery and main field transplant it to main field.
4. Mark the coffee growing regions of the world.
5. Mention various uses of coffee.

FURTHER READING

1. Clifford, M.; Wilson, K. C. Coffee: Botany, Biochemistry and Production. Croom and Helm Publishers, London, **1985,** pp. 453.
2. Wrigley, G. Coffee. Longman Publishers, London, **1988,** pp. 278.

USEFUL WEBSITES

http://www.coffeeresearch.org/agriculture/varietals.htm

PINE PLANTATIONS (*PINUS SPECIES*)

23.1 INTRODUCTION

Pine species are dominant in the temperate regions of the world. They are conspicuous in the sub-arctic region of Northern hemisphere. Pines are widely grown in North American Plains and Hilly terrain. It is predominant in Nordic states such as Norway, Finland and Sweden. Pines are traceable in most parts of Russia and Eastern Europe. They thrive well in the Alpine regions of Germany, Austria and Switzerland. Pines also occur in the Southern hemisphere. Natural stands and plantations are conspicuous in Brazil, South Africa and New Zealand. The lobloby pine is common in the southern hemisphere. They occur in regions with variety of soil types that experience cold climate and moderate rains. They tolerate extreme cold and frosty period better than many other forest species. They also flourish in subtropical and tropical climates (Table 1). Pine plantations put forth rapid growth compared to many other tree species and hence are preferred in plantations meant for paper pulp production and wood. During recent decades, expansion of forest plantations, including pines is preferred because they fix massive quantities of atmospheric CO_2 into biomass and allow its sequestration into soil.

TABLE 1 Pine: Origin, Botany, Nomenclature, Uses—A summary.

Pine trees are predominant in the Northern hemisphere. They flourish on variety of soil types with variable texture and fertility. Pines are traceable in wide range of agroclimatic conditions. They adapt to both low and high rainfall conditions.

Botany and Classification

Kingdom-Plantae; Division-Pinophyta; Class-Pinopsida; Order-Pinales; Family Pinaceae; Subfamily Pinoidea; Genus-*Pinus*; Species-there are several species and subspecies.

There are about 125 species of Pines. Most of the species are localized in the subtropical and cool temperate regions of Northern Hemisphere. Major species that flourish in different parts of the world are as follows:

Pinus halepense – (Jerusalem Pine or Aleppo Pine) is native to Mediterranean zone. It grows to 30–60 ft tall.

Pinus nigra – (Austrian Pine or European Black Pine) native to Southern Europe, Northern Africa, Cyprus and Turkey. It reaches 40–60 ft height.

Pinus aristata-(Bristle cone Pine or Hickory Pine) native to semi arid parts of South-western USA. It grows to 8–30 ft tall.

Pinus caneriensis (Canary Islands Pine) native to Spain. It grows 50–80 ft tall.

Pinus roxburghi (Chir Pine or Imodi Pine) native to Afghanistan, Bhutan, China, India, Myanmar and Nepal. It grows up to 60–120 ft in height.

Pinus coulteri (Coulter Pine or Big cone Pine or Nut pine). Native to Southern and Meso America. It grows up to 150 ft in height.

Pinus strobus (Eastern White Pine or White Pine) native to States of America and Canada. It grows 50–80 ft in height.

Pinus balfouriana (Foxtail Pine) native to California. It grows from 20–50 tall.

Pinus sabiniana (Gray Pine, Foot hill pine, Digger Pine) native to California.

Pinus pinea (Italian Stone Pine or Umbrella Pine). Native to Southern Europe, Lebanon, Turkey and Syria. It grows from 30–60 ft in height.

Pinus radiate (Guadalupe Pine) native to California and introduced into Australia, Argentina, Chile and South Africa. It grows to 50–80 ft height.

Pinus sylvestris (Scots Pine) Native to Europe and most widely spread in the Northern hemisphere. It grow 50–80 ft in height.

Pinus elliotti (Slash Pine) Native to southern USA. Dominant in Georgia and Florida. It grows to 80 ft in height.

Nomenclature

Catalan-Pin; Czech-Borovice; Danish-Pine tree; French-Pin, Conifera; German-Keifer; Hindi-Vudakarchid; Italian-Pino; Polish-Sosna; Portuguese-Pinhal; Romanian-Pionului Brad; Russian-Sosna; Spanish-Pino; Spanish-Pino; Swedish-Buxbom.

Uses

It is said that globally, half of harvested forest plantations are used for industrial purposes and about 25% material is used for nonindustrial end-use. Pinewood is used to prepare variety of furniture and in housing. Pine plantations are a major source of paper production industry in North America and Europe. Pines are used as windshields in many farms. Pines are used to extract variety of industrial chemicals such as tepenes, alcohols and phenolics. Pine wood dust is used as mulch in farms and homes. Pinewood is used as electrical poles in many countries. Pinecone is a decorative item in many homes. Pine needles are used make mats, mattresses and pillows. Pine resin is rich in turpentine.

Pines are rich in Vitamin A and C. Pine flour could be obtained from seeds. Pine seeds are also edible as salads. Pinesap is sometimes used as ointment to reduce pain.

Source: [1, 2]; http://www.2020 site.org/trees/pine.html (November 23, 2010).

23.2 SOILS AND AGROCLIMATE

Natural stands and plantations of pines thrive on very wide range of soil types that occur in the temperate and tropical regions. Pines are deep rooted perennial trees that extract moisture and nutrients from greater depths of soil profile. Roots reach up to 5–10 m depth. They flourish and dominate the landscape on Mollisols, Ultisols and Coastal sands in the North American continent (Plates 1 and 2). The Cambisols, Fluvents and Loess soils support their growth in the Western Europe. In Russia and Eastern European Plains, pines occur on Chernozems, Podzols and sandy loams. In Northern Russia, parts of Nordic region and Alaska, pines thrive on peaty soils very rich in organic matter (20–40% organic-C). In South America, tropical pine is grown on low fertility soils deficient in P and micronutrients. These soils are rich in Al and Fe and need correction for pH if it is extremely acidic for seedling establish and growth. In West Asia, pines occur on Aridisols and Xerasols rich in Ca but deficient for major nutrients. In

Africa and Asia, tropical plantations are usually established on soils with low fertility and not useful for regular farming. Mollisols, Inceptisols, silty/gravely soils derived from shales and secondary minerals support pines in the Himalayan and subHimalyan regions. Pines occur on Inceptisols, Alfisols and Xerasols in Northern China and Far-east. In Australia, pines thrive on sandy loams and rocky loams

Regarding agroclimate, pines tolerate extremes of cold and frost in the temperate regions. Pine plantations occur in regions experiencing subzero temperature –20°C). The average annual temperature in Nordic regions, North America and Russia ranges from −8°C to 12°C, yet pines thrive and put forth biomass and support several normal ecosytematic functions. Temperate Pines grow best and rapidly in most parts of North America and Europe that experience −5° to 25°C. Precipitation may be erratic and range at 200–300 mm annually. The relative humidity is generally high at 50–65%, since loss of moisture via evaporation is not rapid. Diurnal pattern affects pines. Sunshine hours are restricted during winter, but more than 13 during summer. Tropical pines may experience different set of agroclimatic patterns. Temperature regime is warmer and ranges from 5°C–32°C. The diurnal pattern nearer tropics of cancer or equator is different from those found in northern latitudes.

Based on mean temperature and Altitude, we can group the Pine plantations as follows [3]:

	Mean Temperature (°C)	Altitude (m.a.s.l)	Pine species-examples
Tropical	> 24	900	*P. hondurensis, P strobus*
Subtropical	19–24	900–1800	*P. strobus var chiapensis; P.oocarpa*
Warm Temperate	16–19	1800–2700	*P.douglasiana; P.lawsoni; P.patula*
Cold Temperate	10–16	2700–4000	*P.hartwegii, P montezume; P.cooperi*
Arctic	< 10	> 4000	*P rudis*

23.3 EXPANSE AND PRODUCTIVITY

Globally, forest plantations thrive on variety of geographic regions, topography, soils and agroclimatic conditions. They are composed of several species of trees that are broad leaved or coniferous. In the present context, we are interested in Pine plantations that flourish better in temperate and subtropical regions of the world. During past decade, forest plantations occupied 187 m ha. Asia accounted for 62% of forest plantations. The annual increment in forest zone has been conspicuous in Asia and South America. In these two continents it has increased by 4.5 m ha annually [4]. Based on tree species, it is said 40% of forest zone is engulfed by broad-leaved species. Pine species occupy 20% of global forest belts. They are of course dominant in temperate regions of the world. In North America pines make up 88% of forest plantations. Pines are replacing other species at a relatively faster pace [3] (Plate 2). Whereas, in South America and Asia broad leaved tree species occur in greater intensity at 45–47% area. In these areas pines make up 20–28% of area.

Following is the distribution of Pine plantations in different continents during past decade:

Plantation Area (000 ha)

Tree Species	Africa	Asia	Europe	North/Central America	Oceania	South America	World
All Tree	8036	11584732015		17533	3201	10455	118706
Pines	1648	15532	16344	15440	73	4699	37391

Source: Ref. [3].

PLATE 1 Pine Plantation near a National Park in Oregon State of USA. Pine Plantations do coexist and support a variety of flora and fauna under their canopy.

PLATE 2 Tillage operation just before Re-planting of Panderosa Pine seedlings.
Note: Back ground shows up mature Pine trees fit for harvest and transport to paper industries for processing. Location: Austin Carey Forest Plantations near University of Florida, Gainesville, Florida, USA.
Source: Krishna, K.R., Gainesville, Florida, USA.

The composition of pine species in different regions differs markedly. It is easily attributable to genetic traits and adaptability of pine species to agroclimate and agronomic procedures adopted. In Asia, pine plantations are predominantly developed using following six species. They are *P exelsa, P.longifolia, P gerardania, P.khasya, P.insularis* and *P.merkusi*. They negotiate the tropical climate, altitude and soil fertility conditions of Asia better and still contribute wood and biomass. In tropical Africa, *P. canaeriensis, P.caribea, P. excelsa* and *P halepnse* dominate the pine landscape. In America, several pine species fill up the forest regions. A few examples are *P.strobus, P.sylestris, P. elliottti, P.radiata, P.coulteri, P.hondurensis,* etc. (FAO, 2010). In several locations within Southern USA, production of pine plantations ranged from 252–532 MAI ft. 3/ac [4]. Productivity of tropical pines may range from 16 to 56.6 m^3/ha and specific gravity from 0.35 to 0.44 g cm^3 in Brazil [5].

During recent years, South African forestry department has envisaged intensive pine production strategies. Improved land and soil management techniques and nutrient supplies, planting highly productive varieties and careful management of nursery and field have led to better returns. Pine species such as *P. radiata* and *P. pinaster* are preferred [6].

Forest plantations in South and South-east Asia are most often filled with *P khasya* and *P. merkusii*. They occur from 300–2700 m.a.s.l. The annual growth rate is 0.8–1.9 cm in diameter and 54–142 cm in height. They produce quality resin (gum) of 560 kgm^3 at 12% moisture. On an average, 20 year old trees may provide 1.8–2.4 kg resin/tree. Trees also provide turpentine of good quality [7].

In Australia, softwood pine plantations occur on approximately 1.0 m ha. Pinus radiate, locally known as *P. insignia* dominates the plantation ecosystem. They thrive on low fertility soils and in regions with 250–800 mm annual precipitation. Mean annual increments range from 2.21–8.09 m^3/ha.

KEYWORDS

- **P merkusii**
- **P. pinaster**
- **P. radiata**
- **Pinus**

REFERENCE

1. Louhrey, R. E.; Kossuth, S. Slash Pine-Pinus elliottii. http://www.na.fs.fed.us/spfo/pubs/silvics_manual/volume_7/pinus/ellioti.htm **2010,** 1–18 (October 20, 2010).
2. Myers, R. Forty Pine trees from around the world. http://www.treesandshrubs.about.com/od/selection/tp/PineTreees.htm **2010,** 1–5 (November 23. 2012).
3. FAO Global data on Forest Plantations Resources. Food and Agricultural Organization of the United Nations. Rome, Italy. http://www.fao.org **2010,** (November 24, 2012).
4. Moorhead, D. J.; Dangerfield, C. W.; Beckweth, J. R. Opportunities for Intensive Pine Plantation Management. http://www.bugwood.org/intensive/98–002.html **2002,** (November 23, 2012).

5. Lima, W. Soil moisture regime in tropical pine plantations and in Cerrado vegetation in the state of Sao. **1983,** pp. 23–38.
6. Zwalenski, J.; Groenewald. W. Natural regeneration of Pine plantations in South Africa as a cost-effective way of stand reestablishment. Forestry **2004,** *77,* 483–493.
7. ICRAF, Agroforestry Tree Database. World Agroforestry Center, Nairobi, Kenya, **2002,** 1–3 (October 20, 2012).
8. Paulo, Brazil. IPEF 23, 5–10 WRM, USA: Losing forests to Pine Plantations. World Rain Forest Movement Bulletin No 60, **2002,** 1–3.

EXERCISE

1. Mention at least 10 Pine species used in different plantations.
2. Describe the spread of Pines in Europe and North America.
3. Mark distribution and dominance of different Pine species on a map.
4. List top Pine producing nations. State the area, production and productivity of Pine plantations in these nations.
5. What are the various uses of Pine trees?
6. Explain nursery management and Pine tree management procedures briefly.

FURTHER READING

1. ABARE. Australia's Plantations. Bureau of Rural Science, Canberra, Australia. **2006,** pp. 223.
2. Daoust, G.; Beaulieu, J. Genetics, Breeding, Improvement and Conservation of Pinus strobus in Canada. USDA Forest Service Proceedings RMRS-P-32, **2004,** 1–11 http://www.fs.fed.us/rm/pubs/rmrs_p032/ rmrs_p032_003_011.pdf (November 23, 2004).
3. MacDonald, E. Sustainable management of Scotspine in the Northern Periphery: Special Overview. Northern Periphery Program, European Union, Ross-Shie, United Kingdom, http:/www.forestry.gov.uk/fr/timberproperties.**2010,** 1–14 (November 23, 2012).
4. Nunez, M. A.; Medley, K. M. Pine invasions: climate predicts invasion success; something else predicts failure. Diversity and Distribution. **2011,** *10,* 1–11.

USEFUL WEBSITES

http://www.ehow.com/grow-pine-trees/ (October 15, 2012)
en.wikipedia.org/wiki/Pine_plantation (November 23, 2012)
http://smallfarms.ifas.ufl.edu/environment_and_recreation/forestry/pine_production_and_straw.html (November 23, 2012)
www.maplandia.com/united-states/…/pine-forest (November 23, 2012)
http://www.fs.fed.us/database/feis/plants/tree/pinard/all.html (November 23, 2012)
http://www.fao.org/forestry/fo/fra/index.jsp (November 23, 2012)

PART II

PRINCIPLES OF NUTRIENT DYNAMICS

CHAPTER 24

ATMOSPHERE INFLUENCES NUTRIENT DYNAMICS IN AGROECOSYSTEMS

CONTENTS

24.1 INTRODUCTION

Agroecosystems are in constant confluence with atmosphere. The crop and soil phase interact with atmosphere in terms of water, gaseous exchange, dust and particle distribution. Atmosphere is a source of water. Precipitation events add to soil water. Precipitation from atmosphere that accumulates in soil profile is then absorbed into crops. Precipitation carries a certain amount of dissolved nutrients, dust and small particles depending on the geographic location. Atmospheric deposits and dusts may carry relatively more nutrients, if the location is nearer to industries that emanate large amounts particulate/gaseous material into atmosphere. For example, industries that emanate carbonaceous and sulfur containing fumes/particles may pollute atmosphere. The rainfall events that occur in such areas will obviously add more of sulfur into agricultural fields located nearby. Generally, cropping belts closer to seacoast receive greater amounts of nutrients through atmospheric deposits. There are geographic locations, such as in West African Sahel, or dusty plains in North Africa, West Asia or North West India. In these regions, dust storms (e.g., *Harmattan* in Sahel, *Lu* in Gangetic Plains) shift sand particles and nutrient attached along with it, from one place to another. Atmosphere aids shift of nutrients through storms. In the general course, in most locations if not all, morning dew carries a certain amount of water and dissolved nutrients into crop phase. The atmospheric deposits of nutrients may not be conspicuous in many regions, yet it occurs and affects nutrient dynamics of individual field. The extent of nutrients added may be small. It may often range from 5–10 kg N, 0.3–0.5 kg P and 2–3 kg K/ha annually. Atmospheric deposits may also contain S and Zn, rather regularly. It is a sizeable amount of nutrients shifted into soil and crop phase, if large agroecosystems are considered. Even marginal reductions in nutrient received from atmosphere may become perceptible in few years. Yet, many of the fertilizer schedules prescribed for individual crop/field and even large expanses grossly neglect the nutrients added from atmosphere. Let us consider a few specific examples that relate to nutrients received from atmosphere.

24.2 NUTRIENT INPUTS TO AGROECOSYSTEMS FROM ATMOSPHERE

24.2.1 MAJOR NUTRIENTS-NITROGEN, PHOSPHORUS AND POTASSIUM

Nitrogen is impinged into soils of the rice ecosystem through different sources. Firstly, rain water and atmospheric deposits are known to provide rice fields with 5–8 kg N/ha annually. Of course, extent of N deposits depends on precipitation pattern and geographic location. Nitrogen input via atmospheric deposition is generally considered negligible, while deciding N fertilizer schedules for high yield crops. Reports from Eastern United States of America that supports large expanses of peanut crop suggest that farmers generally do not consider amount of N received through deposits and precipitation. Nitrogen received from atmosphere is relatively negligible considering that fields that support peanuts or maize are supplied with high doses of fertilizer-based nutrients. Hodges [1] states that farms in Georgia that support peanuts and maize receive 2–3 kg N/ha via atmospheric deposits.

Reports from Citrus belt of Central Florida indicate that during a year, a sizeable amount of soil-N that gets emitted as NH_3, NO_2, gaseous N_2 and N_2O to atmosphere is returned along with precipitation. The extent of N derived from atmosphere may depend on location of the citrus groove, precipitation pattern, intensity of rainfall and N compounds dissolved in the rain drops. According to researchers at Citrus Research and Experimental Centre (CREC) at Lake Alfred, Florida, a citrus groove may receive up to 6 kg N ha from atmosphere via depositions and precipitation.

There are several reports and summaries about the extent of N derived from atmosphere in the European agrarian zones [2]. For example, in wheat cropping zones of Germany, atmosphere may contribute 20–40 kg/ha,/yr. Estimations by Weigel [3], at Bad Lauschstatd indicate that crops may derive 50–58 kg N/ha/yr from atmosphere. Soil and plant analysis based on ^{15}N techniques shows that N derived from atmosphere could be as high as 60 kg N/ha/yr in certain locations of Western Europe. Reports emanating from Rothamsted Agricultural Experimental Station, indicates that cereal crops may derive 45 kg N/ha/yr through atmospheric deposits [4]. Wheat and barley fields in the Eastern European plains around Prague in Czeckhoslavakia receive 30–46 kg N/ha/yr through atmospheric depositions and precipitation.

Nutrients derived via atmospheric deposits could be relatively small in the humid tropics that support large acreages of rice and other tropical crop species. Yet, atmosphere does supply a definite amount of nutrients to the cropping zones. Kellman and Tackaberry [5] state that ordinarily 0.3 kg N, 0.2 kg P, 3.7 kg K, 2.2 kg Ca and 3.4 kg Mg/ha/yrare received by crop fields in humid tropics of Africa. There are several reports about the extent of sand shifts, movement of sand particles and nutrients attached to sand particles. Storms and sand movement can transfer nutrients from a location to other in sizeable amounts. It depends on factors like geographic location, topography, size of sand particles, soil type, carrying capacity of sand particles for nutrients, precipitation pattern and wind velocity. *Harmattan* is an important atmospheric phenomenon that adds significant amount of dust and nutrients attached with it into a millet/cowpea field. Reports by researchers at ICRISAT Sahelian Center suggests that *harmattan* dust storms, on an average add up 53 kg N/ha in a crop season ([6]; Plates 1 and 2). A few pearl millet farmers have derived greater advantage from *harmattan* dust by placing mulches in the field that traps more of sand and nutrients [7].

PLATE 1 An imminent Sand storm and Precipitation event near Bernin-en-Konni, Niger in West African Sahel.

Note: Sand storms are common to Sahel. Here, sand and dust particles shift from one location and another, causing soil/nutrient erosion in one field and creating deposits in another region a distant apart. Soil erosion control measures are important to preserve the cropping belt.

Source: Krishna, K.R., Bangalore, India.

PLATE 2 A Sand/dust storm in progress in Sahelian, West Africa. This location is near Dosso, in Southern Niger.

Note: Highly sandy Oxisols are prone to wind and water erosion.

Source: Krishna, K.R., Bangalore, India.

Estimation of nutrient fluxes in the peanut growing Sudano-Sahelian region of West Africa, specifically in Senegal; indicate that fallow periods may enrich soils with mineral nutrients. During fallow, atmospheric N deposits may add 5 to 10 kg N/ha. Capture of bird droppings estimated at 40 kg dry matter/ha/yr adds 2.2 kg N, 0.4 kg P and 0.7 kg K/ha. '*Harmattan*' dust from the Sahara is known to contribute 3.0 kg N, 1.0 kg P and 15 kg k/ha/yr ([7]; Plates 1 and 2). Applying mulches or crop residues to trap nutrient rich *harmattan* dust is a traditional practice in West African peanut zone [8].

The sorghum agroecosystem in South India is vast and expansive. It receives nutrients from atmosphere through precipitation events, settling dusts and storms that occur periodically. In general, extent of nutrients received by Alfisols and Vertisols from atmosphere is low and often negligible. About 5–6 kg N, 0.8–1.0 kg P and 2–3 kg K/ha is added annually into sorghum fields [9]. Yet another evaluation in South India indicates that sorghum belt receives 6–12 kg N/ha from atmosphere, but it depends on precipitation levels [10].

24.2.2 SULFUR

Sulfur that occurs in the ambient atmosphere is an important source of this essential element to crops. Sulfur occurs in gaseous state and this could become available to crops. Sulfur deposits could be directly absorbed via hydathode in dissolved state as dew, rain droplets or even water vapor. In the humid tropics, rain water adds 1.58–3.8 kg S/ha, but in dry regions S derived from atmosphere could be low at <1.0 kg/ha.The natural sources of S could provide 2–4 kg S/ha, which is generally sufficient, depending on cropping intensity and yield goals. Reports suggest that in the Great Plains of North America, S deposits from atmosphere ranges from 1.8–10.7 S/ha. In most agrarian regions, S derived from industrial emissions could add to S in crop fields. Globally, anthropogenic S could be as much as 220 × 10^6 kg/yr. Similarly, crops grown in the seacoasts may actually a certain share of S through deposits and precipitation. Seawater is richer in S and this gets recycled via rainfall on the crops. It is said that ambient air near seacoast contains 10 times more S than that in interior landscapes [2].

Over all, nutrients received from atmosphere could be constant, timely and useful, although it is mere small fraction of total nutrients absorbed by crops. However, agro-ecosystems are vast. Individual cropping patches range from few to over 10,000 ha in one stretch. The nutrients received by such vast zones needs to be collected, conserved and used efficiently.

KEYWORDS

- **Citrus Research and Experimental Centre**
- ***Harmattan***
- **Seawater**
- **Sudano-Sahelian region**

REFERENCES

1. Hodges, S. C. Environmental guidelines for Plant Nutrient used. Reports of University of Georgia Extension Services, Athens, Georgia, USA, **1991,** 116.
2. Krishna, K. R. Agrosphere: Nutrient Dynamics, Ecology and Productivity. Science Publishers Inc.: Enfield, New Hampshire, USA, **2003,** 346.
3. Weigel, A.; Russow, R.; Korschens, M. Quantification of Airborne N-input in long-term field experiments and its validation using ^{15}N isotope dilution. Journal of Plant Nutrition and Soil Science **2000,** *163,* 261–265.
4. IACR Annual Report Institute of Arable Crop Research. Rothamsted, England, **1997,** 35.
5. Kellman, M Takaberry, R. Tropical Environments. Routledge Publishers, New York and London, **1997,** 375.
6. Bationo, A.; Traore, Z.; Kimetu, J.; Bagayoko, M.; Kihara, J.; Bado, V.; Lompo, M.; Tabo, R.; Koala, S. Cropping systems in the Sudano-Sahelian zone: Implications on Soil fertility Management. Syngenta Workshop, Bamako, Mali, West Africa, **2004,** 2009.
7. Romheld, D, Marschner, H.; Becker, K.; Schecht, E.; Hulsebusch, C.; Burkert, A. Effects of site factors on the efficiency of soil amendments and on nutrient fluxes in subsistence-

oriented cropping systems of the West African Sahel. http:/www.troz.uni-hohenheim.de/research /sfb308/Ebcont/b13.pdf **2000,** 1–3.
8. Keitchi, H.; Toshiyuki, W. Sustainable soil fertility management by indeginous and Scientific Knowledge in the Sahel zone of Niger. Proceedings of 17 World Congress of Soil Science. Symposium No *15,*Paper No 1, **2002,** *251,* 611.
9. Murthy, K. S.; Sahrawat, K. L.; Pardhasaradi, G. Plant nutrient contribution by rainfall in Patancheru area of Andhra Pradesh. Journal of the Indian Society of Soil Science **2000,** *48,* 803–808.
10. Venkateswarulu, J. Rain fed Agriculture in India. Indian Council of Agriculture, New Delhi, **2004,** 566.

EXERCISE

1. Describe the effect of sand/dust storms on nutrient inputs to agricultural fields.
2. Discuss at least three studies relevant to nutrient deposits from atmosphere.
3. What is carrying capacity of sand/soil particle?
4. What is the effect of heavy precipitation events and storms on soil structure, erosion and nutrient loss from fields?

FURTHER READING

1. Emerson, A. A. *Atmospheric Inputs and Plant Nutrient Uptake along a 3 million year Semi-arid belt.* Northern Arizona University: Arizona, USA, **2010,** pp. 226.
2. Flower, C. A. *Atmospheric Inputs of Nutrients and Pollutants to Uplands in North-east Scotland.* University of Aberdeen: Aberdeen, United Kingdom, **1987,** pp. 398.

USEFUL WEBSITES

http://www.icrisat.org
oar.icrisat.org

CHAPTER 25

NUTRIENT SUPPLY INTO AGROECOSYSTEMS

CONTENTS

Nutrient supply is an integral process that occurs in all agroecosystems. Nutrient supply occurs both through natural process and extraneous additions by farmers. In any well-stabilized agroecosystem, the extent of nutrients stored and that supplied from external sources decides the intensity of cropping and productivity levels.

25.1 NUTRIENT SUPPLY THROUGH NATURAL FACTORS

Agroecosystems may receive nutrients dissolved in water through precipitation, irrigation water, percolation, seepage from neighboring fields, riverine and atmospheric deposits, etc. Nutrients are carried on sand and dust particles that float in atmosphere. The amount of nutrients derived depends on geographic location, topography and source of the deposits. For example, fields located within large cropping belts may receive negligible or small amounts of major nutrients in a year. Whereas, fields situated near seacoast or industries with large effluent discharge or emissions may accrue larger amounts of nutrient.

Through the ages, farmers have stabilized soil fertility aspects of their fields using inherent mineral nutrient in soil and that recycled as organic residues plus small additions of FYM. The nutrient dynamics under such supply schemes have allowed subsistence level grain/forage yield or at best average levels. Soil types that are naturally fertile, for example, Chernozems in Northern Great Plains or Eastern European plains, Peaty soils of Northern Europe, or Black clayey soils with high nutrient buffering capacity have supported relatively higher grain/forage yield. Quite often, nutrient turnover in such fertile zones have been significantly higher than in other areas with low or moderately fertile soils.

Farmers supply nutrients to their fields through a variety of inorganic chemical sources, ores, partially treated ores, soluble inorganic fertilizer, amendments and organic manures. Farmers use several types of formulations such as granules, grits, powders, pastes, liquids, pressurized gas (e.g., NH_3) and sprays, etc.

25.1.1 TRENDS IN NUTRIENT SUPPLY AND SOIL FERTILITY MANAGEMENT PRACTICES

Agronomic procedures standardized using folklore, recorded experiences, chemical and microbiological assays, crop response trials; multilocation trials have all played crucial roles in sustaining the growth, productivity and perpetuation of agroecosystems. Each technique or procedure might have imparted proportionate effects on nutrient dynamics, ecosystematic functions and grain productivity. The current soil fertility status and productivity of any of the agroecosystems world-wide is obviously the net result of effects of nutrient management, weather pattern and cropping systems adopted by farmers through the ages. As a corollary, we ought to realize that basic soil fertility concepts adopted, nutrient schedules devised and techniques employed during supply of nutrients have immense effects on nutrient dynamics and productivity. Any error in judgment regarding nutrient quantities, formulations, timing, placement methods, recycling through residues and cropping systems could show up as enlarged malady that engulfs whole agroecosystem. Even minor errors may get accentuated, if practiced through each year for long time and in large expanses. Hence, accurate judgment of soil fertility and devising most appropriate nutrient supply schedule is an essential step during farming. Actually, at every stage of crop development, nutrient supply should match the crop's demand based on yield goals set by the farmer and that possible in a given location. Any excess or deficit will affect the agroecosystem productivity.

Let us quote a hypothetical example, although real situations abound in the literature. Suppose that an agroecosystem of 1.0 m ha is supplied with 3 kg N/ha in excess of crop's ability to extract through roots. It results in 3 million kg N extra in that area. This excess N in the crop belt finds its way into irrigation channels, lower horizons of soil, ground water and via emissions to atmosphere. We should note that this excess N input may after all be forgotten by the farmer while scheduling nutrients for next crop. Incidentally, farmers adopting crude nutrient schedules, not thoroughly based on soil fertility maps and cropping history are easily prone to errors of margins greater than 3–4 kg N/ha. Spread of fertilizer-based nutrients too may not match the soil fertility maps. We must also realize that critical limits derived based on soil-test crop response studies are at times highly specific to small areas and soil types. They do not really help farmers in judging exact needs of nutrients at different stages of the crop growth and yield formation. In other words, errors imbibed into soil analysis techniques, crop response studies, fertilizer formulation and soil reactions may all affect net availability and utilization of nutrients. These factors too add to errors in nutrient supply into agroecosystems. We ought to know that a minor error in calibration of instruments, computer based growth/fertilizer supply models adopted, soil/nutrient availability tests adopted and their accuracies may all have a lasting effect on nutrient dynamics, ecosystematic functions and productivity of any agroecosystem.

Let us consider a situation where in nutrient supplied is lower than crop's requirements, For example, a high intensity cropping expanse of 1 m ha, if supplied with a deficit of even 2 kg N/ha per season, it may result in mining of 2 million kg N/season from the belt. Obviously, in a few seasons the whole agroecosystem may experience nutrient deficits and result in proportionately lower grain/forage yield. Reduction in grain productivity has further consequences on organic carbon and mineral nutrient recycling. Over all, accuracy in soil sampling, soil analysis techniques, computer programs and nutrient supply schedules is necessary to ensure greater grain/forage productivity of agroecosystem.

Now, let us consider soil fertility techniques that were devised to maintain crop productivity and ecosystematic functions. Farmers have adopted soil fertility restoration methods since many centuries. These nutrient management techniques generally involved *in situ* measures like residue recycling, fallows and refurbishment using animal or farmyard manures. Soil fertility was held at optimum level, to the extent possible, depending on the nutrient content of the organic manures. Variation in nutrient availability in soil was removed only to a certain extent and it persisted. Repeated cropping and insufficient crop residue recycling resulted in lower grain/forage yield. In many cases, nutrient deficiencies were conspicuous. As a consequence, obtaining potential grain/forage yield in a given location or environment was not possible. Further, repeated cropping exhausted soil nutrients and consequently crop yield decreased.

Knowledge gained through mineral theory of crop growth, several other soil fertility concepts and fertilizer technology helped us to develop a series of different soil nutrient management procedures, all aimed at restoring soil nutrient status and maximizing crop yield. Earliest of the soil fertility management methods involved visual score of crop, identification of nutrient deficiencies and matching it with a supply of fertilizer-based nutrient. Although, nutrient deficiencies were overcome to a certain extent, it did not ensure optimum yield. The nutrient supply was not based on a yield goal. It only ensured removal of nutrient deficiencies transitorily. Further, nutrient ratios were also not at all maintained and this could have led to nutrient imbalance. In many situations, it actually led to lack of yield response due to expression of Liebig's Law of Minimum.

During early 20th century, a concept more generally applicable and based on previous evaluation of crop's response to different levels of fertilizer supply was adopted. It was called 'Critical Nutrient Level.' Critical nutrient availability in soil is a level at which at least 95% of maximum grain/forage yield potential in the given location or environment could be produced. The critical nutrient level varies depending on several factors mostly related to soil, crop, environmental parameters and yield goals. During past 2–3 decades, farmers cultivating perennial crops have been exposed to fertilizer recommendations based on plant tissue analyzes. It involves elaborate sampling and analysis of all essential nutrients in leaf, leaf blade or petiole or any other portion of a plant. It is more commonly known as DRIS. Since the method uses only a small portion of plant tissue and just a few samples, it is nondestructive, if we consider the entire crop. Fertilizer recommendations could be altered periodically as the crop grows or seasons progress, based on nutrient status of the crop. Adoption of critical nutrient level techniques based on soil or plant tissue or both had its specific effect on nutrient dynamics, crop growth and ecosystematic functions [1–4].

Next, a concept called 'Soil Test Crop Response (STCR)' was envisaged. It involved series of field trials that evaluated a crop's response to different levels of nutrients supplied using fertilizers. Farmer's traditional practices were also in vogue in many agricultural belts. Farmers adopt several practices that maintain soil fertility and maximize crop yield. Such agronomic measures were standardized through several decades, drawing knowledge from folklore, recorded notes and based on recent experiences. Collectively, these approaches are called 'Farmer's Traditional Practices.' Farmers may apply soil test values before prescribing fertilizers. Farmer's Traditional practices envisage supply of nutrients through both inorganic and organic sources. The quantity of fertilizers supplied is guided by the soil type, crop genotype, season and yield goal. Nutrients supplied may not suffice to achieve maximum possible yield. Also, fertilizer dosages may not be the best in terms of economic advantages. Nutrient ratios in soil may or may not get satisfied. Usually application of organic manures removes dearth for micronutrients. Currently, farmer's practice in a moderately fertile belt involves inorganic fertilizers, FYM, bio-fertilizer and amendments to correct soil pH, if required.

Agricultural administrators and policy makers situated in different nations/regions do contemplate on crop production strategies, nutrient supply trends, fertilizer application and yield goals in their areas of jurisdiction. They formulate nutrient schedules considering the agricultural development of entire region. 'State Agency Recommendation (SAR)' is formulated to guide farmers in an agricultural belt or large cropping expanse. Fertilizer recommendation is primarily guided by State Agricultural Programs related to land management, cropping systems, intensity and annual yield goal. Fertilizer supply rates are stipulated for a soil type, crop, region or yield goal. State Agency Recommendations do not consider within-field variations. Therefore, nutrient supply could often be higher than required or sometimes insufficient depending on variations in soil fertility in a given field. Nutrient dynamics in a large area is optimized through state agency stipulations.

'Best Management Practice (BMP)' is a term more commonly used to denote a collection of soil fertility management methods that result in high grain/forage yield and offers best economic advantages in a given location. Fertilizer supply is held at high levels so that deficiencies are not expressed. Grain yield is generally high. Soil fertility is held optimum using inorganic and organic sources. Bio-fertilizers are also used. These procedures are common in intensive cropping zones.

'Maximum Yield Technology' is a concept that envisages fertilizer supply, so that grain/forage productivity is highest in a given locality. Yield maximization involves application of high rates of inorganic fertilizers and FYM. Basically, aspects like fertilizer quantities, their timing, ratios, and formulations are all aimed at maximizing biomass production. It does not consider soil fertility variations within a single field. Often, a certain quantity is held in subsurface layers of soil as residual nutrients.

'Integrated Nutrient Management (INM)' envisages supply of nutrients based on soil tests, crop's demand for nutrients and yield goals. Most importantly, it considers environmental issues like soil deterioration, exhaustion of nutrients, recycling and soil quality. Therefore, under INM, farmers are asked to supply nutrients using as many different sources. Both, organic and inorganic sources of nutrients are used at different ratios. In addition, bio-fertilizers are also used. Crop yields are generally optimum, but not the best or maximum in a given locality. Again, INM does not consider soil fertility variations that may occur within a field. Nutrient accumulation or depletion in soil based on a given locality and cropping pattern is a clear possibility [3, 4].

Site-Specific Nutrient Management (SSNM) or Precision Farming (PF) is perhaps the most recent technique among the series of nutrient management methods that is known to farmers situated worldwide. It considers relatively minute variations in soil, even in a small area, say 1–2 m^2 within a small field. Often, it involves use of detailed grid sampling, soil nutrient estimations; preparation of soil maps, GIS, computer models and GPS-guided soil fertility distribution using variable applicator (Plate 1). Of course, SSNM is also amenable manually. Precision farming is a relatively new procedure. Basically, it considers soil fertility and crop's demand for nutrients as accurately as possible in time and space. Since nutrient supplies are exact, undue accumulation in the soil profile is avoided. It also avoids soil and ground water deterioration. Crop yield is optimum and economic benefits are generally slightly more than that achieved using other procedures. Ecosystematic functions are not altered or affected to any great extent [4].

PLATE 1 Fertilizer based Nitrogen input into Maize fields using GPS-guided Precision Farming equipment at Nelson Farms in Iowa.
Source: Mr. David Nelson, Nelson Farms, Iowa.

25.2 VARIATIONS IN NUTRIENT SUPPLY TO A CROP SPECIES

As stated earlier, quantities and procedures adopted during nutrient supply to a particular crop species or even a genotype varies enormously, within and between continents. There may be innumerable examples possible to prove the above fact. As examples, wide differences in N supply procedures adopted by maize and peanut farmers situated in different continents have been discussed in the following paragraphs. Maize cropping zones extend into highly fertile areas in North America and Europe. A maize crop produces between 9 and 12 t grain/ha in the area. Moderately fertile soils and nutrient schedules followed by farmers in Nigeria, Brazil, Argentina and India allows only medium levels of productivity reaching 4–5 t grain/ha. Farmers in subsistence farming zones, who supply relatively low quantities of inorganic fertilizer, organic manure and water, obviously harvest only 2–3 t grain/ha (Fig. 1).

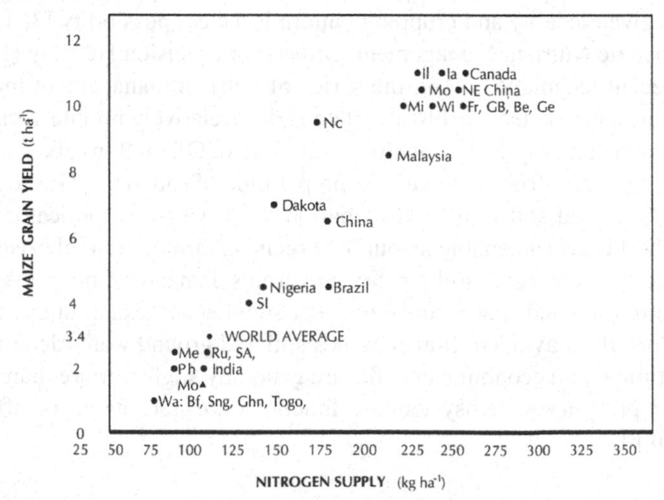

FIGURE 1 Response of Maize crop to Nitrogen supply in different countries.

Note: Nitrogen supply level to Maize cropping zone varies markedly in different countries/ regions. The N-use efficiency ranges from 20–22 kg grain/kg N supplied in high input intensive farming zones to 40–45 kg grain/kgN in subsistence farming regions. Fertilizer-N use efficiency is generally higher in areas with low N input trends and it lessens as fertilizer-N supply increases. In the above graph, fertilizer-N use efficiency is 40–45 kg grain/kg N in countries practicing subsistence or low input technology and fertilizer-N supply is around 25–60 kg N/ ha. As fertilizer-N input is increased to 140–180 kg N/ha, fertilizer-N use efficiency decreases to 30–35 kg grain/kg N and in some of the highest N input (180–280 kg N/ha) regions such as US Corn Belt or European plains, fertilizer-N use efficiency decreases further to 20–22 kg grain/kg N. High grain yield reaching up to 20 t/ha have also been reported from Corn Belt of USA. (Abbreviations: Il-Illinois, Ia-Iowa, Mo-Missouri, Mi-Michigan, WI-Wisconsin, Fr-France, GB-Great Britain, Be-Belgium, Ge-Germany, Nc-North Carolina, SI-South India, Me-Mozambique, Ru-Ruwanda, SA-South Africa, Ph-Philippines, Wa-West Africa, Bf-Burkina Faso, Sng-Senegal, Ghn-Ghana).
Source: Ref. [3].

25.3 NITROGEN SUPPLIED THROUGH CHEMICAL FERTILIZERS

During recent years, N dynamics in almost every major cropping zone has experienced a definite change. Nitrogen supply has increased compared to previous levels. For example, N turnover in cereal fields indeed has enhanced remarkably with the advent of chemical fertilizers [3]. Urea, Ammonium nitrate, Di-ammonium phosphate are some of the common N fertilizers used by farmers situated in different parts of the world. Let us consider 'Corn Belt of USA' as an example. In the 'Corn Belt of USA,' N inputs were originally derived predominantly from residue recycling and FYM. Legumes grown in rotation added small amounts to soil-N status. Nitrogen supply into Corn Belt increased gradually initially in 1940s. In due course, fertilizer-N supply increased rather sharply since 1950s. Nitrogen inputs were progressively revised upwards to suit the fertilizer responsive high yielding maize composites and hybrids. Several types of soil and plant tests were used to ascertain crop's need for N more exactly. In fact, N has been considered as one of the best inputs in the 'Corn Belt of USA.' It is estimated that about 3.6 Tg of N fertilizer was impinged to 'US Corn Belt' at a cost of over 800 m US$ during 2008 [5]. During recent years, computer-based programs that simulate N requirements versus yield goals have been generously used to recommend N inputs, split dosages, their timings and to predict grain yield.

Agronomic procedures, especially N inputs have depended on soil, crop genotype and environment. However, with the advent of computer-based simulations, nutrient dynamics in maize belts have also been influenced by economic considerations rather routinely. Field trials in the Corn Belt of USA, conducted during 2000 to 2006 has revealed that economically optimum nitrogen rates (EONR) for maize was 135–145 kg N/ha if it is soybean-corn rotation and 190–210 kg N/ha if it is corn mono-crop grown year after year [6, 7]. The EONR is a good example of how economic benefits affect nutrient dynamics and other natural processes within maize agroecosystems. The consequence of such decisions needs to be forecasted and evaluated through computer simulations.

Next, let us consider variation in N supply between continents. In the Rolling Pampas, maize mono-crops planted under No-till systems are provided with 50 kg N/ha. The side dress N dosages are dependent on soil fertility and yield goals.

In the Nigerian humid tropics, maize farmers supply 60–70 kg N/ha. However, in Savanna, 100–120 kg N/ha seems economically optimum. In the Southern Guinea Savanna, a maize crop that yields 3 t ha grains is provided with 50 kg N/ha [8]. It is said that N inputs > 50 kg/ha removes N dearth, if any, during hybrid maize production. Split application of fertilizer-N is also in vogue in African maize belts. Kenyan farmers supply 50 kg N/ha to achieve optimum grain harvests of 2–5 t grains/ha [9]. Clearly, geographical aspects, soils and climate have a say while devising fertilizer-N schedules.

A maize silage crop grown in Indus plains with a yield goal of 20 t/ha is supplied with 75 kg N, 60 kg P and 60 kg K/ha. Nitrogen is supplied in split dosages and the second split is usually supplied at 45 DAP [10]. Fertilizers such as P and K are applied at sowing. In the dry regions of South India, maize is cultivated under rain fed conditions. The crop receives 20–40 kg N/ha plus a top dress of 20 kg N/haat knee high stage. Rainy season crop is supplied with 40–80 kg N/ha depending on soil tests. Irrigated maize with relatively higher yield goal of 4–5 t grain/ha is supplied with 80–120 kg N/ha. Often, N is supplied as both inorganic fertilizers and organic manures. Usually 30–50% of total N envisaged is supplied as organic manures. Maize is rotated with legumes. In that case, it is preferable to calculate N requirements of entire sequence and apply N to fields. Legumes may contribute a certain amount of N [2, 3].

25.4 NITROGEN SUPPLY TO A LEGUME AGROECOSYSTEM: EXAMPLE-PEANUT BELTS

Peanut, a legume has been chosen as an example because its nitrogen needs are met through inorganic fertilizer-N, organic matter and most importantly through symbiotic association with *Bradyrhizobium*. As a thumb rule legumes are not supplied with high quantities of N because they derive a portion through biological nitrogen fixation (BNF). Further, high soil-N is supposedly detrimental to nitrogenase activity that has an important role in BNF. Biological nitrogen fixation contributes a portion of peanut crops' need for N. Nutrients received through atmospheric deposition and precipitation in peanut belt may not be significant. Regarding N, reports from peanut belt in Georgia indicate that rainfall supplies 2 to 3 kg N/ha [11]. Like other crops, N supplies to peanut cropping zones that occur in different continents vary enormously. As a consequence, productivity and profitability of peanut belts too differ significantly.

Influence of fertilizer-N on peanuts has been studied for several decades. Reviews by Reid and Cox [12] had concluded that most American research found no response of runner peanuts to N (up to 120 kg/ha) supplied as inorganic fertilizer. Kidder (1994) stated that innumerable studies emanating from Coastal Plains in Georgia, Florida and Alabama show no perceptible increase in pod yield as a consequence of fertilizer-based N inputs. In a few cases, where $(NH_4)_2SO_4$ was supplied, peanut response seems to be due tos as a factor in the chemical fertilizer. Most studies concluded that, practically N inputs from 20 kg N to 120 kg N per/ha did not induce either higher N recovery or pod yield. Foliar spray of N as urea 9 kg/ha too did not result in higher pod yield. As a consequence, fertilizer-N is generally absent from fertilizer recommendations in the United States of America. In many locations within Southern Great Plains, response to N inputs may not be significant, especially if peanut crop is inoculated with *Bradyrhizobium* and soil is limed. Hence, N inputs through chemical fertilizers are feeble, if any. However, N fertilizer inputs are recommended, if peanuts have not been in that field for the past 5 years. In North Carolina, peanuts are known to respond to N inputs up to 300 kg N/ha, especially if sandy soils are deficient for N and crop is not well nodulated. Trostle [13] has reported that in Texas plains nearly 20 to 25% of peanut fields are under-nodulated. Poor nodulation appears to be correlated with high pH >8.0 that curtails effectiveness of rhizobium. Detection of poor nodulation early in the crop season is necessary. Poorly nodulated fields are compensated for loss in fixed N by applying inorganic N fertilizer. Regarding N inputs from Bradyrhizobium, Burkhart and Jones [14] reviewed several sources and suggested that annually 75 to 145 kg N/ha is added via BNF. Much of the fixed N remains in root nodules or in soil after crop harvest. It represents a net source of N available for mineralization or that vulnerable to leaching. Peanut belt in New Mexico is too small in area. Nitrogen dearth is mostly felt at seedling stage before nodule activity begins. Hence, peanut fields are primed with a starter N of 11 to 22 kg/ha, in order to obtain vigorous seedlings and optimum crop stand. If sufficient carry-over N from preceding crop is suspected, priming and additional N inputs are curtailed [15].

In the semiarid tropics of India, crops such as Rice, Wheat, Sugarcane and Cotton garner quite a large share of fertilizer-based nutrients. Further, nutrient inputs are mostly confined to irrigated zones. Consequently, a rain-fed crop like peanut is relegated appreciably in terms of nutrient supply [16, 20].

In uplands of Indonesia, peanut farmers apply 50 to 70 kg urea, along with 75 to 100 kg super phosphate and 0 to 50 kg K/ha. Supply of 50 kg N/ha is almost mandatory, if the crop is irrigated. In China, N input into peanut field is accomplished through chemical fertilizers as well as BNF. Major fertilizer-based N sources that improve soil-N status during peanut cultivation are ammonium sulfate (20% N), ammonium nitrate (3–34% N), ammonium chloride (24% N) and Urea (44–46% N). Fertilizer-based N applied at 180 kg N/ha improves peanut productivity by 20% over control [17]. In fields suspected for low fertility and containing <0.045% N, 90 to 100 kg N input suffices. In soil with medium fertility status (0.045–0.065% N) N inputs are curtailed to 60 kg N/ha. In high soil fertility zones with > 0.065% N, peanuts do not receive N inputs.

Legumes vary markedly with regard to amount of N added via BNF. It depends on genotype, bradyrhizobial strain and soil environment. Genotypes that recover higher N and fix higher quantities of atmospheric N will be useful during cropping. Clearly, peanut genotype and Bradyrhizobial strain, both affect N dynamics in a peanut field, especially extent of N inputs.

25.5 NITROGEN SUPPLIED THROUGH FYM, CROP RESIDUE AND GREEN MANURES

Farm Yard Manure has been the main organic source of N in most of the farming zones. As stated earlier, before the advent of inorganic source, rates of FYM applied almost decided N supply to a crop. Nutrient dynamics that ensued in the ecosystem was solely regulated by FYM inputs. The rapidity of FYM decomposition actually dictated the release of N and other nutrients into soil solution. However, availability of inorganic fertilizer, high yielding genotypes and high yield goals have necessitated higher inputs of N. Globally, farmers have used a wide range of organic manures derived mostly from crop residues, vegetation, animal wastes, industrial wastes and effluents. Often, it is preferable to apply organic manure a couple of weeks ahead of planting to allow proper mineralization. The nature of organic manure, its physical form, N concentration and ratio of other nutrients are crucial. The extent of N derived also depends on soil traits such as texture, moisture content, microbial load, temperature, season and timing of organic manure supply. Following are few examples of organic manures used during maize farming and their percentage N contents on dry matter basis:

Percent N on dry matter basis in Organic Manures:

Crop and Tree residues: Paddy straw = 0.36; Rice hulls = 0.3–0.5; Sorghum straw = 0.41; Pearl Millet straw = 0.65; *Cassia auriculata* = 0.98; *Careya arborea* = 1.67; *Terminalia chebula* =.46; *Terminalia tomentosa* = 1.39; Cowpea = 0.71; Dhaincha = 0.62; Guar = 0.34; Kulthi = 0.33; Mungbean = 0.72; Black gram = 0.85; Sun hemp = 0.75; Groundnut husks = 1.0–1.8; Farm Yard Manure (general) = 0.4–1.5. Compost= 0.4–0.8; Groundnut cake = 7; Coconut cake = 3.

Animal Manures: Cattle dung = 0.3–0.4; Sheep Dung = 0.5; Rural Compost = 0.5–0.76; Vermi-Compost = 1.6; Piggery Slurry = 0.8–1.2; Guano deposits = 0.80–1.4; Poultry manure =1.08–1.2.

Industrial Byproducts, Effluents and Wastes: Coir Pith = 0.26–0.35; Coir Pith Compost = 1.3; Press mud =1.12; Distillery Yeast Sludge = 1.45; Paper Industry wastes = 1.25.
Source: Ref. [2, 18].

Reports suggest that in high productivity belts, crop derives about 45–55 kg N via organic residues. Farmers tend to apply 10–15 t FYM/ha. Crops grown in high rainfall, tropical savanna may derive up to 15 kg N/ha from residues, if planted after long fallow and about 5 kg after a short fallow. Maize planted in low rainfall tropical savanna may derive 2–4 kg N/ha from the organic residues. Crop residue recycling trends are important.

Supply of N via organic manures is common to all agrarian regions. The extent to which N is supplied via FYM or other organic manures depends on inherent soil fertility status, availability of organic manures, timing, agronomic procedures and yield goals. Organic farming may stipulate supply of entire N via FYM or animal manures. The N supplied to soil depends on quality of organic manure, its C:N ratio, succulence, microbial load of soil, soil temperature and other environmental parameters.

The amount of FYM or other organic manures applied varies widely depending on region, season, cropping intensity, soil type, its inherent fertility, cropping systems, and yield goals. In some of the low input subsistence farming zones, availability of crop residues or other organic sources decides extent of N derived. For example, FYM inputs are low and range between 2–5 t/ha in Sahelian West Africa. In this region, crop residue is precious and used to feed farm animals. Animal manures recycled are also small. Natural bird dropping is known to add 2.2 kg N/ha annually. In the dry lands, farmers again recycle only small quantities of crop residues ranging from 2–5 t FYM. Nitrogen derived may range 10–20 kg N/ha. Hence, grain yields are low at 0.8–1.5 t/ha in such subsistence belts [3].

Inorganic fertilizers are supplied to fields using variety of formulations, in different concentrations, ratios and at different timings. They are broadcasted, spot or band placed near the crop, applied in furrows or supplied through irrigation water in dissolved state. The timing, split dosages and method of application of inorganic fertilizers are important factors that affect crop production. Fertilizers formulations used are mostly in powder or granule form. Liquid formulations are also common.

Nitrogen is also supplied to a cereal or cash crop using foliar sprays. Most commonly, a solution of Urea (0.2%) or KNO_3 (1–2.5%) in water is sprayed at seedling/tasselling stage. In a foliar spray, N sources are absorbed through leaf tissue and it avoids contact with soil phase of the agroecosystem. Many of the soil related transformations that fertilizer-N undergoes belowground in soil is absent with foliar fertilization. Nitrogen is required in relatively lower dosages if sprayed on foliage. Therefore, fertilize-N efficiency is generally high. Foliar sprays are also common on maize grown for forage.

The chemical fertilizers containing N are also channeled into crop belts using irrigation water. It is referred as 'Fertigation.' Fertigation is a process where in fertilizer-N is dissolved and allowed to spread across the field in a dissolved state along with irrigation water. Fertilizer-N required is relatively small compared to soil application (broadcast or placement). A study in the Central Platt Valley of Nebraska, has shown that utilization of N supplied using irrigation depended on crop's need and amount of N supplied using other systems. The number of side dress-N envisaged and concentration of N in irrigation water also affected N recovery. Due care is needed to avoid excessive leaching or seepage to ground water. It is interesting to note that generally, irrigation source may itself contain a certain amount of dissolved N in it. The N content in irrigation water may be dependent on source and intensity of farming practiced in the area. For example, in western parts of Corn Belt of USA, Halvorson et al. [19] have reported that irrigation water contained 2.8–3.6 mg NO_3-N/L. It amounted to supply of 12–15 kg N/ha.

In some farms of USA, Europe and Australia, extraneous N is also supplied into crop fields using liquid NH_3 dissolved in irrigation water. Pressurized NH_3 is dissolved into irrigation water source and allowed to mix with soil. Due care is needed to avoid undue loss of NH_3 through volatilization. Liquid ammonia is usually added in cool conditions and during night to avoid evaporation. Fertilizer-N efficiency is higher if liquid NH_3 is used compared to broadcasting or banding of powder or granules (Plate 2).

PLATE 2 Transport of liquid NH_3 for application into fields. Liquid-N input via irrigation is usually done during cool nights to avoid excessive loss of N as emissions. Fertilizer-N inputs as liquid NH_3 is efficient compared to granules.
Source: Nelson Farms, Iowa, USA.

25.5.1 PHOSPHORUS SUPPLY TO AGROECOSYSTEMS

Soils vary enormously for inherent P content, various forms of P and availability to crop roots. Total soil P could be sufficient in many regions, but it the available fraction in soil solution that is absorbed by roots. Like soil-N tests, there are several soil-P tests that predict P availability to crop roots. Soil P tests based on extraction solutions such as Olson's-P, Mehlich's and Brays is popular. The critical soil P levels and/or crop response studies are used to devise fertilizer-P schedules. Fertilizer-P is usually added as basal dose at the time of planting. Fertilizer recommendations often consider various avenues of fertilizer-P loss and chemical fixation. The fertilizer-P inputs are guided by inherent soil-P, available-P pool, and yield goals. Nutrient ratios such as N:P and C:P is also considered in some regions. Plantation crops and oil seed crops are often supplied with larger amounts of P. A subsistence crop of cereal in dry land ecosystem may be given 5–10 kg P/ha. An intensive crop of maize in North America receives 40–60 kg P/ha. Plantations such as grapes, citrus or mango that yield over 10–15 t fruits/ha are provided with proportionately high P at 80–120 kg P/ha. Farmers employ various methods to supply fertilizer-P. They are broadcasting, band placement, point placement, fertigation, etc. Extraneous P is also supplied using Rock Phosphate ores or Guano deposits or Bone meal. Residue recycling and FYM supply also adds to soil P based on nutrient content of crop residues.

25.5.2 POTASSIUM INPUTS TO AGROECOSYSTEMS

Agroecosystems in different continents thrive on soils that are often sufficient in total K. Soil-K exists in different forms such as soluble-K, exchangeable and matrix-K. Soil tests adopted to ascertain its availability to crop root is usually based on soluble-K and exchangeable-K. There are several tests that estimate available-K pool. Soil test-crop response studies are also used to decide on fertilizer-K schedules. Due attention is bestowed to ratios of nutrients such as N:K. In case of plantation crops, tissue analysis is used to arrive at fertilizer-K dosages. Potassium supply is avoided in subsistence farming zones. Crop residue recycling and FYM supplies K to the extent found in them. In the dry arable regions of the world, crops such as maize, wheat or legumes may receive 40–80 kg K/ha depending on soil test and yield goal. Intensive farming zones are supplied with 80–120 kg k/ha to match high yield goal.

KEYWORDS

- ***Bradyrhizobium***
- **Critical Nutrient level**
- **Farmer's Traditional Practices**
- **Liebig's Law of Minimum**
- **Precision Farming**
- **Soil Test Crop Response**

REFERENCES

1. Krishna, K. R. Soil Fertility and Crop Production. Science Publishers Inc.: Enfield, New Hampshire, USA, **2002,** 463.
2. Krishna, K. R. Agroecosystems of South India: Nutrient Dynamics and Productivity. BrownWalker Press Inc.; Boca Raton, Florida, USA. **2010,** 543.
3. Krishna, K. R. Maize Agroecosystem: Nutrient Dynamics and Productivity. Apple Academic Press Inc.: New Jersey, USA, **2013a,** 346.
4. Krishna K. R. Precision Farming: Soil Fertility and Productivity Aspects. Apple Academic Press Inc.: New Jersey, USA, **2013b,** 210.
5. Stanger, T. F.; Lauer, J. G. Corn grain yield response to Crop Rotation and Nitrogen over 35 years. Agronomy Journal **2008,** *100,* 643–650.
6. Sawyer, J. Nitrogen Fertilization for Corn following Corn. http://www.ipm.iastate.edu/ipm/icm /2007/2–12/nitrogen.html **2007,** 1–4.
7. Sawyer, J. E.; Mallarino, A.; Killorn, R.; Barnhart, S. K. A General guide for Crop Nutrient and Limestone Recommendations in Iowa. Iowa State University Extension Bulletin **2008,** 1–18.
8. Kogbe, J. O. S.; Adediran, J. A. Influence of Nitrogen, Phosphorus and Potassium application on the Yield of Maize in the Savanna zone of Nigeria. African Journal of Biotechnology **2003,** *2,* 345–349.

9. Smaling, E. M. A, Toure, M.; Ridder, N. D.; Sanginga, N.; Brenan, N. Fertilizer Use and the Environment in Africa: Friends or Foes. Background paper on African Fertilizer Summit. NAPED-IFDC, Abuja, Nigeria **2006,** 1–26.
10. Niaz, A.; Ibrahim, M.; Ishaq, M. Assessment of Nitrate leaching in Wheat-maize Cropping systems: A Lysimeter study. Pakistan Journal of Water Resources **2007,** *7,* 1–6.
11. Hodges, S. C. Environmental guidelines for Plant Nutrient use. The University of Georgia Cooperative Extension Services, Athens, Georgia, **1991,** 1–16.
12. Reid, H.; Cox, R. R. Soil properties, mineral nutrients, and fertilization practices. In: Peanuts, Culture and Uses. American Peanut Research and Education Society, Stillwater, Oklahoma, USA, **1973,** 22–31.
13. Trostle, C. In-Field correlation of Bradyrhizobium nodulation with soil parameters and peanut yield in West Texas. Precision Agriculture Initiative for Texas High Plains. 2002-Annual Report. Texas Agricultural Experiment Station at Lubbock, Lubbock, Texas, USA, **2002,** 1–4.
14. Burkhart, M. R.; Jones, D. E. Agricultural Nitrogen contributions to hypoxia in the Gulf of Mexico. Journal of Environmental Quality **1999,** *28,* 850–859.
15. Baker, R. D.; Taylor, R. G.; McAlister, R. Peanut Production Guide. New Mexico state University, Albuquerque, New Mexico, USA, **2003,** 1–3.
16. Pal, S. S.; Gangwar, B. Nutrient Management in Oilseed-based Cropping systems. Fertilizer News **2004,** *49,* 37–45.
17. Wenguang, H.; Shufen, D.; Qingwei, S. High Yielding Technology for Groundnut. International Arachis News letter **1995,** *15,* 1–22.
18. Krishna, K. R. Agrosphere: Nutrient Dynamics, Ecology and Productivity. Science Publishers Inc.: Enfield, New Hampshire, USA, **2003,** 208–240.
19. Halvorson, A. D.; Del Grosso, S. J.; Reule, C. A. **2008,** Nitrogen, Tillage and Crop Rotation effects on Nitrous oxide emissions from irrigated Cropping systems. Journal of Environmental Quality *37,* 1337–1344.
20. Krishna, K. R. Peanut Agroecosystem: Nutrient Dynamics and Productivity. Alpha Science International Pvt Ltd, Oxford, England **2008,** 145–216..

EXERCISE

1. Discuss various methods of fertilizer supply to farms in different continents.
2. Mention most popular soil tests for N, P and K.
3. Mention various formulations of fertilizer-N used in North America.
4. What is split dosage. What are the advantages.
5. Mention at least five different sources of P used by farmers in different continents.

FURTHER READING

1. DEFRA, *Fertilizer Manual (RB 209).* 8th Edition. Department of Environment, Food and Rural Affairs, London, United Kingdom, **2010,** pp. 254.
2. FAO. Fertilizer Use by Crops. Food and Agricultural Organization of the United Nations, Rome Italy, FAO Fertilizer and Plant Nutrition Bulletin 17, **2006,** pp. 124.
3. Gruhn, P.; Goletti, J.; Udelman, M. Integrated Nutrient Management, Soil Fertility and Sustainable Agriculture: Current Issues and Future challenges. International Food Policy Research Institute, Washington, D.C. USA, Paper No 32, **2002,** pp. 29.
4. Havlin, J. L.; Beaton, J. D.; Tisdale, S. L.; Nelson, W. L. Soil Fertility and Fertilizers 6th, Edition. Prentice Hall Inc., **2010,** pp. 542.

5. IFDC, Training Manual on Fertilizer Statistics in Africa. International Fertilizer Development Centre, Muscle Shoals, Alabama, USA, **2012,** pp. 78.
6. Jones, B. J. Plant Nutrition and Soil Fertility Manual. Second Edition. CRC Press, Boca Raton, Florida, USA, **2012,** 270.
7. Kaizer, D.; Lamb, J.; Elinson, R. Fertilizer Guidelines for Agronomic Crops in Minnesota. University of Minnesota Extension Service. St Minnesota, USA, **2011,** pp. 48.
8. Krishna, K. R. Precision Farming: Soil Fertility and Productivity Aspects. Apple Academic Press Inc. New Jersey, Toronto, **2013,** pp. 210.

USEFUL WEBSITES

ftp://ftp.fao.fao/agl/docs/fpnb17.pdf (August 24, 2012)
AfricaFertilizer.org (October 14, 2012);
www.africafertilizer.org (October 14, 2012);
Africa Union www.africa-union.org (October 14, 2012)
FAO STAT http://faostat.fao.org/site/575/default.aspx#ancor (October 14, 2012)
IFA www.fertilizer.org (October 14, 2012)
IFDC www.ifdc.org (October 14 2012)
NEPAD www.nepad.org (October 14, 2012)

CHAPTER 26

NUTRIENT TRANSFORMATIONS IN ARABLE CROPPING ZONES

CONTENTS

The parent material of soils, extent of weathering, characteristics like bulk density, texture, structure, aeration, precipitation pattern and soil moisture conditions dictate various physicochemical aspects of soil that have direct bearing on nutrient dynamics in the agroecosystem. The physicochemical properties of each soil type markedly influences nutrient dynamics in the profile and cropping system. Regarding soil structure, it is difficult to define the most appropriate characteristics. Soils should be crust free, loamy and congenial for rapid root growth. A well-aggregated soil contains aggregates of 1–5 mm size in dry and wet conditions. Soil aggregation could be improved by adopting suitable mulching and organic matter supplements. For example, recycling straw in the dry lands is known to improve both size and percentage aggregates. The *in situ* decomposition of sorghum residues improved soil aggregates to 0.38 mm compared to 0.11 mm in control. The infiltration rate improved from 2.59 cm/hr in control plots to 4.75 cm/hr in straw treated fields and to 8.10 cm/hr in legume haulm treated fields. The bulk density of soil affects root proliferation in the profile. Root penetration and activity is least in soils with bulk density 1.8 or above. The mechanical impedance to root spread is high. Presence of high bulk density subsurface layers restricts infiltration of water and root penetration. Hence, deep tillage and chiseling up to 20 cm or below is beneficial in the dry lands. High bulk density is uncongenial to sorghum root growth, moisture infiltration and nutrient availability. The percentage of pores > 0.5 mm in diameter decreases as bulk density increases. Further, hydraulic conductivity and water diffusivity reduced due to high bulk density. Hence, infiltration of water and nutrients to root lessened. Vertical mulches are known to improve water infiltration and nutrient conservation. Run off and soil erosion also affects soil nutrient conservation. The subsoil salinity is yet another factor that affects chemical transformations, nutrient availability and finally it reduces mineral acquisition rates by crops grown on Vertisols. A major share of the explanations on soil physicochemical properties, reactions and nutrient availability are drawn from several publications [1–4].

26.1 PHYSICOCHEMICAL TRANSFORMATIONS OF MAJOR NUTRIENTS—NITROGEN, PHOSPHORUS AND POTASSIUM

Soil moisture and aeration are important factors to crop production in dry lands. Actually, soil moisture versus nutrient interactions, play a crucial role in deciding nutrient dynamics and productivity. Most of the physicochemical transformations of nutrients are dependent on soil moisture status. Fluctuations in precipitation, intermittent drought or severe lack of moisture may all affect nutrient transformations in soil to different intensities. Let us consider a few important nutrient transformations relevant to crop production in arable soils. Discussions here are confined to basic knowledge on transformations of major nutrients N, P, and K during arable or dry land cropping.

26.1.1 PHYSICOCHEMICAL TRANSFORMATIONS OF NITROGEN IN SOIL

Nitrogen is an important mineral nutrient that has marked effect on soil fertility, its quality and crop growth. Nitrogen affects below-ground biomass accumulation through its effect on root growth. The crop growth and canopy formation above-ground is also influenced by N supply. Further, crop residue recycling trends are dependent on influence of N on vegetative biomass formation by crops. There is a need to understand different forms of soil-N, their physicochemical aspects, transformations and availability at different stages during crop production. Major forms of N encountered in soils are classified into inorganic and organic forms. They are:

- NH_4N is found in exchangeable form adsorbed to clay surface and colloids or it is traced as fixed $NH_{4\text{-}}N$ in the crystal lattice;
- NO_3N is found in soil solution, as thin layers on crystals and in rhizosphere. It is highly mobile and prone to loss via leaching and percolation;
- N_2O-N is feebly traced and is confined pockets of reduced layer in soil. Whereas, N_2 and N_2O is found in the soil air space and solution.
- Organic form of N is traceable in amino acids, amino sugars, nucleic acids and humans.

Source: Refs. [2–4].

Nitrogen transport in soil occurs through diffusion and/or mass flow. It can be an important factor that affects N dynamics in the root zone. Movement of NO_3 is controlled by mass flow and that of NH_4N is regulated by diffusion. Generally, NO_3 N translocates faster in soil. Incidentally, in dry lands, cereal crops like sorghum or wheat extracts N predominantly in the form of NO_3. Whereas, in wetlands, diffusion of NH_4N is crucial since rice root preferentially absorbs $NH_{4\text{-}}N$. In arable conditions or dry lands, N loss due to down word percolation and runoff are high. Therefore, N gets transported away from root zone, leading to reduction of fertilizer-N use efficiency. Mulching, spot or band placement of fertilizer-N reduces $NO_{3\text{-}}N$ loss from root zone. Volatilization of inherent soil-N or fertilizer-N as NH_3, N_2O or N_2 is an important physicochemical process in dry land soils. Sorption and de-sorption reactions of N compounds and kinetics of such physicochemical phenomenon too affects N dynamics in the roots zone. In most dry land soils, $NO_{3\text{-}}N$ is weakly adsorbed and easily available to plant roots. $NH_{4\text{-}}N$ is relatively tightly adsorbed on to clay surface in the following order: montomorillonite > kaolinite > allophanes. Ammonia volatilization can be rampant in dry lands of South India. Obviously, it increases as soil temperatures reach higher levels during kharif and summer season. It is said that N loss due to volatilization increases by 0.25% per each 1.0°C rise in soil temperature. Ammonia volatilization may reduce fertilizer N-use efficiency by 20–40% depending on soil and environmental factors [5].

Mineralization is an important chemical transformation that affects availability of N in soil. Basically, it is conversion of organic forms of N into mineral forms. Mineralization occurs in both aerobic and anaerobic microsites. In wetlands, under microaerophilic/anaerobic conditions, mineralization steps proceed only up to formation of NH_4^+. Hence, it is termed ammonification. In dry lands, it is oxidative conversion of organic-N to mineral forms. The aerobic microflora mediates conversion of organic-N to NH_4^+-N and then to NO_3N. The mineralization is affected by various factors related to soil (clay, colloids, aggregation, CEC, C:N ratio, pH, salinity, soil moisture, temperature and microbial flora), cropping procedures (tillage, liming, rotations) and environment (precipitation, temperature etc.). Mineralization and release of inorganic-N is rather rapid in dry lands because of warm temperature, oxygenated soil and enhanced microbial activity. Tillage may further accentuate microbial activity and mineralization rates. The N derived from mineralization can be substantial. Therefore, extent of N derived from mineralization reactions (N_{min}) during crop growth season should also be considered while deciding on the exact amount of fertilizer-N to be applied.

Immobilization is a soil microbe mediated transformation, where in inorganic-N (NO_3 N or NH_4^+-N) is converted to organic forms. For example, fertilizer-N is assimilated into soil microbial biomass. In nature, mineralization and immobilization of soil-N are often coupled. Soil microbes assimilate inorganic-N (NH_4^+-N or NO_3-N) depending on its availability in soil solution. Soil properties such as temperature, pH, microbial activity and C: N ratio of organic amendments affects immobilization.

The dry arable soils are generally held in oxidized state during crop production, except, transitorily immediately after a precipitation or irrigation event when soils may get inundated for some time. The aerobic conditions aid rapid conversion of NH_4 N to NO_3. Nitrification is an oxidation process that transforms NH_4^+-N to NO_3. It is mediated by soil microflora mainly Nitrifying bacteria. *Nitrosomonas sp* mediates conversion of $NH_4^+ + 1½ O_2 \rightarrow NO_2 + H_2O$ and *Nitrobacter spp* mediates $NO_2 + ½ O_2 \rightarrow NO_3$. Nitrification supports build-up of NO_3 in soil. However, in flooded paddy soil nitrogen mineralization halts at NH_4^+ step.

In an arable cropping zone, fertilizer-N efficiency is reduced due to natural factors such as leaching, nitrification-denitrification, immobilization and volatilization as ammonia. Hence, nitrification inhibitors are used to restrict or delay conversion NH_4^+-N to NO_3. It reduces accumulation of $NO_{3-}N$ which otherwise is vulnerable to loss via leaching and gaseous loss as N_2O and N_2. An effective nitrification inhibitor (e.g., 2-Ethyle pyridine, Etridiozole, Nitrapyrin) persists for at least a cropping season and inhibits nitrification to the extent of 65 to 86%.

26.1.2 PHYSICOCHEMICAL TRANSFORMATIONS OF PHOSPHORUS IN SOIL

Soil phosphorus is an important factor that affects both nutrient dynamics and productivity of agroecosystem. Phosphorus deficiency is experienced at different intensities. Major soil types encountered across different continents that are loamy, silty or clayey, coarse textured sandy soils found in coasts, all suffer from P deficiency to different intensities. Basically, various physicochemical and biological transformations that soil-P undergoes seem important with regard to P availability and use by crops. Obviously, detailed knowledge about physicochemical transformations helps while deciding on P replenishment schedules.

Firstly, soil P exists both in inorganic and organic forms. Soil inorganic–P is classified into soluble and insoluble forms. Inorganic-P exists mostly as salts of ortho-phosphoric acid. The inorganic forms include Ca-P, Al-P, Fe-P occluded and Fe-P in the order of decreasing solubility. These inorganic-P fractions may account for 50–60% of total P depending on the soil type. Insoluble – P is mostly attached with elements like Al and Fe. The Al-P and Fe-P are usually abundant in acidic soil, but in soils with pH 6.5 or above Ca-P predominates. Organic-P forms encountered in Vertisols and Alfisols of the sorghum belt are inositol phosphates, phospholipids, nucleic acids and their derivatives, and other orthophosphate esters. The total-P, organic-P and available-P contents are usually higher in surface layers than in subsurface. The P content in the soil phase decreases with soil depth.

The behavior of P in most soils could be explained as follows: Solution P ↔ labile P ↔ Non Labile P. Crop plants use P in solution or that found in labile fraction. Non-labile P is found chemically bound to soil matrix or to organic compounds and is not available to plant roots. Soil P undergoes several types of physicochemical transformations. Many of these transformations play a crucial role in supplying P to plant roots and in recycling P within soils. Immobilization of P denotes conversion of fertilizer-P or other inorganic-P sources into organic forms. Soil microbes assimilate inorganic-P and convert them into organic-P. It is generally accepted that if Org-C/Org-P ratio is 300:1 or more, immobilization of P occurs. Whereas, if Org-C/Org-P is less than 200:1, mineralization reactions set in releasing P into soil solution [6]. Mineralization of organic-P compounds into inorganic forms helps in P absorption by crop. The organic compounds such as phospho-proteins, nucleic acids, lecithin and phytin are de-phosphorylated upon addition to soil. Incidentally, rates of C and N mineralization are usually correlated with mineralization of organic-P. Mineralization of organic-P depends on type of compounds. Usually, microbial degradation of nucleic acids in soil occurred much faster than phytates.

Rock Phosphates are inorganic minerals that contain sparingly soluble P source. Acidulating phosphate rocks partially or *in toto* releases P that is soluble and easily used by root system. In nature, soil microbes also aid dissolution and release of P from such a source with sparingly soluble-P. Microbes are known to release a range of organic acids, like lactic, butyric acid, acetic, propionic, gluconic, oxalic and citric, plus several other extra-cellular products. Such microbes found in rhizosphere and in soil are called Phosphate Solubilizing Microbes. The presence of organic acids in the rhizosphere improves release of P from insoluble-P sources. There are alternate mechanisms to explain solubilization of phosphates. It is argued that organic acids and extracellular microbial products get adsorbed to clay surface. Therefore, it decreases sites for P to adsorb. As a consequence, P is retained in soil solution. Another mechanism suggested relates to adsorption of Ca_2^+ causing a shift in mass equilibria of sparingly soluble Ca-P. Most fertilizer-P sources used are inorganic in nature such as SSP, TSP and Ammonium phosphate. Knowledge about fate of fertilizer-P added to arable soils is essential. It is influenced by physicochemical factors such as texture especially clay content, moisture content, pH, temperature, sesquixodes, Ca carbonate and organic matter. Chemical fixation of P into soil matrix is an important factor that affects fertilizer-P schedules and its availability to crops in an agroecosystem.

26.1.3 PHYSICOCHEMICAL TRANSFORMATIONS OF POTASSIUM

Potassium availability in soil influences productivity of crops. Its deficiency could be perceived on most soil types encountered in the agrarian zones. Potassium occurs in different fractions of soil. It is encountered within the crystal lattice of soil minerals as fixed-K or nonexchangeable-K. This fraction is not easily available to roots. The available-K is found as exchangeable-K and solution-K. The availability of K is actually influenced by several physic-chemical factors such as texture, nature of clay, sorption/desorption reactions etc. The interrelations between different forms of soil-K, actually influences the extent of K recovered by crops. The soil solution-K is generally smallest of K fractions. Crop roots explore and acquire K from this fraction. Obviously, rapid intake of K by roots, result in a gradient of availability of solution-K. Steeper gradient induces diffusive flux of K, depending on soil moisture content. In nature, solution-K and exchangeable-K establish rapid equilibrium in soil. The exchangeable-K replenishes the K lost from solution through rapid absorption by roots and leaching. The equilibrium between solution-K and exchangeable-K is dependent on soil parent material, cropping systems and environment. The structural-K is chemically bound. Release of K from nonexchangeable fraction is usually very slow and feeble. Long-term weathering and destruction of crystal lattice induces it.

Several researchers have reviewed aspects of soil K dynamics and their relevance to crop production [7]. Mostly, availability of K in soil is influenced by sorption-desorption mechanisms, pH, soil moisture and buffering. The soil-K available to plant roots is actually indicated by intensity factor. Unlike other essential elements, acquisition of soil-K is immensely influenced by concentration of Ca and Mg in soil solution. Therefore, it is suggested that while depicting availability of soil-K, it is preferable, to state presence of interfering cations like Ca^{2+} and $Mg^{2+.}$ It then allows us to decipher the intensity factor appropriately. The quantity factor of soil-K is actually a measure of sources of K such as exchangeable-K (adsorbed) that replenishes and maintains solution K levels. Together, these factors influence the K-buffering capacity of soil. Soil pH also affects solution-K concentration. Generally, liming to correct acidity or to reduce Al and Mn toxicity (e.g., in laterites) may be detrimental to K availability in soil.

The diffusion of soil K towards plant roots is immensely influenced by soil moisture status. Soil types found in arable belts, support rapid movement of K, both, down the soil profile or laterally into drainage channel. Soil-K depletion accentuates immediately after a precipitation event. There is also greater chance of soil-K depletion, whenever moisture levels are high. Incessant cropping depletes soil-K. However, its impact on K nutrition of crops may not be perceived immediately. It is attributed to K buffering capacity of soils. Sandy soils show poor buffering capacity, whereas clayey soils show up high buffering capacity for K. Fixation of soil-K is an important aspect that affects fertilizer-K efficiency and crop productivity. Soils that possess large clay fraction, such as illite, smectite or montomorillonite are prone to fix high quantities of fertilizer-K applied. The extent of K fixation may depend on percentage clay content and organic matter. The fertilizer-K once fixed into crystal lattice, becomes virtually non-available to plant roots. Intensive farming that occurs in many parts of the world may deplete both exchangeable and available-K and simultaneously increase K-fixation capacity of soil. Together, it leads to rampant K deficiency in soils that could be corrected only by addition of larger doses of fertilizer-K. Albeit, most soil types are moderately rich in K and may not need replenishment even after a couple of cropping cycles. The fraction of K that gets depleted most rapidly needs consideration. In case of exchangeable-K, cereals may deplete about 20% of the 130 ppm K from a Vertisol. It is said that incessant cropping for nine cycles may still leave sufficient exchangeable-K for next crop. While replenishing K, minimal level of exchangeable-K also needs to be ascertained. Inherently, soils contain adequate amount of exchangeable-K. For instance, in the Black Cotton soils of South India, need for K replenishment was felt only after 13th consecutive cropping. Further, there are suggestions that fixed-K should also be considered while devising K replenishment schedules. In most soils, exchangeable-K is as good as K availability index.

KEYWORDS

- **clay fraction**
- **exchangeable-K**
- **Nitrobacter spp**
- **Nitrosomonas sp**

REFERENCES

1. Brady, N. C. Nature and Properties of Soils. Prentice Hall of India, New Delhi, **1975,** 575.
2. Krishna, K. R. Soil Fertility and Crop Production. Science Publishers Inc.; Enfield, New Hampshire, USA, **2002,** 546.
3. Krishna, K. R. Agrosphere: Nutrient Dynamics, Ecology and Productivity. Science Publishers Inc.: Enfield, USA, **2003,** 344.
4. Krishna, K. R. Agroecosystems of South India: Nutrient Dynamics, Ecology and Productivity. BrownWalker Press Inc.; Boca Raton, Florida, USA, **2010,** 565.
5. Hood, R. The use of Stable Isotope in Soil Fertility Research. In: Soil Fertility and Crop Production. Krishna, K. R. (Ed.). Science Publishers Inc.; Enfield, New Hampshire, USA, **2002,** 313–335.
6. Tomar, N. K. Dynamics of Phosphorus in Soils. In: J. N. Mukherjee, ISSS Foundation lectures. Indian Society of Soil Science, New Delhi, **2002,** 244–278.
7. Krauss, A. Potassium-The Forgotten Nutrient in West Asia and North Africa. In Accomplishments and Future Challenges in Dry land Soil fertility Research in the Mediterranean area. Ryan, J. (Ed.) International Center for Agricultural Research in Dry areas, Aleppo, Syria, **1997,** 9–21.

EXERCISE

1. Discuss various Physico-chemical transformations that occur in Arable cropping belts.
2. Mention the various transformations of Nitrogen in an Arable soil held in oxidized state.
3. Describe Nitrification in soils.
4. What is Chemical fixation of Soil Nutrients. List soils that are prone to P-fixation. How is it overcome while preparing fertilizer-P schedule?
5. Mention the different forms of Potassium encountered in soil. How do they affect absorption of K by plant roots.

FURTHER READING

1. Kersebaum, K. C.; Hecker, J. M.; Mirchel, W. Modeling Water and Nutrient Dynamics in Soil-Crop Systems. Application of different models to common datasets. *Proceedings of*

workshop held at Nehberg, Germany. Springer Verlag: Heidelberg, Germany. **2007,** pp. 439.
2. Krishna, K. R. *Soil Fertility and Crop Production.* Science Publishers Inc.: Enfield, New Hampshire, USA, **2002,** pp. 465.
3. Power, J. A.; Prasad, R. *Soil Fertility Management for Sustainable Agriculture.* CRC Press: Boca Raton, Florida, USA, **1997,** pp. 384.
4. Sposito, G. *The Chemistry of Soils.* Oxford University Press: New York, USA, **2008,** pp. 344.

USEFUL WEBSITES

http://www.sssa.org
http://www.eolss.net/Sample-Chapters/C19/E1-05-07-06.pdf
http://www.soils.wisc.edu/frc/
www.journals.elsevier.com/soil-biology-and-biochemistry

CHAPTER 27

NUTRIENT TRANSFORMATIONS IN FLOODED WETLAND RICE BELTS

CONTENTS

Soil submergence with water induces anaerobiosis or microaerophyllic conditions in the upper horizons where crop roots have to grow and absorb nutrients. The soil physicochemical condition during rice production is therefore enormously different from those encountered by roots of crops cultivated under arable conditions. Most striking is the fact that rice crop, that is grown in rotation and soil biota will all have to adapt to rapidly changing, and alternating aerobic and anaerobic conditions, plus have to be versatile in garnering soil nutrients that occur in different physicochemical states, efficiently. Most farmers and agricultural researchers do not pay heed to such massive alteration of soil environment. We need to understand the principle changes that occur in soil physicochemical environment and crop growth in expanses that experience such crop rotations involving flooded rice and an arable crop.

27.1 SUBMERGENCE ALTERS SOIL AND NUTRIENT TRANSFORMATIONS

Flooding and puddling of rice field, both influence physicochemical dynamics of nutrients within soil profile (Plate 1). Submergence affects availability of nutrients to rice seedlings to a certain extent. Physically, a well-puddled, flooded soil possesses an oxidized surface layer; below it is a chemically reduced layer. The plow layer is sandwiched between layers of oxidized subsurface soil. Rice roots proliferate into anaerobic plow layer to absorb nutrients that are held chemically in a reduced state. Actually, plow layer is not entirely anaerobic, but it is interspersed with a mosaic of aerobic pockets. These are called aerobic microsites. Flooding effectively stops flow of O_2 to soil layers within 24 h. It halts gaseous exchange between atmosphere and soil. This leads to a reduction in redox potential, increase in specific conductance and ionic strength. Obviously, as redox potential decreases, aerobic microflora fades leading to a build-up of anaerobic microbes in soil. It is clear that redox potential in soil changes from positive to negative due to submergence of rice field. Sequentially, at redox potential of +800 mv oxygen is reduced to H_2O. As redox potential decreases further reaching +400 to 430 mv, NO_3 is denitrified to N_2, MnO_3 is reduced to Mn_2; at +300 mv $Fe(OH)_3$ is converted to $Fe(OH)_2$; then at –180 mv organic acids and at –200 mv alcohols are reduced. Extremely low redox potential beyond –300 mv could be detrimental to rice roots because sulphites tend to accumulate in soil. Submergence of paddy soil also alters pH. Puddling affects extent of volatilization of nitrogen from paddy soil [1–4]. Following physicochemical reactions relevant to rice plant nutrition occur in the paddy fields due to submergence:

- Quantity of CO_2 released during first few weeks after submergence is larger;
- Usually, de-nitrification of nitrate is accomplished within a month after flooding, without NO_3 accumulation
- NH_4-N^+ is more stable in reduced conditions and it accumulates. Extent of NH_4-N^+ accumulation is dependent on organic matter content and leaching losses;
- Mineralization of organic-C is retarded because of anaerobic conditions, but organic-N mineralization might be hastened due to higher C: N ratio;
- Phosphorus in soil solution increases by at least 0.1 to 0.2 ppm due to flooding;
- Phosphorus fixation reactions due to Al and Fe are also affected;
- Ca, Mg and K are displaced into solution by NH_4^+, Fe_3^+ and Mn_2^+;

— Fe_2^+ increases rapidly initially, but later decreases gradually;
— Mn_2^+ levels increase due to enhanced solubility, but it could also precipitate as $MnCO_3$;
— B, Cu and Mo availability increases due to flooding, but Zn availability decreases;
— In submerged soils, end products of organic matter decomposition are CO_2, NH_4^+, CH_4, NO_2, H_2S, mercaptans and partially humified residues.

PLATE 1 Top: Paddy fields near Dharmasthala, (Western Ghats) in South-west India, kept puddled and inundated to transplant rice seedlings.
Note: Fine tilts allows easy transplantation of rice seedlings. Puddling also aids in stagnation of water upon flooding. This procedure brings about series of physicochemical changes in the soil ecosystem. Puddling induces loss of soil structure. Soil aggregates crumble. Fields are usually inundated with water up to 5 cm depth. Oxygen tension in soil reduces leading to microaerophilic/anaerobic conditions. Microbial flora changes as anaerobiosis sets in. Regarding nutrient dynamics, mineral nutrient availability and soil chemical reactions that ensue too alter enormously due to changes in redox potential. Availability of nutrients like P and Fe increases in flooded soils.
Source: Krishna, K. R., Bangalore, India.

27.2 NITROGEN TRANSFORMATIONS IN SUBMERGED PADDY FIELDS

Soil N status is also affected to a certain extent, by the rates of mineralization/immobilization reactions. These soil N transformations decide the soil available-N pool. Mineralization refers to conversion of organic-N into mineral N via decomposition of organic matter. However, immobilization reactions fix mineral forms like NH_4 or NO_3 into organic matter in soil. The anaerobic conditions that prevail in submerged soils allow mineralization reactions to continue only up to NH_3 formation. Hence, in paddy soils, it is often termed as ammonification. Ammonification is rapid immediately after fertilizer-N application and peaks in about 2 weeks after submergence. However, ammonification rates decrease 8 weeks after transplanting. The rate of ammonification in paddy field is actually dependent on a variety factors related to soil, organic matter and crop management practices [2]. Immobilization means assimilation of inorganic-N forms into organic forms. This process is mediated be soil microbes. Immobilization affects soil N dynamics, especially N accumulation and its availability to rice crop. Soil related factors such as C:N ratio plays an important role in extent of N immobilized.

Nitrification and de-nitrification reactions induce soil-N transformations. The native soil-N in wetlands occurs as NH_4^+. It is oxidized biologically to NO_3. This is a strictly aerobic transformation mediated by soil microbes such as *Nitrosomonas* and *Nitrobacter*. Actually, nitrification occurs in oxidized layers of submerged soils. Thus, nitrification reactions confine mostly to aerobic microsites (site-1). Nitrification also occurs in the rhizosphere, wherever oxygen is not a limiting factor (site-2). Nitrification usually peaks in 7 to 10 days after fertilizer-N supply.

Denitrification occurs in the anaerobic layers available abundantly in submerged soils. Firstly, a gradient of $NH_{4-}N$ gets established between aerobic and anaerobic layers. Next, $NH_{4-}N$ in aerobic layer is converted to NO_3 via nitrification. This NO_3, then diffuses to anaerobic layers in soil, where it is denitrified to N_2O and N_2. A sizeable quantity of N is lost to atmosphere via denitrification reaction in submerged soils. Overall, there are two important sites where nitrification/denitrification reactions could occur. Firstly, in the aerobic microsites then secondly, in rhizosphere of roots/rootlets, wherever oxygen is available. Often, nitrification inhibitors are added with fertilizer-N to suppress formation of $NO_{3.}$ If not, accumulated NO_3 becomes vulnerable to loss via de-nitrification. The soil-N loss due to denitrification at site-2, such as oxidized layers in rhizosphere seems conspicuous. Nitrogen loss from submerged soils due to denitrification at site 2 can be as high as 18% of applied fertilizer. As stated above, efficiency of fertilizer-N is variable but mostly <48%. Firstly, fertilizer-N supplied to paddy fields immediately becomes part of soil N dynamics. It undergoes various physicochemical and biological transformations. Such transformations could be either beneficial or detrimental to fertilizer-N efficiency. Over all, the expected fate of fertilizer-N in submerged soils used for paddy production is as follows [5]:

a) Volatilization of NH_3 from NH_4 based fertilizers;
b) Nitrification and de-nitrification of both NH_4 and NO_3 containing fertilizers;
c) Immobilization of available N (NH_4^+ or NO_3) into biomass;
d) Soluble NO_3 in fertilizers could be leached and lost to anaerobic layers, where, it is further lost via de-nitrification and conversion to N_2O and N_2.
e) Absorption of NH_4/NO_3 by rice roots.

Obviously, recovery of fertilizer-N applied to paddy fields is influenced by several factors. Field trials with ^{15}N tagged fertilizers indicate that in the rice belt of South India, out of 100 kg N applied, 24.3% is traced in grains, volatilization and leaching losses account for 17.2%, about 27.3% is retained in soil and 31% is lost through de-nitrification. Residual effects are minimal at 5%, which is mostly used by succeeding crop [6].

27.2.1 NITROGEN LOSS FROM PADDY FIELDS

Nitrogen loss from paddy fields occurs through ammonia volatilization, de-nitrification, leaching and runoff. Therefore, nitrogen recovery from rice fields is relatively low. It seldom exceeds 30 to 40% of fertilizer N applied. Movement of N in submerged soils is important. In undisturbed submerged fields, N percolates at a rate of 12–14 cm in 4 weeks. Sometimes, N loss via downward leaching can be rampant. Leaching loss is high immediately after a rainfall event. Rainfall events that result in surface flow, leaching, percolation and loss through drainage water may cause another 10 to 12% of N loss from soil. Since, NH_{4}-N and NO_{3-}N are mobile in soil; their loss is accentuated immediately after a heavy rainfall event. Down ward movement that results in loss of N from root zone is also important for yet another reason. It may affect ground water quality. Especially, NO_3 contents may exceed permissible limits. Major suggestions to avoid ground water contamination are to apply N when the crop needs; match N supply rate with need; split N inputs several times in order to avoid undue accumulation; apply in NH_4 forms and adopt a balanced fertilizer input program [7] .

Nitrogen loss from the rice belt can be severe depending on meteorological conditions, soil N status and cropping pattern. Volatilization of N as NH_3, NO_2 and N_2O is dependent on atmospheric conditions, rice genotype and its canopy, soil N status, fertilizer-N supply and farming practices. ^{15}N studies indicate that about 3–8% of fertilizer-N, applied, as prilled-Urea is lost from rice fields via volatilization. However, nitrification inhibitors may curtail loss of N to a certain extent. For example, N loss from neem-coated urea is minimal. Detailed studies on nitrogen dynamics in paddy fields has revealed that, under submerged condition, oxidation of NH_{4-}N derived from NH_4 fertilizers in surface layers and subsequent de-nitrification in subsurface layers may together account for 30 to 49% N loss from rice cropping zones [8]. Generally, volatilization of NH_{3-}N peaks at 3 days after Urea application and later decreases until 15th day. Floodwater, pH, FYM and green manure are major factors that affect NH_{3-}N volatilization from paddy fields.

27.3 PHOSPHORUS DYNAMICS IN RICE FIELDS

Phosphorus inputs via atmospheric depositions are considered negligible, but in some locations it ranges from 0.4 to 0.8 kg P/ha, annually. The inherent, total P in submerged paddy soils may range between 200 to 400 kg P/ha. However, it has little significance because it does not correlate with available-P that roots exploit. Flooded soils contain both organic and inorganic-P forms. Inorganic-P forms encountered in submerged soils are Fe-P, Al-P, Ca-P and occluded-P. Generally, Fe-P forms are dominant accounting for 75–80% of inorganic-P. Submergence enhances P availability in soil. It is attributed to Fe-P fraction. Major reasons for enhancement of P availability are:

— Reduction of hydrous ferric compounds particularly Fe^3 Oxides;

— Under submergence, organic acids released during anaerobic decomposition of organic matter complexes with Ca^{++} to increase P availability. Level of organic matter in soil influences P release from Ferric reduction.
— Mineralization of organic-P is considered less important in terms of P release, because it is too slow;
— PO_4 ion release may occur through exchange of organic anions and PO_4 ions in Fe-P and Al-P containing compounds;
— Flooding enhances P diffusion and buffering of soil P. Increased buffering capacity is attributable to P adsorption from soil solution;
— Microbial activity in the oxidized layers of rhizosphere or microsites can mobilize P;
— In flooded soils, chemically reduced conditions improve P availability via formation of H_3PO_4 from Fe-P and Al-P compounds.

Source: Ref. [9].

During rice-pulse or even rice-rice-fallow rotations, anaerobic and aerobic conditions alternate in soil, for different lengths of time. Submergence induces rapid changes in redox potential of soil. Such alternation of redox potential affects P availability in soil. During flooded condition P availability is relatively higher. Drying a submerged soil reduces P availability. Immobilization of P is also a possibility. Availability of P in soil increases due to submergence during paddy cultivation [10]. It may still be insufficient to grow a high yielding rice crop. This necessitates supply of extraneous P. Therefore, P is applied to paddy crop grown on all major soil types. Vertisols may show high P fixation. Hence, they require still higher quantities of P to satisfy the P-fixation factor first, and then to enhance available-P pool. The extent of chemical fixation ranges from 20 to 60% of fertilizer-based P inputs. Similarly, P – fixation is generally high in lateritic soils. On such lateritic soils, rice crop continues to respond to P inputs up to 260 kg/ha.

Green manures may improve P availability in soils used for rice production. Soil P status improves upon incorporation of green manure and allows sufficient residual effect during rice phase of the sequence. Yet another practice is to apply rock phosphates to green manure crop. It allows residue recycling and improves release of P from rock phosphates. Transformations of P and dynamics of release in neutral soils used for rice-green manure rotation needs to be understood in greater detail [11].

Rock phosphates are mixed along with composts and/or industrial effluents to improve the performance. Phospho-composting is a common practice with farmers in South India. Field tests have shown that kinetics of P solubilization is improved, if acidic effluents were mixed with rock phosphates. For example, Mussoorie rock phosphate mixed with glue waste and distillery effluent improved P availability significantly. A combination of rice straw, rock phosphate and industrial effluent is considered ecofriendly. In addition, this mixture improved soil P availability [12].

27.4 POTASSIUM DYNAMICS IN RICE FIELDS

Flooding affects potassium dynamics in low land paddy fields. Flooding actually increases availability of K to crop roots. It is caused by increased solubility and mobility in soil pores. As a consequence, rice plants are exposed to K ions better and the roots absorb relatively greater amounts of K. Rice farmers add several types of fertilizers. The K concentration in these sources too varies. On average, if farmers apply FYM, upon decomposition it release 2–3 ppm K+, incorporation of Blue Green Algae releases 5.5 ppm, and 3–35 ppm if inorganic-K bearing fertilizers are incorporated into submerged soils [1, 13]. During recent years, there are growing evidences that soil K fertility is declining in the rice agroecosystem of South and South-east Asia. Atmospheric deposits contain K, but it is not significant. Similarly, irrigation water contains traces of K, but it does not match K requirements of crop. At best, K input via irrigation may offset leaching or runoff. Although most soils are endowed with optimum levels of inherent K, incessant cropping, intensification of paddy cultivation and lack of proportionate replenishments have depleted exchangeable-K levels. Therefore, soil test value for exchangeable-K is important while assessing K requirements. Actually, K availability in soil increases immediately after flooding. Obviously, soils with higher background-K are applied with proportionately lesser quantities of K fertilizer. In the Indian subcontinent, paddy crop is supplied with 60 to 140 kg K_2O, if soil test K is 160 kg/ha. If soil test is 260 kg K_2O/ha –40 to 110 kg K_2O is supplied and if soil test is much higher at 380 kg K_2O/ha then K input is reduced to 20 to 60 kg K_2O/ha. In addition to soil test values, K supply to rice agroecosystem is dependent on soil type, location as well as yield goals set by individual farmers. Following are few examples. Potassium inputs to paddy fields are usually split. About 50% K is applied at planting and rest 50% at panicle initiation. In some locations, 50% K is applied immediately after transplanting and 50% before flowering. Sometimes, direct seeded paddy is supplied with 50% K a month after planting and 50% at panicle initiation.

Rice farmers tend to compute crop's response to K addition in single plots or farmer's fields. Actually, knowledge about response of entire rice-based cropping sequence to K input is equally important. Post-harvest availability of K in soil after first crop of rice is also important. It is dependent on irrigation practices followed during crop season. Studies on interaction of K and irrigation schedules has shown that after kharif crop of rice, available K was greatest, if fields were provided with irrigation at 2.5 cm depth 3 days after disappearance of ponded water.

27.5 SULFUR DYNAMICS IN PADDY FIELDS

The rice agroecosystem in South-east Asia and other continents receives S from atmospheric deposits, but it is very small ranging from 4–8 kg S/ha annually. On an average, sulfur content of wetland soils in South-east Asia ranges from 0.1 to 0.6%. Atmospheric deposits containing S could be negligible but at times sizeable, if paddy fields are located nearer to industries or sea. In the paddy cropping zones, atmospheric S is usually encountered as H_2S, SO_2 or particulate S. Irrigation water may contain 3–6 mg S/L. It may suffice to raise a paddy crop. However, intensive rice–rice cropping systems need extra supply of S.

Sulfur occurs both in organic and inorganic forms in submerged paddy fields. The organic-S is generally traceable in ester forms or bonded into S containing amino acids. Organic-S may account for 65% of total S in submerged soils. Majority of S absorbed by rice roots arises from organic fraction of soil. Residual-S is yet another fraction of soil which is resistant to acid and alkali hydrolysis. In flooded soils, $SO_{4-}S$ released via mineralization of organic-S is quickly converted to sulfide form, because of chemically reduced environment. Further, sizeable amount of $SO_{4-}S$ is encountered in forms unavailable to rice roots. Immobilization of $SO_{4-}S$ drawn from fertilizer is another possibility. Knowledge about $SO_{4-}S$ sorption and desorption is also important while devising S input schedules.

Sulfur deficiency has been identified in the Rice belt of South-east Asia. Sulfur deficiency has spread rapidly since past three to four decades. It is attributable to intensive cultivation of rice, season after season without appropriate levels of replenishment. Application of high analysis S-free fertilizers accentuates S deficiency in the agroecosystem. According to Singh [14], sulfur deficiency is felt on 41% of soils in India. About 20 to 30% of soil in the rice growing regions of the world is S deficient. Therefore, response to S supply is significant. Sulfur deficiency is marked in areas supporting rice-arable crop rotations. For most crops, including rice, critical level of S in soil is 10 to 13 mg $SO_{4-}S$/kg soil, if extracted using calcium orthophosphate or calcium chloride [15]. Sulfur status in rice fields could be improved using organic manures, sulfur containing fertilizers, industrial effluents and by-products containing sulfur. Obviously, sulfur supply schedules are guided by soil tests. Soils with < 5 mg/kg S are considered severely deficient. They are supplied with 40–60 kg S/ha. Soils with 10 to 15 mg S/kg are deemed medium in sulfur availability and provided with 30 kg S/ha. Soils containing >20 mg S/kg are not supplied S [14].

Next, we have to understand fate of S-containing fertilizers applied to submerged paddy fields. The release of S due to submergence needs due consideration. Sulfates applied to paddy fields during kharif season undergo several chemical transformations specific to submerged conditions. Most commonly, sulfates are transformed into sulfides, including H_2S. Hydrogen sulfide is detrimental to plant root respiration and aerobic soil microbes. During peak periods of transformation, about 1180 cm^3 H_2S gas/ha may be generated from rice fields. Reduction of sulfate salts in submerged soils also generates meta-stable water insoluble metallic sulfides. There are instances when 229 to 429 mg insoluble metallic sulfides have been recorded in paddy fields. Accumulation of metallic sulfides usually increases with duration of submergence. Extended period of anaerobic condition enhances metallic sulfide formation.

Overall, timing of S bearing fertilizer is important. We need to synchronize S transformation, crop demand and desorption patterns. Basal inputs are preferred for early maturing rice genotypes. For late maturing genotypes S supply could be delayed by 20–30 days after transplantation. A recent compilation by IPNI [15] suggests that rice crop that yields 2.5 to 3.0 t grains/ha removes 7–12 kg S. The S requirement of cereals like rice is about 9–15% of total N absorbed by the crop and it is usually of the same order as P. Mean application of S ranges from 35–40 kg S/ha and response ratio of rice to S input is 16 kg grain/kg S [15].

27.5.1 LIMING PADDY FIELDS

During wetland rice culture, liming is required if soil pH is low, say around 5.5. Lime is applied at 600 kg/ha in two splits if the crop is direct seeded. Basal application is done at plowing and split amount of 250 kg lime/ha is supplied a month after sowing. Transplanted paddy crop too receives lime in two splits of 350 and 250 kg/ha. First split at plowing and other a month after transplanting. The rice ecosystem receives a certain amount of Ca from atmospheric deposits. It ranges from 1–2 kg Ca/ha and is not considered as significant while deciding lime application. Liming the surface soil in the puddle with 1500 to 2000 kg/ha just before transplantation induces mineralization of organic-N in wetland. Liming induces release of at least 30 to 40 kg N/ha of easily available $NH_{4-}N$ [8]. Such an induction of N release through liming improves grain yield by 350 to 400 kg/ha.

27.6 DYNAMICS OF MICRONUTRIENTS IN FLOODED PADDY FIELDS

Like most crops, rice too absorbs only small quantities of micronutrients such as Zn, Fe, B, Cu, Mn and Mo from soil. Yet, incessant cultivation of rice mono-crops and rotations has resulted in deficiency of at least two micronutrients, namely Zn and Fe [16, 17]. Several aspects of micronutrient dynamics such as availability, absorption, accumulation in plant parts and recycling have been affected in the wetlands. Lack of micronutrients in soil often affects response of rice plant to other major nutrients. This is attributable to expression of Liebig's law of Minimum. Micronutrients undergo different physicochemical transformations in submerged soils. Some of them play a crucial role in their availability to rice roots. Therefore, we need to assess micronutrient status of paddy fields periodically and replenish them in apt proportions in order to derive better crop yield. During recent years, micronutrient deficiencies in rice fields have curtailed crop response to fertilizer supply in general. Agronomic efficiencies of major nutrients have declined from 15–20 kg grain/kg nutrient in early 1970s to almost 6 to 8 kg/kg nutrient at present.

Deficiency of Fe is experienced at different intensities in the Rice belt. To a large extent, Fe deficiency is related to intensive cropping procedures. Regular Fe replenishment schedules are guided usually by soil test values taken ahead of sowing or transplanting the crop. Iron requirement of rice crop is usually satisfied by application of Fe salts to soil. For example, supply of 10 to 20 kg $Fe.SO_{4-}7H_2O$/ha plus 10 t FYM improves Fe recovery by rice crop significantly from 1.01 kg Fe/ha to 1.20 kg Fe/ha. Consequently, rice grain production improved from1.5 t/ha in control plots to 3.9 t/ha due to supply of Fe. Rice crop absorbs Fe supplied through foliar spray. Foliar spray of Fe improves grain and straw yield of rice. In addition to fertilizer-Fe supplement, we can explore genetic tolerance of rice crop to Fe deficiency could also be used to thwart Fe deficiency. Rice cultivars such as IET 7613, Akashi, WBPH25 and Prasanna are tolerant to low levels of Fe in soil.

Zinc deficiency appeared in rice belts during mid-1970 s. In general, submergence reduces concentration of different forms of Zn. Therefore, Zn salts are applied to soil or sprayed on rice foliage. It is easier, if Zn is replenished into low lands annually or bi or tri yearly basis. Zinc applied to submerged paddy soils undergoes transformations into different forms. Certain forms of Zn are preferentially used by rice roots and rest retained as residual-Zn in soil. In fact, about 60–70% of Zn salt applied to submerged soils is retained. Such residual-Zn fraction in soil usually contains 3% water soluble (exchangeable) Zn, 47% Zn bearing amorphous sesquioxides and 12–20% Zn complexed with manganese oxides or in crystalline bonded form. We should note that, only 5% of Zn applied to submerged soils is recovered into paddy crop. However, Zn being a micronutrient, amount of Zn-salt supplied and that needed by crop are relatively small. Zinc input to rice fields has ranged from 20–50 kg/ha based on location and soil test value. Deficiency of other micronutrients such as B, Cu and Mn are sporadic. Molybdenum salts are applied, if rice is routinely rotated with N-fixing legumes such as groundnut, cowpea or pigeonpea or a leguminous green manure crop during fallow period.

27.7 SOIL ORGANIC MATTER, CROP RESIDUES AND ORGANIC MANURES IN WET LANDS

Soil organic matter (SOM) is an important component that influences soil fertility. Soil organic matter specifically affects availability of nutrients in wetlands. Paddy fields are supplied at least 5 t/ha FYM or compost as a matter of routine. Otherwise, 10 t/ha rice straw is recycled to stabilize SOC and organic-N [8]. Long-term studies on carbon dynamics have shown that SOC accumulated at greater quantities during rice-rice double cropping. Locations supporting in rice-arable crop rotations possessed lower amounts of SOC. The SOC: N ratio was wider, if soils were continuously used for wetland rice cultivation compared to those under arable cropping system. For example, mono-cropping of rice at Teligi in Karnataka lead to SOC accumulation (1.03%), whereas soils exposed to rice-arable crop sequence possessed only 0.45% SOC. Sahrawat et al. [18] suggest that preferential accumulation of organic matter in wetland is attributable to anaerobiosis in submerged conditions during rice-rice continuous cropping. Prolonged flooding that reduces oxygen tension in soil results in chemically reduced environment. Under these conditions, organic matter gets decomposed slowly. Sometimes, organic matter decomposition is inefficient or even incomplete. In comparison, during arable cropping, oxidized environment aids rapid decomposition of SOC. In fact, accumulation of organic matter is an important aspect considered with regard to C dynamics in wetland [19–21]. Actually, rate of FYM supplied and decomposition is crucial during rice-based cropping systems. Often, in submerged conditions, decomposition rates are slow thus leading to net positive balance for SOC. In other words, rice-rice intensive cropping provided with FYM or organic manures tend to sequester relatively greater quantities of SOC than arable cropping belts.

Recycling of rice straw and stubble plays an important role in maintaining SOC in soil. It builds SOM, improves soil physical properties and recycles C and N in rice fields. Average nutrient content of oven-dry paddy straw that could be recycled is as follows:

Major and Secondary nutrients (%):

N= 0.70 (0.4–1.2); P= 0.07 (0.02–0.17); K= 1.5 (1.0–3.7), Ca = 0.23 (0.03–0.39); M = 0.15 (0.10–0.26); S = 0.08 (0.06–0.10)

Micronutrients (ppm):

Cu = 4.2 (3.0–6.0); Zn = 1380 (30–2860); Mn =720 (560–1090); B = 6. (3.4–8.9); Mo = 2.5; Fe = 200; Ni = 0.04; Na =0.25%; Si = 7.4%; Cl = 0.18%; Al = 600.

Indeed several types of bio-digested organic manures and organic wastes are used to enhance SOC status and rice productivity. Bio-digestion or composting helps in improving quality and nutrient availability of manures. Organic wastes like, banana waste, bio-digested slurry, subabul, neem leaves, and palmyrah leaves are often recycled. Addition of cellulose digesters such as *Trichoderma* or *Pleurotus* hastens organic matter decomposers. Most importantly, nutrients released from organic matter that are recycled are immediately available in soil [22].

27.7.1 GREEN MANURES AND NUTRIENT DYNAMICS IN RICE FIELDS

Rice-rice intensive cropping is in vogue for past three to four decades. Incessant cropping has affected soil quality and organic-C sequestration. In many locations of the rice belt, sequestration of organic-C has declined. Hence, farmers tend to add FYM or grow green manure crops such a *Sesbania rostrata* or *Glyricidia spp* and incorporate it *in situ*. This process is said to improve SOC. Green manure crops are usually grown during fallow after two cycles of rice or as intercrop at 4:1 ratio simultaneously with rice. Traditional rice-rice cropping improves SOC, but it is marginal. Actually, SOC increased by 0.29% compared to initial levels. However, intercropping rice with green manure (*S. rostrata*) increased SOC by 10.6 %. Green manures preferentially improved humic and fulvic fraction of SOC. Repeated incorporation of *S.rostrata* helps in building SOC without significant loss from the ecosystem [23]. Recent trend in the rice belt is to adopt organic farming. It envisages supply of required nutrients through different organic sources. For example, during organic farming a composite of manures comprising FYM (8–10 t/ha), poultry waste 2.5 t/ha, vermin-compost (2.5 t/ha), neem cake (50–60 kg/ha) and fresh cow dung (10 kg/ha) are applied to rice fields prior to transplantation.

Sometimes, succulent weeds are used as green manures. Judicious use of weeds can improve soil-K status. Rice farmers in Godavari region prepare K-rich green manures from weeds that infest the surrounding areas of their rice fields. Most commonly used weeds are *Pitia stratiotes, Salvinia molesta, Gynandropis gynandra, Lantana camara* and *Parthenium hysterophorus*. These K-rich weeds are cut at a succulent stage and incorporated into rice fields to improve soil nutrient status [24]. Field tests show that application of such K-rich weeds improves water soluble-K by 2–4 kg K/ha and exchangeable-K by 10 to 43.5 kg K/ha over control plots. Fluctuations in nonexchangeable-K were not discernible distinctly.

Application of Green leaf manures (GLM) reduces need for inorganic-N input and improves rice grain yield. It improves soil quality and SOC content. The GLMs are prepared by harvesting succulent leaves from trees found commonly in the rice belt. Tree species are cultivated on the bunds and along margins, so that litter and fresh leaves could be recycled easily. Tree species commonly used to prepare GLM are *Cassia auriculata, Azadirechta* (neem), custard apple, *Calotropis, Leucana, Gliricidia, Pongamia* and *Tephrosia*.

In addition to quantity, mechanism of nutrient release from leaf litter belonging to different tree species is important. The soil physicochemical condition at the time of leaf litter incorporation plays vital role in decomposition and release of nutrients into root zone. Leaf litter from certain tree species such as Gliricidia may contain relatively higher levels of N and K. Immediately after incubation of leaf litter, nutrient availability increased in following order K > N > P > S.

At present, Azolla is an important component of Rice Agroecosystem of Southeast Asia. *Azolla* influences C and N dynamics in Paddy fields. *Azolla* is a floating fern that harbors N-fixing BGA. Among the seven species of Azolla traced in paddy fields, *A. pinnata* is common. Azolla needs 5 to 10 cm water in paddy fields to float plus to absorb dissolved nutrients. Azolla requires all macro and micronutrients generally essential to plants. Optimum levels of soil P and Ca are necessary to support growth and N fixation. The growth rate of Azolla decides amount of organic-N and organic-C that could be added to soil in a season. Potentially, biomass of azolla cultures increase 2 to 6 fold in a week and fix up to 75 mg N g/dry mass/day. Succulent azolla contains 3 to 5% N by dry weight or 0.2 to 0.3 % N by fresh weight. In other words, on a yearly basis, about 347 t fresh Azolla could be harvested. It amounts to 86.8 **kg** N fixed via symbiosis and added to paddy fields [8]. In fact, Azolla is said to release greater quantities of C and N into soil compared to leguminous green manures.

Blue Green Algae (BGA) such as *Anabaena, Nostoc, Aulosira, Gleotricha* and several others are regularly used as microbial inoculants in rice fields. They fix atmospheric nitrogen to different extents, depending on strains and soil conditions. Field studies in Southern India have revealed that BGA inoculation was equivalent to 20 to 25 kg N/ha. Actually, algalization improves organic-C, total N and P in soil [8]. At present, intensification of paddy culture is the trend in most parts of South India. It involves application of high doses of N that can interfere with BGA activity and reduce N fixation.

KEYWORDS

- **Blue Green Algae**
- **flooding**
- ***Glyricidia spp***
- **microaerophyllic**
- **puddling**
- ***Sesbania rostrata***
- **soil microbes**

REFERENCES

1. Krishna, K. R. Soil Potassium and its Transformation. In: Soil Fertility and Crop Production. Krishna, K. R. (Ed.) Science Publishers Inc.: Enfield, New Hampshire, USA, **2002b,** 456.
2. Krishna, K. R. The Intensive Rice Culture in south and South-east Asia: Nutrients in Rice land Ecosystem. In: Agrosphere: Nutrient Dynamics, Ecology and Productivity. Science Publishers Inc.: Enfield, New Hampshire, USA, **2003,** 105–140.

3. Ponnemperuma, F. N. The Chemistry of Submerged soil. Advances in Agronomy **1972,** *24,* 29–96.
4. Ponnemperuma, F. N. Chemical kinetics of wetland rice soils relative to soil fertility. In: Wetland soils, Characterization, Classification and Utilization. International Rice Research Institute, Manila, Philippines, **1985,** 71–89.
5. Rosen, C.; Krishna, K. R. Nitrogen in Soil: Transformations and Influence on Crop Productivity. In: Soil Fertility and Crop Production. Krishna, K. R. (Ed.) Science Publishers Inc.: Enfield, New Hampshire, USA, **2002,** 92–108.
6. Sharma, A. R. Fertilizer use in Rice and Rice-based cropping system. Fertilizer News **1995,** *40,* 29–41.
7. IPNI Managing Nitrogen to protect water. Enviro-Briefs. International Plant Nutrition Institute, Norcross, Georgia, USA, **2007a,** 1–2.
8. CRRI, Soil Science and Microbiology Report. Central Rice Research Institute, Cuttack, India, http://crri.nic.in/soilscience.htm **2006,** 1–13.
9. Krishna, K. R. Soil Phosphorus, its transformation and their relevance to Crop Productivity. In: Soil Fertility and Crop production. Krishna, K. R. (Ed.). Science Publishers Inc.: Enfield, New Hampshire, USA. **2002a,** 109–142.
10. Dobermann, A.; Cassman, K. G.; Sta-Cruz,; Adviento, M. A. A.; Panipolon, M. F. Fertilizer inputs, nutrient balance and soil nutrient supplying power in intensive, irrigated rice system. 3. Phosphorous. Nutrient Cycling in Agroecosystems **1996,** *46,* 41–125.
11. Medhi, B. N.; DeDataa, S. K. Phosphorus and availability t irrigated lowland rice as affected by source, application level and green manure. Nutrient Cycling in Agroecosystem **1997,** *40,* 195–203.
12. Singh, K Dhaliwal, R. S. Kinetics and Phosphorus during Composting of Rice straw with Rock Phosphate and Industrial effluents. 18 World Congress of Soil Science, Philadelphia, USA, http://www.cops.confex.com/crops/wc2006 /techprogram/P11927.htm **2006,** 1.
13. Kemmler, G. Potassium deficiency in Soils of the Tropics as a constraint to Food Production. In: Soil Related constraints to Food production. International Rice Research Institute, Manila, Philippines, **1980,** 253–275.
14. Singh, M. Evaluation of current Micronutrient stocks in different Agro-Ecological zones of India for Sustainable Crop Production. Fertilizer News **2001a,** *46,* 25–42.
15. IPNI. Sulphur: An Introduction. International Plant Nutrition Institute. Norcross, Georgia, USA, **2007b,** 1–20.
16. Singh, M. Current status of Micro and Secondary nutrient and crop responses in different Agroecological regions. Fertilizer News **1999,** *44,* 65–82.
17. Singh, M. Importance of Sulphur in Balanced Fertilizer Use in India. Fertilizer News **2001b,** *46,* 13–35.
18. Sahrawat, K. L.; Bhattacharya, T.; Wani, S.; Chandhran,; Ray, Ray, S. K.; Pal, D. K.; Padmaja, K. Long-term Lowland Rice and Arable cropping effects on Carbon and Nitrogen status of some Semi-arid Tropical soils. Current Science **2005,** *89,* 2159–2163.
19. Sahrawat, K. L. Organic Matter Accumulation in Submerged soils. Advances in Agronomy **2004,** *81,* 169–201.
20. Sahrawat, K. L. Fertility and Organic matter in Submerged Rice soils. Current Science **2005,** *88,* 735–739.
21. Purakayastha, T. J.; Ananda Swarup and Singh, D. Strategies to manage Soil Organic Matter for carbon sequestration-Indian Perspective. Indian Journal of Fertilizers **2008,** *4,* 11–22.

22. Velayutham, K.; Alwar, A. A.; Veerabhadhran, V.; Manoharan, S.; Balasubramaniam, R. Effect of Organic wastes on growth and yield of rice (*Oryza sativa*). Indian Journal of Agronomy **1996,** *41,* 584–585.
23. Ramesh, K.; Chandrashekaran, B. Soil Organic Carbon build-up and Dynamics in Rice Cropping Systems. Journal of Agronomy and Crop Sciences **2004,** *190,* 21–23 New Delhi, *1*, 276.
24. Raju, R. A.; Gangwar, B. Utilization of Potassium-rich Green Leaf Manures for Rice (*Oryza sativa*) nursery and effect on Crop Productivity. Indian Journal of Agronomy **2004,** *49,* 244–247.

EXERCISE

1. Mention the various physicochemical changes that occur in flooded Paddy fields.
2. Discuss the influence of submergence on Phosphorus dynamics in Paddy field.
3. List the nutrients whose availability to Rice roots gets enhanced due to flooding.
4. What are aerobic microsites in flooded soils? Discuss the N dynamics in Aerobic microsites and Submerged anaerobic regions in a Paddy field.

FURTHER READING

1. DeDatta, S. K.; Biswas, T. K.; Charoenchampratcheep, C. Phosphorus requirements and Management for Low land Rice. In: Phosphorus requirement for Sustainable Agriculture in Asia and Oceania. International Rice Research Institute, Manila, Philippines, **1990,** 307–313
2. Iyamuremaya, F.; Dick, R. B. Organic amendments and Phosphorus sorption by Soils. *Advances in Agronomy*, **1996,** *56,* 139–185
3. Krishna, K. R. Agrosphere: Nutrient Dynamics, Ecology and Productivity. Science Publishers Inc. Enfield, New Hampshire, USA **2003,** pp. 343
4. Sass, R. L. CH_4 emissions from Rice Agriculture. In Good Practice Guidance and Uncertainty Management in National Green House Gas Inventories. http://www.ipcc-nggip.iges.or.jp/public/gp/bgp/4_7_CH4_Rice_Agriculture.pdf, **2012.**
5. Savant, N. K.; DeDatta, S. K. Nitrogen Transformation in Wetland Soil. *Advances in Agronomy*, **1982**, *35*, 241–300.

USEFUL WEBSITES

http://www.irri.org (October 14, 2012)
http://www.macaulay.ac.uk/MERES/ (October 14, 2012)

CHAPTER 28

NUTRIENT RECOVERY BY CROPS

CONTENTS

28.1 INTRODUCTION

Agroecosystems are endowed with large quantity of mineral nutrients and organic matter that are held in bed rock, soil phase, crop phase, atmosphere, water bodies and aquifers. Nutrients are translocated/retranslocated from each aspect of agroecosystem to others. For example, nutrients are absorbed/recovered from substrate (soil) to crop phase. Similarly, nutrients are transferred from crop to soil through recycling of residues, then from soil to atmosphere via emission and from soil surface to ground water through percolation. Nutrients are also shifted away from an agroecosystem through harvest and transport to long distances from original fields. In the present context, our focus is on the net nutrients recovered from soil to crop phase and its accumulation into different plant parts. The consequences of nutrient recovery from soil on crop growth, forage and grain formation is also important. Nature abounds with variations in terms of crops/mixtures that dominate an agroecosystem, rates of nutrients recovered and total quantity of nutrients held in the crop phase at different stages. Several factors related mostly to geographic location, topography, agroclimate, soils, precipitation and water resources, crop species, nutrient supply rates, and yield goals affect nutrient absorption/recovery rates in a given agroecosystem. Let us discuss a few agroecosystems with regard to above aspects. Geographically, there are regions on earth that support intensive, high input, high yielding agroecosystems. For example, Corn Belt in USA, Wheat cropping zones in Western European plains, Wheat production zones in Northeast China, Rice in Cauvery delta zone of South India, Rice belts in Far-eastern regions etc. High productivity automatically means proportionately massive absorption of soil nutrients into above-ground portions and grains/tubers/fruits. The nutrient absorption rates per time and per unit area are high. Plant tissues accumulate relatively greater amounts of nutrients. To quote a few examples, Wheat grown in Northern Great Plains of United States of America is said to recover at least 81–85 kg N/ha to yield 3.5–4.2 t grains/ha [1]. Over a large expanse of >100,000 ha in Western Europe, a wheat crop that yields 7.5 t grain/ha removes 190 kg N/ha. About 140 kg N/ha could be traced in the grain and the rest 50 kg N/ha in stover that can be either recycled or used to feed animals [2]. In the Australian wheat belt, N recovery rate at different stages of the crop on Zadock's scale has been documented. Usually, about 60–80 kg N/ha is recovered from soil phase. A crop that produces 2.2 t/ha partitions 25–40 kg N/ha into grains. In Argentinean Pampas, a wheat crop that forms 2.27 t grain/ha recovers 45–50 kg/ha. Wheat crop may actually absorb 14–20 kg N/ha to form 1.0 t grain plus forage depending on the harvest index. The exact amount of N recovered, of course depends on the agroecosystem, its soil fertility status, crop genotype and yield goals. Nutrient recovery by a crop may vary depending on general agroclimate, season and crop rotations followed. Whatever is the geographic location, soil fertility status, water supply level, agronomic procedures, or even crop genotype and yield goals set, we ought to realize that physiological genetics of a crop species stipulates quantity of nutrients to be removed to produce a unit, say one ton of grains and foliage.

There are different ways to enhance nutrient recovery from soil phase to above-ground of the ecosystem. A monocrop of cereal like rice or wheat or any other shallow rooted species will be able to scavenge soil nutrient only to a certain depth. The nutrient depletion will be restricted to the root zone. Root system may not reach beyond shallow depths of 1–2 feet. A monocrop of maize with shallow roots may scavenge nutrients from 1.0 m depth. In comparison, a mixture of two or even three crops with different rooting pattern and ability to reach both shallow and deeper layers of soil profile will recover more nutrients. For example, a deep-rooted cotton removes soil moisture and nutrients from depths up to 2–3 m and intercropped maize scavenges nutrients from surface till 1 m depth. Yet another, technique is to stagger and plant intercrops at different dates. Nutrient inputs could be matched with pattern of nutrient recovery by intercrops. Relay cropping helps in removing nutrient from soil more efficiently without leaving large pools of residual nutrients in the soil profile. There are agronomic procedures that allow planting of catch crops or trap crops. Planting a catch crop during off-season or in the inter row space is common. For example, citrus groves allow annuals or short season catch crops to grow and extract soil nutrients not used by the main crop. Such catch crops store the nutrients that otherwise could be lost to ground water. Catch crops (residues) are often recycled to release nutrients at a later date. In certain agroecosystems, nutrients supplied using highly soluble nutrient formulations may rapidly traverse down the soil profile and reach ground water. To avoid it, intercrops with ability to put forth profuse root system that easily traps and absorbs the percolating nutrients are planted.

In nature, nutrient absorption is actually enhanced by factors like soil moisture, warm temperatures, well spread out root system etc. For example, in the Great Plains of North America, nutrient uptake by wheat crop and productivity are generally greater in a wet year with congenial precipitation pattern, compared with low rainfall years [1, 3]. Nutrient recovery by crops could also be enhanced by selecting appropriate genotypes and maturity groups. For example, in case of maize, some of the very early maturing genotypes absorb only 35 kg N/ha and yield proportionately low. However, long duration high yielding genotypes are known to extract as much as 110–120 kg N/ha to 5.5–7.0 t grains/ha. Here, nutrient versus soil moisture interaction is important [4]. Most importantly, crops absorb maximum amount of nutrients during vegetative phase lasting from seedling to tillering/branching then on till seed-fill stage. The period of maximum nutrient absorption may vary depending on crop species and its genotype. We have to use this short window of period and the appropriate physiological stages to maximize nutrient recovery by crop genotypes. The length of physiological stages that support high nutrient absorption rates could be enhanced. If not, rates of nutrient recovery per unit time need to be improved, through genetic selection. It obviously has consequences on nutrient depletion rates from soil. Also, it affects absorption, translocation and accumulation pattern of nutrients in the crop phase.

28.2 NUTRIENTS RECOVERED BY DIFFERENT CROPS

Rice-Lowland, Irrigated, High Input Regions: Examples-North-east China and Fareast; East Coast of India; Cauvery Belt in Tamil Nadu, India
Nutrient Supply: 180–240 kg N; 80–120 kg P; 180–200 kg K/ha
Nitrogen Recovered: 140–190 kg N/ha; @ 19–21 kg N/t grain produced
Total Grain Yield 7–9 t grain and 15–17 t forage/ha

Low land, Rain fed, Moderate Input Regions: Examples Indo-Gangetic belt; South-east Asia: West Africa;
Nutrient Supply: 120–160 kg N; 80 kg P and 120 kg K/ha
Nitrogen Recovered: 85–130 kg N/ha; @ 17–21 kg N/t grain
Grain Yield 4–6 t grain plus 12 t forage/ha
Low Input, Subsistence Systems and Upland Paddy regions:
Nutrient Supply: 60 kg N; 30 kg P; 45 kg K
Nitrogen Recovered 55–70 kg N/ha; @ 20–22 kg N/ha
Grain Yield 2.5–3.0 t/ha

Sources: [4–7].
Note: Average nutrient absorbed by rice ranges from 14.5 to 22 kg N/ha depending on soil fertility regime and genotype.

Nutrient demand by crops may be dependent on accumulation pattern. More precisely the harvest index of a crop genotype needs attention. The total nutrients partitioned into grains and forage is governed partly by harvest index. Rice grown on fertile South-east Asian soils absorbs relatively larger quantity of nutrients. Tropical climate actually induces better biomass formation. Generally, grains accumulate nutrients in relatively high concentrations. The ratio of nutrients recovered into straw and grains depends on several factors, among them; harvest index is a deciding factor. Following is summarized data about nutrient acquisition and accumulation by rice crop, for example:

	N	P_2O_5	K_2O	MgO	CaO	S	Fe	Mn	Zn	Cu	B	Si	Cl
	kg/t[1] grain			kg/t[1] grain			g t/grain						
Straw	7.6	1.1	28.4	2.3	3.80	9.34	150	310	20	2	16	42	5.5
Grain	14.6	6.0	3.2	1.7	0.14	0.60	200	60	20	25	16	10	4
Total	22.2	7.1	32	4.0	3.9	0.94	350	370	40	27	32	52	10

Source: [4–6, 8–10].

Wheat-Reports from several studies suggest that on an average a wheat crop recovers 18–21 kg N/ha to form a t grains. It also absorbs 8–10 kg P_2O_5 and 22 kg K. About 40% of the nutrients absorbed is traced in grains. Nutrient partitioning into grains depends on harvest index of the genotype.

Maize-Maize crop recovers relatively higher quantities of N and K. Reports based on several studies in the South Indian plains suggests that a crop that produces 6.27 t grains removes 168 kg N, 57 kg P_2O_5, 130 kg K_2O and 30 kg Zn. It amounts to 20.7 kg N, 9 kg P_2O_5, 20.3 kg K_2O and 5 kg Zn/ha to produce 1.0 t grains. In Sub Sahara, average productivity of maize is low at 0.8 to 1.2 t grain/ha. It means about 20 kg N, 3 kg P and 18–20 kg K is absorbed/ha from the sandy soils. In Mexican highlands, where about 3 t grain/ha are harvested, proportionately higher quantities of nutrients (60 kg N, 7–8 kg P and 55–60 kg K/ha) are recovered per season. Let us consider nutrient recovery in an intensively cropped zone. On a wider scale, maize productivity in US Corn Belt has increased from a mere 1.25 t/ha in 1900 to 11.5 t/ha in 2007 [11, 12, 31]. This maize belt has graduated from being a subsistence farming zone to one of the highly intensely cropped areas of the world. We know that maize absorbs 18–21 kg N, 2.5–3.0 kg P and 18–20 kg K/ha to produce a ton grain. In 100 years, 'US Corn Belt' has been intensified enormously, almost 10–11 folds to be accurate. The nutrient recovery by a corn crop in US Mid-West has increased from 25 kg N, 4 kg P and 24 kg K/ha in 1900 to 235 kg N, 50–60 kg P and 240 kg K/ha due to higher productivity [11].

Sorghum-In the semiarid tropics of India, a crop of hybrid sorghum CSH1 that produces 4.4 t/ha removes 98 kg N, 35 kg P_2O_5, 117 kg K_2O, 44 kg Ca, 28 kg mg, 705 g Fe, 447 g Mn, 132 g Zn, 37 g Cu [4, 13]. Nutrient absorption is greatest during seedling stage that begins at 17–21 days after planting and lasts for 75–90 days until end of seed filling stage. Nutrient recovery is reduced at crop maturity and senescence.

Finger millet-Physiologically, nutrient requirement of finger millet crop depends on total productivity, including both stover and grains. The harvest index may alter the nutrient extraction and accumulation to a certain extent. On an average, a crop that produces 1.0 t grains recovers 42 kg N, 5.1 kg P and 46 kg K. There are reports that in dry lands, finger millet grown under subsistence conditions without any extraneous source of nutrients, either through FYM or inorganic fertilizers, naturally yields low, say 0.6 to 0.8 t/ha. Such a low input crop may still recover 36 kg N, 3.8 kg P and 38 kg K to produce 1.0 t grain/ha. Incessant nutrient depletion without matching refurbishments will lead to soil nutrient mining and deterioration.

Chickpea: Chickpea crop that yields 1.5 t grains recovers 91 kg N, 6 kg P, 49 kg K and 13 kg S [14–16]. A few other reports indicate that chickpea crop may recover 60 to 140 kg N/ha, 5 to 10 kg P/ha and 60 to 100 kg K/ha depending on productivity that may range from 1.0 to 1.8 t grains/ha. Chickpea crop grown under rain fed condition, that yields 1.0 t grain/ha, recovers 46.3 kg N, 8.4 kg P_2O_5 and 49.6 kg K. Chickpea removes 8–9 kg S and 7.5 kg Mg per ton of grain produced [17–19]. Regarding micronutrients, crop that yields 1.5 t/ha removes 96 mg Zn, 1302 mg Fe, 105 mg Mn, 25 mg B [20].

Cowpea: According to Shrothriya and Phillips [21], cowpea grown on Alfisols and Vertisols extract on an average 260 g Zn, 89 g Mn, 53 g B, 17 g Zn, 11 g Cu and 1.3 g Mo to produce 1.0 t grains.

Pigeonpea: A compilation by Masood Ali [22] suggests that depending on soil fertility status, a pigeonpea crop that yields 2 t grains/ha could remove 93.5 kg N, 25.4 kg P_2O_5 and 65.6 kg K_2O. A few other studies indicate that on an average 85 kg N, 18 kg P_2O_5 and 75 kg K_2O, 1440 g Fe, 128 g Mn 38 g Zn and 31 g Cu is absorbed from soil to produce 1.2 t grain/ha.

Black gram: Black gram crop that produces 600 kg grain/ha absorbs 45 kg N, 8 kg P, 18 kg K and 5 kg S [16]. In other words, to produce 1.0 t grain, black gram crop has to remove 63 kg N, 11.5 kg P, 25.2 kg K and 7 kg S. Trials on Inceptisols have shown that P and S requirement to produce 1.0 t grain (plus 3–3.5 t haulms) ranges from 4.7–6.4 kg P and 6.1–6.5 kg S [23]. According to Masood Ali et al. [15], most pulse crops including black gram removes 30 to 50 kg N, 3–7 kg P, 12–30 kg K, 1–5 kg Ca, 1–5 kg Mg, 1–3 kg S, 200–500 mg Mn, 5 g B, 1 g Cu, and 0.5 g Mo to produce 1.0 t biomass (not grains). Masood Ali and Mishra [24] state that black gram that yields 0.9 t grains/ha recovers 70 kg N, 5.6 kg P, 50 kg K and 5.1 kg S.

Green gram: According to IPNI [16], green gram crop absorbs 92 kg N, 15 kg P, 100 kg K and 8 kg S to produce 1.0 t grains. Masood Ali and Misra [24] stated that mung bean grown on Alfisols capable of procuring 1.0 t grain/ha removes 106 kg N, 21 kg P, 90.4 kg K and 6.5 kg S. Whereas, those grown on sandy, coarse textured soils in Coastal Andhra Pradesh may recover 82 kg N, 12.5 kg P, 89 kg K and 6.0 kg S to produce 0.8 t grain/ha. Regarding micronutrients, Shrothriya and Phillips [21] state that, mung beans grown in South India extracts about 170 g Fe, 18 g Mn, 32 g B 13 g Zn, 11 g Cu and 1.5 g Mo to produce 1.0 t grain.

Sunflower: Sunflower crop that produces 2.3 t grains/ha removes 114 kg N, 26 kg P, 141 kg K and 17 kg S [16]. In the semiarid tracts of India, sunflower extracts 63.3 kg N, 19.1 kg P, 72 kg K, 11 kg S, 27 kg Mg, 47 g Zn, 1075 g Fe, 177 g Mo and 30 g Cu.

Groundnut: Groundnuts are grown in south and south-eastern regions of USA, with moderate yield goals. A groundnut crop that produces 4.5 t pods/ha recovers 110 kg N, 18 kg P_2O_5, 90 kg K_2O, 12 kg Mg and 9.5 kg S. On an average, 20.5 kg N, 5 kg P_2O_5 and 8 kg K_2O are extracted to produce 1.0 t pods [25]. In the semiarid regions of South Asia, groundnut crop that produces 1.2 t pod/ha extracts 58.5 kg N, 17.3 kg P_2O_5, 23.9 kg K_2O, 7.9 kg S, 20.5 kg Ca, 13.2 g Mg, 109 g Zn and 2284 g Fe [26]. According to Srinivasa Rao et al. [27] groundnuts grown in dry lands under rain fed conditions recovers 58 kg N, 19 kg P_2O_5 and 30 kg K_2O per ton grain formed.

28.3 NUTRIENT REMOVALS MAY LEAD TO MINING OF AGROECOSYSTEMS

Crops remove quiet a large quantity of nutrients from soil, season after season. It results in mining, if proper refurbishment schedules are not in place. Harvest and transport of grains/forage result in removal/mining of nutrients from fields. Soil fertility depletion due to recovery of nutrients by crops may become a major problem. Soil fertility loss becomes conspicuous in regions that are cropped incessantly. However, nutrient mining is not a major problem in all the farming zones. For example in many of the intensive farming zones, nutrient depletion is not perceived at all because replenishment schedules are commensurate or at times more than that required by the crop. Yet, sizeable nutrients are removed and transported out of the ecosystem. For example, consider nutrient removal by maize grown in the Corn Belt of USA. At the end of each season farmers' fields loose nutrients in grains and forage depending on extent of recycling and that apportioned to farm animals. On an average, maize fields in the Corn Belt of USA, lose 147 kg N, 26 kg P, 35 kg K, 3 kg Ca, 14 kg mg and 12 kg S in grains. An addition, forage removes 78 kg N, 15 kg P, 179 kg K, 39 kg Ca, 38 kg Mg and 18 kg S/ha/season [28].

'Soil Nutrient Mining' could be explained as removal of soil minerals, so much that it results in loss of soil fertility in general. It happens if quantity of soil nutrients removed is beyond that recycled or replenished. In many areas, fertilizer and/or organic manure replenishment is either nil or meager. Soil mining is felt in cropping zones of Asia, Sub-Sahara, Brazilian Cerrados and Pampas of Argentina. Cereal belts in Sub-Sahara is supposedly most prone to nutrient mining since farmers practice subsistence farming without much nutrient replenishment [10]. Vander Pol and Traore

[29] have pointed out that nutrient balance is negative for most part of cereal fields in Mali, West Africa. In tropical West Africa, nutrient removal by maize is consistently higher than amount of fertilizers supplied to fields. Maize is said to mine 25 kg N, 2.5 kg P, 20 kg K and 5 kg Mg per each ton of grains formed. The annual loss of major nutrient in Sub-Sahara, mostly due to removal into forage/grain is said to be 22.5 kg N, 2.5 kg P and 15 kg K/ha. Such a depletion of soil nutrients is attributed to removal by crops, surface runoff, erosion, leaching and volatilization. Crop productivity decreases as a result of rampant mining of nutrients. Nutrient mining is major problem in many other regions of Africa. Reports from International Maize and Wheat Centre suggest that wheat crop in East Africa suffers due to soil nutrient mining [30]. Small farms are prone to nutrient mining more than large commercial farms, where fertilizer schedules are well structured. In India, nutrient mining trends have been studied. Currently soil maladies connected with nutrient mining seems corrected in most areas. Nitrogen was the first nutrient to get depleted, rather mined out rapidly. Soil-N was mined in greater quantities. Deficiencies of P and K appeared next, followed by secondary and micronutrients. These deficiencies were attributed to rampant mining without appropriate replenishment schedules. Trends in soil nutrient mining and appearance of deficiencies were similar in many agrarian regions. Let us consider cereal crop production trends in South India. The arable cereal fields showed negative balance for N (–8.9 kg N/ha), and K 91.2 kg K/ha). Generally, soil-P mining is nil or marginal for P, because of P-buffering capacity of soil. The nutrient depletion by crops and loss due to natural factors is pronounced during rapid growth of crop. To thwart such depletion, that may also create nutrient imbalance, farmers are generally suggested to add fertilizers commensurate with depletion trends. We should avoid incessant cropping without fallows and nutrient replenishments. Soil nutrient mining may also lead to inappropriate nutrient ratios. It affects crop productivity.

KEYWORDS

- **black gram**
- **chickpea**
- **cowpea**
- **green gram**
- **groundnut**
- **pigeonpea**
- **sunflower**

REFERENCES

1. Rasmussen, E.; Rohde, C. N. Tillage, Soil depth and Precipitation on Wheat response to Nitrogen Agronomy. Agronomy Journal **1991,** 121–124.
2. Jarvis, S. C.; Stockdale, E. A.; Sheperd, M. A.; Powlson, D. S. Nitrogen mineralization in temperate Agricultural soils. Process and Measurement. Advances in Agronomy **1996,** *57,* 187–239.

3. Peterson, G. A.; Schlegal, A. J.; Tanaka, and Jones, C. R. Precipitation use-efficiency as affected by cropping and tillage systems. Journal of Production Agriculture **1996,** *9,* 180–186.
4. Krishna, K. R. Agroecosystems of South India: Nutrient Dynamics, Ecology and Productivity. BrownWalker Press Inc.: Boca Raton, Florida, USA, **2010,** 456.
5. De Datta, S. K. Advances in Soil Fertility Research and Nitrogen fertilizer management for Lowland rice. International Rice Research Institute, Manila, Philippines, **1987,** 236.
6. Krishna, K. R. Soil Fertility and Crop Production. Science Publishers Inc.; Enfield, New Hampshire, USA, **2002,** 453.
7. Sharma, S. K.; Gangwar, K. S.; Pandey, D. K.; Tomar, O. K. Increasing Productivity of Rice-based systems for Rainfed Upland and Irrigated areas of India. Indian Journal of Fertilizers **2006,** *2,* 29–40.
8. DeDatta, S. K.; Biswas, T. K.; Charoenchamratcheep, C. Phosphorus requirements and sustainable management in Asia and Oceania. International Rice Research Institute, Manila, Philippines, **1990,** 307–323.
9. DeDatta, S. K.; Broadbent, F. E. Development changes related to Nitrogen use efficiency in Rice. Field Crops Research **1993,** *34,* 47–56.
10. Krishna, K. R. Maize Agroecosystem: Nutrient Dynamics and Productivity. Apple Academic Press Inc.: New Jersey, **2013,** 341.
11. Abendroth, L Elmore, R. Demand for more Corn following Corn. http:// www.agronext.iastate.edu/corn/production/ management/cropping/demand.html. **2007,** 1–2.
12. Sawyer, J. Nitrogen Fertilization for Corn following Corn. http://www.ipm.iastate.edu/ipm/icm/2007/2-12/nitrogen.html **2007,** 1–4.
13. ICRISAT A review of Fertilizer Use Research on Sorghum in India. International Crops Research Institute for the Semiarid Tropics, Patancheru, India. Research Bulletin No 8, **1984,** 1–59.
14. Masood Ali, Ganeshmurthy, A. N.; Srinivasa Rao, Ch. Role of Plant Nutrient Management in Pulse Production. Fertilizer News **2002,** *47,* 83–90.
15. Masood Ali, Singh, A. A.; Saad, A. A. Balanced Fertilization for Nutritional quality in Pulses. Fertilizer **2004,** *49,* 43–56.
16. IPNI. Sulphur: An introduction. International Plant Nutrition Institute. Norcross, Georgia, USA, **2007,** 1–20.
17. Hegde, D. M.; Sudhakar Babu, S. N. Nutrient Management Strategies in Agriculture; A future Outlook. Fertilizer News **2001a,***46,* 61–72.
18. Hegde, D. M Sudhakar Babu, S. N. Nutrient mining in Agro-climatic zones of Karnataka. Fertilizer News **2001b,** *46,* 55–72.
19. Shrothriya, G. C. Promoting Potash Application for Balanced Fertilizer Use. Indian Journal of Fertilizers **2007,** *1,* 31–43.
20. Prasad, B. Conjoint use of Fertilizers with Organics, Crop Residue, Green manures for the Efficient use in Sustainable Crop Production. Fertilizer News **1999,** *44,* 67–73.
21. Shrothriya, G. C.; Phillips, M. Boron in Indian Agriculture. Retrospect and Prospect. Fertilizer News **2002,** 95–102.
22. Masood Ali, A. Tropical Pulse Crops. World Fertilizer Use Manual. International Fertilizer Association, IFA, Paris, http://www.fertilizer.org/ifa/publicat/html/pubman/tpulse.htm. **2004,** 1–4.
23. Singh, Y. Role of Sulphur and Phosphorus in Black gram Production. Fertilizer News **2004,** *49,* 3–36.
24. Masood Ali and Mishra, J. Nutrient Management in Pulses and Pulse-based Cropping systems. Fertilizer News **2000,** *45,* 57–69.

25. PPI Nutrient uptake and harvest removal for Southern crops.Potash and Phosphate Institute. Norcross, Georgia, USA, **2004,** 1–3.
26. Ghosh, K. Growth, yield, competition and economics of groundnut/ceral fodder intercropping system in semiarid tropics of India. Field Crops Research **2004,** *88,* 227–237.
27. Srinivasa Rao, C.; Prasad, J. V. N. S.; Vital, K. P. R.; Venkateswarulu, B.; Sharma, L. Role of optimum plant nutrition in Drought management in rain fed Agriculture. Fertilizer News **2003,** *48,* 105–114.
28. Patzek, T. W. Thermodynamics of Agricultural sustainability: The case of US maizeAgriculture. http://petroleum.berkeley.edu/papers/Biofuels/816patzek4–8-08.pdf **2008,** 1–50.
29. Van Der Pol, F.; Traore, B. Soil nutrient depletion by agricultural production in Southern Mali. Nutrient Cycling in Agroecosystem **1993,** *36,* 1385–1394.
30. Nkonya, C.; Kaizzi, C.; Pender, J. Determinants of Nutrient Balances in a Maize farming system in Eastern Uganda. Agricultural Systems **2004,** *85,* 155–182.
31. Sawyer, J. E.; Mallarino, A.; Killorn, R.; Barnhart, S. K. A general guide for Crop Nutrient and Limestone Recommendations in Iowa. Iowa State University Extension Bulletin **2008,** 1–18.

EXERCISE

1. Describe nutrient recovery rates and patterns for five major crops.
2. What is Soil mining? How is it avoided in Practical agriculture?
3. Mention nutrient ratios found in different cereal crops.
4. Mention soil factors that influence nutrient recovery rates by crops.

FURTHER READING

1. Krishna, K. R. Peanut Agroecosystem. Nutrient Dynamics and Productivity. Alpha Science International Inc. Oxford, England, **2008,** pp. 292.
2. Krishna, K. R. Maize Agroecosystem. Nutrient Dynamics and Productivity. Apple Academic Press Inc. New Jersey, USA, **2013,** pp. 341.

USEFUL WEBSITES

http://www.agronomy.org
http://www.soilscience.org

CHAPTER 29

NUTRIENT LOSS FROM AGROECOSYSTEMS

CONTENTS

Soil and nutrient loss from agroecosystem is a phenomenon traced in all regions of the world. It may affect farm productivity detrimentally, both immediately with short-term effects and for long periods. This natural phenomenon needs careful consideration, while devising soil conservation practices, fertilizer schedules and cropping systems. Nutrient loss is incessant and intense in some parts of the world. In such areas, nutrient loss and soil deterioration has to be thwarted by adopting a range of agronomic procedures. In other words, integrated approaches are needed to restrict loss of soil fertility. Nutrient loss is negligible or small if fertilizer schedules are split and carefully matched with crop's demand. Nutrients that accumulate in soil become vulnerable to natural processes that hasten their loss from agroecosystem. Soil nutrients are also lost through leaching, percolation, erosion and gaseous emissions. It is marked in case of soil-N and C. There are several agronomic procedures that halt loss of soil nutrients. For example, crop residue recycling *in situ* and adoption of soil conservation practices are essential to thwart loss of nutrients.

29.1 LOSS OF SOIL AND NUTRIENTS DUE TO EROSION AND LEACHING

Soil erosion is defined as a process that causes dislodgement and transport of soil particles, mainly by two agents wind and water. Wind erosion involves gradual removal but water mediated erosion is rapid. The erosive power of wind increases exponentially with velocity but, unlike water, it is not affected by gravity. Wind velocities less than 12–19 km/hr at 1.0 m above soil surface do not dislodge and cause movement in soil particles [1]. Leaching refers to loss of nutrients in dissolved state from soil profile to locations away via surface flow, percolating water, seepage horizontally to channels. Emissions involve conversion of plant nutrients into gaseous phase and its loss via diffusion into atmosphere. Emission of different forms of N is most common in agricultural fields. Loss of soil carbon to atmosphere via emissions as CO_2, CH_4 and CO is also major cause of concern in agricultural belts.

In the 'US Corn Belt' and other parts of Great Plains, nutrient loss via natural processes is severe. For example, average soil erosion during past 2–3 decades has fluctuated between 12.0–16.9 t/ha/yr. Loss of surface soil depreciates fertility perceptibly, depending on nutrient carrying capacity of soil particles [2]. There are innumerable reports about nutrient loss from crop fields due to natural factors like leaching to lower horizons, rapid percolation to vadose zone and seepage laterally into adjacent fields or irrigation channels. Nutrient loss due to above factors varies depending on geographic location, topography, soil type, cropping history and current crop standing in the field, season and general weather pattern. The extent of nutrient lost is often proportionate to that supplied. Excessive application makes fertilizers more vulnerable to loss via leaching. Therefore, fertilizer schedules are shrewdly tailored so that nutrients supplied into soil profile is scavenged and exhausted rapidly by crop roots. For example, one of the comments by Westfall et al. [3] indicates that traditionally standardized procedures in Wheat agroecosystem of Great Plains suggest addition of N, based on exact requirements at different stages of the crop. Yet, it is said application of only 40 kg N/ha at mid-stage of the crop, still allows perceptible quantities of fertilizer-N to be leached. In fact, ^{15}N balance studies suggest that in certain locations, about 10–90% of total N lost from field or an area is attributable to leaching and percolation away from root zone. Reports indicate that in the Central Plains there are instances wherein 66–88 kg N is lost via leaching and percolation. Leached N may often reach vadose zone and then on to ground water. Kucharik and Brye [4] state that in United States Midwest region, farmers have to enhance maize grain yield using higher amounts of NO_3 fertilizer, but they have to control leaching loss, so that ground water contamination is avoided. It has been observed that for an increase of fertilizer-N by 30% to 234 kg N/ha, NO_3 leaching increased by 56% of normal levels. N concentration in leachate could increase to 30 μg NO_3/L. The permissible limit is only 10 μg NO_3/L.

Nitrogen leaching is substantially higher in peanut growing regions of Georgia and Florida due to higher total precipitation. During La Nina or neutral crop seasons preceding the El Nino phase, especially bare fallow periods show highest leaching [5]. High N leaching is a possibility during neutral fallow periods following El Nino growing seasons. Among ENSO phases, La Nina seasons induce greater N leaching. Simulations indicate that management of planting dates and irrigation schedules to suit ENSO phase can decrease N leaching by 10% [6]. For example, delayed planting during El Nino reduces N leaching. Within cropping zones, nitrate movement into groundwater is more likely in sandy soils characterized by greater permeability or if soils are shallow. In the Piedmont region of Georgia, North Carolina and Virginia, red clayey soils are supposedly prone to NO_3 leaching. Nitrate retention occurs in some acid subsoils of Georgia. Soils are comparatively less prone to erosion, but careful management can further minimize loss of soil and nutrients. In certain places, leaching of NO_3 has not been a major problem, because a thick relatively impermeable layer restricts down word movement of water and nutrients. However, NO_3 may still percolate away from peanut root zone and accumulate in the shallow ground water. Lateral movement of NO_3 via ground water is also a possibility. Extent of NO_3 lost via percolation may vary depending on soil, fertilizer inputs and rate of absorption by crop roots. Loss of nitrate and a few other nutrients like SO_4 and PO_4, through percolation and later accumulation in lower horizons is a phenomenon that occurs all through the peanut belt. It is more frequent in highly weathered red soils [7].

29.1.1 SOIL EROSION AND NUTRIENT DYNAMICS IN WEST AFRICAN SAHELIAN ZONE

In the Sahelian agricultural zone, control of soil nutrient loss that results due to erosion needs special attention (*see* Plates 1 and 2). Soil erosion, it seems increases due to natural factors and as a function of human population density. For example, in the Sudano-Sahelian region, if population exceeds 20 to 40/km^2, intensification of cropping ensues and fallow period gets shortened, sometimes making it ineffective in terms of amelioration of soil fertility. Thereafter, soil degradation may soon begin. Wind erosion can be a severe problem when soil is loose, dry and wind velocity is beyond threshold. In the sub-Sahara, *harmattan*, a hot dusty wind sets in during October to April. During the year, severe erosion is experienced during May and July. It coincides with early rainy season, when thunderstorms move westward through Sahelian zone. Wind erosion becomes severe during dry season lasting from November to April in Mali and Senegal [8]. Dust storms in West Africa lasts 10 to 30 min at a time, but they can cause severe loss of topsoil and nutrients. Loss of nutrients and resulting low soil fertility seems to be more important than just soil particle loss that wind erosion causes. Analysis of nutrients removed via erosion has revealed that during the two storms that a crop suffers, as much as 76 kg C/ha, 18 kg N/ha, 6 kg P/ha and 57 kg K/ha could be lost. Average enrichment ratio of dust sample for dust storms at ICRISAT's experimental station in Sadore, Niger is shown in Table 29.1. The extent of nutrients lost is dependent on height of wind (Plates 1 and 2). However, during a *harmattan* storm in Sahel, apart from nutrient losses, nutrient rich deposits are also received. Therefore, balance between inputs and outputs due to *harmattan* is to be considered, while computing effects of wind erosion. Soil erosion control methods adopted commonly by Sahelian farmers are soil ridging, application of surface and crop residues, mulches, strip cropping that exposes only reduced area of field to high wind velocities, crop rotation, intercropping, etc. (Table 1).

PLATE 1 Sheet erosion caused due to excessive surface flow immediately after a storm in the Sahelian zone near Dosso, Niger, West Africa.

Note: Surface soil loss is rampant leaving behind large gravel and stones. It makes Pearl millet establishment difficult. Loss of nutrients depends on thickness of surface soil lost from a field. Quite often planted pearl millet/cowpea seedlings are swept away, if the rooting is not deep. It leads to poor crop stand and plant density. Soil fertility becomes uneven due to large-scale movement of suface soil and nutrients contained in the sand particles. Nutrient accumulation in a far away low-lying area depends much on the nutrient carrying capacity of sand particles, particle, size and velocity of storms.

Source: [12, 13].

PLATE 2 Landscape in Sahelian West Africa showing rampant loss of topsoil and erosion leading to formation of gullies and ravines.

Note: Soil erosion literally reduces chances for even a subsistent level yield of pearl millet. Massive loss of soil actually induces its transport to other locations where soil may actually play a detrimental role by submerging young seedlings. The above picture depicts topsoil erosion and gulley formation at a location near Dosso, in Southern Niger, West Africa.

Source: Krishna, K. R., Bangalore, India.

TABLE 1 Sand storm, Nutrient enrichment ratios of Dust particles and Erosion material encountered in South-west Niger during a Rainy season June/July, 1988).

Duration	Wind Speed and Direction	Mass Transport	Storm height	Nutrient Enrichment Ratio			
				C	N	P	K
S	m/S	kg/m	m				
1481	10.3 (SE)	102.7	0.05	1.18	1.33	0.83	0.98
			0.26	1.91	2.39	1.42	1.10
			0.50	2.06	3.02	2.08	1.14

Source: [1, 9].
Note: Mean wind speed was measured at 2 m above soil surface. Nutrient enrichment ratio is defined as Ratio of the total nutrient element content of trapped sediment to the total Nutrient element content of top soil. S = seconds, SE = South-east.

Among various environmental factors that affect nutrient dynamics in peanut cropping zones of West Africa, wind erosion ranks 3rd, after drought and soil fertility [10]. Bielders et al. [10, 11] conducted field trials to measure soil mass fluxes and monitor sand storms for 3–4 months during each year from May/June to August. Both, soil loss and deposition was measured. Results indicate that wind erosion resulted in considerable soil loss in farmer's fields maintained under traditional management. About 27 t/ha soil was lost in first five storms and deposition amounted to 24 t/ha in the fallow, corresponding to a layer of 1.5 to 2 mm of topsoil. Ridging was less effective in thwarting erosion compared to mulching. Reduction in soil erosion was perceptible only for first few storms, if ridging was used. Incorporating residues into ridges constituted effective measure to control erosion for lengthier period than usual. Ridging early in the season and ridging in the mid season is a common practice with farmers. These agronomic methods reduce soil and nutrient loss via wind erosion. As stated earlier, while assessing wind effect on peanut fields, both loss and influx of sediments need to be considered. There may be large seasonal/yr variation in these natural processes. Let us consider an example. In Western Niger, during the first year after clearing vegetation, crop fields received a net sediment deposition of 5.4 t/ha. In the next year, about 5.0 t/ha of sediment was lost through erosion. Beilders et al. [14] argue that gradual disappearance of ground cover resulted in excessive sediment loss during next year. The magnitude of reduction depended on the extent of mulching. At a mulch rate of 2 t/ha, the enrichment ratios of sand particles decreased to 0.9–2.3 for organic matter, 0.9–1.8 for N, 1.5–1.8 for P and 1.5–2.9 for K [15].

A study by Michels et al. [8] at ICRISAT Sahelian Center, Niamey, indicated that soil flux from above the ground was significantly reduced at 2000 kg/ha crop residue application but not at 500 kg residue/ha. Farmers in Sahel use '*Zai*' system of soil rehabilitation. This agronomic procedure aims at reducing soil erosion and nutrient loss. The *Zai* system involves use of small quantities of organic matter, either manures or crop residues that is placed in small pits. Seeds are sown in these pits. Seedlings are benefited from the available nutrients and run-off water collected in the pits. Fatondji et al. [16], therefore, argue that a *zai* system combines both water and nutrient management. On farm trials indicate a 40% increase in crop produce due to *zai* system. Sometimes, as much as 3 t/ha organic residues are used during *zai* system.

29.1.2 SOIL EROSION AND LEACHING IN SEMI-ARID REGIONS OF ASIA

Annually, soil erosion due to wind and heavy precipitation events, may range from 1.95 to 3.2 t/ha in the arable cropping zone of South India [12]. This eroded soil carries away precious nutrients and reduces the depth of top soil. Such loss of soil nutrients may depreciate crop productivity in a field, but elsewhere, down the stream or gullies, eroded soil and nutrients may collect and help cropping. According to Sharda and Rattan Singh [17], all three predominant soil types found in the Peninsular India, namely Alfisols, Vertisols and Red Laterites are prone to soil erosion (Plate 3). Soil leaching is rampant in most locations of semiarid tropics. In addition to loss of surface soil and clay, it induces loss of 50–70% of NO_3-N applied to fields. The extent of soil loss varies from slight (5–10 t/ha), moderate (10–20 t/ha) to severe (20–40 t/ha). The depreciation in grain yield due to soil loss is perceptible. Nutrient depletion is proportionate to soil erosion. Remedial measures such as contour planting and bunds may reduce soil and nutrient loss from cropping zones.

PLATE 3 A Groundnut field in the Semi-arid region of South India, showing soil and nutrient loss from fields.
Note: Soil tillage and land preparation, especially contouring the furrows and ridges is important. In the above case, a large storm has eroded furrows and ridges. It has caused sheet erosion, destroying groundnut seedlings. Soil and nutrients move rapidly with surface runoff and accumulate at a distance in the field. Location is near Kanakpura, Bangalore Rural district in Karnataka State of India.
Source: Krishna, K. R., Bangalore, India.

Soil erosion induced by high volumes of runoff significantly affects nutrient dynamics in the agroecosystem [18]. The soil erosion rates may range from 5–20 t/ha annually. Actually, amount of soil loss may not be high at 7 t/ha, but enrichment ratio seems greater than 2. Therefore, total nutrient loss via sediments reaches 27 kg N/ha and 178 kg C/ha. The nutrient loss also depends on topography, tillage and cropping system. In case of pigeonpea/soybean intercrops, N loss could be 14 kg N/ha in fields with broad-bed and furrow and up to 17 kg N/ha on flat surface fields [19]. Long-term trials on Alfisols indicate that application of straw mulches; FYM or intercrops reduced runoff and soil erosion by 50 to 87% compared to control (Plate 4).

PLATE 4 Mulched Soybean (left) and Maize (Right).
Note: Soybean fields are usually mulched with maize residues obtained from previous season to avoid soil erosion and nutrient loss. Mulching regulates soil temperatures, percolation of water and rooting pattern to a certain extent. Mulching adds to soil carbon pool, since the mulch degrades and releases organic compounds into soil. Mulching affects several other processes like soil microbial flora and their activity.
Source: Alvarez, University of Beunos Aeries; Dr Albert Quirogog and Bono, INTA, Argentina.

Nutrient leaching can be severe in fertile soils. A good portion of nutrients could be lost due to leaching, surface runoff and percolation, if inherent soil-N or residual-N is high. Immobilization of fertilizer-N into soil organic fraction is yet another avenue for loss of N. It depends on C: N ratio of the soil but it is transitory. In due course, N lost due to immobilization becomes available to crop roots as mineralization proceeds. Mineralization of organic fraction is rapid in arable soils found in semiarid regions.

Nitrogen input to crops as urea or other formulations is not efficiently used. Crops usually recover only 35 to 48% of urea-N. Residual-N that accumulates is vulnerable to leaching and volatilization. Loss of N through leaching can be significant, especially when NO_3 fertilizer supply to field is high. Actually, high mobility of NO_3 accentuates leaching loss. Factors such as rainfall, initial moisture status and water management methods affect rate of NO_3 leaching. Autil et al. [20] suggest that frequent, low rates of irrigation avoids excessive $NO_{3-}N$ leaching down the soil profile, and reduces ground water contamination. The down word leaching of NH_4 is comparatively rapid than NO_3, if it is applied as urea. Soil texture influences N loss through down word leaching. In coarse textured soils 19% of urea-N leached below 7 cm within 48 h, but only 11% NH_4N traversed below 7 cm.

On coarse textured soils of South India, 19% of N could be lost below the root zone. The loss of N from soil profile is high when NO_3 containing fertilizers are used. Nitrates are highly mobile in soil profile. A high intensity rainfall may further accentuate movement of NO_3 to lower horizon, thus making it unavailable to roots. Usually, nutrient loss via runoff, leaching and percolation is greatest immediately after a precipitation event. Regulating irrigation is also essential.

Let us consider an example from Australia. Nitrate leaching is rampant in the arable wheat belts of Australia. The NO_3 leaching is induced by high rainfall events, irrigation and seepage of water into canals. For example, a 112 mm precipitation may induce loss of 12 kg N/ha from wheat fields. Nitrate leaching actually becomes conspicuous at the start of winter rains in Queensland. Soil texture and crop phase are important factors to consider fertilizing wheat fields. Nitrate leaching is highly dependent on rate at which fertilizer-N is absorbed into crop phase. Fertilizer-N applied at late stages and not used efficiently are prone more to leaching.

29.2 NUTRIENT LOSS FROM AGROECOSYSTEM DUE TO GASEOUS EMISSIONS

Loss of soil nutrients as gaseous emission has been reported from most of the agrarian regions of the world. Fields with annuals, cash crops, long duration plantations and forest species are all prone to loss of N due to volatilization. The extent of N lost as NH_3 and N_2 may vary depending on several factors related to crop, topography, soils, and environment during the year. Experiments in the citrus groves of Central Florida suggest that potential NH_3 volatilization could account for 10–40% of N lost from grooves. The volatilization of N increased with temperature.

European farmers are used to intensive cropping on heavily fertilized fields. This induces massive loss of soil-N and fertilizer-N through emissions. Reports indicate that in southern France, total loss of soil-N through volatilization and emissions fluctuates between 30–110 kg N/ha per season. About 40% of N loss occurs as NO_2, <1% as NH_3, 14% as N_2O and 46% as N_2 [21]. Experimentation in the Broadbalk fields of England has shown that N loss of de-nitrification reactions accounted for 17% of total n loss. In the Spanish wheat producing regions, ammonia volatilization could account for 20 kg N loss to atmosphere. Nitrous oxide emissions are said to be the major avenue of N loss from arable cropping zones of Europe. Nitrous oxide emissions are proportionate to high fertilizer-N supply into wheat belt. Loss of N as N_2O ranges from 0.5 kg N/ha to 16.5 kg N/ha. Estimations in Germany have shown that for every 60 kg N/ha absorption into wheat crop, about 2 kg N/ha is lost as N_2O into atmosphere [13]. Nitric Oxide emissions are also significant in European cropping zones.

Wood et al. [22] state that dry weather combined with high soil temperatures experienced in the wheat/barley growing zones of West Asia and North Africa, induces rampant loss of N via volatilization. About 9–27% fertilizer-N could be lost as NH^3 from the fields. Loss of N via percolation too could be severe. Hence, these factors reduce fertilizer-N efficiency significantly. Loss of fertilizer-N due to nitrification/de-nitrification reaction in soil is supposedly sizeable in the Northern Chinese peanut belt [23]. Application of higher doses of N increases N_2O emission. The N_2O emission was 0.33 kg N/ha in the absence of urea application, but emission increased to 2.91 kg N/ha if urea was broadcast and 2.5 kg N/ha if urea was point placed. Obviously, appropriate timing and fertilizer disbursement method reduces loss of N from peanut cropping zone.

Major avenues of N loss from pigeonpea ecosystem are NH_3 volatilization, de-nitrification, soil erosion, surface runoff, leaching, percolation beyond root zone and immobilization. Ammonium volatilization is severe in pigeonpea intercropping zones because of high temperatures in the semiarid tropics. Although N supply is meager on pigeonpea, it could still be vulnerable to loss via volatilization. Hence, banding and incorporation of fertilizer-N into soil is recommended. Loss of N via de-nitrification is negligible. It accounts for small amounts up to 8 kg N/ha. It is believed that loss of N due to de-nitrification from arable cropping zones in South India rarely exceeds 28 kg N/ha [24].

Groundnut fields maintained under rain fed conditions in Peninsular India are also prone to N loss through ammonia volatilization. Deeper placement of fertilizer-N reduces volatilization. However, during rabi, both to enhance availability of N and to avoid its loss, fertilizer-N must be placed deep in the root zone [25]. We may note that under irrigated conditions, peanut fields are prone to N loss both, through ammonia volatilization and leaching. Vertical leaching of N moves it to lower layer of soil well below the root-zone. Split application of N is a good option to avoid both excessive leaching and volatilization loss. It improves fertilizer-use efficiency. Use of slow-release N fertilizers and nitrification inhibitors too reduce N loss, so they improve fertilizer-N efficiency.

Soils used for groundnut production may also suffer significant loss of N via N_2O emission. Both biological and abiotic N transformations in soil may lead to N_2O emission. Mostly, denitrification, nitrification, dissimilatory nitrate reduction and chemo-denitrification reactions all lead to N_2O formation. Experiments at Indian Agricultural Research Institute, New Delhi has shown that total emission of N_2O-N during 51 days ranged from 0.21 to 0.5 mg/kg soil. Emission of N_2O increased almost linearly, with enhanced N fertilizer inputs. About 0.2% of applied N fertilizer could be emitted as N_2O-N in the peanut fields [26].

Nitrogen loss from cropping belts could be severe, even if carefully designed crop sequences were adopted. For example, ammonia volatilization is an important mode of N loss from cropping zones. ^{15}N studies on a crop sequence involving legumes and sorghum has shown that out of 100 kg N applied to sorghum crop, about 23% was lost as ammonia. Such a loss of N via volatilization increased, if high temperatures continued during crop season. The extent of volatilization is also high, whenever sorghum crop is supplied with high levels of inorganic N. Inefficient absorption of N by sorghum and other crops in sequence is yet another factor that increases ammonia volatilization. Unused soil N retained is vulnerable to volatilization and /or leaching.

However, incorporating fertilizer-N into soil at a depth and splitting its application into at least 2–3 times could reduce volatilization. Such a precaution minimizes volatilization to less than 5% of the amount applied. In addition to loss of N through volatilization of NH_3, denitrification too induces massive loss of N from agroecosystem. Loss of N via denitrification is accentuated in anaerobic conditions. Loss of N occurs when denitrifying bacteria convert NO_3-N into NO_2 and N_2 that escape into atmosphere. Nitrification inhibitors are used to reduce conversion of fertilizer-N to NO_3 that may accumulate and become vulnerable to leaching and denitrification. Coating urea with 'Nimin' or 'Nitrapyrin' and other neem-based products suppresses nitrification. Sometimes, sulfur coated urea is also used. These are called slow-N release fertilizers. They reduce loss of N and improve fertilizer-N use efficiency. For example, application of Nimin-coated urea reduced loss of fertilizer-N through leaching and denitrification by 30–35%. It resulted in 5–19% higher N uptake and 20% more grain production. It is believed that phenolic components in 'Nimin' inhibit nitrifying and denitrifying bacteria.

Several types of organic manures are applied to fields, often a couple weeks prior to seeding. This is to allow mineralization and rapid availability of nutrients to sorghum roots. However, detailed analysis of decomposition patterns of FYM or other manures is essential to understand the pattern of N loss via ammonia volatilization and the fraction recovered by crop. For example, during 75-day period in sorghum fields at Coimbatore in Tamil Nadu, loss of N via volatilization as NH_3 was greatest in plots applied with poultry manure (50 kg N/ha) compared to FYM (21.2 kg N/ha). Clearly, type of organic manure used during sorghum production affects extent of N loss. A certain amount of N is also lost via conversion of NO_3 to N_20 and N_2. Loss of N via de-nitrification can be significant depending on soil N status, fertilizer-N applied, type of fertilizer and other agronomic procedures adopted during crop production.

29.3 NUTRIENT LOSS FROM SUBMERGED, LOWLAND PADDY FIELDS

Nutrient loss from submerged paddy fields could be significant. It may reduce grain/forage yield. Paddy fields that are submerged provide different physico-chemical conditions in the field. Nitrogen loss from the rice belt can be severe depending on meteorological conditions, soil N status and cropping pattern. Volatilization of N as NH_3, NO_2 and N_2O is actually dependent on atmospheric conditions, rice genotype and its canopy, soil N status, fertilizer-N supply and farming practices. ^{15}N studies indicate that about 3–8% of fertilizer-N applied, as prilled-Urea is lost from rice fields via volatilization. Generally, volatilization of $NH_{3-}N$ peaks at 3 days after Urea application and later decreases until 15th day. Nitrification inhibitors may curtail loss of N to a certain extent [27].

Nitrogen recovery from rice fields is relatively low. It seldom exceeds 30 to 40% of fertilizer N applied. In undisturbed submerged fields, N percolates at a rate of 12–14 cm in 4 weeks. Leaching loss is high immediately after a rainfall event. Rainfall events that result in surface flow, leaching, percolation and loss through drainage water may cause another 10 to 12% of N loss from soil. Since, $NH_{4-}N$ and $NO_{3-}N$ are mobile in soil; their loss is accentuated immediately after a heavy rainfall event. Down ward movement that results in loss of N from root zone affects ground water quality. Nitrate contents may exceed permissible limits. Major suggestions to avoid ground water contamination are to apply N when the crop needs; match N supply rate with need; split N inputs several times in order to avoid undue accumulation; apply in NH_4 forms and adopt a balanced fertilizer input program [28].

Phosphorus fertilizers applied to a crop or a cropping zone is not entirely absorbed and used to produce crop biomass formation. Sizeable quantity of P is chemically fixed into soil lattice structure. It becomes unavailable to crop roots immediately. For example, lateritic soils are prone to high chemical-P fixation ranging from 20–60%. Alfisols too fix high amounts of P. Across different locations, efficiency of fertilizer-P added to soil is only 18–22%. Rest is either held in soil as residual-P or lost via erosion, runoff and seepage through irrigation channels [7, 29]. Phosphorus in soil may be lost in several ways, such as leaching, chemical fixation, surface flow, erosion, etc. Transport of P through subsurface flow is an important mechanism, especially from manured fields. Obviously, while assessing P leaching potential from a peanut field, it is necessary to accumulate sufficient knowledge about soil profile and P translocation rates. Kleinman et al. [30] reported that quantity of soil P lost via subsurface did not alter much based on P fraction. Actually, total amount of P fertilizer and P containing manures added decides the extent of subsurface P flow. Leaching rates of different fractions of P is also dependent on tillage [31].

Knowledge about P saturation thresholds or the 'change point' in soil is useful. It helps in judging extent of P inputs possible and onset of leaching. It helps in avoiding loss of P through leaching. The P saturation threshold or change point depends on soil factors such as texture, chemical nature, especially Al and Fe or P/ [Al + Fe] ratio [32]. Agronomically, optimum P level may often be less than 'change point.' Still, an understanding about change point in soils used for crop production can avoid risk of P loss via leaching.

Potassium is a relatively mobile element in soil. It occurs in different forms such as nonexchangeable-K, exchangeable-K and solution-K. Loss of solution-K could be rampant in many agrarian regions. Large basal dosages of fertilizer-K, induces accumulation of exchangeable and soluble-K in soil. These fractions are vulnerable to loss via surface runoff, seepage and percolation. Splitting fertilizer-K is not common, but it is a useful procedure to thwart loss of soil-K. Splitting actually helps us in matching demand for K with its supply levels. It avoids undue accumulation of K in subsurface soil that could become vulnerable to loss via erosion.

Finally, a general suggestion applicable to most agrarian regions, especially those prone to rampant loss of soil and nutrients, is to include soil conservation practice mandatorily. It is most essential to conserve soil, water and nutrient resources *in situ* in any farm [33]. This suggestion applies even to large expanse of crop.

KEYWORDS

- **erosion**
- **harmattan**
- **leaching**
- **zai system**

REFERENCES

1. Sterk, G.; Stroosnijder, L.; Raats, A. C. Wind Erosion processes and Control techniques in Sahelian zone of Niger. http://www.wu.ksu.edu/symposium/ proceedings /sterk.pdf **1998,** 1–14.
2. Patzek, T. W. Thermodynamics of Agricultural Sustainability: The Case of US Maize Agriculture. http://petroleum. Berkeley. Edu/papers/Biofuels/816patzek4–8-08.pdf **2008,** 1–50.
3. Westfall, D. G.; Havlin, J. L.; Hargest, G. W.; Raun, W. R. Nitrogen management in Dry land Cropping System. Journal of Production Agriculture **1996,** *9,* 192–199.
4. Kucharik, C. J.; Brye, K. R. IBIS Yield and Nitrate Loss Predictions for Maize Agroecosystems receiving varied N-fertilizer. American Geophysical Union, Fall Meeting, Abstract No 31 adsabs.harvard.edu/abs/2001AGUFM. B31B0096K **2001,** July 11, 2012.
5. Mavromatis, T.; Jagtap, S. S.; Jones, J. W. ENSO effects on Peanut yield and Nitrogen leaching. Climatic Research **2002,** *22,* 129–140.
6. Mavromatis, T.; Boote, K. J.; Jones, J. W.; Irmak, A.; Shinde, D.; Hoogenboom, G. Developing genetic coefficients from crop simulation models using data from crop performance trials. Crop Science **2001,** *41,* 40–51.
7. Hodges, S. C. Environmental guidelines for Plant Nutrient use. The University of Georgia Cooperative Extension Services. Athens, Georgia, USA, **1991,** 1–16.
8. Michels, K.; Sivakumar, M. K.; Allison, B. E. Wind erosion control using crop residue. 1. Effects on Soil flux and Soil particles. Field Crops Research **1995,** *40,* 101–110.
9. Shivakumar, M. K. Agroclimatic aspects of rain fed agriculture in the Sudano-Sahelian zone. In: Soil, Crop and Water management in the Sudano-Sahelian zone. International Crops Research Institute for the Semi-Arid Tropics, Patancheru, AP, India **1989,** 17–38.
10. Bielders, C. L.; Alvey, S.; Cronyn, N. Wind Erosion: The Perspective of Grass Roots communities in the Sahel. ICRISAT@org CCER.htm **2002a,** 5 (August 24, 2012).
11. Bielders, C.; Michels, K.; Rajor, J. On-farm evaluation of wind erosion control technologies. ICRISAT@org CCER.htm **2002b,** 5 (August 24, 2012).
12. ICRISAT Annual Report 1985, International Crops Research Institute for the Semi-arid Tropics. Patancheru, A.; India **1986,** 489.
13. Krishna, K. R. The Temperate Wheat Cropping zones of European plains, Australia and Pampas in Argentina: Nutrient Dynamics and Productivity. In: Agrosphere: Nutrient dynamics, Ecology and Productivity. Science Publishers Inc.: Enfield, New Hampshire, USA, **2003,** 58–104.
14. Bielders, C.; Rajot, J.; Amadou, M.; Skidmore, E. On-farm quantification of Wind Erosion under Traditional Management Practices. International crops research Institute for the Semi-arid tropics, Patancheru, India, A.; Internal Reports 1–6 ICRISAT@org CCER.htm **2002c,** 6.
15. Krishna, K. R. Peanut Agroecosystem: Nutrient Dynamics and Productivity. Alpha Science International Inc.: Oxford, England **2008,** 290.
16. Fatondji, D.; Bielders, C.; Bationo, A.; Vlek, A. Evaluation of *Zai* technology for rehabilitation of Degraded Land. ICRISAT@org CCER.htm **2002,** 8.
17. Sharda, N.; Rattan Singh. Erosion control measures for improving productivity and Farmers' profitability. Fertilizer News **2003,***48,* 55–68.
18. Cogle, A. L.; Rao, K. C.; Yule, D. F.; Smith, G. D.; George, J.; Srinivasan, S. T.; Jangwad, L.; Soil management for Alfisols in the semiarid tropics: Erosion, Enrichment ratio and runoff. Soil Use and Management *18,* 10–17.
19. Piara Singh Productivity and water balance of Soybean-based systems on a Vertic Inceptisols. http://www.icrisat.org/gt-aes/researchbriefs18.htm **2002,** 1–4.

20. Autil, R. S.; Kumar, V.; Gangwar, M. S. Leaching of Nitrate in a typic Ustipsamment as influenced by Water regimes. Journal of Indian Society of Soil Science, **2002,** *50,* 209–212.
21. Jambert, C.; Serca, D.; Delmas, R. Quantification of N losses as NH3, NO N2O N2 from fertilized a Maize fields in Southern France. Nutrient Cycling in Agroecosystems **1997,** *48,* 91–104.
22. Wood, S.; Sebastian, K.; Scherr, S. J. Pilot analysis of Global Ecosystems: Agroecosystems. International Food Policy Research Institute. Washington, D.C. **2000,** 109.
23. Ding, H.; Cai, G.; Wang, Y.; Chen, D. Nitrification and De-Nitrification loss and N_2O emission from urea applied to crop-soil systems in the North China Plains. Proceedings of 17 World Soil Science Congress Symposium No 7 Paper No 214, **2002,** 214–218.
24. Prasad, R.; Blaize, B. Soil Nitrogen dynamics in Cropping systems. In: Roots and Nitrogen in Cropping Systems of Semi-arid Tropics. Ito, O.; Johansen, C. J.; Adu Gyamfi, S.; Katayama, K.; Kumar Rao, J. D. K.; Rego, T. J. (Eds.). Japan International Research Center for Agricultural Sciences. Series **1996,** *3,* 429–440.
25. Hegde, D. M.; Sudhakar Babu, S. N. Management Strategies in Agriculture: A Future Outlook. Fertilizer News **2001,** Nutrient *46,* 61–72.
26. Majumdar, D.; Rastogi, M.; Sushil Kumar, S.; Pathak, H.; Jain, M. C.; Upendra Kumar, K. Journal of Indian Society of soil Science **2000,** *48,* 732–741.
27. CRRI Soil Science and Microbiology Report. Central Rice Research Institute, Cuttack, India **2006,** 1–3.
28. IPNI Managing Nitrogen to protect water. Enviro-Briefs. International Plant Nutrition Institute, Norcross, Georgia, USA, **2007,** 1–3.
29. Hodges, S. C.; Goshco, G. J.; Kidder, G. Research-based soil testing information and fertilizer recommendations for Peanuts on Coastal plains. Southern Cooperative series Bulletin **1994,** *380,* 1–12.
30. Kleinman, J. A.; Needelman, B. A.; Sharpley, A. N.; McDowell, R. W. Using Soil Phosphorus profile data to assess Phosphorus leaching potential of manured soil. Soil Science Society of America Journal **2003,** *67,* 215–224.
31. Tarkelson, D. D.; Mikkelson, R. L. Runoff Phosphorus losses as related to soil test phosphorus and Phosphorus saturation on Piedmont soils under conventional and No-tillage system. Communications in Soil Science and Plant analysis **2004,** *35,* 1532–1538.
32. Maguire, R. O.; Sims, J. T. Measuring Agronomic and Environmental Soil Phosphorus saturation and predicting Phosphorus leaching with Mehlich-3. Soil Science society of America Journal **2002,** *66,* 2033–2039.
33. Zougmore, R.; Mando, A.; Stroosnijder, L.; Ouedraogo, E. Economic benefits of combining Soil and Water conservation measures with nutrient management in Semi-arid Burkina Faso. Nutrient Cycling in Agroecosystems **2005,** *70,* 261–269.

EXERCISE

1. What are the different soil and atmospheric process that induce loss of nutrients from fields/agroecosystems?
2. Explain mechanisms by which nutrient is lost from a field due to sand and rainstorms.
3. What are the methods employed by farmers in different agroecosystems to thwart soil erosion?
4. What are the permissible concentrations of major nutrients N, P and K in the ground water?
5. Mention a few catch crops used to reduce Loss of nutrients from plantations.

FURTHER READING

1. Brye, K. R.; Norman, J. M.; Bundy, L. G.; Gower, S. T. Nitrogen and Carbon leaching in Agroecosystems and their role in De-nitrification potential. *J. Environ. Qual.* **2001,** *30*, 58–70.
2. Godone, D.; Stanchi, S. Soil Erosion Issues in Agriculture. Intech Publishers, New York, USA, **2011,** pp. 347.
3. Huang, P. M.; Li, Y.; Sumner, M. E. *Hand Book of Soil Sciences Volumes 1 and 2.* CRC Press, Taylor and Francis Company: Boca Raton, Florida, USA. **2011,** 2272.
4. Palmer, J. J. Soil and Nutrient conservation for Slope land areas. http://www.agnet.org/htmlarea_file/library/ 20110808143338/bc48003.pdf, **2011.**

USEFUL WEBSITES

http://www.sssa.org
http://www.icrisat.org
http://www.iita.org

CHAPTER 30

NUTRIENT CYCLES IN CROP FIELDS

CONTENTS

30.1 INTRODUCTION

Nutrient cycles are integral to almost all geographic locations. Nutrient cycles are essential phenomena within each crop field, a farm, a small cropping zone or an entire agroecosystem. Within the context of this book, nutrient cycles in nature, especially those occurring in vast natural vegetation and agroecosystems deserve greater attention. Knowledge about nutrient cycles that occur in different agroecoregions or cropping belt has been accumulated consistently for the past several decades, using wide range of techniques. Yet, there are lacunae. Several aspects of nutrient cycling specific to different regions need attention. Long term effects of priming and altering nutrient cycles, on steady states of soil nutrients and crop productivity need due attention.

Nutrient cycling in a field is accomplished through variety of natural and man-made processes that act at various intensities, duration, rate and at different times during a crop season or year. Nutrients move from atmosphere to soil via rainfall, deposition and wind. Natural process like precipitation brings in a certain amount of nutrients in dissolved state into a crop field. Dust storms and wind also add to nutrients in a field. It actually depends on extent of surface soil removed from a field versus that added. Sand and dust particles may add sizeable quantities of nutrients depending on parent material and carrying capacity of particles. Irrigation water too adds nutrients based on the source and nature of soils on which it moves before reaching the stipulated field. For example, in areas with rampant seepage loss (sandy soils), irrigation water may bring in more of dissolved nutrients such as NO_3 or SO_4. Nutrient inputs via atmospheric deposits and precipitation are relatively small or even negligible in many regions.

Next, addition of nutrients through inorganic and organic manure is a major source of nutrients to most fields. This is a crucial step, since it primes nutrient cycles in farmer's fields. If extrapolated to a cropping zone or agroecosystem, supply of manures is a major phenomenon that induces and regulates the intensity of nutrient cycles and crop productivity. Fertilizer supply is actually an important point in the nutrient cycle that replenishes nutrients and improves nutrient turnover in a field. Nutrients are also recycled into a crop field through crop residues that are left to decompose in a field. Crop residues are also carefully collected and incorporated into soil. This procedure returns a portion of nutrients that moved from soil to crop phase during previous season.

Soil is a major repository of nutrients in any agroecosystem. This pool of nutrients is almost always amenable for transportation to lower horizon or laterally through seepage or runoff. It is also prone to physicochemical and microbiological transformation. Soil nutrient transformation is a natural phenomenon that regulates nutrient cycles both in the surface and subsoil. Most importantly it has a major say in the conversion and reconversion of soil nutrients, so that they are soluble or exist in chemical state easily available to crop roots. Nutrient availability to roots partly decides the extent of nutrients absorbed by crops. It is nothing but movement of nutrients from soil to crop phase. Nutrient transformations are either rapid or slow depending on the nutrient element and soil physicochemical environment.

In case of N, which is a major soil nutrient and required by crops in relatively larger quantities, soil processes (nutrient transformations) regulate its various forms and availability to roots. Important soil processes are N-mineralization, N-immobilization, N transport in soil through seepage, percolation, nitrification/de-nitrification, N loss to groundwater, N-emissions and N absorption by roots and incorporation into plant tissue. Incidentally, biological nitrogen fixation (BNF) adds to soil-N and contributes to plant N status. Aspects like rate of BNF, its duration and timing during a crop season may be important, since they decide the quantum of N added to crop field.

Phosphorus cycle in a field is regulated by atmospheric deposits, supply of inorganic fertilizer-P and organic manures, inherent soil-P and its availability to roots, soil chemical fixation of P, loss of P through seepage, percolation and surface erosion of soil, as well as mineralization and immobilization of P in soil. The removal of soil-P by crop roots, its transport to aboveground and recycling through crop residue are again important aspects. Potassium is also required in relatively higher quantities by crops. Potassium cycling in agricultural belts is regulated by phenomena like atmospheric deposits, storms, precipitation and winds that bring in a certain amount of nutrients into field. Supply of K from atmospheric processes is often small or even negligible. Supply of potassium through inorganic fertilizers and manures is an important aspect that adds to soil-K status of a field. It often primes K turnover at higher rates. Soil-K undergoes several physicochemical transformations that affect its availability to roots. Soil transformations actually affect accumulation of various forms of K in soil. A portion of soil-K is fixed into soil matrix through chemical-fixation. It may not be easily amenable for recycling. Potassium uptake by roots is rapid since it is a highly mobile and soluble nutrient in the soil profile. Irrigation water too adds to soil-K in the crop field. Soil-K is lost to lower horizon and ground water through percolation, seepage and drainage. Crops absorb relatively large quantities of soil-K. This process actually moves large amount of soil-K into above-ground crop phase in a short span of 6–10 weeks in a crop season. The duration of rapid K uptake varies with crop species. For example, a wheat crop may transport as much 130–160 kg K in a matter of 6–8 weeks from soil to above ground biomass [1].

30.2 FACTORS THAT INFLUENCE NUTRIENT CYCLES

Nutrient cycling in a crop field, a cropping zone or an agroecosystem is dependent on several factors related to geographic location, topography, soil, crop, environment, yield goals and economic considerations. They can be grouped and identified individually as follows:

Soil related factors that affect nutrient cycling and its intensity are soil type, its depth, texture, structure, moisture holding capacity, temperature, pH, EC and redox conditions. Soil fertility characteristics such as available nutrients, buffering capacity, optimum fertilizer supply, soil organic matter and biotic components, like soil microbes are essential. Optimum levels of soil nutrient transformation rates. Any of these factors operating singly or in combination at different intensities and for variable stretch of time can affect nutrient cycling in a cropping zone. Most important factor that can either induce or reduce nutrient recycling is the inherent soil fertility and fertilizer/organic manure supply trends in a given field or area.

Crop related factors immensely affect nutrient cycling in an agroecosystem. They are rooting depth and its pattern; nutrient recovery pattern; nutrient translocation, its re-translocation and accumulation pattern within the plant tissue, quantity of forage and grain produced; harvest index, nutrient partitioning ratios, emissions from crop, if any; succulence of tissue and susceptibility to degradative processes in soil. Leaf Area and photosynthetic efficiency and CO_2 fixation pattern also affect nutrient cycling.

Environment related factors that regulate nutrient transformations and turnover are atmospheric conditions like temperature, moisture, relative humidity, diurnal pattern; season and precipitation pattern; emissions of CO_2, N_2, NO_2 and N_2O from fields.

Yield goals and economics aspects too affect nutrient supply, recycling and utilization aspects. For example, cost of inorganic fertilizers and FYM are often weighed with net return. In other words, economic advantages dictate the extent of nutrients supplied to fields or cropping expanses. The value of the produce (grains, forage, fruits) is crucial with regard to extent of nutrients impinged into fields, extent recovered into produce and that recycled into fields. The turnover of nutrient in most areas is dependent on intensity of farming, amount of produce transported out of the location and extent recycled *in situ*.

30.3 NUTRIENT CYCLE IN DIFFERENT AGRO-ECOREGIONS

Nutrient cycles in different agroecosystems are mainly dependent on natural ability of the agro-ecoregion to generate biomass/grains, nutrient turnover rates possible and recycling trends followed by farmers. Aspects like inherent soil fertility, nutrient supply, cropping systems and weather pattern affect nutrient cycles in a given agroecoregion. Let us consider a few examples from different continents. The North American Great Plains, European Plains or Agrarian zones of North-east China all support intensive cultivation of crops such as corn, wheat, soybean, cotton and forest species. Farmers in Great Plains prime their fields with high amounts of fertilizer-based N, P and K. They also supply 15–20 t FYM/ha/season. Corn or wheat removes relatively large quantities of nutrients to support high yield goals. The productivity of crops in the temperate Great Plains is high at 5–10 t grain/ha and 8–15 t forage depending on crop species. A large share of stover is recycled and a certain amount is used as silage. The nutrient recycling procedures allow excellent nutrient turnover rates (Fig. 1). The high soil fertility and nutrient supply may induce loss of nutrients. Therefore, in such agroecosystems, accuracy in quantity of nutrient supplied, its timing, selection of appropriate genotype that rapidly scavenges nutrients is essential. We have to match nutrient turnover rates with crop productivity, in order to avoid undue loss of nutrients to atmosphere. Despite it, nutrient loss or movement to ground water or atmosphere could be massive, if an entire agroecosystem is considered. Soil-N saturation is observed frequently in the intensive cereal cropping zones. Undue soil-N accumulation is a problem. It induces greater loss of N from the system. There are indeed several ways to regulate nutrient (N) cycles in crop fields. McSwaney et al. [2] believe that we could immobilize the excess fertilizer-N applied to soil using cover crops. Such N-immobilization helps in conserving soil-N and avoiding loss via emissions. It can be recycled at a later date as crop residue.

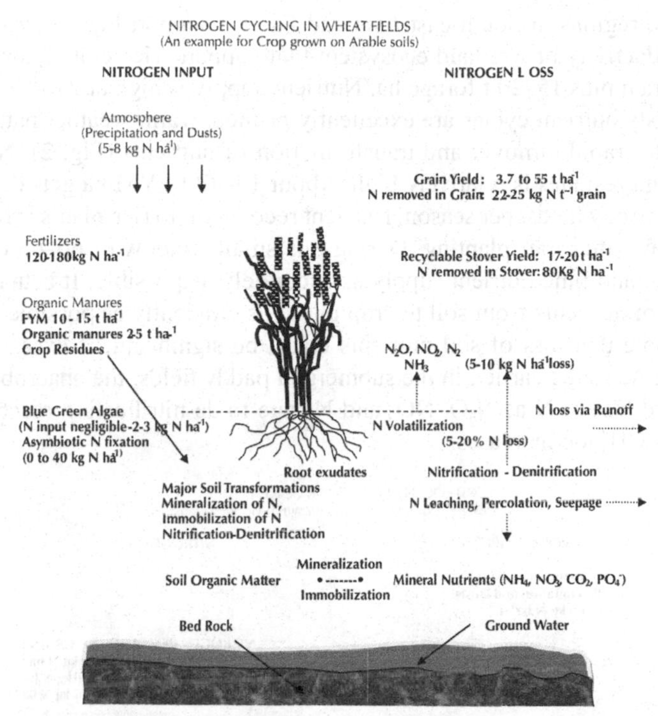

FIGURE 1 Nitrogen Cycle in a Wheat field.
Note: Nitrogen priming into the ecosystem is commensurate with high grain/forage yield. Nitrogen removal too is relatively high. Intensification of cropping requires higher N inputs. However, high N inputs may lead to higher amounts of N loss from ecosystem. Nitrogen loss values in parenthesis refer to percentage of fertilizer-N lost through volatilization.
Source: Modified based on Krishna [3, 4].

The semiarid regions that occur in Latin America (Pampas and Cerrados), Sub-Saharan Africa and Indian subcontinent support nutrient cycle at commensurately moderate levels. The productivity and profitability of crops decides the nutrient supply into cropping zones. The agroecosystem is primed with nutrients using both inorganic and organic manures. Nutrient priming is strictly dependent on inherent soil fertility and yield goals. Nutrient input may range from 40–80 kg N, 20–40 kg P and 60 kg K depending on crop and soil fertility status. Nutrient recycling through crop residues is relatively smaller. Farmers tend to allocate a larger share of crop biomass to farm animals. The grain harvest is low or moderate at 0.8 to 2.5 t/ha. Nutrients recycled are commensurate with grains and forage productivity. The nutrient transformation and turnover is rapid owing to warm temperatures experienced in tropics. Soil characteristics such as aeration, moisture, temperature and high microbial load allow rapid mineralization of nutrients and CO_2 evolution. Nutrient loss to atmosphere or to ground water could be significant. In the absence of appropriate fertilizer input schedule, nutrient accumulation in soil profile is minimal, if any.

Tropical regions in South-east Asia and Fareast support high intensity cropping. Again, productivity of wet land ecosystem that supports rice cultivation here is high at 7–10 t grain plus 15–20 t forage/ha. Nutrient supply is high at 270–300 kg N, P, K/ha. Obviously nutrient cycles are excellently primed, warm weather pattern is highly congenial for rapid turnover and transformation of nutrients (Fig. 2). Nutrient recycling through residues is relatively high. About 15–12 t FYM/ha gets decomposed in submerged paddy fields, per season. Nutrient recovery into rice plants is rapid between 21 to 85 days after transplanting. During this small window of 60–65 days, farmers try to match and time nutrient supply as accurately as possible. It actually improves movement of nutrients from soil to crop phase as efficiently as possible without loss. We must note that loss of soil nutrients could be significant, if excessive nutrients are primed. As stated earlier, in the submerged paddy fields, the anaerobic conditions induce rapid loss of N as N_2O, NO_2 and N_2 due to de-nitrification reactions. Carbon emission as CH_4 too increases.

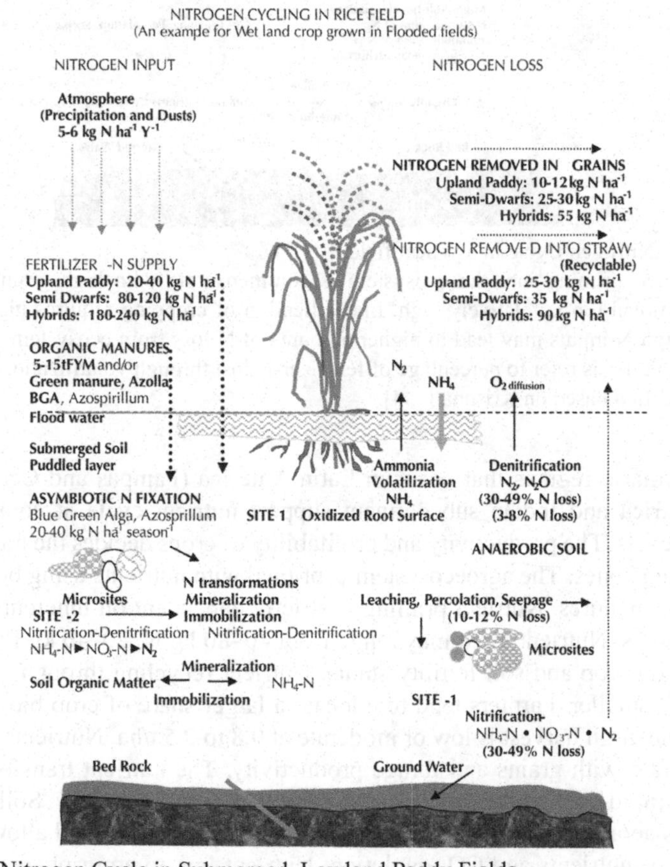

FIGURE 2 Nitrogen Cycle in Submerged, Lowland Paddy Fields.

Note: Nitrogen removed in Grains and Straw has been computed based on the fact that to produce. 1.0 t grains, 22–23 kg N, 8–12 kg P and 32 kg K are removed from soil by a paddy crop.

Source: Modified based on Krishna [5].

Legumes are an important component of cropping belts found in humid and semi-arid regions of different continents. Legumes are often intercropped or grown in rotation with cereals. This procedure adds to soil-N through biological nitrogen fixation. It allows better turnover of N in the crop field. Legume cultivation actually primes N cycles in fields with 20–40 kg N/ha that is derived from atmosphere. Fertilizer-N supply to legumes such as lentils, groundnut (Fig. 3) or pigeonpea is much less compared to a cereal or a cash crop.

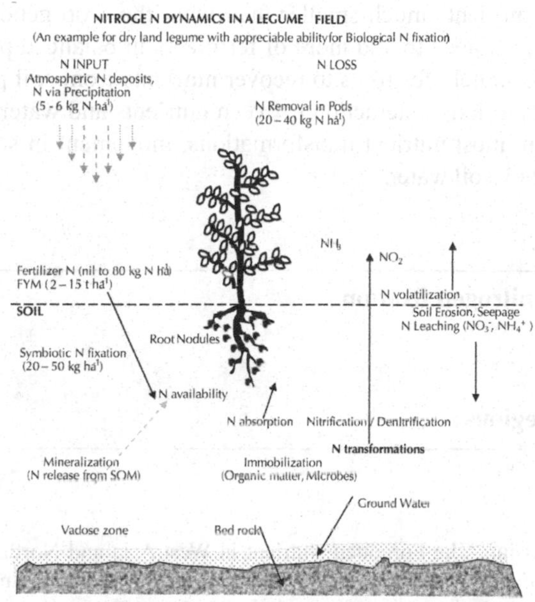

FIGURE 3 Nitrogen Dynamics in Legume (Chickpea) fields in Semi-arid Tropics of India. *Source:* Modified based on Krishna [6].

Study of nutrient cycles is important since it allows us to trace the major lacunae that affect crop nutrition and productivity. It actually allows us to accurately judge problems related to quantum of nutrient supply, its timing, cropping systems and genotypes most suitable to a given agro-environment. We can manipulate various aspects of nutrient cycling preferentially and regulate crop productivity. For example, a field could be prone to accumulate nutrients in the lower horizons or is susceptible to ground water contamination. It means nutrient supply is greater than that removed to crop phase in a season. We have to judge nutrient need accurately, lessen nutrient supply and time it exactly so that crop roots scavenge effectively. This system overcomes undue nutrient accumulation that may be vulnerable to loss into lower horizons, ground water or to atmosphere via emissions. Next, if the movement of nutrients from soil to crop phase is inadequate to support the yield goals envisaged. It means we should mend the nutrient cycle by planting a new genotype that has rapid growth rate and absorbs more of nutrients, so that grain yield is higher. It is fairly common to mend nutrient cycles by manipulating cropping sequences and genotypes. Now, if the crop field is located near an industrial zone that causes S and micronutrient deposits. It is preferable to plant an oil seed crop that requires greater quantities of soil-S to support fat accumulation in grains. Such a step regulates S cycle effectively. In subsistence farming zones, if we suspect nutrient turnover to be low, it is preferable to plant hardy crops that withstand paucity of natural resources like soil nutrients and water. These crops may sustain nutrient cycles at appropriately low level, but still offer optimum grain productivity. In most cases, the grain/forage yield potential of crop genotype seems much higher than that derived routinely. Nutrient cycle is obviously primed with quantities of nutrients much smaller than what the crop genotype can handle. In such a situation, we need to add more of fertilizers in balanced proportions, place and time it most accurately for roots to recover nutrients from soil phase and deliver it to above ground portions. Interaction between nutrients and water is rather crucial. Nutrient absorption, most nutrient transformations, movement in soil and emissions are highly affected by soil water.

KEYWORDS

- **biological nitrogen fixation**
- **Chickpea**
- **nutrient**
- **semiarid regions**

REFERENCES

1. Krauss, A. Potassium-The Forgotten Nutrient in West Asia and North Africa. In Accomplishments and Future Challenges in Dry land Soil fertility Research in the Mediterranean area. Ryan, J. (Ed.) International Center for Agricultural Research in Dry areas, Aleppo, Syria, **1997,** 9–21.
2. McSweney, C.; Snapp, S. S.; Gentry L. L. Use of N immobilization to tighten the N cycle in Conventional. Agroecosystems. Ecological Applications **2009,** *20,* 648–662.
3. Krishna, K. R. Agrosphere: Nutrient Dynamics, Ecology and Productivity. Science Publishers Inc.: EnField, New Hampshire, USA, **2003.**
4. Krishna, K. R. Maize Agroecosystems: Nutrient Dynamics, Ecology and Productivity. Apple Academic Publishers Inc.: Toronto, Canada, **2013,** 343.
5. Krishna, K. R. Agroecosystems of South India: Nutrient Dynamics, Ecology and Productivity. BrownWalker Press Inc.: Boca Raton, Florida, USA. **2010,** 312.
6. Krishna, K. R. Peanut Agroecosystem: Nutrient Dynamics and Productivity. Alpha Science International Inc.: Oxford shire, United Kingdom, **2008,** 292.

EXERCISE

1. Describe Nitrogen cycle in a cereal and a legume crop.
2. What are the major transformation reactions of soil-N that relate to its availability to plant roots?
3. What is Nitrification? Mention at least two important measures that reduce loss of N caused by nitrification reactions.

4. Describe carbon cycle in a mixed farm involving intensive cereal production and farm animals.
5. Describe methods to sequester more of C into soil.

FURTHER READING

1. Nutrient Cycling in Agroecosystems, Journal Series. Springer Inc.: Heidelberg, Germany.

USEFUL WEBSITES

en.wikipedia.org/wiki/Nutrient_cycle
http://link.springer.com/journal/10705
http://www.sssa.org
http://www.asa.org
www.soils.wisc.edu/~ruark/
www.journals.elsevier.com/soil-biology-and-biochemistry/
www.journals.elsevier.com/agricultural-systems/

PART III: FACTORS INFLUENCING NUTRIENT DYNAMICS IN AGROECOSYSTEMS

CHAPTER 31

SOIL TILLAGE AND MULCHES INFLUENCE NUTRIENT DYNAMICS

CONTENTS

31.1 INTRODUCTION

Land and soil management in any agroecosystem begins with clearing of natural vegetation, weeds, volunteers, stubbles, stones and large sized gravel (Plate 1). Soil tillage and land preparation are perhaps earliest of agronomic procedures that has bearing on productivity. Soil tillage has distinct effects on soil nutrient dynamics and moisture availability to crops. Tillage actually involves opening or loosening soil. Deep plowing using disks turns and mixes the top layer of soil thoroughly (Plate 2). Land preparation encompasses marking ridges and furrows, flat beds, broad beds, contour ridges etc. Tillage basically aims at obtaining better soil structure and tilth, so that seed germination, seedling establishment and a good crop stand is achieved (Plates 3 and 4). Most of the tillage operations, implements, traction vehicles and timing of tillage or inter culture steps have been improvised periodically. During recent period, a major trend is to reduce tillage or even adopt No-till system in order to thwart loss of SOC. Aspects such as preservation of soil structure; conservation of SOC and better rates of C-sequestration seems to be the prime aim. Tillage procedures used in different agroecosystems involves use of different types of equipments. Each of these implements is designed to suit a specific purpose. The effect of tillage operation on soil characteristics and nutrient dynamics invariably depends on the kind of implement and intensity of its use. Commonly used primary tillage implements are Mould board plows, Disc plows, Chisel plows and Rippers. Secondary tillage implements such as rotary tillers, spiked or tined cultivators, clod crushers and harrows (Plate 4) are also common in all agroecosystems. Farmers in different continents adopt a wide range of tillage systems. Over all, they can be grouped into No or Zero-tillage systems, Stubble-Mulch Tillage systems, Reduced or Restricted tillage systems and Conventional Tillage systems. In general, conventional tillage systems involve thorough plowing, turning and exposing subsurface layers of soils to atmosphere. This results in loss of soil organic fraction and making soil vulnerable to surface erosion and loss of nutrients. Hence, during recent years, the trend is to adopt No-tillage or Zero tillage systems. No tillage systems are preferred because it protects soils from massive loss of SOC as CO_2 and avoids erosion to a great extent [10]. Land preparation systems adopted are equally important. They protect fields from erosion, reduce loss of soil moisture and provide better crop stand (Plates 5 and 6).

PLATE 1 Field clearance through Bull dozer and Disc plows before sowing Soybean or Maize in the Cerrados of Brazil.

Note: Such heavy tillage is necessary to clear fields and manage a crop of soybean/maize. It induces loss of surface vegetation cover, opens up soil to nature's vagaries. It causes rampant loss of soil and nutrients via surface runoff, percolation and emissions. Procedures such as contour planting, bunding, intercropping and planting windrows are some suggestions that reduce loss of top soil and nutrients.
Source: EMBRAPA, Sao Paolo, Brazil.

PLATE 2 A Red Alfisol field is often deep plowed (heavy disking) and left to fallow.
Note: This procedures affects soil aeration, turns-up soil, brings in consistency to soil pH and nutrient distribution in the profile. Since deep plowing induces soil aeration and creates an oxidized state, it induces soil aerobic microflora, leading to rapid loss of C through soil respiration. Soil carbon sequestration is therefore reduced in arable soils.
Source: Krishna, K.R., Bangalore, India.

PLATE 3 Left: Deep red soils common in Southern Indian plains are used to produce dry land cereals, millets and oil seeds such as sunflower and groundnut. Right: A close up of plowed field with clods. The above red soil fields are located near Mahakoota, North Karnataka, India.
Source: Krishna, K.R., Bangalore, India.

PLATE 4 Land leveling and clod crushing prior to seeding Wheat in the Great Plains of North America.
Source: Nelson Farms, Fort Dodge, Iowa, USA.

PLATE 5 Land preparation and Bund formation to conserve soil moisture and control erosion from Oxisols of West Asia.
Source: ICARDA, 2012

PLATE 6 Ridging and construction of furrows prior to transplantation of Pine seedlings in the main fields at Austin Carey Forest Plantations near Gainesville, Florida, USA.
Source: Krishna K.R. University of Florida, USA.

31.2 TILLAGE AND SOIL ORGANIC CARBON

Tillage causes physical disturbance of soil, either marginally or drastically depending on the tillage intensity, implements and traction vehicles used. In addition, tillage induces a series of physicochemical and microbiological changes. There are indeed innumerable reports about the small or large-scale changes in soil that occur as a consequence of tillage. Tillage enhances aeration and oxidative state of arable soils. Tillage affects nutrient dynamics in soil. It affects nutrient transformation steps and their rates. Tillage influences nutrient availability to plant roots. As a consequence, tillage may affect rooting, nutrient recovery, crop growth and productivity.

Now, let us examine the influence of tillage on C dynamics. Tillage actually affects soil-C transformations rather markedly. Tillage has direct impact on C sequestration and emissions such as CO_2 and CH_4 [10]. Tillage affects rates of mineralization of organic matter and immobilization of C in soil. In arable soil, excessive tillage induces oxidative process improves microbial load and respiration. Excessive microbial respiration results in loss of C as CO_2. In fact, to avoid loss of soil organic carbon (SOC), farmers adopt restricted or reduced tillage practices during crop production. In other words, C sequestration procedures are preferably adopted. Conventional tillage affects soil up to 30 cm depth. Conventional tillage affects soil structure, bulk density, redox potential and microbial activity to a greater extent. It induces rapid transformations of soil C and mineral nutrients. Fields kept under conventional tillage are prone to rapid loss of SOC. Tillage and land preparation practices do affect extent of soil erosion, loss of surface soil, percolation trends, seepage and loss dissolved nutrients. Clearly, overall nutrient dynamics in soil gets affected based on type and intensity of tillage adopted.

Long term trials in North America indicate that chisel plowing of fields for cultivating maize induces greater loss of SOC compared with strip tillage or no-tillage systems. Actually, chisel plowing induces greater loss of surface residues. Fields retain less amount of surface organic residue, if kept under chisel plows. Strip tillage and no-tillage system retain relatively greater amount of organic matter on the surface. Soil loss is rampant in fields kept under conventional tillage systems (4.67 t/ha) compared to strip tillage (0.28 t/ha) or no-tillage systems (0.05 t/ha) [1]. Field trials in Missouri during 1980s and 1990s have shown that if no-tillage system retains 4.5% SOC, under similar environment and inputs, chisel and disk plows retain mere 3.4% SOC [2]. Field trials in Wisconsin, USA, have shown that tillage firstly induces higher levels of drainage loss of water received through precipitation. This has direct impact on loss of dissolved nutrients. Following is an example:

	Loss of Water	**Leaching Loss of Nutrients**	
		Nitrogen	**Soil Organic Carbon**
	mm	**(kg N/ha)**	**(kg C/ha)**
No-Tillage	116	201	435
Chisel Plow	1575	179	502

Source: [3].

The effect of tillage on drainage loss of water (soil moisture) and nutrients in dissolved state, especially N and C is enormous, if entire cropping zone or an agro-ecosystem is considered. Tillage could easily become an important factor affecting both nutrient and water dynamics. The interaction between tillage system and cropping sequences adopted may also have its own specific influence on C sequestration, loss and turnover. For example, during a maize-corn rotation followed in the Northern Great Plains, no-tillage plots supporting soybean induced better decomposition of organic matter. Tillage depth is an important factor affecting C sequestration. Over all, an integrated approach is preferred since tillage interacts with many factors related to soil and cropping systems. In fact, no-tillage, mono-cropping with large amounts of residue recycling is preferred in some areas.

31.3 ORGANIC MULCHES AND NUTRIENT DYNAMICS

Mulching is a common procedure in many agrarian regions. Mulching is an important agronomic procedure that usually follows tillage and is largely aimed at protecting soil and crop from environmental vagaries. In addition, mulching adds to soil organic matter and improves residue recycling. Mulching adds to soil mineral nutrition upon decomposition of organic matter. Mulching has a series of advantages in different agro-ecosystems. They are as follows: a) Mulches protect soil from rains, sun and wind; b) Mulches reduce soil erosion. They actually protect topsoil from being eroded rapidly; c) Mulches reduce runoff and conserve water that would otherwise be lost as surface flow; d) Mulches reduce weed growth. They also suppress weed germination and growth; e) Mulches reduce evaporation of soil moisture; f) Decomposing organic matter and crop residues used as mulch, generally improve soil structure and aggregates; g) Mulches enhance water infiltration and protect soil from sealing and crusting; h) Mulching regulates soil temperature. It protects seeds and seedlings from frost damage and low temperature induced seed dormancy. Mulching also protects seeds/roots from high temperature, especially in areas prone to high temperature conditions, for example in Savannas of Africa or Dry lands of South India; i) Mulching improves soil microbial activity; j) Mulching improves rooting and root activity; k) Mulching aids in recycling nutrients with in the maize fields. It allows nutrients held in residues to percolate to top 40–60 cm soil layer.

Farmers use a variety of materials such as farm residues, crop residues, fine gravel, industrial by-products, plastic sheets, cardboards, etc. The nature of material used as mulch has an impact on extent of protection against soil erosion, nutrient loss and addition of nutrients. For example, in the maize cultivating zone of Eastern Nigeria, legume on a bare-soil control yielded 2.3 t grain/ha. In comparison, application of legume husks as a mulch soon after seeding gave 4.4 t grain/ha, soybean tops 4.2 t grain/ha, cassava stems 3.8 t grains/ha, Andropogan straw 3.5 t grain/ha and oil palm leaves 3.2 t grain/ha. Inorganic materials are also used frequently as mulches. For example, application of fine gravel resulted in 3.1 t grain/ha, placing black plastic sheets gave 3.0 t grain/ha and translucent sheets gave 2.7 t grains/ha [4].

The soybean-maize cropping zone in Cerrados of Brazil is largely developed by converting natural savannas. Massive clearing and tillage during land conversion resulted in loss of surface vegetation (Plate 1). In addition, farmers practiced deep tillage using disks and chisel plow. These procedures induce loss of surface soil and nutrients. Therefore, during recent years, farmers in this belt consistently adopt mulching with organic residues and industrial wastes or by-products. Generally, farming in the Brazilian Cerrados involves soil tillage, direct seeding, maintenance of maize residues in between crop rows and following most appropriate crop sequence [5]. During 1980s, mulching became popular and mulched area improved from 1.0 m ha to 15 m/ha by 2001 [5, 6]. Farmers' experience suggests that direct seeding and mulching enhances C sequestration and soil-N. It avoids nutrient loss through percolation. Mulching seems to improve both soil quality and grain productivity. Mulching is a common procedure in almost all cropping zones situated in cold temperate regions. Here, mulching protects seeds and germinated seedlings from cold damage. Rooting and nutrient acquisition also get improved due to mulching. Advantages of mulching can also be perceived on final grain/forage yield. Mulching is equally important in hot, semiarid tropics. Here, mulching thwarts excessive ambient heat and protects seeds from exposure to high soil temperature. Mulching regulates soil temperature and stops loss of soil moisture.

31.4 TILLAGE AFFECTS NUTRIENT DYNAMICS IN SUBMERGED, LOWLAND PADDY FIELDS

Soil puddling is a preferred procedure in lowland paddy growing regions. Puddling produces a soft soil tilth, so that transplantation of paddy seedlings is achieved with ease. It helps seedlings to attain anchorage. Puddling affects rooting and nutrient acquisition that may ultimately affect seedling establishment and better crop stand. Puddling affects various soil traits such as porosity, aggregation, bulk density, hard pans, etc. Puddling reduces seepage and percolation of impounded water. Puddling intensity affects extent percolation down the profile. For example, improving tillage and puddling reduces loss of water by half from 5.6 mm/h to 2.6 mm/h, if puddling intensity is enhanced. Puddling actually helps in attaining standing water in a flooded field. Puddling is said to be an important agronomic procedure in the rice agroecosystem of South-east Asia. It has direct influence on water requirement, nutrient recovery and as a consequence grain/forage yield. For example, rice grain yield improved from 4.1 t/ha to 5.0 t/ha if number of passes by puddlers are increased from 1 to 4 [7]. Puddling intensity also affects a series of physiological characteristics like tiller number, leaf area index and biomass formation. Intensive puddling and consistent water logging improves rice grain yield by 20–22%2over control [8].

In the wetland ecosystem, tillage affects three important aspects namely soil structure, nutrient transformations and rooting. Puddling to greater depth of soil reduced loss of N to lower horizons. Shallow puddling results in relatively higher loss of N. Preplanting tillage and intensity of puddling affects distribution of $NH_{4\text{-}}N$ and $NO_{3\text{-}}N$ in the root zone [9]. Puddling affects distribution of water down the profile. Therefore, distribution and availability of all soluble nutrients too are affected by puddling intensity.

We should note that even in intensive rice cropping zones, a short spell of fallow or dry spell occurs. This short fallow period allows soil to get aerated and oxidative processes to set in. The alternate flooding (anaerobic/microaerophillic) and arable period actually causes drastic upheavals in soil physico-chemical process, microbial population, nutrient transformations and availability. Puddling and impounding of water induces anaerobic conditions, anaerobic microflora and low redox condition. Dry spells in soil and tillage induces oxidative process, rapid microbial build up and activity, and loss of CO_2 due to soil respiration.

KEYWORDS

- **flooding**
- **soil organic carbon**
- **Stubble-Mulch Tillage**
- **Zero-tillage**

REFERENCES

1. Walkowski, R. Using Recycled Wallboard for Crop Production. University of Wisconsin-Extension Bulletin. http://www.soils.wisc.edu/extension/publications/ **2003,** 1–8. (December 6, 2012).
2. Quarles, D. Effects of 10 years of continuous conservation tillage crop production and infiltration for Missouri Claypan soils. **1994,** 12, 1–28.
3. Brye, K. R.; Norman, J. M.; Bundy, L. G.; Gower, S. T. Nitrogen and Carbon leaching in agroecosystems and their role in denitrification. Environmental Quality **2001,** 30, 58–70.
4. Okigbo, B. N.; Lal, R. Residue mulches. inter-cropping and agriculture potential in tropical Africa. In: Basic Techniques in Ecological Farming, S. Hill (Ed). IFOAM Conference. Montreal, Canada **1980,** 54–69.
5. Corbeels, M.; Scopel, E.; Cardoso, A.; Douzet, J. M.; Neto, M. S.; Bernoux, M. Soil carbon sequestration and mulch-based cropping in the Cerrado region of Brazil. Proceedings of the four International Crop Science Conference. Brisbane, Australia. www.cropscience.org **2004,** 1–8.
6. Evers, G.; Agostini, A. No-tillage farming for sustainable land management: lessons from the 2000 Brazil Study tour. TCI Occasional Paper Series No 12, **2001,** 1–26.
7. Sharma, S. K.; Gangwar, K. S.; Pandey, D. K.; Tomar, O. K. Increasing Productivity of Rice-based systems for Rainfed Upland and Irrigated areas of India. Indian Journal of Fertilizers **2006,** 2, 29–40.
8. Subramanyam, D.; Srinivaulu Reddy, D.; Raghava Reddy, C. Influence of Integrated Weed Management practices on growth and yield of transplanted rice (Oryza sativa). Crop Research **2007,** 34, 1–5.
9. Kar, S.; Sahoo, S. Dynamics of water and fertilizer-N as influenced by dry and wetland tillage in low retentive permeable soil. 17 World Congress on Soil Science, Bangkok. Thailand, paper No 1, **2002,** 356, 1–10.
10. Duxbury J. M. Reducing Greenhouse warming potential by carbon sequestration opportunities, limits and tradeoffs. In: Climate Change and Global Food security. R. Lal (Ed.) Taylor and Francis, Boca Raton, Florida, USA, **2005,** 435–450.

EXERCISE

1. Describe conventional, restricted and zero tillage systems in few sentences.
2. How does tillage affect soil organic matter component?
3. What is the influence of tillage on soil microbes?
4. Briefly mention methods that improve C-sequestration in soil.
5. List various land preparation systems and mention their advantages with regard to soil moisture, nutrient conservation and crop productivity.
6. How does intensity of tillage affect soil nutrients, C-sequestration and crop productivity?

FURTHER READING

1. Landers, J. N. How and Why the Brazilian Zero tillage occurred. **1999.** http://topsoil.nserl.purdue.edu/nserlweb-old/isco99/pdf/ISCOdisk/SustainingTheGlobalFarm/P037-Landers.pdf (December 6, 2012).
2. Titi, A. E. *Soil Tillage in Agroecosystems.* CRC Press: Boca Raton, Florida, USA **2002.** pp. 384.

USEFUL WEBSITES

www.fao.org/docrep/004/yr2638E/yr2638e08.htm (December 6, 2012)
http://www.cirad.fr/en/research-operations/research-results/2009/no-tillage-with-covercrops-for-the-brazilian-cerrados
www.journals.elsevier.com/soil-and-tillage-research/ (December 6, 2012)
http://soilquality.org/practices/tillage.html (December 6, 2012)

CHAPTER 32

SOIL ORGANIC MATTER AND NUTRIENT DYNAMICS

CONTENTS

32.1 INTRODUCTION

Agroecosystems that support wide range of field crops and plantations were all derived or developed on soils that supported natural vegetation, forests, prairies or wasteland. The dynamics of plant species, their growth pattern and most importantly carbon and mineral nutrient dynamics in these natural habitats were actually result of wide range of interactions through the ages. Such established natural ecosystems had to be removed progressively or at times drastically to make way to large scale cropping. The nutrient dynamics that ensued in the developing agroecosystems was markedly different from those known for several decades. Conversion of natural vegetation, prairies and scrubland to cropland almost always affects carbon dynamics, rather immensely. Obviously, the inherent soil fertility, biomass production, its recycling and weather pattern that induce changes in soil carbon pools are important aspects to note. There are several studies that have aimed at understanding the effects of changes in vegetation pattern on carbon dynamics. Let us consider a few examples from the Great Plains of North America. Here, pine forests, shrub land, natural prairies and waste land have given way to large scale production of cereals like wheat, maize, sorghum, millets or cotton and plantation crops. The cropping system adopted in different regions has specific influence on carbon dynamics. Long term effect on soil-C sequestration pattern, carbon recycling and loss due to emission need attention. In the Midwestern USA, C and N declined in regions that supported maize. The decline in soil-C was greater at 38% of the original level recorded 60 years ago. In case of natural vegetation, decline of soil-C was marginally low at 32% of original [1]. Hass et al. [2] reported 39–42% decline in soil organic carbon (SOC) due to maize cultivation in the Corn Belt. Reeder et al. [3] found a 16–28% decline in SOC due to continuous cropping of cereals. Incessant cropping of Chernozems found in Russian plains induced 38–42% decline of SOC in a matter of 50 years [4]. Several reports about decline in SOC actually pertain to upper horizon of soil that holds large portion of roots system of the crops. However, there are clear suggestions, that entire soil profile that gets explored by crop roots need to be considered. We should note that fields or even large expanses of crops reach a steady state level regarding soil-C transformations, loss/gain of SOC. Once, this steady state is reached, measurements about SOC need careful interpretation. Actually, factors such as crop rotation, residue recycling and rates of C-emissions may play a vital role in stabilizing SOC of a field. Pikul et al. [5] report that monocrop of corn induced a loss of 2.3 t C/ha but a rotation with corn-alfalfa-wheat resulted in only 0.3 t C/ha loss. Generally, a well-diversified cropping system conserves SOC better than a mono-cropping stretch.

Cereals such as maize, sorghum and sugarcane are preferred for producing biofuels. This aspect induces farmers to either replace crops previously grown with these cereals or convert prairies or wasteland in the vicinity to improve production of biofuels. We ought to realize that conversion of natural grasslands has immediate effect on C dynamics in the fields and ecoregion *per se*. In North America, it seems regions marked as conservation programs too have been converted into cereal fields, whose produce is destined for biofuel production [6]. In many regions of Southern USA, conservation with Brome grass has been replaced by lush high productivity maize crop meant for biofuels. Long term trials by Follet et al. [6] indicated that despite sizeable return of maize residues, the loss of C due to emissions and removal for biofuel reduces *c* sequestration immensely. Fields with maize consistently sequestered low amounts of C compared to Brome grass held as conservation.. There are suggestions to adopt No-tillage systems while producing maize, so that C sequestration improves. In the European plains, farmers are advised to follow stringent residue recycling patterns and suitable tillage systems that conserve SOC [7].

Soil quality is an aspect that depends on SOC, nature of organic fraction, microbial load, its activity, availability of nutrients and ultimately the ability of soil to generate a rich crop. Cropping systems play a vital role in conserving SOC and soil quality. It is said that the extent to which soil quality indicators are affected adversely or accentuated depends on cropping systems adopted. According to Karlen [8], crop rotations or intercrops of maize-soybean which dominates the Northern Great Plains has positive influence on several parameters that decide soil quality. A maize-soybean rotation improves C sequestration compared with maize sole crop. Indeed, evaluating various cropping systems and practices that aid better C sequestration is a preoccupation among crop scientists. Currently, we know that there are several computer-based cropping models and simulations that provide us with forecasts about C sequestration trends. For example, computer models such as CENTURY, Cquester or CERES-maize can help us with trends and forecasts about C sequestration in crop fields. No doubt, intensive cropping belts generate relatively larger amounts of biomass and recyclable residues. Whereas, subsistence farming approaches common in dry land leads us to low-C recycling trends. It results in progressive soil deterioration and reduction in crop productivity. Basically, biomass production and recycling is of lower order in dry lands. Hence, C accumulation is marginal. For example in Northern Indian plains, a crop rotation involving dry land maize-cowpea sequestered only 3 t SOC, despite supplying inorganic nutrients and farm yard manure [9]. We should also consider aspects like the extent to which C can be sequestered in soil. We know that soils differ for SOC status, its saturation levels and ability to sequester C (*see* Table 1). There are C-rich soils found in the Northern latitudes. The peaty soils of Northern Russia and Nordic region possesses 25–38% C. Loamy Mollisols that support large stretches of maize, wheat, soybean or cotton are termed relatively rich for SOC (> 3–5%). Silty or loamy Inceptisols found in Indo-Gangetic plains with 1–3% SOC are considered marginal in terms of SOC status, but possess greater potential to sequester C through residue recycling. Sandy soils in Sub-Sahara are relatively poor with regard to inherent SOC. Subsistence farming too does not allow residue recycling. Hence, potential to improve C sequestration is high. There is a large gap between actual SOC content and saturation levels. At this juncture, we should also note that soil characteristics such as texture, clay contents, moisture holding capacity, microbial load and activity also affects C sequestration trends and saturation point.

TABLE 1 Soil types encountered, Soil Organic Carbon content and Soil pH encountered in the Cropping Zones of different continents.

Cropping Zone	Soil Type	SO C %	pH	Reference
Corn Belt of USA				
Sterling, Nebraska	Cumulic Hapludoll	1.6	6.2	[10]
Fort Collins, Colorado	Aridic Haplustalfs	2.1	7.6	[11]
Dekalb, Illinois	Aquic Argidoll	2.4	7.2	[12]
European Cropping Zone				
European plains	Cambisols	2.6	6.3	[13]
Rothamsted, England	Chernozems	2.2	7.4	
	Luvisols	1.8	6.2	
Sophia, Bulgaria	Pelic Vertisols	3.4	6.7	[14]
South American Plains/hills				
Santa Caterina, Brazil	Hapludox	0.4	5.9	[15]
Botucatu, Sao Paulo	Latosols	0.2	6.5	[16]
Mossoro RN, Brazil	Red Yellow Argisol	0.2	6.8	[17]
Colombian Savannas	Acid Oxisol	2.4	5.5	[18]
Balcarce, Argentinean Pampas	Arguidoll	0.6	5.8	[19]
Eastern Colombia	Acidic Oxisol	2.3	4.8	[20]
West African Savannas				
Togo	Ferralsols	0.07–1.00	5.2–6.8	[21]
Ghana	Acrisol 0.97–1.56	5.5–5.6	[22]	
Ghana	Haplic Lixisols	0.88–1.12	5.5–5.8	[23]
Northern Nigeria	Arenic Haplustalfs	0.37–0.67	6.0–6.3	[24]
Ibadan, Nigeria	Oxic Paleustalfs	1.11–1.51	6.5	[25]
Borno State (Azir), Nigeria	Argillic Alfisols	0.82–0.94	6.1–6.5	[26]
Ile-Ife, Nigeria	Oxic Tropudalfa	0.85–1.12	6.5	[27]
Cameroon (highlands)	Palehumults	1.32–1.45	5.3–5.7	[28]
Other Sub Saharan locations	Aridisols	1.3–4.9	[13]	
South and East African Cropping zones				
Mpangala, Tanzania	Volacanic soils	2.2	5.9	[29]
Chitedze, Malawi	Ferruginous soil	2.1	6.2	[30]
Bembeke, Malawi	Ferraltic latosols	1.8	5.9	[31]
Harare, Zimbabwe	Lixisols	0.4	4.6	[32]
Mediterranean Region				
Hebron, El-Khalil region	Clay Loam	2.1	8.2	[33]

Central Valley, Jordan	Calcareous Loam	2.3	7.8	[34]
South and South-east Asia				
Dharwar, India	Vertisols	0.65	7.2–7.6	[35]
Bangalore, India	Alfisols	0.35	6.5–6.8	
Coimbatore, India	Haplustalf	0.40	8.7	[36]
Ludhiana, India	Ustrochrept	0.42	8.0	[37]
Bikaner, India	Sandy Alfisols	0.24	8.2	[38]
Faisalabad, Pakistan	Calciferous Inceptisols	0.85	7.7–7.8	[39]
Kampong Chang, Cambodia	Loamy Alfisol	2.4	6.7	[40]
Berili, Philippines	Sandy Alfisols	2.1	7.8	[41]
Chinese and Fareast Maize belts				
Pinglian, Gansu, China	Calcarid Regosol	0.9	8.2	[42]
Hainan, China	Ferralsols	0.9	4.8	[43]
Suwon, South Korea	Sandy soil	0.8	6.1	[44]
Kyoto, Japan	Alluvial soil	1.9	5.7	[45]
Australian Maize Growing Regions				
Comet in Queensland	Vertisols	1.31	8.8	[46]
Quirindi in New South Wales	Vertisols	1.19	8.9	[46]

Source: Ref. [69].

Carbon emission from natural vegetation, agricultural expanses and urban settlements is an important aspect of global C cycle. Carbon emission from agricultural soils is sizeable. It has consequences on SOC content, soil chemical process, soil microbial load and activity. Soil and crop respiration together result in emission of large amounts of C that was otherwise held sequestered in soil. Soil respiration may contribute about 25% of C exchange between terrestrial and atmosphere phases of ecosystem [47]. It amounts to exchange of 75 Pg C annually. There is no doubt that factors like geographic location, topography, cropping pattern and residue recycling trends, immensely influence the extent of C emissions from agricultural expanses. It is believed that small changes in soil properties, fertilizer inputs and cropping intensity may affect net C-emission from that location. However, it is difficult to monitor and estimate the minor changes in C emissions. Carbon emission is also influenced by the stage of the crop and type of vegetation. There are reports that C emissions may increase with crop age. Soil temperature, soil oxidative condition, pH, microbial load are few other factors that affect soil respiration and hence C emission from a field [47]. Soil respiration and C emission could increase exponentially with temperature. Fertilizer input, especially N has impact on seasonal changes in C emissions. For example, for an increase in N supply from 0 to 250 kg/ha, C emissions increased from 294 g C/m to 539 g C/m. Soil moisture and soil biota play a vital role in C emissions from agricultural fields. It is said that coarse textured soils with low soil moisture and microbial activity emit less CO_2. Fine textured soils and those with higher SOM content result in proportionately enhanced C emissions. Tillage is an important agronomic process that directly influences soil respiration and C emission. Conventional tillage often induces greater loss of C from soil. Hence, C sequestered into soil is small in fields kept under conventional tillage [48].

Crop production involves large-scale generation of biomass that encompasses plant tissues such as roots, stem, branches, leaves twigs, fruits/grains. Farmers allocate a portion of it to farm animals while a sizeable part may be recycled. Stover is used enhance to organic-C in soil; rather recycle organic-C from above-ground systems back into soil. It ultimately helps in improving soil quality and productivity (*see* Plates 1, 2, and 3).

PLATE 1 Storage of Cereal hay in a Farm near Doddaballapur, Bangalore District, South India.
Note: Large fraction of cereal forage is consumed by farm animals. Therefore, organic matter recycled is proportionately small.
Source: Krishna, K.R., Bangalore, India.

PLATE 2 Maize crop residue ready for incorporation into soil.
Source: Krishna, K.R., Bangalore, India.

PLATE 3 Sunflower stover applied on soil surface. It adds to soil-C.
Source: Krishna, K.R., Bangalore, India.

Generally, roots are allowed to decompose *in situ* and release nutrients. The mineral nutrient content and organic-C present in the tissue, biochemical nature of plant tissue, C:N ratio of residue/soil, age, moisture content, succulence of the organic matter and environmental parameters like temperature, relative humidity and soil type are some of the factors that influence decomposition rates and extent of nutrients recycled [49, 50]. Crop species differ widely for mineral nutrient content (Table 2). The C:N ratio of crop residue is an important indicator helpful in forecasting decomposition rates. There are several computer-based models and simulations (e.g., Century, Daisy, and C-quester) that help us in predicting crop residue decomposition rates and release of organic-C and mineral nutrients into soil.

TABLE 2 Nutrient contents of Crop Residue, Organic Manures and Green Manures used during Crop Production.

Source	Nutrient Content (%)		
	N	P	K
Paddy straw	0.36	0.08	0.12
Rice hulls	0.40	0.33	0.45
Sorghum straw	0.41	0.23	2.1
Pearl Millet straw	0.65	0.75	0.4
Cassia auriculata	0.98	0.12	1.8
Careya arborea	1.67	0.40	2.3
Terminalia chebula	1.46	0.35	1.2
Terminalia tomentosa	1.39	0.40	1.2
Crotolaria spp	2.89	0.29	0.7

Tephrosia spp	3.73	0.28	1.8
Water Hyacinth	2.04	0.37	3.4
Azolla spp	3.68	0.20	0.2
Cowpea	0.71	0.15	0.22
Dhaincha	0.62	0.41	0.81
Guar	0.34	0.51	0.68
Horse gram	0.33	0.25	0.18
Mungbean	0.72	0.18	0.15
Black gram	0.85	0.18	0.12
Sun hemp	0.75	0.12	0.62
Cattle Dung	0.3–0.4	0.1–0.2	0.1–0.3
Sheep Dung	0.5	0.4	0.3
Rural Compost	0.5	0.8	0.8–1.0
Farm Yard Manure	0.4–1.5	0.3–0.9	0.3–1.2
Vermi-Compost	1.6	5.0	1.7

Source: [51].

32.2 CROP RESIDUE RECYCLING

Crop residue recycling has direct influence on grain yield of cereals. For example, maize responds with grain/forage yield increase if crop residue is recycled. Partial removal of residue from crop fields and recycling the remaining biomass led to an increase of grain production by 5–13% over control. Crop residue recycling is also known to reduce fertilizer-N requirement by maize. Most importantly, recycling crop residue improves soil quality, SOC content and turnover of nutrients in the agroecosystem. Crop residues could be spread on the surface of soil, incorporated into soil at a depth or mixed thoroughly with soil (Plates 2, 3, and 4).

PLATE 4 Agroforestry tree species generate a large amount of crop residues that is collected as leaf litter, twigs and branches. Recycling such residues *in situ* or applying it to crop fields is beneficial in terms of sequestration of organic-C and supply of mineral nutrition. Nutrients are released into soil upon decomposition of such organic residues.
Source: Krishna, K.R., Bangalore, India.

Let us consider a few more examples that depict importance of crop residue recycling and organic matter inputs. Maize is cropped intensively in the Corn Belt of USA. Residue recycling enhances SOC content. Farmers, here add small amounts of fertilizer-N to prime decomposition of residue. Bundy [52] states that C:N ratio status of maize residue recycled and soil nutrient status are important factors that need attention. Next, in the Piedmont region of North-east United States of America, residue recycling supposedly enhances C sequestration. Tillage systems adopted during cotton/maize production interacts with extent of residue recycling creating better soil conditions. Removal of stover and avoiding residue recycling has deleterious influence on crop productivity. For example, maize silage production decreases from 18.5 t/ha to 17.5 t/ha. According to Amos et al. [53], improving plant population, irrigation and soil fertility results in more biomass. Consequently residue recycled too increases, thus enhancing C sequestration in the soil. We can quantify C sequestration by measuring soil surface CO_2 fluxes. We should note that soil emits CO_2 due to microbial respiration. In addition, other gases like CH_4 and N_2O that gets emitted also possess green house effects.

Studies on Mollisols of Northern Great plains have shown that recycling legume residue improves soil-N status. This is in addition to its influence on organic-C and microbial load/activity. Succulent legume residues decompose rapidly. In Spanish farms, over 20 t/ha of crop residues or cotton gin compost is added along with usual inorganic fertilizers. This aids rapid decomposition of organic matter and release of nutrients to crop roots [54].

The sandy soils of West African Sahel and savanna zones are highly prone to loss of organic matter component. This happens due to variety of factors such as loss of topsoil, erosion, leaching and percolation to subsurface. Tropical conditions induce rapid loss of soil-C through brisk microbial activity. West African farmers often find it difficult to recycle organic residues into soil. They actually apportion most of it to farm animals. Yet, West African farmers add organic matter just prior to rainy season and achieve perceptible increase of soil-C content so that crop productivity is enhanced. We can improve soil-C build up by adding sufficient amounts of N, P and K, so that rooting is improved. This leads to enhanced soil-C in the below-ground portion of ecosystem. In some parts of Northern Nigeria, farmers' burn scrub vegetation and weeds prior to adding inorganic fertilizers. Burning organic residues adds to mineral nutrients, mainly N, P and K. In East Africa, farmers tend to recycle N-rich pigeonpea and maize residue to improve soil-C contents [55]. Priming with fertilizer-N improves crop productivity.

Farmers in the Indo-Gangetic plains generate large quantities of crop biomass each season. Residue recycling is stringently followed during rotations like maize-legume, rice-wheat and oilseed-wheat. There are indeed several studies and routine reports that stress the need for residue recycling to improve nutrient recycling *in situ*. For example, retaining crop residues *in situ* enhanced soil-N by 11.5–38.5 kg N/ha under intercropped conditions and by 17.5 to 83.5 kg N/ha under sole [56] (Plate 2). Legume residues added 45–56 kg N/ha, if recycled into Inceptisols. Residue recycling during maize-groundnut rotation improved soil-C, if crop residue were recycled. Crop residue recycling is a common practice almost all agrarian zones of China. In Northern

China, where cropping is intensive, N recovery from residues ranges from 28–56% depending on soil moisture and environmental parameters [57].

There are indeed large number of reports and manuals that deal with ways to collect crop residues in plantations. Organic residue recycling is essential in order to sequester organic-C into soil within a plantation. Plantations generate a large amount of biomass in the form of new branches, twigs, leaves, inflorescences and fruits, each year. The amount of crop residue generated depends on a range of factors related to weather, crop, genotype, age of the plantation, agronomic practices, fertilizer supply and irrigation. For example, in Central Florida, young citrus groves of 4–7 years age generate 2,300–4,650 kg residue/season. Roots are major contributors to SOC. Roots add 1,890–4,200 kg residues to soil under the canopy. Leaf litter adds 377–410 kg residue/ha. The age of the citrus grove is important. A 20-year-old citrus grove in the same geographic area is known to add 6,100–10,287 kg organic residues/ha to soil under the citrus tree canopies. Cover crops such as Rhizoma peanuts add to soil-C upon incorporation at succulent stage (68).

32.3 GREEN MANURES AND NUTRIENT DYNAMICS IN AGROECOSYSTEMS

Agroecosystems thrive on variety of soils that differ enormously with regard to traits such as cation exchange and nutrient buffering capacity, organic matter content and crop productivity. Several of these cropping belts have actually evolved on highly weathered soils that are not endowed with characteristics optimum for crop production. Quite a few soils actually experience loss of surface soil, organic fraction and nutrients. Farmers try to overcome such soil deterioration effects by supply good quality organic matter in the form of green manure. Green manures are supplied periodically to improve SOC and soil quality. Slowly decomposing organic matter adds SOC rather slowly. Therefore, it is relatively less preferred in intensive cropping zones. They are apt to derive long-term effects on SOC content. Green manures add a certain quantity of mineral nutrients in addition to organic-C. In fact, leguminous green manure with a degree of succulence is preferred. It adds both N derived from atmosphere through BNF and organic-C to fields.

The soil types encountered in the savannas of Latin America are not richly endowed with organic-C, N and P that actually contribute to stable crop yield. Therefore, farmers in this region adopt agronomic procedures such as green manure production and incorporation *in situ*. This may first stabilize SOC content. They also avoid repeated tillage which otherwise leads to loss of SOC. They adopt cropping systems that systematically interject green manure production, during a short span of time between two major crops. Farmers in Brazilian Cerrados grow sun hemp, cowpea or pearl millet and incorporate biomass into soil to enhance SOC and soil quality. Sometimes, a mixture of cereal plus legume is grown, so that it adds to soil fertility especially C and N upon incorporation. Reports indicate that about 173 kg N/ha contained in the aboveground biomass of sun hemp and 89 kg N/ha found in pearl millet could be incorporated into soil, for the succeeding maize crop to use. We should note that inclusion of a leguminous green manure crop in the sequence, easily improves the soil-N budget by 10–15%. It also improves the C:N ratio of green manure and soil. Field trials on acid Oxisols of Cerrados have shown that incorporation of soybean-based green manure enhances maize grain yield by 1–1.5 t/ha. The maize-green manuremaize rotation improved SOC, N and P substantially [18, 58]. The amount of N derived from atmosphere through soybean-*Bradyrhizobium* and the extent to which it is channeled to soil are crucial aspects of crop sequence in this part of the world.

The Southern African farming zones are prone to soil deterioration, erosion of surface soil and nutrients. Therefore, farmers try to overcome the problem by growing leguminous green manure crops and recycling large quantities of SOC and N. Green manure species common to this zone are *Macuna, Crotolaria* and *Lablab*. It is said that timing of harvest of green manure is crucial. Aspects like selective pruning of twigs, young leaves and succulence of green manure need attention. Reports suggest that C content of green manure ranges from 0.1% to 0.4%, N ranges from 4.5 to 5.5% and lignin content from 2.6 to 16%. The C:N ratio of green manures ranges from 7.4 to 30.8. The average gain in maize grain harvest due to green manure crops is 2.7 to 2.9 t/ha for *Macuna*, 2.8–3.0 t/ha for *Crotolaria* and 3.0 t/ha for *Labalab* [30, 31]. Mtambanengwe and Mapfuma [59, 60] state that in Zimbabwe, green manures improve nutrient holding capacity and quality. Otherwise, sandy soils are prone to deterioration. Farmers in East Africa regularly cultivate green manure species and incorporate the succulent tissues to regain soil quality. They recycle a certain amount of nutrients. *Tithonia, Luecana, Sesbania* and *Cajanus* are popular as green manure crops. Farmers in East Africa tend to use green manures along with inorganic fertilizers. For example, supply of urea plus green manure improves N recovery by maize crop. Synchronizing availability of N using green manure species seems important. Long-term trials too have indicated that green manure plus fertilizers improves maize forage production by 3–9 t/ha. Residual effect of green manure species that may last for 2–3 years is an added advantage noticed by farmers in East Africa [61].

South-east Asian farmers, who specialize in wetlands and those situated in dry lands consistently, practice green manure production. Green manures are usually grown during short summer season and incorporated well ahead of main crop. *Leucana, Crotalaria, Macuna* and *Sesbania* are common green manure species in this part of the world. Reports from Indonesia suggest that N supplied by incorporation of green manure may reach 71–178 kg N/ha depending on the green manure species. The extra N recovered into succeeding maize crop fluctuates between 121–147 kg N/ha. Generally, South-east Asian farmers prefer to collect succulent green manure tissues, incorporate them just a couple of weeks prior to sowing the main crop. This allows timely release of nutrients to feed crop roots. Priming with inorganic nutrients is an important agronomic procedure.

32.4 ORGANIC MANURES

Farmers use a variety of organic wastes derived from animals to improve SOC content and soil quality. Farmers recycle organic-C and mineral nutrients via organic manures. Some of the most frequently used farm manures are dairy wastes, cattle dung, cattle slurry, piggery and poultry wastes. Let us consider a few examples. Application of dairy wastes improves soil microbial processes, enhances nutrient release and improves soil quality. For example, Habtesellaize et al. [62] found that dairy wastes improved SOC, soil-N, available-P and available-K. Soil organic pool almost doubled in 5 years. The soil microbial activity that is generally responsible for various nutrient transformations improved perceptibly due to supply of dairy wastes to soil. Population of Nitrifying bacteria increased 3 folds in dairy waste treated plots. The grain and silage yield of maize from dairy waste treated fields was almost always higher than those not provided with organic manures. Hountin et al. [63] have shown that application of piggery waste known as Liquid Piggery Manure (LPM) resulted in accumulation of soil nutrients such as N and P. Long term supply of piggery wastes had beneficial effect on soil nutrient availability. The field treated with LPM accumulated nutrients such as C, N and P in perceptible quantities. Cattle manure is an important organic amendment in most regions of the world. It has several beneficial effects on soil physicochemical traits and microbial component. However, in major farming belts of North America and European plains, farmers tend to supply both cattle manure plus inorganic fertilizers. This procedure is known to be significantly more effective in improving crop productivity. Cattle slurry or manure supply is known to recycle sizeable quantities of C and N within a farm. Cattle slurry incorporation in deeper layers of soil avoids undue loss of C, via leaching and emissions. The extent of nutrients added to soil depends on pattern of nutrient release from organic manures. Obviously, nature of organic manure affects nutrient dynamics in the root zone. Farmers in southern Europe, actually supply about 200–300 kg N/ha through cattle manure or slurry [64]. Generally, soil nutrient status of fields could get depleted rapidly, if organic manure such as cattle slurry is not added. The nutrient recycling due to supply of cattle manure is an important aspect of farming in Southern Europe. There are many instances that suggest that addition of cattle slurry could lead to build-up of soil nutrients. Poultry litter and wastes are most useful in enhancing soil-N and P status. Fields provided with poultry wastes are more productive. For example, poultry litter supply enhanced maize silage production by 2–3 t/ha [65].

In West Africa, cereal production depends on organic manure and residue recycling trends. Supply of organic manures is essential. On the sandy Oxisols, organic amendment improves nutrient and moisture buffering capacity of soils. It allows better rooting, nutrient acquisition and biomass formation by crops. In the humid tropics of West Africa, recycling animal wastes is an important procedure that ensures optimum nutrient dynamics. A combination of animal wastes, leaves from shrubs and crop residue may enhance soil-N by 120 kg N/ha. Such a procedure is known to improve N acquisition into crops by 31% over untreated control plots [66]. There are several types of organic manures used by Asian farmers. Many of them possess large amounts of nutrients essential to plants. However, they are often bound to organic components that need predigestion in order to release mineral nutrients. Mineralization of organic wastes helps in releasing nutrients to crop roots rather rapidly. Soil processes such as mineralization/immobilization of nutrients is actually dependent on several characteristics related to organic manures, soil and environment. For example, C:N ratio of organic manure, inherent soil nutrient status, soil moisture, atmospheric conditions may all influence nutrient release into soil. Succulence of crop residues mixed with animal manures is an important factor that affects decomposition and nutrient release rates in soil.

32.5 INDUSTRIAL BY-PRODUCTS

Industrial by-products rich in organic content are also recycled into soils. For example, farmers situated in different agrarian regions often recycle rice mill wastes, effluents from yeast distilleries, press mud, paper board wastes, timber and wood wastes, coir by products etc. The extent of organic-C and mineral nutrients added to soil and that available to crop roots depends on physicochemical nature of industrial by-products. In South-east Asia, rice mill waste (RMW) is an important industrial waste available close to farming zones. Organic manures prepared using RMV can be effective in enhancing crop productivity. In West Africa, field trials with RMW have shown that RMW adds soil nutrients and improves soil quality. Soil chemical analysis has suggested that 55% N, 40% C and 90% P held in RMW was degraded and released as soluble nutrients that gets easily absorbed by crop roots. The rice grain yield improved 0.5 t/ha, if the plots were treated with RMW. Supply of 30 t RMW/ha increased soil-C from 0.7% to 1.3%. Improvement in SOC has direct impact on crop growth and yield.

Yeast distillery Waste and Press mud are other by-products most frequently used by farmers. They improve soil-C and N status. Yeast sludge and Press mud contain about 1.5–2% N that can be effectively used to substitute fertilizer-N inputs. The list below compares nutrient composition of Yeast Distillery Sludge, Press Mud and Farmyard manure.

	pH	EC	Org-C	N	C:N	P	K	Fe	Cu	Zn	Mn
		dSm	%	%	ratio	%	%	ppm			
Yeast Distillery Waste	7.4	15.8	40.7	1.4	28.1	0.21	2.14	148	38	134	76
Press mud	6.4	3.12	39.4	1.1	35.2	1.40	0.95	35	45	35	25
Farmyard Manure	7.1	0.82	48.8	0.6	84.1	0.18	0.40	20	34	12	28

Source: [35].

Biofert is an industrial by-product derived during fermentation of lysine and other amino acids. It contains 6% N and several other ingredients that include amino acids, vitamins, enzymes, micronutrients, etc. [67]. Basically, *Biofert* affects soil N dynamics. It contributes large amounts of N based on dosages adopted. The agronomic efficiency is almost similar to N fertilizers. Interaction between soil moisture status and *Biofert* seems important. On Chernozems, application of 120–190 kg N/ha using *Biofert* was optimum and it produced 11.0 t maize grain/ha. Farmers use several types of organic preparations. Some of them contain a combination of organic wastes, microbes and earthworms. They are sometimes termed as vermin-composts. For example, 'Revital Maize' is an organic formulation based on vermicasts. It is said that such formulations play a vital role in releasing nutrient slowly to soil. Application of vermin-composts has resulted in enhanced maize grain and silage productivity. A typical organic formulation made of worm casts and soil microbes may contain following microbial ingredients:

Total Bacterial mass (mg/g):	249
Total Fungal mass (mg/g):	238
Flagellates (number/g):	11772
Amoeba (number/g):	94235
Ciliates (number/g):	2836

Source: http://www.revitalfertilizers.co.nz/files/11852%20 Maize.indd2005–6.pdf

The extent of nutrients supplied by such organic formulations may vary based on a range of factors related to source, composition, soil fertility status etc. In case of 'Revital Maize,' application of 3 t/ha releases 50 kg N, 49 kg P, 26 kg K, 7 kg S, 12 kg Mg and 171 kg Ca. In general, application of such organic formulations improves soil nutrient contents and quality. It enhances root growth and activity. It mainly improves nutrient cycling in the field.

KEYWORDS

- ***Biofert***
- ***Bradyrhizobium***
- ***Crotolaria***
- **Lablab**
- ***Macuna***
- **Mollisols**
- **soil organic carbon**

REFERENCES

1. Jenny, H. Factors of Soil formation. McGraw Hill, New York, USA, **1941,** 345.
2. Haas, H. J.; Evans, C. E.; Miles, E. A. Nitrogen and Carbon changes in the Great Plains soils as influenced by cropping and soil treatments. United States Department of Agriculture, USDA Technical Bulletin No **1957,** *1164,* 1–87.
3. Reeder, J. D.; Schuman, G. E.; and Bowman, R. A. Soil C, N changes on conservation reserve program lands in the Central Great Plains. Soil Tillage Research **1998,** *47,* 339–349.
4. Mikhailova, E. A.; Bryant, R.; Vassenev, J. I.; Schwager, S. J.; Post, C. J. Cultivation effects on Soil Carbon and Nitrogen contents at depth in Russian Chernozems. Soil Science Society of America Journal **2000,** *64,* 738–745.
5. Pikul, J. L.; Johnson, J. M. F.; Schumacher, T. E.; Vigil, M.; Riedell, W. E. Change in Surface soil Carbon under Rotated Corn in Eastern South Dakota. Soil Science Society of America Journal **2008,** *72,* 1738–1744.
6. Follet, R. F.; Varvel, G. E.; Kimble, J. M.; Vogel, K. No-Till corn after Brome Grass: Effect on Soil Carbon and Soil aggregates. Agronomy Journal **2009,** *101,* 261–268.
7. King, J. A.; Bradley, R. I.; Harrison, R.; Carter A. D. Carbon sequestration and saving potential associated with changes to the management of Agricultural soils in England. Soil Use and Management **2004,** *20,* 394–402.

8. Karlen, D. L.; Hurley, E. G.; Andrews, S. S.; Camberdella, C. A.; Meek, D. W.; Duffy, M. D.; Mallarino, A. Crop rotation effects on Soil Quality at three Northern Corn / Soybean belt locations. Agronomy Journal **2006,** *98,* 484–495.
9. Purkayastha, T. J.; Rudrappa, I.; Singh, D.; Swarup, A.; Badhraray, S. Long-term impact of fertilizers on soil organic carbon pools and sequestration rates in maize-wheat-cowpea cropping system. Geoderma **2008,** *144,* 370–378.
10. Lindquist, J. L.; Arkebauer, T. J.; Walters, D. T.; Cassman, K. G.; Dobermann, A. Maize Radiation use efficiency under optimal growth condition. Agronomy Journal **2005,***97,* 72–78.
11. Halvorson, A. D.; Mosier, A. R.; Reule, C. A.; Bausch W. C. Nitrogen and tillage effects on irrigated Continuous Corn yields. Agronomy Journal **2006,** *98,* 63–71.
12. Coulter, J. A.; Nafziger, E. D. Continuous corn Response to Residue Management and Nitrogen fertilization. Agronomy Journal **2008,** *100,* 1774–1780.
13. Smalling, E.; Toure, M.; Ridder, N. D.; Sanginga, N.; Breman, H. Fertilizer Use and the Environment in Africa: Friends or Foes. Background paper African fertilizer summit, NEPAD-IDFC, Abuja, Nigeria **2006,** 1–26.
14. Alexieva, S.; Stoimenova, I. A model concerning to the yield loss of maize from weed density or dry biomass. http://www.toprak.org.tr/isd/isd_82.htm **1998,** 1–5.
15. Sangoi, L.; Ender, M.; Guidolin, A. F.; Almeida, M. L.; Konflancze, A. Nitrogen fertilization impact on Agronomic traits of Maize hybrids released at different decades. Pesquisa Agroepcuaria Brasiliera **2001,** *36,* 1–14.
16. Theodoro, B. L.; Ferreira, E. Growth and osmotic adjustment of maize plants as influenced by Potassium and Water stress. http://natres.psu.ac.th/link/ soilCongress/bdd/symp 142219-t.pdf **1995,** 1–10.
17. Silva, S. L.; Silva, E. S.; Mesquita, S. S. X. Weed Control and Green ear yield in Maize. Planta Daininha **2004,** *22,* 1–9.
18. Basamba, T. A.; Barrios, E.; Amexquita, E.; Rao, I. M.; Singh, B. R. Tillage effects on Maize yield in a Colombian Savanna Oxisol: Soil Organic Matter and P fractions. Soil and Tillage Research **2006,** *91,* 131–142.
19. Sainz Rozas, Echevarria, H. R.; H. E.; Picone, L. I. De-nitrification in Maize under No-tillage. Soil Science Society of America Journal **2001,** *65,*1314–1323.
20. Oberson, A.; Buneman, E. K.; Friesen, D. K.; Rao, I. M.; Smithson, Turner, C.; B. J.; Frossard, E. Improving Phosphorus fertility in tropical soils through Biological Interventions. http://ciat-library.ciat.cgiar/articulos_Ciat/DK3724_CO37.pdf **2009,** 1–19.
21. Sogbedji, J. M.; Van Es, H. M.; Agbeko, K. L. Cover cropping and Nutrient Management strategies for Maize Production in Western Africa. Agronomy Journal **2006,** *98,* 883–889.
22. Yeboah, E.; Ofori, P.; Quansah, G. W.; Dugan, E.; Sohi, S. Improving Soil Productivity through Biochar amendments to Soil. African Journal of Environmental Science and Technology **2009,** *3,* 34–41.
23. Abunyewa, A.; Asiedu, E. K.; Nyameku, A. L.; Cobbina, J. Alley cropping *Gliricidia sepium* with maize: 1. The effect of hedgerow spacing, pruning height and phosphorus application on maize yield. Journal of Biological Sciences **2004,** *4,*81–86.
24. Kogbe, J. D. S Adedira, J. A. Influence of Nitrogen, Phosphorus and Potassium application on the yield of maize in the Savanna zone of Nigeria. African Journal of Biotechnology **2003,** *2,* 345–349.
25. Babalola, O.; Jimba, S. C.; Maduakolam, O.; Dada, O. A. Use of Vetiver Grass for Soil and Water Conservation in Nigeria. http://www.vetiver.org/NIG_SWC.pdf **2009,** 1–10.
26. Kamara, A. Y.; Ekeleme, F.; Chikoye, D.; Omoigui, L. O. Planting Date and Cultivar Effects on Grain Yield in Dry land Corn Production. Agronomy Journal **2009,** *101,* 91–98.

27. Tijani, F. O.; Oyedele, D. J.; Aina, O. Soil moisture storage and Water Use Efficiency of maize planted in succession to different fallow treatments. International Agrophysics **2008,** *22,* 81–87.
28. Yamoah, C.; Ngueguim, M.; Ngong, C.; Dias, D. K. W. Reduction of fertilizer requirement using lime and macuna on high P-sorption soils of North West Cameroon. African Crop Science Journal **1996,** *4,* 441–451.
29. Lisuma, J. B.; Semoka, J. M. R.; Semu. Maize yield Response and Nutrient uptake after micronutrient application on a Volcanaic soil. Agronomy Journal **2006,** *98,* 402–406.
30. Sakala, W. D.; Cadisch, G.; Giller, K. E. Interactions between residues of Maize and Pigeonpea and mineral N fertilizers during decomposition and N mineralization. Soil Biology and Biochemistry **2000,** *32,* 679–688.
31. Sakala, W. D.; Kumweda, J. D. T.; Saka, A. R. The Potential of green manures to increase Soil fertility and Maize yields in Malawi. www.ciat.cgiar.org/tsbf_institute/managing_ nutrient_cycles/AfnetCh26.pdf **2009,** 1–16.
32. Mapfumo, P.; Mtambanengwe, F. Base nutrient Dynamics and Productivity of Sandy soils under maize-pigeonpea rotational systems in Zimbabwe. www.ciat.cgiar.org/tsbf_institute /managing_nutrient_cycles/Afnetch16.pdf **2008,** 1–19.
33. Al-Bakeir, H. M. Yield, Growth rate and Nutrient content of Corn (*Zea mays L.*) hybrids. Hebron University Research Journal **2003,** *1,* 2003.
34. Khattari, S. Phosphorus levels in Soils of different vegetable crops in Jordan Valley. In: Plant Nutrient management under pressurized irrigation systems in the Mediterranean region. Ryan, J. (Ed.). World Phosphate Institute, Casablanca, Morocco, **2000,** 223–228.
35. Rajeshwari, R. S.; Hebsur, N. S.; Pradeep, H. M.; Bharamagoudar, T. D. Effect of Integrated Nitrogen Management on Growth and Yield of Maize. Karnataka Journal of Agricultural Science **2007,** *20,* 399–400.
36. Praharaj, C. S.; Shankarnarayanan, K.; Khader, S. E. S. A.; Gopalakrishnan, K. Sustaining Cotton productivity and soil fertility through *in situ* management of green manure and crop residues in Semi-arid irrigated condition of Tamil Nadu. Indian Journal of Agronomy **2009,** *54,* 415–422.
37. Dhaliwal, S. S.; Sadana, U. S.; Khurana, M. S.; Dhadli, H. S.; Manchanda, J. S. Enrichment of rice grains with zinc and iron through ferti-fortification. Indian Journal of Fertilizers **2010,** *6,* 28–35.
38. Sharma, K. C. Integrated Nitrogen Management in Fodder Oats in hot arid ecosystem of Rajasthan. Indian Journal of Agronomy **2009,** *54,* 459–464.
39. Iqbal, Z.; Latif, A.; Sikander Ali and Iqbal, M. Effect of fertigated Phosphorus on P use efficiency and Yield of Wheat and Maize. Songkalakarin Journal of Science and Technology **2003,** *25,* 697–702.
40. Belfield, S.; Brown, C. Field Crop manuals: Maize-a guide to Upland production in Cambodia. Australian Center for International Agricultural Research, Canberra, Australia. http://www.aciar.gov.au/publication/CoP10 **2008,** 1–75.
41. Comia, R. A. Soil and Nutrient conservation oriented practices in the Philippines. http://www.agnet.org/library/eb/472a/.htm **1999,** 1–13.
42. Fan, T.; Stewart, B. A.; Payne, W. A.; Yong, W.; Luo, J.; Gao, Y. Long term fertilizer and water availability effects on Cereal yield and soil chemical properties in North-west China. Soil Science Society of America Journal **2005,** *69,* 842–855.
43. Wu, W.; Chen, M.; Sun, B. Effects of Land use changes on Soil chemical properties of sandy soils from tropical Hainan, China. http://www.fao.org/docrep/010/ag125e/AG125E13.htm **2002,** 1–23.

44. Hur, S.; Kim, W. T.; Jung, K. H.; Ha, S. K. Study on soil and nutrients loss with soil textures and two crops during rainfall. Proceedings of International Crop Science Conference. http://www.cropscience.org.au/icsc2004/poster/1/6/1298_huros.htm **2004,** 1–4.
45. Li, K.; Shiraiwa, T.; Saitoh, K.; Takeshi, H. Water use and Growth of Maize under water stress on the long-term application of Chemical and/or organic Fertilizers. Plant Production Science **2002,** *5,* 58–64.
46. Carter, M. A.; Singh, B. Response of Maize and Potassium dynamics in Vertisols following Potassium fertilization. http://www.regional.org/au/au/asssi/supersoil2004/s13/ oral/1603_ carterm.htm **2004,** 1–12.
47. Ding, W.; Cai, Y.; Cai, Z.; Yagi, K.; Zheng, X. Soil respiration under Maize Crops: Effects of Water, Temperature and Nitrogen fertilization. Soil Science Society of America Journal **2007,** *7,* 944–951.
48. Duxbury J. M. Reducing Greenhouse warming potential by carbon sequestration opportunities, limits and tradeoffs. In: Climate Change and Global Food security. R Lal (Ed.) Taylor and Francis: Boca Raton, Florida, USA, **2005,** 435–450.
49. Johnson, J. M. F.; Barbor, N. W.; Weyers, S. l. Chemical composition of Crop Biomass impacts its Decomposition. Soil Science Society America Journal **2007,** *7,*155–162.
50. Machinet, G. E.; Bertrand, I.; Chabbbert, B.; Recous, S. Role of cell wall components on the decomposition of maize roots in soil: Impact on carbon mineralization. Proceeding of 18 World Congress of Soil Science, Session 2.2A Philadelphia, Pennsylvania, USA. **2006,** *138,* 1–2.
51. Krishiworld. The Pulse of the Indian Agriculture: Maintenance of Soil fertility. http://www.krishiworld.com/html/soil_ferti3.html **2002,** 1–9.
52. Bundy, L. G. Nitrogen application and Residue decomposition. Area Fertilizer Dealer Meeting. University of Wisconsin-Madison, Wisconsin, Soil Science Extension papers. **2001,** 1–2.
53. Amos, B.; Arkebauer, T. J.; Doran, J. W. Soil Surface Fluxes of Green House gases in an Irrigated maize-based Agroecosystem. Soil Science Society of America Journal **2005,** *69,* 387–395.
54. Tejada, M.; Gonzalez, J. L. Crushed cotton gin compost effects on soil biological properties, nutrient leaching and maize yield. Agronomy Journal **2006,** *98,* 749–759.
55. Mupangwa, W.; Twomlow, S.; Walker, S.; Hove, L. Effect of Minimum tillage and mulching on maize (*Zea mays*) yield and water content of clayey and sandy soils. Physics and Chemistry of the Earth **2007,** *32,* 15–18.
56. Sharma A. R.; Behera, U. K. Recycling of legume residues for Nitrogen Economy and Higher Productivity in Maize (*Zea mays*) – Wheat (*Triticum aestivum*) cropping system. Nutrient Cycling in Agroecosystems **2008,** *83,* 197–210.
57. Wang, X. B.; Cai, D. X.; Hoogmooed, W. B.; Perdock, U. D.; Oenema, O. Crop residue, Manure and Fertilizer in Dry land Maize under Reduced tillage in Northern China 1. Grain Yields and Nutrient Use efficiencies. Nutrient Cycling in Agroecosystems **2007,** *79,* 1–16.
58. Adriano, P Santos, R.; Urquiaga, S. S.; Guerra, J, G. M.; Freitas. Sun hemp and millet as green manure for tropical maize production. http://biblioteca.universia.net/ficha.do?id=10255750.htm **2006,** 1–3.
59. Mtambanengwe, F.; Mapfumo, P. Effects of Organic resource quality on soil profile N dynamics and maize yields on Sandy soils of Zimbabwe. Plant and Soil **2006,** *28,* 1132–1138.
60. Mtambanengwe, F, Kosina, P.; Jones, J. Maize Crop Residue Management-mulch feed or burn. IRRI-CIMMYT Cereal knowledge bank. Knowledgebank.cimmyt.org **2007,** 1–3.

61. Zingore, S.; Mafogoya, P.; Myammgufata,; Giller, K. E. Nitrogen mineralization and maize yields following application of tree prunings to a sandy soil in Zimbabwe. Agroforestry Systems **2003,** *57,* 199–211.
62. Habtesellassie, M. Y.; Miller, B. E.; Thacker, S. G.; Stark, J. M Norton, J. M. Soil Nitrogen and Nutrient Dynamics after repeated application of treated Dairy–waste. Soil Science Society of America Journal **2006,** *70,* 1328–1337.
63. Hountin, J. A.; Couillard, D.; Karam, A. Soil Carbon, Nitrogen and Phosphorus contents in maize plots after 14 years of Pig slurry applications. The Journal of Agricultural Science **1997,** *129,* 187–191.
64. Grignani, C.; Zavatarro, L.; Saco, D.; Stefano, M. Production, Nitrogen and Carbon balance of maize-based forage systems. European Journal of Agronomy **2007,** *26,* 442–453.
65. Juan, H.; Walter, I.; Undurraga,; Cartagena, M. Residual effects of poultry litter on silage maize (*Zea mays*) growth and soil properties derived from volcanic ash: fertilizers and soil amendments. Soil Science and Plant Nutrition. **2007,** *53,* 480–488.
66. Ikpe, Ndegewe, F. N.; Gbaraneh, N. A.;, L. D.; Torunana, J. M. A.; Williams, T. O.; Larbi, A. Effects of sheep browse diet on fecal matter decomposition and N P cycling in the humid lowlands of West Africa. Soil Science **2003,** *168,* 646–659.
67. Pepo, **2001,** Nitrogen-fertilization using 'Biofert' in sustainable Maize production. http://www.date.hu/ acta-agraria/2001–01/pepo.pdf 1–8.
68. Rouse, Muchovej, R. E.; R. M.; Mullahey, J. Guide to using Perennial Peanut as a cover crop in Citrus. University of Florida IFAS Extension Publication No HS-805 http://edis.ifas.ufl.edu/ch180 **2012,** 1–2 (June 25, 2012).
69. Krishna, K. R. **2013,** Maize Agroecosystem: Nutrient Dynamics and Productivity. Apple Academic Press Inc.: New Jersey, pp. 348

EXERCISE

1. Give examples of sources of Organic matter and C content in at least three Cropping Areas.
2. What is the difference between incorporation and surface spreading of organic matter in terms of improvement of SOC content in soil?
3. Mention at least 10 soil types and their SOC content and what is effect on crop production.
4. Why should we apply FYM or Organic Manure 4–7 weeks ahead of sowing dates?
5. Define Cellulose Digesters and mention few cellulose-digesting microorganisms.

FURTHER READING

1. Magdoff, F. In: *Soil Organic Matter in Sustainable Agriculture*. CRC Press: Boca Raton, Florida, USA, **2004,** pp. 416.
2. Kucharik, C. J.; Bryce, K. R.; Norman, J. M.; Foley, J. A.; Gower, S. T.; Bundy, L. R. Measurements and Modeling of Carbon Nitrogen cycling in Agroecosystems of Southern Wisconsin. Potential for SOC sequestration during the next 50 years. Ecosystems, **2001,** *4,* 237–255.
3. Partey, S. T.; Preziosi, R. F.; Robson, G. D. Effects of Organic residue chemistry on Soil Biogeochemistry: implications for Organic Matter Management in Agroecosystems In: Adewuyi, B.; Chukwu, K., Eds., Soil Fertility: Characteristics, Processes and Management, Nova Publishers, NY, USA, **2012**, pp. 1–28.

USEFUL WEBSITES

http://www1.agric.gov.ab.ca/$department/deptdocs.nsf/all/aesa1861 (October 14, 2012)
http://landresources.montana.edu/NM/Modules/Module8.pdf (October 14, 2012)
http://www.eolss.net/sample-chapters/c19/E1–05–08–02.pdf (October 14, 2012)
http://www.dpi.nsw.gov.au/__data/assets/pdf_file/0016/41641/Organic_matter.pdf (October 14, 2012)
http://naldc.nal.usda.gov/download/50219/PDF (October 14, 2012)
http://www.licor.com/env/newsline/2007/10/measuring-soil-co2-flux-in-maize-soybean-agro-ecosystems/

CHAPTER 33

SOIL MICROORGANISMS AND NUTRIENT DYNAMICS IN AGROECOSYSTEMS

CONTENTS

33.1 MICROBES IN SOIL

We may take note that soil forms a major ingredient of agroecosystems that sprawl different continents. The soil phase of an agroecosystem sequesters mineral nutrients, organic matter, moisture and harbors soil microbes. Soil microbial flora may constitute only a small fraction of total organic matter, yet its significance to soil biotic phase and nutrient transformations is immense. Soil microbes are crucial to crop growth and productivity. Soil microflora that occur on the root surface i.e., rhizoplane microbes; those existing close to roots within 0.2–0.3 mm from root surface i.e., rhizosphere and soil microflora in the bulk soil, all have their important role to play in nutrient transformation. Soil quality is an aspect of concern in most regions that support moderately productive or intensive cropping belts. Soil microflora is an equally crucial component of dry land soils that are poor in fertility traits. In fact, ensuring optimum soil microbial component, mainly its population and activity is needed even for subsistence crop production. Soils are particularly endowed with richly diverse microflora. The agroclimate may have its share of influence on soil microbial component. The dry lands may show relatively low microbial populations and activity. Actually, low water tension and high temperatures seem to curtail microbial activity. Loamy arable soils may show up higher microbial density of soil microbes and greater diversity. Soils that are rich in organic matter or those supplied with optimum levels of organic manures and continuously cropped with staple crops, may however show high microbial counts. We should note that almost every factor that affects crop also affects soil microflora found in the cropped field.

In addition to general activity, soil microflora is known to perform certain specialized functions in soil that are of great relevance to crop growth and production. Until date, we have accrued large body of information about the activities of microbes that fix atmospheric-N, either symbiotically with legumes or saprophytically in the rhizosphere and bulk soil. There are several genera of soil microbes that ensure availability of soil nutrients by transforming them into available forms. Microbes that dissolve partially available soil-P called Phosphate Solubilizing Microbes (PSM) have their share of impact on soil-P availability to crops. Mycorrhizas, different types that occur in symbiotic association with forestry trees and annual field crop species, partly ensure P acquisition by crops. There are still other groups of soil microflora that affect nutrient dynamics in the agrecosystem by releasing plant growth promoting substances. They are called Plant Growth Promoting Rhizobacteria (PGPR). All soil microbial groups are influenced by environmental factors.

The soil microbial component is important in many ways to the development, sustenance and perpetuation of agroeocsystems. Soil microbes are highly adaptable biological entities and are part of agroecosystems that thrive in high or low fertility conditions. They negotiate almost all the congenial or harsher factors that crops perceive during a season. Soil microbes are diverse with regard to genera, species, subspecies, and biotypes or even isolates. They perform diverse functions that contribute to sustenance of soil biotic activity and plant growth.

The soil microbes traced in an agroecosystem can be classified using several criteria, for example, based on their preference to environmental parameters [1]. They may be classified based on morphological traits and biochemical aspects of nutrition. Microbes are also classified based on their need for nutrients into groups such as autotrophic, chemo-autotrophic, lithotrophic, heterotopic, saprophytic, facultative and obligate pathogens etc. Then, microbes are also grouped based on ecosystematic functions performed. The extent to which a particular group or class of microbes flourish and function depends on variety of factors that operate within an agroecosystem. Soil microbes are important as mediators of several different nutrient transformations in the soil profile. Many of these nutrient transformations are essential to make nutrients available to plant roots. Several other microbial processes that add to soil-N (BNF), solubilize P, enhance P acquisition by crops (mycorrhizas) and those which release plant growth promoting substance are also equally important.

33.1.1 FACTORS AFFECTING DISTRIBUTION, QUALITY AND QUANTITY OF SOIL MICRO-ORGANISMS

Soil factors and environmental parameters that influence the soil microbial component and its functions are as follows: Soil texture and structure; soil moisture, soil temperature, soil aeration, soil reaction (pH), soil organic matter, agronomic procedures including cropping systems adopted, soil fertility and supply of extraneous nutrients.

Nature of soil has its impact on the crop that develops above ground, its roots and microbial component. The soil physical aspects like texture, structure, pores and aggregates all affect soil microbes and their distribution. The soil mineralogical and chemical features such as extent of chemical weathering, lattice structure, nutrient contents and their availability and SOM also influence soil microbial population and their activity. Microbial activity may in turn influence crop growth in a given agroecosystem. Usually, soil with good aggregation supports microbial growth and activity better than highly gravely or sandy soils, prone to percolation and leaching. Soil aggregates allow better adhesion and niche for soil microbes to operate.

Soil moisture is among the most important factors that affects soil microbial flora, its diversity, total population and activity. Soil water is essential to dissolve and ensure diffusivity of nutrients and minerals. Almost all soil microbial activity needs a certain level of soil moisture, for example to absorb nutrients, effect biological transformations and derive energy for growth and activity. This aspect has direct bearing on soil microbial component of the agroecosystem. Soil microbes proliferate and function at optimum levels when soil moisture ranges from 20–60%. In the crop field, soil water held at one-third water holding capacity is said to be optimum for crop root as well as aerobic microflora. In flooded soils, soil microflora differs enormously. Since only anaerobic and microaerophillic microorganisms survive, grow and function. The anaerobic flora that negotiate low-redox potential, inundated water and low oxygen tension have major role to play in the vast rice agroecosystem of Asia.

Soil temperature is an important factor that affects optimum physiological activity, growth, multiplication of soil microbes. Soil temperature optimum for microbes encountered in different agroecosystem may vary. For example, those encountered in temperate wheat belts of North America and European Plains may be active at relatively low temperature ranges of 12°C–24°C. Psychrophilic microbes may dominate the soil microbial component of temperate agroecosystems. Where as those in tropical cropping zones of Africa or Asia may with stand higher temperature at 18°C–32°C and still multiply and function normally. Mesophyllic soil microbes thrive and flourish in the soil of cropping zones found in tropical and subtropical regions of the earth. Soil microbial species may tolerate even higher temperatures in the dry lands. For example, they thrive in the Xerasols of West Asia or Vertisol belts of South India, where soil temperature during crop season may reach 38°C–41°C. The characteristics of soil microbes, such as production of spores or propagules that stay dormant or over winter or hibernate; if they encounter unfavorable soil temperature is rather important. Seasonal fluctuation of soil temperature too has its effect on soil micro-biota and its activity.

Soil aeration and availability of O_2 during respiration is essential for all aerobic microflora. Often, activity of soil microbes is detected or gauged quantitatively using the respiratory quotient i.e., ratio of O_2 absorbed versus CO_2 liberated. Gaseous exchange in the soil micropores is necessary for microbes to proliferate and function. Whenever, soil O_2 tension diminishes, perhaps due to inundation or flooding or saturated moisture conditions in soil, there is possibility for anaerobic microflora to appear, grow and function in the soils. For example, in the low land rice belts, flooded ecosystem supports anaerobic microflora during paddy cultivation. However, when farmers switch to arable crop in the next season, microflora undergoes a massive and drastic change in terms of diversity and function. The aerobic microflora functions in the ecosystem. Clearly, aeration has immediate effect on plant roots, their respiratory activity and growth. Equally so, it affects soil microflora, their number and species diversity as well as ecosytematic functions [2, 3]. Soil microbial diversity and their preferences need due attention. For example, most soil fungi and symbiotic species such as arbuscular mycorrhizas are aerobic. Arbuscular mycorrhizas proliferate and improve P uptake in soils that are aerobic. Agroecosystems that exist perpetually on aerobic soils excepting during heavy precipitation events support AM fungi, their growth and multiplication. Agroecosystems that experience heavy rains, flooded conditions and low aeration will obviously support anaearobic microflora and their activity. Soil aeration, if it is feeble with low O_2 tension allows micro-aerophillic microflora. In summary, soil aeration affects soil microflora. As a consequence, soil nutrient availability and nutrient transformations mediated by soil microbes are also influenced.

Soil reaction or pH (H^+ activity) affects proliferation and activity of soil microbes. The soil pH optima and extremes that soil microbes can withstand differs based on soil type, cropping systems, soil amendments, etc. Agroecosystems developed on acid soils in the Cerrados, for example, will allow proliferation of soil microbes that with stand acidic pH. Let us consider another example. Vertisols in South India are alkaline by nature. They predominately harbor microbes that have soil pH 7.0–8.2 as optima. Root exudates released may affect soil pH and microbial component in the soil close to roots and rhizosphere. Aspects such as quantity, quality and period for which root exudates are released and their diffusivity need due attention. Root sloughings, small sized organic debris and exudates that are energy rich may have direct impact on soil microflora and its activity in the root zone. If extrapolated to large expanses of soil and the crop species that exists in an agroecosystem, soil pH and its fluctuations as well root exudates may together influence microbial flora, root activity and nutrient acquisition enormously.

Soil organic matter is an important fraction that again varies enormously based on soil type, profile depth, aeration, cropping systems, and natural processes such as erosion, percolation and seepage. Soil microbes depend directly on SOM for food and energy. Microbial population is often high in soils that are rich in SOM. For example, SOM rich Mollisols found in Great Plains may show high microbial count compared to dry sandy soil in Sahel or West Asia with very low SOM component. Soil microbes bring about several types of transformations to SOM components. Microbial flora responsible for each or a few steps of nutrient transformations may differ. Microbial succession that matches the nutrient transformation process in soils needs due attention. For example, SOM decomposers first liberate mineral-N forms such as NO_3. Nitrifying bacteria (*Nitrosomonas, Nitrobacteria*) may dominate during next step. There are indeed several soil microbe mediate transformations of SOM that relate to lignins, polysaccharides, sugars, proteins, fats and complex organo-metals. The above aspects affect soil phase of any agroecosystem.

Agronomic procedures do alter soil properties, nutrient composition and microflora in the upper layers of soil. Tillage is an activity that has direct consequences on soil aeration, microbial respiration and population. Tillage induces soil microbes to release CO_2 and reduces organic component in soils. Tillage provides a chemically oxidative condition that induces CO_2 loss. However, higher soil microbial population that results improves soil nutrient availability. Mineralization process gets induced. It releases mineral forms of nutrients useful for plant roots. Supply of fertilizer-based nutrients is an important procedure during crop production. It may initially induce immobilization of mineral nutrients into soil microbial fraction. This process may be transitory. Application of large amounts of FYM and other organic manures induces mineralization of soils. The soil C:N ratio is said to be a factor that influences mineralization/immobilization reactions effected by soil microbes. No doubt, many of these microbial functions have direct impact on soil nutrient availability to crops. Soils, that are fertile with long history of cropping and rich in organic matter, harbor high microbial population density per unit soil mass/volume. Hence, farmers try to build up SOM in their fields.

Tillage affects several types of soil borne fungi that reside in the rhizosphere and nonrhizosphere soil. There are indeed several reports on soil fungi that reside in root zone soil. In the Argentinean maize belt, Nesci et al. [4] have made an evaluation of different tillage systems such as no-till, reduced-till and conventional-till on various fungi that colonize soil under a maize crop. No-tillage plots without grazing had highest fungal density at 5.7×10^3/g soil. Major genera of soil fungi noted were *Aspergillus, Fusarium*, *Penicillium, Trichoderma*, *Cladosporium* and *Alternaria*. Several species of each genus were detected in the rhizosphere of maize. There are innumerable reports about effects of tillage on soil bacteria, actinomycetes, protozoa etc. Clearly, tillage operations do affect soil microbiota. Consequently, it may have an impact on soil quality and nutrient transformations in soil.

TABLE 1 Influence of Tillage on Soil Microbial Diversity, Soil aggregation and Grain Yield of Maize grown at Nyabeda in Kenya.

Tillage	Bacterial Diversity	Fungi	Soil Aggregate	Maize Grain Yield
	H	H	MWD	t/ha
Restricted Tillage	2.02	1.56	1.81	3.51
Conventional Tillage	2.04	1.67	1.47	3.71
SE±	0.075	0.057	0.080	0.156

Source: [5].

Microbial diversity in soil was assessed using PCR-DGGE technique.

MWD = Mean Weight Diameter; H = Shanon's index of Diversity.

33.1.2 'RHIZOSPHERE EFFECT' AND MICROBIAL POPULATIONS IN AGROECSOYSTEMS

In any agroecosystem, crops are among the prime inducers of biomass formation and consequent nutrient transformation and dynamics *per se*. The crop roots that form an integral part of below-ground portion of the agreocosystem has dominant impact on soil microflora, their composition, diversity and function. There is a phenomenon known as 'rhizosphere effect' that markedly affects soil microbes. Soil microbial population is several times greater in the rhizopshere (0.2–0.3 mm) compared to bulk soils away from roots. This phenomenon is crucial to crops, since it enhances soil microbial activity. The volume of rhizosphere encountered in a field is proportionate to crop species, especially it's rooting traits and rooting pattern. Aspects such as size of field, soil profile depth, root growth/senescence, rooting density, and physiological activity will govern the microbial population below-ground in an agroecosystem. A virgin field or large patch without cropping history and root activity and lack of rhizosphere effect will harbor only low intensity of soil microbes.

During past years, the topic of rhizosphere, its physicochemical characteristics in different cropping belts, microbial composition and nutrient transformation that get accentuated in the rhizosphere have been reviewed in detail [6]. According to McCully [7], rhizosphere is an important zone in the soil-plant continuum that regulates nutrient movement from soil to below-ground root portion. Rhizosphere mediates transport of almost all of the nutrients and other chemical substances from soil to plants. The value of rhizosphere and its impact on nutrient acquisition by crops may differ based on intensity of cropping. We may realize that in low fertility subsistence farming zones, nutrient acquisition from soil to root is relatively small compared to a densely planted high fertilizer input crop field. The rhizosphere and microbes in it will have to mediate transfer of relatively massive amounts of nutrient within the same given crop duration in a high intensity cropping zone. We need to study the mechanisms that aid absorption of larger quantity of nutrients within the same period i.e., from seedling to grain filling stage in case of most cereals and legumes.

Rhizosphere usually has higher microbial count compared with bulk soil. The root exudations rich in C and energy bearing compounds and sloughings are known to influence C dynamics in the rhizosphere favorably to microbes. There are reports that 'rhizosphere effect' influences accumulation of major nutrients. For example, NH_{4}-N and NO_3 accumulated in greater quantities compared with bulk soil. Fertilizer-N input too affects N accumulation in the rhizosphere. Rhizosphere microbes that mediate transformation of N may get accentuated based on forms of N available in soil.

In case of soil-P, it is said that P gets depleted in the rhizosphere rapidly compared to bulk soil. The diffusivity of soil-P is slower than the rate at which plant roots deplete P from the rhizosphere. Regarding K, rhizosphere accumulates slightly greater quantity of K compared with bulk soil. Soil amendments, fertilizer and FYM supply may all impart their effect on K accumulation and microbes in rhizosphere [8]. John et al. [9] state that knowledge about spatial and temporal changes in nutrient accumulation and microbes are essential to gauge the importance of rhizosphere to nutrient absorption in soil.

Cereal/legume intercrops are perhaps most common across different continents. It offers a unique situation regarding rhizosphere and microflora found in the root zone. The roots and rhizospheres of cereal and legume intercrops overlap and cross over. Root exudations, rhizosphere microflora and nutrient accumulation pattern of both crop species interact and overlap. The interactions could detrimental or beneficial at varying intensities to each intercrop. Of course, we may perceive only the net result when observe the roots, soil or the crops. According to Song et al. [10], intercrops improved microbiological component of root zone. The diversity and population increased due to intercropping. The total microbial biomass was higher in the root zones of intercrops than sole crops. Intercropping enhanced organic-C in the rhizospheres. There could be combinations of crops that might exhibit antagonistic effects on rhizospheres and microbial populations. Crop mixtures of several different species occur in different agrarian zones. The interactive effects of rhizospheres of crops that make up the mixture need attention.

Rhizosphere consists of an assortment of microbes that are useful, symbiotic, antagonistic, or commensalistic. For example, survey of microbial population in the rhizosphere of maize has shown bacterial species such as *Streptococcus pyogenes*, *Bacillus subtilis*, *Psuedomonas aeruginosa* and *Micrococcus* species. These were dominant bacterial species. On the other hand *Fusarium chlamydosporum*, *Aspergillus sp* and *Rhodosporum* were common among fungal species. Regarding symbiotic microbes, *Rhizobium japonicum*, *R.leguminosarum* and *R melilotti* were traced frequently. Interactions among certain Bradyrhizobial species and fungi were antagonistic. No doubt, alteration in Bradyrhizobial population and AM fungi, especially suppression could reduce beneficial effects of these organisms [11]. Such interactions may have direct bearing on the nutrient dynamics in the rhizosphere. Soil-N status and P absorption by maize roots could be influenced. Populations of plant growth promoting bacteria (PGPR) may also affect nutrient availability to maize roots. Therefore, studying microbial diversity and interactions may be useful. Molecular marker and restriction fragment length analysis (RFLP) based techniques could help in assessing microbial species in the rhizosphere [12–16].

In summary, we should note that root exudations, rhizosphere microbes and nutrient transformations that occur in soil may appear minute. However, if extrapolated to large expanse of crops or an agroecosystem in general, it is a massive quantity of nutrients that get transformed and/or transported across rhizosphere. Literally, every unit of nutrient that appears in the above-ground crop phase moves across rhizosphere, unless the crop is supplied with nutrients via leaves.

33.2 BIOLOGICAL NITROGEN FIXATION

33.2.1 LEGUME-RHIZOBIUM SYMBIOSIS AND SOIL MICROBES

The cropping systems adopted in different continents have often included a legume. This is done mainly to take advantage of its ability to form symbiotic association with Bradyrhizobia and fix atmospheric N. This process adds to soil-N. A legume crop in sequence offers proteinaceous grains and fodder. Lupwayi and Kennedy [17] have made some interesting observations regarding wheat-fallow system followed in the Northern Great Plains that got shifted to wheat-legume system. They state that historically, cereal–fallow system has been in vogue in the Canadian Prairies and Northern Great Plains. Nearly 80 years of consistent wheat-fallow with conventional tillage has caused loss of soil fertility, induced moisture deficits and progressive depreciation of productivity [18]. However, during recent years, with the advent of conservation/zero tillage systems and introduction of wheat-legume system, the fallow region in USA, decreased from 1.55 m ha to 0.83 m ha. In Canadian Prairies, fallows have shrunk from 1.08 m ha to 0.6 m ha. Legume belt filled with field peas, lentil and chickpea increased significantly in the Canadian Prairies. The field pea cultivating zone expanded markedly from 200,000 ha in 1992 to 1250,000 ha in 2005 [17]. This shift to wheat-legume system and inoculation with N-fixing Bradyrhizobia has influenced soil characteristics and microbial component.

Following is a comparison of effect of cropping systems on soil microflora and enzyme activity:

	Fallow-Wheat	Legume-Wheat Rotation			
		Lentil-Wheat	Flat pea-Wheat	Vetch-Wheat	Field Pea-Wheat
Bacteria (× 10^6)	17	73	54	68	63
Fungi (× 10^3)	58	130	103	112	142
Dehydrogenase (µg TPF/g/h)	47	109	85	98	89
Urease (µg Uh/g/h)	37	57	49	56	47

Source: [17].

Note: Soil microbial population and enzyme activity is vastly improved in soils grown to legume-wheat compared to fallow-wheat.

The contribution of grain legume-rhizobium symbiosis to soil-N and N dynamics of the system is important, since it reduces on fertilizer-N consumption. Estimates of

N derived from atmospheric-N fixation have varied depending on wide range of factors related to crop, rhizobial isolates, soil and environmental conditions. For example in the Canadian Prairies and Great Plains, extent of N fixed by legumes inoculated with seed inoculants are as follows: Chickpeas 0–141 kg N/ha, dry bean 0–165 kg N/ha; lentil 5–191 kg N/ha; Vicia faba 12–330 kg N/ha; soybean 0–360 kg N/ha; Alfalfa 44–308 kg N/ha; Cowpea 44–132 kg N/ha [17, 19]. Reports by International Plant Nutrition Institute at Norcross in Georgia, USA, suggest that naturally, soybean derives N from three sources namely that fixed by rhizobium, NO_3 and NH_4 derived from soil and lastly via N supplied using fertilizers. Inoculation with Bradyrhizobium and maximization of atmospheric-N fixation will help us to reduce on fertilizer-N application. Repeated inoculations are practiced to build a permanent rhizobial population that helps in symbiotic association. We can improve efficiency of rhizobia by using better strains of inoculant [19, 20]. There are clear indications that inoculation of legumes with rhizobia affects soil microbial component and soil-N dynamics in the cropping belt. For example, legume stover recycled into soil improves soil-N cycling. Natural processes such as gaseous-N emissions do occur from legume fields. The extent of N loss could be more or less depending on N atmospheric N fixed by the legume-Bradyrhizobium association. We can decrease loss of this N fraction by efficiently scavenging it using appropriate catch crops. Legume catch crops or fodder crops such as alfalfa are common in Northern Plains. It affects soil microflora, rhizobial population as well soil-N dynamics in the vast pasture zones [18].

Legume crops are integral to many of the agroecosystems that flourish in the African continent. Legume-Rhizobium associations have greater relevance to subsistence farming belts in the sub-Sahara. Nitrogen contributed by legume-rhizobium association is highly important in the absence of fertilizer-N supply. For example, cowpea is intercropped in large expanses of West Africa. On sandy Oxisols, cowpea is strongly dependent on biological N fixation. Reports suggest that nearly 66% of their N requirements is satisfied through symbiosis with rhizobium [22]. In some locations, almost 90% of N requirements of the legume crop could be derived via biological N fixation. In Botswana, inoculation of cowpea with rhizobium produced 1500 kg grain/ha and in South Africa it was 2600 kg/ha. Selection of cowpea genotype and rhizobial isolates for higher efficiency could impart consistency to grain yield. Cereal-legume intercrops are common in the semiarid regions of Africa. The legume component that adds to soil-N via bradyrhizobial symbiosis, again affects N dynamics in the field [23]. Bean cultivation as an intercrop is in vogue many areas of tropical Africa. Bean, again adds to soil-N through symbiosis with rhizobium. The extent of N derived may vary depending on several factors related to crop, soil and environment. Estimates may show that N derived from BNF could be 25–50 kg N/ha. Now, if we consider the entire cropping belt, N added to the region is of enormous advantage [24]. There are indeed innumerable reports and reviews about contribution of legume-rhizobium symbiosis to agroecosystems of South Asia and Fareast. Major legumes such as chickpea, pigeonpea, cowpea, bean, peas etc., fix N ranging from nil to 360 kg N/ha. The legume-rhizobium association may also affect multiplication and activity of several other soil microbes. Rhizobium interacts beneficially with several other soil microbes. Leguminous forages such as *Trifolium spp* are common in Australia. Estimations using natural abundance of ^{15}N suggest that annually clovers may add 101–137 kg N/ha to pasture soils. The extent of N derived from atmosphere and accumulated in plant tissue ranged from 65–74% depending on clover genotype [25]. Clearly, N derived from atmosphere could affect the N dynamics in the pastures. It can reduce fertilizer-N supply to agroecosystems.

33.2.2 ASSOCIATIVE NITROGEN FIXING ORGANISMS

The soil phase of most agroecosystems in tropics supports a large population of microbes termed associative N-fixing microorganisms. These soil microbes thrive on the soil organic matter and other energy yielding substances. They are traced in rhizoplane, rhizosphere and even in the bulk soil. These associative N fixing microbes add N to soil. Following are most commonly encountered associative N fixing bacterial species *Azospirillum brazilense* and *A. lipoferum*. There are six species of *Azotobacter* namely *A.chroococcum, A.vinelandi, A.beijerincki, A. nigricans, A. armeniacus* and *A. paspali* that are termed free-living, symbiotic N fixing microbes. In addition to the above two major genera, microbial species such as *Azoarcus sp, Gluconacetobacter diazatrophicus, Herbaspirillum sp., Achromobacter sp., Acetobacter sp., Arthrobacter sp., Azomonas sp., Bacillus sp., Beijerinckia sp., Derxia sp., Enterobacter sp., Klebsiella sp., Psuedomons sp., Rhodospirillum sp.,* and *Xanthobacter sp* traced in soils of different agroecosystems are known to fix atmospheric -N through nitrogenase enzyme activity (Saharan and Nehra, 2011). The extent of N derived from asymbiotic N fixation may vary depending on the crop species, soil type, and other environmental parameters. Farmers practicing integrated management methods invariably inoculate seeds with N fixing microbes. There are several brands of microbial inoculants that contain efficient strains of associative N fixing microbes. Often a combination of N fixers is found in the microbial inoculants (e.g., Symbion-N; Rhizoagrin; Biofix-N, Biobact etc.). These microbial inoculants ensure optimum population of associative N fixers in the root zone soil. The amount of N derived from atmosphere and added on to soil may vary from nil to 25 kg N/ha depending on crop and microbial isolates. For example, in case of barley, 23–32% of N absorbed by crop was derived through associative N fixing microbes [26]. There are indeed several factors that affect amount of N added to the soil ecosystem by these associative N-fixers. We may note that in cereal-based subsistence belt, N added to soil via associative N fixers is important. The fertilizer-N saved may be small but considering the vast expanses of subsistence farming zones in semiarid regions, the contribution of associative N fixers could be most useful. Clearly, asymbitoic N fixing microbes do affect N dynamics in the agroecosystems. It is said 90% of tropical cropping zones and 60% of those in temperate agrarian regions derive a certain amount of N from such asymbiotic soil microflora [27]. There are reports that inoculation with *Azospirillum* increase N uptake and grain yield of cereals such as rice, pearl millet, sorghum, foxtail millet and several other crop species [28].

33.3 PHOSPHATE SOLUBILIZING BACTERIA

In an agroecosystem, Phosphate Solubilizing Microorganisms (PSM) occur in greater density in the rhizosphere. However, proportion of PSM is small compared to total counts of general microflora. The population of PSM may depend on soil type, its P status, forms of P encountered, root traits of crop species and environmental parameters. In soils of poor fertility, population of phosphate solubilizing bacteria (PSB) was 0–107 cells/g soil. About 4% of total bacterial counts were contributed by PSB.

There are several species of bacteria that are capable of solubilizing organic-C and partially soluble phosphate sources. Following are few examples: strains of *Psuedomonas, Bacillus, Enterobacter* and several other genera are endowed with ability to solubilize P in soil. Isolates of fungi such as *Aspergillus* and *Penicillium* also release soluble P forms in soil. We may note that PSB may constitute between 2–50% of total microbial count in a soil sample. However, counts of P solubilizing fungi (PSF) are low at 0.1–0.5% of total microbial count. Phosphate solubilizing microbes are encountered in greater number in agricultural soils than in virgin uncropped zone [29]. The species diversity, population and efficiency of P solubilizers are influenced by soil type and cropping history. Bowsar [30] has recently listed several species of bacteria such as *Achromobacter, Acetobacter, Bacillus cereus, B. circulans, B. maegaterium, B polymixa, B. pumilis, B.subtilis, Psuedomonas putida, P. striata, P.liquifans, P fluorescence* and *Serratia phosphaticum*. Actinomycetes such as *Nocardia, Streptomyces* and *Micromonospora* also solubilize P in soil. A few of the fungal species that solubilize P in soil are *Aspergillus awamori, A.carbonum, A flavus, A fumigatus, Candida albicans, Penicillium bilagi, P digitatum, Rhizoctonia sp, Sclerotium rolfsii, Pythium sp,* and *Fomitopsis sp*. Certain blue green algae such as *Anabaena sp, Aulosira sp, Nostoc sp* and *Anacystis* are also known to exhibit P solubilizing ability in soils.

There are indeed several mechanisms recognizable by which PSMs bring about solubilization of insoluble-P fraction in soil. A large fraction of inorganic-P is actually complexed or held tightly with crystal lattice of soil particles. The release of this fraction is an important process leading to availability of P to roots. Firstly, soil bacteria may enhance availability of Ca-Phosphates by releasing acids that reduce soil pH in microsites. The reduction in soil pH leads to increased solubility of Ca-phosphates. Phosphate solubilizing microbes are known to release organic acids. Acidification of soil medium surrounding the microbes helps in releasing P from sources like apatite. On the other hand, release of carboxylic acids that have higher affinity for Ca, solubilizes more of P. Formation of complexes of organic acids with cations is another mechanism that helps in release of P from insoluble-P sources in soil. Phosphorus desorption from Al and Fe-bound P is another mechanism that provides soluble-P to roots. Mineralization of organic-P is an important phenomenon that releases inorganic-P moiety through the action of soil phosphatases. Plant roots and microbes release phosphatase enzymes active at both acidic and alkaline soil pH. Rock phosphates are ores of P that possess insoluble and small quantities of partially soluble-P in them. During preparation of soluble-P fertilizers, acidification using sulfuric or phosphoric acid of rock phosphates leads us to super phosphate fertilizers. In agricultural soils, rock phosphates cannot be used directly. Instead, inoculation with P solubilizing bacteria that liberate relatively higher levels of organic acids may be effective. In some flooded soils with anaerobic conditions prevailing in the soil profile, release of H_2S that reacts with insoluble ferric PO_4 yields solubilized ferrous-PO_4. Chelation is another mechanism by which P gets released into soil solution. Phosphate solubilizing microbes found in arable soils release large amounts of α-keto-glutaric acid into soil solution. In soil, α-keto glutaric acid is a powerful chelator of Ca ions. It can easily remove Ca containing moiety from P sources such as Flour-apatite, Chlor-apatites and Hydroxy-apatities aid easy availability of P in soil.

In nature, ability of PSM to effectively release soluble-P that finally gets translated to better P nutrition varies depending on several factors related to soil, microbial species and environmental factors. The PSMs may inherently vary with regard to production of organic acids, phosphatases and siderophores that help in enhancing P availability to roots. Let us consider an example. Oliviera et al. [16] assessed over 350 bacterial colonies derived from Oxisols of Brazilian Cerrado region that are low in P content. They classified microbes based on their ability to solubilize P found in various organic and inorganic sources. Further, they identified and typed microbes using molecular markers. Greatest P solubilizing effect was noticed on medium containing Ca_3PO_4. Bacterial strains like B17 and B5 identified as *Bacillus sp* and *Burkholderia sp* were most effective among the hundreds of species screened. The above species mobilized 57–68% of Ca_3PO_4 applied in 10 days period in the rhizosphere soil. Strom et al. [31] have shown that mobilization of P in the rhizosphere is mediated by elaboration of organic acids. The extent of organic acid exudation has direct impact on release of P in rhizosphere. Reports suggest that P solubilizing ability of PSMs may range from 25–42 μg P/L from inorganic sources and 8–18 **μg**/L from organic sources [32]. Addition of rock phosphates may improve release of P into medium. Evaluation of bacterial strains has shown that a few of them can release between 51 mg P/L to 156 mg/L from Ca-phosphate sources [29]. Screening of several different fungal and bacterial species has shown that interactive effects between soil. Plant and PSM governs the net P solubilizing effect. In case of certain Vertisol regions of Southern India, soil fungi showed better P solubilizing ability than bacteria. For example, fungal species of genera such as *Aspergillus, Pencillium* and *Fusarium* solubilized greater amounts of P than bacterial species such as *B subtilis* or *B. megaterium* [33]. We encounter high degree of diversity among P solubilizers with regard to establishment and proliferation in soil, in addition to P solubilizing capacity. Microbial inoculant containing mixtures of P solubilizers are preferred. There are several commercial microbiological formulations that contain phosphate solubilizing bacteria and fungi. They are being used worldwide in different agroecosystems to obtain better efficiency from insoluble P ores.

33.4 MYCORRHIZAS

Knowledge accrued during past 3 decades has shown that mycorrhizas are involved in several different ecosystematic functions and aspects of agroecosystems. Mycorrhizas have been attributed to play a wider role in the agroecosystems and forests [34–37]. Mycorrhizas mobilize N and P from soil to plant roots. Mycorrhizas help in improving P uptake from rock phosphates that possess P in sparingly soluble condition. Mycorrhizas are intimately linked to C cycle in soil. Mycorrhizas are known to interact with variety of soil microbes and bring about changes in soil biogeochemistry. Mycorrhizas have also been attributed to affect absorption of water from soils. They extract water from soil phase at levels generally not congenial for crops, if they are afflicted by drought. We should note that nutrients are mostly absorbed in dissolved state. Therefore, mycorrhizas may actually aid both water and nutrient acquisition under drought prone conditions. Mycorrhizas are known to play an important role bringing about interconnections between plants sown in a field. Roots from different plants, species or genera are interconnected via mycorrhizal hyphae. Nutrient transfers that occur between plants may be of great value in agricultural or natural vegetation. Mycorrhizas may affect soil microflora by changing the root exudation pattern. The carbon and energy rich compounds released by mycorrhizal roots improve microbial density and activity in the root zone. Mycorrhizas are capable of playing an important role in organic farming [38]. Most important function attributed to mycorrhizas is their ability to enhance root surface area and improve ability of plants to scavenge nutrient held in soil solution better. Mycorrhizal roots explore greater volume of soil than roots without them.

Mycorrhizas have potential to perceptibly affect P dynamics in crop field. They may enhance transfer of P from soil phase to above-ground stem and foliage, increase microbial biomass and activity in soil and also induce interplant transfer P and other nutrients. Consequently, crop growth and biomass could be increased in many instances. One of the advantages of microbial inoculants is that after a few repeated inoculations into soil, the said symbiont may establish itself and allow roots of crops that are grown in sequence to be infected without extra effort. Compatibility is not a major concern with AM fungi. They do form symbiotic associations with most annual crop species. Mycorrhizal benefits could be conspicuous in soils that are sandy, low in nutrient availability, low in organic matter and nutrient buffering capacity (Plate 1). Mycorrhizal fungal species or even isolates are to differ in their ability to improve P acquisition by crops. Hence, it is preferable to survey different agrarian regions, collect, isolate and multiply fungal isolates. Then inoculate those with better P acquisition capacity (Plate 1). In case of AM fungi, spores and propagules have to be multiplied on a living host; their roots and surrounding soil is extracted and applied as inoculum. Perennial grass host or legumes have been employed to develop AM fungal inoculum. Sand or peat based inoculum is most common. Mycorrhizal fungi occur naturally in most crop fields. The extraneous inoculum that contains efficient strains is supposed to enhance its population activity and help in P acquisition better. There are innumerable reports that deal with improvement of P acquisition by crop plants [8].

In a given crop field or a vast expanse (e.g., maize) AM fungi exhibit enormous diversity with regard to species, isolates, their ability to proliferate and improve P absorption of crops. Specific genotype may have overriding influence on AM fungal flora [39]. The maize genotype that dominates the field or cropping expanse affects the AM fungal flora, perhaps more perceptibly [40]. In Guatemala, for example, a change from local landraces to maize hybrids affected AM fungal colonization and propagule number. Omar [41], reports that AM fungus (*Glomus constrictum*) improved P uptake and accumulation by maize grown on low-P soils of Egypt. Reports from Kenya and Eastern Tanzania state that indigenous AM fungal strains are efficient in P uptake. Their accentuation and inoculation may improve P dynamics in the fields [42, 43]. In Central Anatolia (Turkey), inoculation with *Glomus mosseae* and *G etunicatum* improved uptake of P and Zn by maize [44]. Karasawa et al. [45] state that soil moisture affects the extent of P benefits derived from mycorrhizal inoculation. For example, on Andosols found in Japan, improvement of P recovery by maize roots was influenced by soil moisture status. Such reports clearly indicate that AM fungi potentially affect P dynamics in the crop field/zone. Estimations of P benefits derived could be small, but given the vast crop belts that harbor natural AM fungal flora the total of fertilizer-P or inherent soil-P that transits through AM fungi could be large. Hence, accentuation of AM fungal component in soil using efficient inoculant strains may help us manipulate AM fungi and P dynamics better. In case of ectomycorrhizas, forest tree species are usually inoculated at the seedling stage and transplanted so that the crop derives optimum benefits in terms of P dynamics.

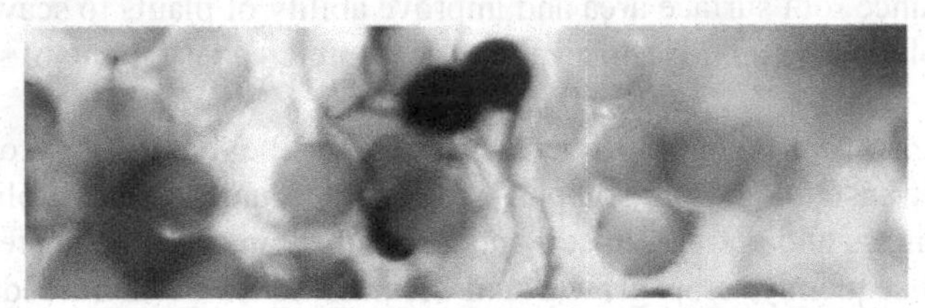

PLATE 1 *Glomus aggregatum*, a pure culture of Arbuscular Mycorrhizal fungus maintained on a perennial grass host.
Source: Krishna, K.R. Bangalore, India, ICRISAT, Hyderabad, India

33.5 PLANT GROWTH PROMOTING RHIZOBACTERIA

Plant growth promoting rhizobacteria (PGPR) are beneficial soil microbes that reside on rhizoplane, in the rhizosphere and bulk soil surrounding roots. They help the plant in different ways such as: release of siderophores that improve nutrient availability, induction of rapid root growth by releasing hormones, increase of nutrient uptake and shoot growth. A few of them may show antifungal properties and reduce disease incidence in root region. There are a few other PGPRs that could have a role in overcoming drought stress. Plant root growth stimulation that leads to better nutrient scavenging ability, has also been attributed to PGPRs [46]. In the present context, we are interested in the PGPRs that elaborate siderophores and improve nutrient uptake. Most commonly traced PGPRs in the rhizosphere are *Azotobacter, Azospirillum, Arthrobacter, Enterobacter, Bacillus* and *Pseudomonas* species. Many of these species are also active N-fixing microbes in free-living state in soil. The use of PGPRs may help us replace chemical fertilizers to a certain extent. They also enhance soil-N status. Obviously, PGPRs could be affecting a series of nutrient related functions in soil. There are several examples where PGPRs are known to improve nutrient uptake by maize and other cereals. Let us consider an example. In the dry regions, evaluation of several strains of PGPRs such as *Azospirillum lipoferum*, *A.brasilense*, *Azotobacter species* and *Bacillus sp* has shown that PGPRs improve absorption of N, P, K, Fe, Zn, Mn and Cu [47, 48].

The PGPRs seem to have a role to play in making Fe available to plant roots by releasing siderophores. The siderophores are high affinity, chelating compounds secreted by soil microbes that reside in the root zone. PGPRs are found to improve availability of Fe by elaborating siderophores. According to Paskiewicz and Berthelin [49], rhizosphere of plants like maize that adopt strategy II liberate siderophores that help in improving Fe availability to roots. The phyto-siderophores can dissolve iron from gothites and other similar sources. Experiments with maize have shown that, on Ferralsols, rhizosphere microflora enhances weathering of Fe and Mn hydroxides. It alters availability of Fe, Mn, Ni, Cr and Co. Several factors related to soil type and microbes affect the siderophore production.

It is important achieve higher population density of PGPRs in agricultural expanses. We may select them for higher efficiency and introduce them into the soil ecosystem. However, they have to multiply into optimum levels in the root zone and function. There are several factors that actually influence establishment, proliferation and efficiency of PGPRs. The ineffectiveness of PGPRs is often attributed to inability to establish in the roots zone. Aspects such as low motility, root exudate production and rapid multiplication are aspects related to PGPRs that affect their establishment. Over all, PGPRs affect nutrient dynamics in the roots zone of crops. The extent to which nutrient dynamics and crop productivity is influenced by them may be marginal or at times significant based on several soil/plant related factors.

32.6 AZOLLA AND BLUE GREEN ALGAE

Azolla is a floating fern found frequently in flooded paddy fields (Plate 2). They are well integrated with the wetland ecosystem of South Asia. Azolla put forth biomass rapidly and possess the ability to garner atmospheric-N through Blue Green Algae that are epiphytic. The extent of N fixed through BNF may vary from 20–40 kg N/ha in a season. Farmers often recycle and incorporate the succulent Azolla just prior to transplanting. Since Azolla is easily decomposable, it releases nutrients for roots to absorb in a matter of 7–10 days. There are several species of Azolla and Blue Green Algae that flourish in the wetland rice belts.

PLATE 2 *Azolla sp* common to wetland rice fields of India.
Source: Krishna, K.R., Bangalore, India.

KEYWORDS

- **A. lipoferum**
- **Azospirillum brazilense**
- **bradyrhizobial symbiosis**
- **Phosphate Solubilizing Microbes**
- **rhizoplane microbes**
- **soil fungi**

REFERENCES

1. Higa, T.; Parr, J. R. Beneficial and effective Microorganisms for a Sustainable Agriculture and environment. International Nature Farming Research Centre, Atami, Japan, **1994,** 1–20.
2. Krishna, K. R. Agrosphere: Nutrient Dynamics, Ecology and Productivity. Science Publishers Inc.: Enfield, New Hampshire, USA, **2003,** 346.
3. Krishna K. R. Agroecosystems of South India: Nutrient Dynamics, Ecology and Productivity. BrownWalker Press Inc.: Boca Raton, Florida, USA, **2010,** 548.
4. Nesci, A.; Baros, G.; Castillo, C.; Etcheverry, M. Soil Fungal Population in preharvest Maize Ecosystem in different Tillage practices in Argentina. Soil and Tillage Research **2006,** *91,* 143–149.
5. Kihara, J.; Vlek, P, Matius, C.; Amelung, W.; Bationo, A. Influence of conservation tillage on soil microbial diversity, structure and crop yields in sub humid and semiarid environments in Kenya. **2008.**
6. Drinkwater, L. E.; Snapp, S. S. **2006,** Understanding and Managing the Rhizosphere in Agroecosystems.http://ecommons.library.cornell.edu/bitstream/1813/3470/2/ Drinkwater-Snapp%20Revised%20Chapter-3–2006-JLW-LED-Final%20version-2rtf **2006,** 1–16.
7. McCully, M. The Rhizosphere: The key functional unit in plant/soil/microbial interactions in the Filed. Implications for the understanding of allelopathic effects. Journal of Agricultural Science, Cambridge **2005,** *130,* 1–7.
8. Krishna, K. R. Maize Agroecosystem: Nutrient Dynamics and Productivity. Apple Academic Press Inc.: New Jersey, USA, **2013,** 341.
9. John, K.; Kelly, J.; Schroeder, P.; Wang, Z. *In situ* dynamics of our macronutrients in the rhizosphere soil solution of Maize, Switch grass and Cottonwood. In: Rhizosphere: Perspectives and Challenges – A tribute to Lorez Hiltner. Neurenberg, Germany, **2005,** p. 192.
10. Song, Y. N.; Zhang, F. S.; Marschner, P.; Fan, F. L.; Gao, H. M.; Bao, X. G.; Sun, J. H.; Li, L. Effect of intercropping on Crop Yield and Chemical and Microbiological properties in Rhizosphere. Biology and Fertility of Soils **2007,** *43,* S14–15.
11. Liasu, M. O.; Shosanya, O. Studies of Microbial development on mycorrhizosphere and rhizosphere soils of potted maize plants and the inhibitory effect of rhizobacterial isolates on two fungi. African Journal of Biotechnology **2007,** *6,* 504–508.
12. Chang, C. Y.; Chao, C. C.; Chao, W. L. An evaluation of the Diversity of Fluorescent Psuedomonads in Maize Rhizosphere using 16S-23 rDNA intergenic spacer region restriction fragment length polymorphism and the Biology GN plate method. Taiwanese Journal of Agricultural Chemistry and Food Science **2007,** *45,*67–75.
13. Krishna K. R. Mycorrhizas: A Molecular Analysis. Science Publishers Inc.: Enfield, New Hampshire, USA, **2005,** 343.

14. Martin-Laurent, F.; Benoit, B.; Isabelle, W.; Severine, P.; Devers, M.; Guy, S. Phillipot, L. Impact of the Maize rhizosphere on the genetic structure, the diversity and the atrazine-degrading gene composition of cultivable-degrading communities. Plant and Soil **2006,** *282,* 99–115.
15. Payne, G. W.; Ramette, A.; Rose, H. L.; Weightman, A. J.; Jones, T. H.; Tiedje, J. M.; Eswar, M. Application of a rec-A gene-based identification approach to maize rhizosphere to assess novel diversity in *Burkholderia* species. FEMS Microbiology Letters **2006,** *259,* 126–132.
16. Oliviera, C. A, Alves, V.; M. C.; Mariel, I. E.;.; Gomes, E. A.; Scotti, M. R.; Carneiro, N.; Guimareas, C. T.; Schafert, R. E. Sa, N. M. H. Phosphate solubilizing microbes isolated from rhizosphere of maize cultivated in an Oxisol of the Brazilian Cerrado Biome. Soil Biology and Biochemistry **2009,** *41,* 1782–1787.
17. Lupwayi, N. Z.; Kennedy, A. C. Grain Legumes in the Northern Great Plains: Impacts on Selected Biological Processes. Agronomy Journal **2006,** *99,* 1700–1709.
18. Grant, C.; Entz, M. Crop Management to Reduce N Fertilizer Use. http://www.umanitoba.ca/afs/agronomists_ conf/proceedings/2005/grant_crop_management. **2005,** 1–6 (September 18, 2012).
19. Erker, B.; Brick, M. A. Legume Seed Inoculants. Colorado State University Extension Service. http://www.ext.colostate.edu/pubs/crops/00305.html. **2006,** 1–4 (September 19, 2012).
20. Stewart, D. M. Soybeans and N Fertilizer-Do they go together. International Plant Nutrition Institute. Plant Nutrition Today. **2012,** 1–3.
21. McVicar, R.; Panchuk, K.; Pearse, P. Inoculation of Pulse crops. Agricultural Knowledge Centre, Saskatchewan Agriculture, Canada, **2012,** 16.
22. Pule-Muellenberg, F.; Belane, A. A.; Krasova-Wade, T.; Dakora, F. D. Symbiotic functioning and Bradyrhizobial biodiversity of Cowpea (*Vigna unguiculata*) in Africa. BMC Microbiology **2010,** *10,* 89–90.
23. Musa, E. M.; Elshiek, E. A. E.; Mohammed, I. A. M.; Babiker, E. E. Intercropping Sorghum and Cowpea: Effect of Bradyrhizobium inoculation and Fertilization on Minerals composition of Sorghum seeds. ISRN Agronomy **2012,** *12,* 1–9.
24. Musundu, A. O.; Joshua, O. O. Response of Common Bean to Rhizobium inoculation and Fertilizers. The Journal of Food Technology in Africa **2001,** *6,* 121–125.
25. Denton, M. D.; Coventry, D. R.; Belloti, W. D.; Howieson, J. G. Nitrogen fixation in annual *Trifolium* species in alkaline soils as assessed by the ^{15}N natural abundance method. Crop and Pasture Science **2011,** *62,* 712–720.
26. Zawalin, A. Agroecological assessment of Associative Nitrogen fixation in Barley cultivation. In: Miedzynarodowe Sympozjum Ekologiczne Aspekty Mechanizacji Produkcji Roslinnej. Warszawa, Polska, **2001,** 304–313.
27. Swedrzynska, D.; Sawicka, A. Effect of Inoculation with *Azospirillum brasiliense* on Development and Yielding of maize 9Zea mays) under different cultivation conditions. Polish Journal of Environmental studies. **2000,** *9,* 5050–509.
28. Rafi, M. D.; Varalakshmi, T.; Charyulu, B. B. N. Influence of Azospirillum and PSB inoculation on growth and yield of Foxtail Millet. Journal of Microbiology and Biotechnology Research **2012,** *2,* 558–56.
29. Khan, A. A.; Jilani, G.; Akhatar, M. S.; Naqvi, S. S. M.; Rasheed, M. Phosphorus Solubilizing Bacteria: Occurrence, Mechanisms and their Role in Crop Production. Journal of Agriculture and Biological Sciences **2009,** *1,* 48–58.
30. Bowsar, S. Microbial Phosphate Solubilization www.biotecharticles.com **2011,** (October 10, 2012).

31. Strom, l.; Owen, A. G.; Godbold, D. L.; Jones, D. L. Organic acid mediated P mobilization in the rhizosphere and uptake by maize roots. Soil Biology and Biochemistry **2002,** *34,* 703–710.
32. Tao, G. S.; Tian, M.; Cai, M.; Xie, G. Phosphate solubilizing and mineralizing abilities of bacteria isolated from soils. Pedosphere **2008,** *18,* 515–523.
33. Sanjotha, P.; Mahanthesh, P.; Patil, C. S. Isolation and screening of efficiency of Phosphate solubilizing microbes. International Journal of Microbiology Research **2011,** *3,* 56–58.
34. Cardoso, I. M Kuyper, T. W. Mycorrhizas and Tropical Soil Fertility. Agriculture, Ecosystems and Environment **2006,** *116,* 72–84.
35. Turk, M. A.; Assaf, T. A.; Hameed, K. M.; Al-Tawaha, A. M. Significance of Mycorrhiza. World Journal of Agricultural Science **2006,** *2,* 16–20.
36. Finlay, R. D. Ecological aspects of Mycorrhizal symbiosis: with special emphasis on the functional diversity of interactions involving the extra radical mycelium. Journal of Experimental botany **2008,** *59,* 1115–1126.
37. Muchovej, R. M. Importance of Mycorrhizae for Agricultural Crops. University of Florida, IFAS Extension. Edis.ifas.ufl.edu/ag116 **2012,** 1–4 (September 18, 2012).
38. Mahmood, I.; Rizvi, R. Mycorrhiza and Organic Farming. Asian Journal of Plant sciences **2010,** *9,* 241–248.
39. Douds, D. D.; Millner, D. Biodiversity of Arbuscular Mycorrhizal fungi in Agroecosystems. Agriculture, Ecosystems and Environment **1999,** *74,* 77–93.
40. Hess, J. L.; Shiffler, A. K.; Jolley, D. Survey of mycorrhizal colonization in native, open pollinated and introduced hybrid maize in villages of Chiquimula, Guatemala. Journal of Plant Nutrition **2005,** *28,* 1843–1852.
41. Omar S. A. The role of rock phosphates-solubilizing fungi and vesicular arbuscular mycorrhiza in growth of maize plants fertilized with rock phosphate. World Journal of Microbiology and Biotechnology **1998,** *14,* 211–218.
42. Waceke, J. W. Use of Arbuscular Mycorrhizal fungi for improved crop production in sub-Saharan Africa. Journal of Tropical Microbiology **2002,** *1,* 14–21.
43. Yamane, Y.; Highuchi, H. Function of Arbuscular Mycorrhizal fungi associated with maize roots grown under indigenous farming systems in Tanzania. http://www. cababstractsplus. org/abstracts/ Abstracts.aspx?AcNo=20036796725.htm, **2003,** 1–2.
44. Ortas, I.; Kaya, Z.; Cakmak, I. Influence of arbuscular mycorrhizas inoculation on growth of maize and green pepper plants in Phosphors and Zinc deficient soil. Plant Nutrition **2001,** *92,* 632–633.
45. Karasawa, T.; Takebe, M.; Kasahara, Y. Arbuscular Mycorrhizal (AM) effects on maize growth and AM colonization of roots under various soil moisture conditions. Soil Science and Plant Nutrition. **2000,** *46,* 61–67.
46. Lugtenberg, B.; Kamilova, F. Plant Growth Promoting Rhizobacteria. Annual Review of Microbiology **2009,** *63,* 541–556.
47. Biari, J. A.; Gholami, J. A.; Rahmani, A. Z. Growth Promotion and Enhance nutrient Uptake of Maize (*Zea mays*) by application of Plant Growth Promoting Rhizobacteria in the Arid Regions of Iran. Journal of Biological Science **2008,** *8,* 1015–1020.
48. Saharan, B. S.; Nehra, V. Plant Growth Promoting Rhizobacteria: A Critical Review Life Sciences and Medicine Research 2011, **2011,** 1–30 http://www.astonjournals.com/lsmr (September 21, 2012).
49. Paskiewicz, I.; Berthelin, J. Influence of maize rhizosphere and associated microflora on weathering of Fe and Mn oxides and availability of trace elements in a New Caledonia Ferralsols. Geophysical Research Abstracts **2006,** *8,* 762.

EXERCISE

1. Discuss Legume-Rhizobium symbiosis with reference to Atmospheric-N fixed, Quantity of N contributed to soil.
2. Discuss the influence of legumes on succeeding cereal crop with regard to N nutrition.
3. Discuss different types of Mycorrhizas encountered by crops and plantations.
4. Explain how plant benefits from Mycorrhizas.
5. What is the influence of Mycorrhizas on P dynamics in Agroecosystem?
6. Give the names of at least five major genera of Arbuscular Mycorrhizas.
7. Mention the amount of Nitrogen gained by rice crop due to Azolla-Blue Green Algae.
8. Discuss the mechanism of P solubilization effected by Phosphate Solubilizing Bacteria.
9. Mention names of few Plant Growth Promoting Bacteria.

FURTHER READING

1. Abbot, L. K. Soil Biological Fertility: A key to sustainable land use in Agriculture. Springer: New York, **2010,** pp. 264.
2. Cheeke, T. E.; Coleman, D. C.; Wall, D. H. Microbial Ecology in Sustainable Agroecosystems. CRC Press: Boca Raton, Florida, USA **2012,** pp. 308.
3. De Vries, F.; Bardgett, R. D. Plant-Microbial Linkages and Ecosystem Nitrogen retention: Lessons for Sustainable Agriculture. Frontiers in Ecology and the Environment, **2012,** 10, 425–432.
4. Dixon, G. R.; Tilston, E. L. Soil Microbiology and Sustainable Crop Production. Springer: Heidelberg, Germany, **2010,** pp. 340.
5. Dixon, G. R.; Tilston, E. L. Soil Microbiology and Sustainable Crop Production. Springer Ltd: Dordrecht, Heidelberg, **2010,** pp. 451.
6. Khan, M. S.; Zaidi, A. Phosphate Solubilizing Microbes for Crop Improvement. Nova Publishers Inc.: New York, USA, **2011,** pp. 452.
7. Van Elsas, J. D.; Jansson, J. K.; Trevors, J. T. Modern Soil Microbiology 2nd Edition CRC Press: Boca Raton, Florida, USA, **2006,** pp. 672.

USEFUL WEBSITES

www.sssa.org
mycorrhiza.ag.utk.edu
www.microtrop.ird.sn/anglais/organization/organisateur.htm
www.nrel.colostate.edu/
www.ctahr.hawaii.edu/bnf/

CHAPTER 34

WEEDS AND NUTRIENTS IN AGROECOSYSTEMS

CONTENTS

34.1 INTRODUCTION

Weeds are worldwide in distribution. They are traced at different intensities in all agroecosystems of the world. The infestation of weed species differs enormously based on geographic location, season, weather pattern, soil type and its fertility status, crop species, its growth stage, and agronomic practices adopted by farmers. There are indeed several families of plants that behave as weeds in cropping expanses. At times volunteers from previous crop too become weeds. This situation is common in no-tillage plots. The weed flora in a crop field is often dynamic. The weed species that become dominant at different times and their potential to thwart crop growth needs to be understood properly. Weeds are no doubt part of a cropping expanse. They need to be controlled, if not, they divert significant quantities soil nutrients and moisture. Weeds could be deep rooted. Weed roots compete for underground soil space, nutrients and water with crop plants growing closely. They are often fast growing and in such situations weeds compete and intercept photosynthetic radiation. Weeds may reduce photosynthetic light reaching the crop canopy and therefore reduce biomass formation. Weeds multiply rapidly. The crop stage at which weeds become rampant seems important. Weeding at early stages of crop is most effective. The intensity and species of weeds flourishing in a field and the crop species are factors that affect biomass and grain/forage production.

34.2 METHODS TO CONTROL WEEDS IN DIFFERENT CROPPING ZONES

Weed control measures adopted depends firstly on weed species that infest a region and their impact in terms of grain/forage yield. Weeds could be easily classified as annuals, biennials and perennials. Weeds could also be classified as monocot grasses, dicots, sedges, and parasites. There are weeds that are specific to certain crops and locations. The weed control strategy should also consider location, growth pattern and life cycle of the weed(s). To quote a few examples, in a cereal field, application of herbicides specific to dicot and leguminous weeds is feasible, but not a chemical that works against cereals. Farmers may aim at suppression of weeds by applying pre-emergent herbicides. During mid-season, it is better to spray post-emergent herbicides. There are several cultural practices devised specifically to reduce or control weeds. Farmers often tend to restrict weed growth and keep them at levels below threshold. Agronomic measures that regulate weeds in a farmers' field are important. Foremost, weed propagules such as seeds, stolons, rhizomes, bulbs and tubers have to be cleaned and removed from crop fields. Dispersal agents responsible for spread of weed should be controlled. Major dispersal agents of weed seeds are wind, water, animals, man, farm machinery, FYM contaminated with seed, seeds, etc.

Tillage is one of the earliest and most important agronomic measures that farmers adopt to control weeds. Tillage may induce seed germination making weeds prominent. Such germinated weeds could be removed easily. Deep plowing is needed if weeds are well established and roots are widely spread. Intercultural operation with tined implements reduces weeds in the inter-row space. Weeds could also be controlled by sowing intercrops. There are also options such herbicide application and biological control of weeds. Weed control measures help the farmers in reducing loss soil moisture and nutrients to weeds. Worldwide, some of the most common herbicide brands used are Atrazine, Butachlor, Alachlor, Trifluralin, Simazine, etc. The quantity of weedicide (active ingredient), timing, method and number of applications are important aspects. During recent years, we have crop genotypes that are herbicide tolerant. Planting an herbicide-genotype will allow farmers to apply more of herbicides without affecting the crop and its physiological manifestations.

34.3 WEEDS AND NUTRIENT DYNAMICS IN DIFFERENT AGROECOSYSTEMS

Weeds are an important factor during crop production, anywhere in different agroecosystems. As stated earlier, weed flora that appear in the fields vary enormously with location, soil fertility, dominant crops grown in the area, agronomic practices and herbicide sprays. Tillage is a major procedure that reduces weed infestation in cereal fields. Pre-emergent soil application of herbicide is a useful procedure in almost all agrarian regions. Post-emergent sprays are also adopted by farmers. Intercropping and practicing cover crop during fallow period are important. These procedures reduce weed intensity, loss of soil moisture and nutrients. Timing of weed control procedure may have significant impact on nutrient dynamics of a field.

In the Northern Great Plains, crops such as wheat, maize, legumes and vegetables are all affected by weed population of different intensities. It is said that cultural weed control begins with selection of vast areas or fields with low weed emergence. Kochia, Russian thistle, wild buckwheat, wild oats, and several other grass species may all be troublesome to the growth and development main crop. On a large scale, reduction in wheat production due to weeds could be significant. Procedures such as conventional tillage or deep plowing once in 2–3 seasons, inter culture with tined implements, spray of weedicides such as glyphosate, sulpentrazone, atrazine, simazine, treflan and others may reduce weed flora considerably. Generally, integrated weed management procedures are preferred.

In Southern and Eastern regions of United States of America, crops such as cotton, groundnut and legumes are predominant. They are often rotated with cereals such as maize, sorghum and wheat. The weed flora in Southern Plains is diverse. Weed species such as *Cyperus rotundus, C.esculentus, Amaranthus palmae, Edipta prostrate, Achypa ostrifolia, Digitaria ciliaris, Panicum maximum, Brachiaria sp, Ipomea lacuna* and *I. hederaceae* are common to Southern Plains of United States of America. The weed infestation in a field could vary with regard to diversity of species, intensity and stage in a field could vary within each field or agrarian zone. Most of these factors affect extent of soil moisture and nutrients diverted to weeds. The quantum of crop yield reduction due to weeds and their activity may also vary. For example, total control of weeds using hand-weeding as a procedure could provide 687 kg groundnut pods/ha more than a weedy check plots. Spraying with Metachlor could increase groundnut pod yield by 277 kg/ha over a weedy check plot. In Georgia, strip tillage plus pre-emergent herbicide application is very effective in restoring nutrient dynamics in the groundnut fields. Nutrient diversion to weed flora is almost nil. Reports compiled based on over 100 field trials conducted for several years during past decade have shown that weeds may reduce crop yield by 7–12%, if stringent weed control procedures are not in place and adopted timely during early stages of crops [1]. Series of studies on weed infestation of pastures has shown that monocultures of pastures are vulnerable to weeds. Hence, farmers in the Midwest of USA, have been advised to establish mixed pastures using grass-legume combinations. The diversity of weed flora and intensity seems to depend on the grass-legume combinations sown and its ratios. They say growing mixed pastures helps in reducing application of herbicides. Grass-legume combinations based on Orchard grass (*Dactylisglomerata*), quack grass (*Elytrigia repens*), legumes like clover (*Trifolium repens*) and alfalfa (*Medicago sativa*) have performed well to thwart weed infestation in Wisconsin [2].

Weeds commonly encountered in the rice and legume-growing regions of South-east Asia are diverse and belong to several different families and genera. Weed species such as *Digitaris ciliaria, Cynodon dactylon, Cyprus rotundus, C.iria, Portulacaoleracia, P.indica, Agetarumconyzoides* and *Physalis minia* are predominant in South-east Asia (Plates 1 and 2). The weeds that infest wetlands do garner a sizeable quantity of nutrients that otherwise was meant for the main crop-rice. The extent of nutrients diverted from rice, of course depends on weed intensity and weeding procedures adopted. Farmers in South-east Asia adopt mild tillage and mulching to avoid germination of weed seeds. This procedure gives effective control over loss of nutrients to weeds, right from seedling stage. Cultural practices like crop rotation, intercropping and post-emergent herbicide spray is also in vogue. Weed biomass gets effectively reduced due to cultural practices. However, grain yield advantages are mostly dependent on types of weed control measures, their timing and intensity [3, 4].

PLATE 1 *Cassia occidentalis*is frequently traced in the Dry lands of South India and elsewhere in different continents.

Note: It is a strong shrub with deep roots and difficult to uproot and eradicate. Deep plowing and culling is essential.

Source: Krishna, K.R. Bangalore, India.

PLATE 2 *Parthenium* species are weeds common to many Agroecosystems across different Continents.
Note: Parthenium is deep rooted, prolific in growth, flowers profusely and propagates rapidly. Hence, it is a formidable weed for farmers to control. Eradication and burning prior to flowering is important. Deep plowing is essential to remove the weed roots and stolons from regenerating into fresh plants.
Source: Krishna, K.R. Bangalore, India.

Wheat is a major cereal grown in Indo-Gangetic belt. This wheat belt may suffer due to weed infestation of various intensities. There are several species of weeds that infest. Species such as *Philaris minor* is predominant in the plains. It is said that nearly 5 m ha of wheat belt is severely affected and rest mildly due to weeds. A few of the weeds are supposedly herbicide tolerant and may cause loss of soil nutrients and moisture. Integrated weed management measures such as suitable cropping system, herbicide application, hand weeding and deep plowing periodically seems important. There are suggestions to monitor and forecast weed infestation by studying the weed seed diversity and seed distribution in soil during off season [5].

Agrarian regions in South and South-east Asia support a very large intensive rice cropping belt. Farmers adopt continuous rice, with a rather short fallow. Weeds that proliferate during rice culture may reduce nutrient recovery by the main crop. It is said that in a single season, about 18 kg N, 3 kg P and 16 kg K/ha could be lost to weeds that grow in low land rice fields [6, 7]. Weed management is essential to reduce diversion of nutrients. Transplanted paddy seems to be affected less due to prolific weed growth. Generally, a small portion of fertilizer nutrients supplied to fields could be diverted to weeds. On an average, if rice crop has absorbed 35–55 kg N/ha, then during the same period and given similar conditions weeds in the field may have absorbed 2–12 kg N/ha. Maintaining rice fields weed-free throughout the season is important. For example, if a weedy rice field absorbs 35 kg N, 15 kg P and 45 kg K/ha. During the same period a thoroughly weeded plot may absorb 62 kg N, 26 kg P and 80 kg K/ha. Over all, a weed-free duration of 45 days is essential, if interference due to weeds has to subside [8].

The Dry land Agroecosystem of South India supports cropping systems that include several different cereals, legumes and oil seeds. The growth and yield formation of these dry land crops is affected by weeds (Plates 3 and 4). Crop loss assessments suggest that grain productivity may decrease by 20–31% due to weeds. In fact, there are suggestions that depreciation in grain/forage yield due to weeds is much higher than that attributable to disease and pestilence. Weeds directly affect nutrient dynamics in the agroecosystem. According to Venkateswarlu [9] nutrient loss to weeds may depreciate grain productivity by 20–40% depending on crop species and location.

PLATE 3 Weeds in between two plants of Finger Millet grown on Alfisols found in the Southern Indian Plains.
Note: The intensity of weed infestation could be high in certain patches of a Finger Millet cropping zone.
Source: Krishna, K.R. Bangalore, India.

PLATE 4 A thoroughly weeded Groundnut field and culled weeds heaped by the side of the plot.
Note: Groundnuts are weeded at least twice during first 30–45 days after emergence. If not, grassy weeds appear rather rapidly and over grow the small groundnut canopy. This leads to reduction in photosynthetic light interception by groundnut. As a consequence crop biomass formation gets reduced.
Source: Agricultural Experiment Station at Gandhi Krishi Vignana Kendra, near Bangalore, South India.

The effect of weeds on legume production in the subsistence farming zones of semiarid tropics needs due attention. Timely weeding is almost essential in every field that supports legume crops (Plate 4). If weeds are unchecked within first 30 days of crop growth, then they affect grain/forage productivity of legumes. In the semiarid tropics, where soil moisture and nutrients are major constraints, impact of weeds on crop growth and yield could be severe. For example, in a green gram field within Gangetic belt, if weeds were unchecked for first 30 days then grain yield decreased by 30–68% [10]. Weeds affect pigeon pea grown on Vertisols of Central Plains of India. Farmers in this region adopt various intercropping mixtures to thwart weed growth. For example, a sole crop of pigeon pea allowed 168 kg weeds/ha. This is equivalent to diversion of 4.2 kg N, 1.6 kgP and 4.0 kg K/ha from crops. Whereas, an intercrop of pigeonpea allowed only 142 kg/ha weed biomass formation, which is equivalent to 3.6 kg N, 1.16 kgP and 3.5 kg K/ha. Clearly, intercropping is a useful procedure to reduce deleterious effects of weeds [11]. In the Southern Indian plains, it is said that hand weeding timed at seedling stage of crop can effectively avoid loss of 50 kg N, 9 kg P and 45 kg K/ha in any field that supports a crop for 100–120 days. In some groundnut fields, two weeding exercises at appropriate interval can reduce weed effect to negligible level.

The Chinese agrarian region is vast and supports several cropping systems. Crop species encountered in different regions of China varies enormously. Weed flora that infest crop fields and reduce their productivity too vary proportionately. The distribution and infestation of weeds in China, like any other agrarian zone is dependent on natural factors like temperature, water resources, soil fertility, light distribution, and cropping system [12]. Weed distribution in China is dependent on agroclimate. For example, in tropics, weeds such as *Dactyloctenium aegypiacum. Heydiotis costata, Ageratum conyzoides, Paspalum conjugatus, Alopecarus acqualis, Leptochloa chinensis* and *Alternanthera philoxeroides* are predominant. Their influence on nutrient dynamics and crop productivity could be severe. In the temperate regions, weeds such as *Descurania sorphia, Acalypha australis, Amaranthus retrofluxes, Avena fatua, Polygonum convolvulus* and *Galeopalus bifid* are common. *Echinochloa crus-galli* and other related species are important grain yield reduces. Generally, degree of infestation by weeds ranges from light (5–10%), or medium damage (10–30%), or serious damage (30–50%) and very serious damage (50–80%). The extent of soil nutrients and moisture diverted to weeds is important at each level of infestation. Over all, it is said that 53 m ha of Chinese cropping belts are susceptible to weed infestation. Crop loss due to weeds in each of the different agroecosystem could reach millions of tons per year. Timely application of weedicides such as Propanil, Butachlor, Oxadiazon, Atrazine and Simazine is most essential to reduce loss of nutrients to weeds.

Australian farmers grow cereals like wheat, maize and sorghum in rotation with legumes such as Lupin, Lentils, Groundnuts and Trifolium. Weeds occur throughout the year and affect nutrient recovery by main crops. Pre-emergent application of herbicides is practiced to thwart weed infestation. Glyphosate is a common herbicide used by farmers. Yet, procedures such as deep plowing once in three years, adopting suitable rotations and intercrops are needed. Cover crops during fallow season also avoid loss of residual nutrients to weed flora. Weeds that grow rapidly into large sized canopies should be effectively removed during early stages of crop. If not, it may affect photosynthetic light interception and biomass formation by the main crop [13].

Aquatic and floating weeds are well distributed across different agroecosystems of the world. They infest water bodies that are used for irrigation, such as lakes, ponds and channels (Plate 5). It has been observed that weeds such as Water Hyacinth (*Eichhornia species*), water fern (*Salvinia molesta*) and water lettuce (*Pistia stratiotes*) are observed in lakes and streams in most tropical and subtropical agrarian region [14]. Submerged weeds such as *Elodea spp, Hydrilla verticillata* and *Myriophillum spicatum* are also frequently found in cropping regions. Integrated measures such as mechanical removal of water weeds, culling the weeds that occur in channels, adoption of biological control and application of herbicides to crop fields are essential. Aquatic weeds clog the channels and hinder movement of irrigation water. They may remove large amount of dissolved nutrients from soil. In lowland paddy fields aquatic weeds are a major hindrance to transplanting, establishment and productivity of rice plants. Aquatic weeds could affect photosynthetic light interception in addition to diversion of soil nutrients.

PLATE 5 Excessive weed growth has clogged the lake.

Note: Weeds are prolific in regions with lakes and channels. Weeds may grow rapidly and clog water sources and channels. They reduce water-holding capacity of water bodies and hinder easy movement of water in irrigation channels. Location: Vishwanathpura near Bangalore, South India.

Source: Krishna, K.R. Bangalore, India.

KEYWORDS

- ***Cassia occidentalisis***
- ***Parthenium***
- ***Philaris minor***
- **preemergent herbicides**
- **weed propagules**

REFERENCES

1. Grichar, W. J.; Sestak, D. C. Herbicide systems for Golden Crown Beard (*Verbesina encilodes*) control in peanut fields. *Peanut Science* **2000,** *27,* 23–26.
2. Sanderson, M. A.; Brink, G.; Ruth, L.; Stout, R. Grass-Legume Mixtures suppress weeds during establishment better than Monocultures. *Agronomy Journal* **2012,** *104,* 36–42.
3. FFTC, Weed control for Peanut (*Arachis hypogaea*). Food and Fertilizer Centre. Institute for Agricultural Technology, Taipei, Taiwan **2002,** 1–2.
4. Krishna, K. R. Peanut Agroecosystem: Nutrient Dynamics and Productivity. Alpha Science International Inc.: Oxfordshire, United Kingdom, **2008,** 234–235.
5. Srivastava, R. Diversity of weed soil seed bank in Indian Dryland and Irrigated Agroecosystems. *Indian Journal of Fundamental and Applied Life Sciences* **2012,** *2,* 3–37.
6. Rao, K. V.; Rao, B. P.; Ramarao, K. Weed control techniques in transplanted rice (*Oryza sativa*). *Indian Journal of Agronomy* **1995,** *38,* 474–475.
7. Choubey, N. K.; Tripathi, R. S.; Ghosh, B. C. Effect of Fertilizer and Weed management of Direct seeded Rice on Nutrient utilization. Indian Journal of Agronomy **1999,** *44,* 313–315.
8. Singh, U.; Cassman, K. G.; Ladha, J. K.; Bronson, K. R. Innovative nitrogen management strategies for low land rice Systems. In: Fragile Ecosystems. International Rice Research Institute, Manila, Philippines, **1995,** 236.
9. Venkateswarlu, J. Rain fed Agriculture in India. Indian Council of Agricultural Research, New Delhi, India, **1988,** 383.
10. Rana, K. S.; Mahendra Pal and Rana, D. S. Nutrient depletion by Pigeonpea (*Cajanus cajan*) and weeds as influenced by intercropping systems and weed management under rainfall conditions. *Indian Journal of Agronomy* **1999,** *44,* 267–270.
11. Patil, B. M.; Pandey, J. Chemical weed control in Pigeonpea (*Cajanus cajan*) intercropped with short-duration grain legumes. Indian Journal of Agronomy **1996,** *41,* 529–535.
12. Yuan, T. H. Weed Distribution and Infestation in China.Shanghai Academy of Agricultural Science, Shanghai, China www.caws.org.au/awe/1993/awe199310231.pdf **2012,** (July 8, 2012).
13. Prostko, E. Down Under: New Ideas for control of Tropical Spiderwort. Peanut Farmer **2004,** *40,* 14–18.
14. Labrada, R. Present trends in Weed Management. Food and Agricultural Organization of the United Nations. www.fao.org/docrep /006/yr5031e/yr5031e0j.htm **2006,** 1–15 (July 8, 2012).

EXERCISE

1. What are weeds? Mention at least 10 major weeds from different Agroecosystems considered important in Your Continent.
2. Photograph and mention scientific names of at least 10 major weed species common to Agricultural Belts of your area.
3. Explain the phenomenon by which weeds compete with crops and reduce crop grain/forage formation
4. Mention various cultural methods popular in your area that aim at controlling weeds.
5. Mention at least five most popular and important herbicides known in your area. Collect information on chemistry and mode of action of Active Ingredient found in each herbicide.
6. Give a few examples of crop loss assessments due to weeds.

FURTHER READING

1. Caton, B. P.; Mortimer, M.; Hill, J. E.; Johnson, D. E. Weeds of Rice. International Rice Research Institute: Manila, Philippines. **2010,** pp. 124.
2. Pott, A.; Pott, V. J.; de Souza, T. W.; Pasture weeds in the Brazilian *Cerrado* region. EMBRAPA Gado de Corte. Campo Grada. Brazil. **2006,** pp. 336.

USEFUL WEBSITES

http://www.weeds.iastate.edu/mgmt/qtr97–1/weedid.htm (July 8, 2012)
http://www.notill.org/links/pest_links.htm (July 8, 2012)
http://extension.missouri.edu/explorepdf/agguides/pests/ipm1014.pdf (July 8, 2012)

CHAPTER 35

IRRIGATION SYSTEMS INFLUENCE NUTRIENT DYNAMICS IN AGROECOSYSTEMS

CONTENTS

35.1 INTRODUCTION

Water resources are integral to development, productivity and perpetuation of Agro-ecosystems. Water affects a series of ecosystematic functions in a cropping zone. To a certain extent, water decides nutrient dynamics and productivity of various agroecosystems. Knowledge about water resources is essential to all agricultural agencies and individual farmers. Satellite aided assessment of precipitation pattern and irrigation resources are in vogue in many agrarian regions of the world. It helps in planning appropriate planting dates, cropping systems, fertilizer application schedules and forecasting grain/forage yield.

At present, major providers of satellite imagery systems data regarding irrigation potential are Aquastat, FAO STAT, Actinich, etc. The Global Map of Irrigation consists of a spatial database for each continent i.e., North America, South America, Europe, Africa, Asia and Oceania. It includes sizeable details on locations, variations and conditions that affect irrigation. Currently, data on irrigation potential is based on countries. Political boundary of each nation is considered during satellite imagery and computing. The top 10 countries with better irrigation potential are as follows: India has largest irrigated area in the world at 501,020 km^2, followed by China (460,030 km^2), United States of America (234,938 km^2), Pakistan (172,000 km^2), Iran (72,640 km^2), Mexico (61,000 km^2), Russian Federation (53,600 km^2), Thailand (50,040 km^2), Indonesia (45,800 km^2) and Turkey (41,600 km^2). India, China and United States of America together constitute 47% of global irrigated cropping area [1]. Maps depicting irrigation potential could also be based on fluctuations in boundaries of various agroecosystems. Matching irrigation trends with crop species and its expanses may help the policy makers in drawing more accurate inferences. Perhaps, channeling irrigation resources could then become more efficient. A recent trend is to mark production zones of a particular crop based on water resources and soil fertility. For example, in case of major cereals like wheat, researchers recognize mega-environments in different continents. Their boundaries are recognized based on precipitation pattern, irrigation potential, soil fertility status and grain productivity [2].

According to Faures et al. [3], globally there is still untapped potential for expanding land and water use. They have cautioned us that extent of development in irrigation and crop production during past three decades should not mask the limits imposed by water and land to agricultural expansion in the world. The current trend among agricultural researchers and policy makers is to adopt frame work known as 'World Agriculture: Towards 2015/2030 (AT 2030).' The anticipated enhancement in arable land, cropping intensity, total harvested area and crop production in the rapidly developing zones of the world, based on projections using Aquastat and projections of AT 2030 is as follows:

Percent expansion/increase anticipated by the year 2030 compared to 1999/2000 A.D.

	Arable Land Expansion	Increase in Cropping Intensity	Harvested Land	Yield Increase
Developing countries				
Rain fed	25	11	36	64
Irrigated	28	15	43	57

Source: [3].

The agroecosystems that thrive on arable land under rain fed conditions seems to hold the key for higher crop production caused mainly through expansion of cropping area. Whereas, within agroecosystems thriving under irrigated conditions, both intensity of cropping and expansion of harvested land seem to generate more food grains during next 20 years. The expected grain/forage yield increase in the developing world is about 57–64% more compared to 1999/2000 levels (Table 1) [3]. Globally, cropped area could increase by 130 m ha in the period from 1999 to 2030. Yet, it is believed that much (80%) of increased grain/forage would be derived by intensification effected through irrigation and fertilizers. About 20% increase in crop production could be due to land expansion. Therefore, in future, irrigation and its impact on soil, nutrient dynamics and crop production needs greater attention. Fluctuations in area, intensity and cropping pattern within an agroecosystem could be largely guided by the irrigation potential of the region. Globally, irrigated area may expand from 202 m ha in 1999/2000 A.D. to 242 m ha by 2030 A.D. The influence of irrigation on ecosystem functions, nutrient dynamics and crop productivity could be simulated and studied periodically. It could help us to avoid any detriment to nature.

TABLE 1 Agricultural Cropping Area and Irrigation Trends in Different Continents.

Continent	Total Geographic Area M ha	Arable and Perennial Crop Area M ha	% APC	Irrigated Area M ha	% Irrigated Area within APC	Population Million	Food Production MT
Asia	3077	544	18	211	39	4071	1182
Americas	3816	376	10	44	12	860	629
Europe	2174	288	13	22	8	699	446
Africa	2247	225	10	13	8	863	148
Oceania	803	45	6	3	7	28	36
World	13428	1533	9	299	18	6521	2494

Source: International Commission on Irrigation and Drainage (ICID); http://www.icid.org/imp_data.pdf
Note: Population, Geographic and Irrigated area data pertain to mid 2010. APC = Arable and Perennial Area.

We ought to realize that water resource is an important ingredient of every agroecosystem. Water, in fact, influences the type of cropping belt possible, crop species and productivity. Therefore, it would be a useful exercise to demarcate various major cropping ecosystems of the world and evaluate the irrigation potential of the region. In any geographic region, cropping zone may actually be supported by water derived from several different sources. For example, the Rice Agroecosystem in South Asia comprises variety of crop production strategies like rain fed, rain fed upland, rain fed low land, irrigated lowland, flooded etc. Similarly, Corn Belt in North America includes areas that are rain fed, irrigated, irrigated with high fertilizer inputs etc. Sorghum belt in South India, includes rain fed dry land cropping zones, rain fed moderately intense cropping zones, irrigated high fertility cropping zones etc. Irrigation pattern and intensity may also vary based on the crop species, genotype and water requirements. In a given region, we often perceive a mosaic of irrigation systems adopted, water depletion pattern, cropping pattern and grain productivity trends.

35.2 PRECIPITATION PATTERN INFLUENCES AGROECOSYSTEMS, NUTRIENT SUPPLY AND PRODUCTIVITY

Agroecosystems have several stringent requirements. Their characteristics like composition of crop species, their genotypes, intensity or supply of inputs and productivity are highly regulated by soils and weather parameters. Precipitation pattern and total moisture received during a crop season is a potent factor that decides crops, cropping sequences and productivity in any geographic region. The crops that flourish in high rainfall zone are often those that need more quantity of water. Water may be derived either via rains or extraneous sources. Rice grown under lowland flooded condition is an excellent example for crops that are preferred in high rainfall zones or those with sufficient irrigation facility. The humid tropics of South-east Asia, especially in South-eastern India and Southern China that receives 1,500–2,500 mm precipitation supports two or even three crops annually. Rice demands relatively higher quantity of water to grow and form panicles. When matched with high inputs of nutrients and FYM rice productivity are also high. In China, high precipitation pattern (2,000–3,000 mm annually) allows intensive cropping and high yield goals for maize (7–10 t grain/ha or wheat 5–8 t grain/ha). The maize-soybean or wheat-maize or wheat-fallow sequences yield high amounts of grains and forage. In the Northern European plains, wheat monocrops are provided with relatively larger quantities of fertilizers to use high rainfall pattern efficiently (see Table 2).

TABLE 2 Agroecosystems that flourish in different Precipitation Zones, their Cropping Systems and Nutrient Supply levels.

Precipitation Pattern /Agroecoregion	Cropping Sequences (NPK/ha/season)	Fertilizers Impinged
High Precipitation Zone (1500–3000 mm/yr)		
South-east Asian Rice belt,	Rice-Rice-fallow	180–240
Maize, Wheat in North-east China	Maize-wheat or Wheat-Fallow	140–180
North-west European Plains	Wheat –Fallow; Wheat-Berseem	240–300
Maize in the Chile	Maize-vegetables-fallow	140–180
Coffee in Western Ghats	Coffee with shade tree or citrus	140–220
Medium (700–1450 mm/yr)		
Corn Belt of USA,	Corn continuous, Corn-soybean	80–160
Wheat in Great Plains of North America	Wheat-fallow; Wheat-legume-fallow	80–120
Soybean in Cerrados	Soybean continuous; soybean-cereal-fallow	60–100
Soybean in Central Plains	Soybean-maize; soybean continuous	60–100
Wheat, Maize and soybean in Argentinian	Pampas Maize/corn; Wheat/soybean	40–80
Wheat-Rice in Indo-Gangetic Plains	Rice-wheat-fallow; rice-wheat-vegetable	80–120
Sorghum-Cotton in Indian Vertisol Plains	Sorghum/cotton; Sorghum	80–120

Sunflower in Central/South Indian Plains	Sunflower-fallow; sunflower continuous	60–120
Low (350–600 mm/yr)		
Pearl Millet in Sahel	Pearl millet-fallow	nil-40
Pearl millet in North-west India	Pearl millet-fallow	nil-40
Wheat/Barley in West Asia	Wheat/Barley-fallow; Wheat-vetch-fallow	20–40
Sorghum in Cerrados	Sorghum-soybean-fallow; sorghum-fallow	40–60
Groundnut in West Africa	Pearl millet/Groundnut; Groundnut fallow	Nil-40
Groundnut in Southern Indian	Groundnut-Groundnut-fallow;	20–40
Alfisol regions	Groundnut-fallow	

Note: Annually, agroecosystems situated in high rainfall mountain forests and plantation zones receive 2500–3000 mm; high rainfall tropical agricultural zones receive 2000–3000 mm; subtropical plantation and arable cropping belts receive 900–1500 mm; semiarid cropping zones receive 700–1100 mm; dry land regions receive 350–650 mm and arid zones receive 300–450 mm. Most importantly, we should note that fertilizer-based nutrients supplied into fields/cropping zones is dependent on precipitation and water resources augmented. Interaction between precipitation and fertilizer-N is important. Cropping sequences that include leguminous crop species require relatively lesser quantity of fertilizer-N.
Source: Compiled from several publications.

35.3 WATER RESOURCE AND MODE OF IRRIGATION

The water source and mode of irrigation adopted by farmers residing in an agrarian region has its share of impact on farming enterprise, crop species, intensity of cropping, fertilizer and other inputs, and finally the productivity of individual fields. Through the ages, human ingenuity has led us to variety of water resources. They have been exploited to different levels year after year. Methods that allow refurbishment of such water resources have also been shrewdly devised and practiced. Yet, some methods do exhaust the water resource beyond threshold. Water resources available worldwide can be classified into at least two main categories-namely; low or surface irrigation and lift irrigation. Each of the irrigation systems and local variations, if any, have played crucial role in formation, sustenance, productivity and perpetuation of agroecosystem. Water resource and its interaction with soil type, especially fertility has major impact on intensity and expanse of an agroecosystem.

Basically, if the farmer's land is situated at higher level than location of water resource, then lift irrigation, mediated via Reciprocal lifting systems, Persian wheels (Noria) or Electric pumps need to be used [4]. Water sources like wells, ground water (bore well); low land lakes need lift irrigation. Irrigation from wells is common almost everywhere in semiarid, wet tropical or temperate regions. The rainwater sinks deeper in soils that are porous and gets accumulated at a depth. This accumulated ground water is available for lifting through pumps, artisan wheels or manually. Small areas in fields can be irrigated using wells. Wells serve excellently when drought occurs at crucial stages of the crop and farmers direly need to supply moisture, though in small quantities. Of course, wells also support large fields in individual farms (Plate 1). For example, in the Gangetic belt and Southern Indian plains, wells support large expanses of cereal/legume intercrops during post-rainy and summer seasons. Ground water and tube wells could play a role in development of cereal/soybean based cropping expanses in Cerrados and Pampas of South America. Surface irrigation is feasible when water reservoirs, tanks are situated at levels higher than cropping zones, for example, large lakes, dams and rivers. Water flows naturally based on gravity through the canals, channels, pipes or even irrigation equipment, allowing farmers to feed water to crop lands. Most agroecosystems are endowed with several different types of water resources. The impact of each on the agroecosystem may vary based water source/ irrigation method, factors related to geographic location, crop, environment and economic advantages (Plates 2–5).

PLATE 1 Ground water (Bore well) in use to irrigate a Paddy crop. Farmers adopt flood irrigation method to supply water.
Source: Agricultural Experimental Station, at Gandhi Krishi Vignana Kendra near Bangalore, India.

PLATE 2 Top: Flat beds of small or larger dimensions that hold seedlings of Cole crops are provided water through flood irrigation system. Bottom: A close-up view of surface-flow of water in each flat bed. Location: Doaddaballapur near Bangalore, South India.
Source: Krishna, K.R. Bangalore, India.

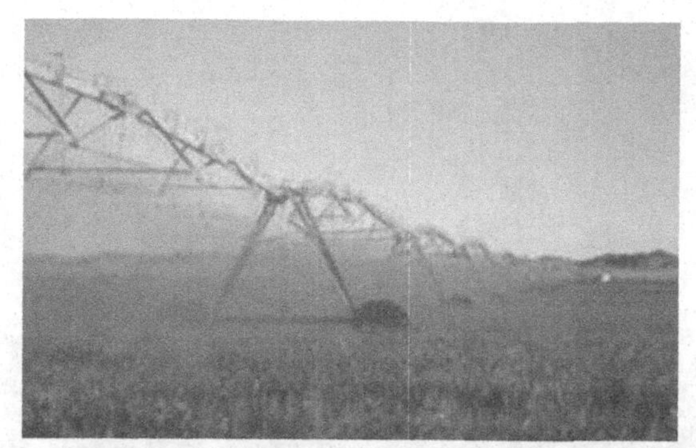

PLATE 3 Sprinkler Irrigation System adopted on a Wheat crop in West Asia.
Note: Sprinkler irrigation system is an efficient method of water distribution in the dry land zones of West Asia, especially large farms. It reduces loss of water via percolation, seepage and undue evaporation. Moisture gets absorbed both through roots and hydathodes of leaf. Farmers may also use this system to supply dressings of nitrogen in dissolved state. Fertilizer consumption is perceptibly very low if supplied via irrigation water.
Source: ICARDA, Aleppo, Syria; Plow Creek Farm, Illinois, USA.

PLATE 4 River Niger near Niamey in West Africa.
Note: The river is in full-flow just ahead of a small barrage. River Niger and human civilization on its banks has domesticated and gifted many a crop species that feed not only the local populace, but the spread of these crop species into other continents has provided food to human populations in other regions and continents. For example the upper reaches of river Niger, around Mali and Northern Niger is a major center of genetic diversity for crops like sorghum, pearl millet and cowpea. These crops were domesticated during 3rd millennium B.C. on the banks of river Niger. Later they spread to regions in Southern Africa, Indian peninsula and China. At present, river Niger is the major riverine irrigation source for crop production in the Sahelian West Africa. Volta and Senegal are other rivers that supply water to West African crops that thrive on their banks.
Source: Krishna, K.R., Bangalore, India; ICRISAT, Hyderabad, India, 1985.

PLATE 5 A view of Lake Nasser created by Aswan High dam across river Nile.
Note: This dam irrigates a large cropping zone in Egypt. This river converts an arid expanse into productive agroecosystem.
Source: http://www.ilec.or.jp/database/afr/afr-19.html; www.dams-info.org/en

Large water bodies like lakes, small or large masonry dams have almost always transformed the agrarian zone that occurs in its region (Tables 3 and 4). In fact, in many developing nations, construction of a large or small dam across a river/rivulet has consistently changed the cropping pattern and ecosystematic functions within the command area. Often, dry land ecosystems have been converted to intensive arable or wetland ecosystems. The assured water resource has induced farmers to revise crop species. High yielding genotypes of field crops or plantations with better yield goals that demand enhanced supply of fertilizer-based nutrients and pesticide usage have been preferred. There are indeed too many examples, where in dams and canal systems have altered that agroecosystems and improved farmers' economic status. A few examples are river and dam systems in Central and Southern Great Plains of USA, River-based irrigation in Amazon region and Eastern Brazil (Cerrados), Nile valley in North-east Africa, Niger in Central Nigeria, Indo-Gangetic plains in South Asia, Yangtze and Huang He in China, Euphrates and Tigris in Mesopotamia, Riverine zones in Central and Eastern Europe, Mekong in South-east Asian tropics, Darling and Murray rivers in Australia have all induced marvelous transitions of landscape from dry, low intensity cropping belts to intense high yielding and luxuriant agricultural zones (Table 3). Irrigation and nutrient supply systems have played a dominant role in creating these agricultural marvels termed generally as cropping expanses or agroecosystems.

TABLE 3 Major Riverine Irrigation systems of the World, Geographic regions or Agroecosystems supported by them and Crop species grown in that Area.

Major River(s)	Agroecosystem/Region	Crops/Cropping Systems
North America		
Mississippi/Missouri and tributaries	Great Plains of USA	Wheat-Soybean, Maize-Soybean
Platte, Arkansas, Red,	Central and Southern Great Plains	Wheat-Cotton, Wheat-Soybean
Pecos, Rio Grande	Southwest cropping zones	Maize-cotton, Sorghum-Cotton
California Irrigation System		Citrus, Grapes, Vegetables
Amazon and Tributaries		
Amazon, Branco, Negro, Putmayo, Ucayali, Jurua, Purus, Madeira, Roosevelt, Tapajos	Amazonia, Matto Grasso	Wheat, Maize, Soybean in mono /intercrops and rotations, Sorghum,
Brazilian Riverine systems		
Araguala, Tocantins, Parnaiba	Matto Grasso, Cerrados, Catingas	Soybean, Cotton, Maize, Forages
Sao Francisco Parana, Uruguay	Cerrados, Paraguay	Maize, Sorghum, Cowpea, Alfalfa
Argentinean Riverine Zones		
Parana, Bermejo, Pilcomayo, Uruguay Salado, Colorado, Negro	Rolling Pampas and Cordoba region	Wheat, Maize, Soybean, Groundnut,
European Riverine Systems		
Guadiniano, Guadalquivir, Ebro	Spanish Cereal belt	Wheat, legume, oilseeds
Dordogne, Loire, Seine	French Plains	Wheat, Barley, Oats, lentils, Brassica
Rhine, Danube, Weser	Central Plains of Europe	Wheat-legume/Oilseed zones Wheat, Barley, Lentils, and Brassica
Vistula, Dniester,	Polish Cereal belt	Wheat, Barley, Oats
Drava, Sava	Southern European Cropping zones	Wheat, Oats, Lentils, Olives
Dnieper, Volga, Prut, Oka, Kama	Russian Agricultural zones	Wheat, Barley, Oats, legumes
Brassicas and Vegetables		
African Riverine Agricultural Belt		
Niger, Senegal, Volta,	Sahelian and Tropics of West Africa	Pearl millet, Sorghum, Cowpea, Groundnuts, Maize, Bananas
Congo, Orange, Limpapao, Zambezi	Central and Southern African Farming	Maize, Sorghum, Finger millet, Cowpea, Groundnut
Nile and its tributaries	Cereal and legume regions of Sudan and Egypt	Wheat, Barley, Lentils, Vetch, Olive
Southern Asia		
Indus and Ganges plus tributaries	Rice-Wheat belt of Indo-Gangetic Plains,	Wheat- Rice-legume/Rye
Brahmaputra, Irrawaddy	Tropical Forestry, Agroforestry,	Rice/Legume belt, Forest Plantations
Narmada, Godavari, Krishna, Tungabhadra, Cauvery	Deccan, Vertisol and Alfisol Plains and Coastal Plains of South India	Sorghum, Maize, Rice, Pearl millet, Pigeon pea, Cowpea, Beans, Sunflower, Groundnuts, Brassicas, Cotton, Sugarcane
China and Fareast		
Huang He	Farming zones of Shanxi, Hubei and Beijing	Maize-wheat, Wheat-fallow, Rice, Groundnut
Chiang Jiang, Huangpu, Xiang, Yalong,	Cropping belts of Wuhan, Hunan, Anhui, Zhejiang,	Rice, Wheat, Maize, millets, groundnut vegetables,
Wu, Gan, Zijiang,	Hangzhou, Shandong, Jiongx, Fujian	
Hongshui He, Bei Jiang, Pearl river	Cropping belts in Yunnan and Guangxi,	Rice, maize, groundnut
Tarim He, Qargen He, vegetables	North-western dry basins of Tarim	Millets, legumes, groundnut and
Heilong Jiang, Songhua, Ussuri	Northeast cold regions	Wheat, maize, rice, legume, forage
Australian		
Ashburton, Fortescue, De Gray, Murchison	Western Australian Farming Zones	Wheat, legumes, pastures
Darling, Murray, Barwan	South-eastern Australian Cropping Belts	Wheat, Barley, Canola, Pastures
Finders, Mitchel	Queensland Farming zones	Wheat, Canola, Groundnuts, legumes, vegetables and pasture

TABLE 4 A few examples of large lakes across different continents that support agricultural belts, in addition to other services like recreation, drinking water, fisheries and power generation.

Lake/Country	Geographic Area Irrigated	Crops/Cropping Belt
Chilwa, Malawi	Southern Malawi and Mozambique	Rice, tobacco, groundnut
Sibaye, South Africa	Natal Province of South Africa	Pine and eucalyptus plantations,
Albert, Zaire	Western Uganda and Zaire	Coffee, maize, cassava, millet, cowpea
Chad, Chad	Cameroon, Chad, Nigeria, Niger	Pastures, pearl millet, cowpea
Nyasa, Malawi	Tanzania, Malawi, African Rift Valley	Sorghum, maize, millet, cowpea, pigeonpea
Victoria, Tanzania	Northern Tanzania, Kenya, Uganda	Sorghum, millet, maize, legumes, trees
Aswan High Dam, Egypt	Northern Egypt, parts of Sudan	Wheat, cotton, lucerne, groundnut
Guiers, Senegal	Senegal	Sugarcane, pearl millet, groundnut, cowpea,
Ozero Baykal, Russia	Buryat and Irkutsk region	Coniferous forests, barley and wheat
Krasnoyarsk Reservoir, Russia	Krasnoyarskaye district	Forest plantations
Changshou Hu, PR China	Sichuan province	Horticultural crops, wheat and soybean
Titicaca, Peru	Regions in Peru and Bolivia	Maize, beans, vegetables, forest tree species
Ypecarai, Peru	Cordillaria and Central Peru	Maize, potato, groundnut
San Roque, Argentina	Cordoba and Piniella region	Wheat, soybean, sunflower, groundnut
Sobrandhino, Brazil	Sobrandhino, Bahia in Brazil	Sorghum, maize, soybean

Source: www.ilec.or.jp/afr/afr-03.html and several others.
Note: This list is not exhaustive. There are many water bodies and lakes in different continents that have induced and sustained cropping belts and ecosystems. Space and context are insufficient to mention and discuss all of them.

35.4 METHODS OF IRRIGATION ADOPTED IN VARIOUS AGROECOSYSTEMS

As stated earlier, factors like precipitation and irrigation supplements have influenced the development, perpetuation and productivity of different agroecosystems. Since ages, farmers have devised and used a wide range of irrigation techniques, some with great advantage in terms of water-use efficiency [5]. Water resource available at the disposal is a crucial factor that decides irrigation quantum and method adopted. Farmers have often matched crops, more accurately their water requirements with type of irrigation. Knowledge about various irrigation methods, water sources that could be exploited and crops that are feasible in a given region or season is essential. Generally, factors such as soil type, slope, climate, water availability and quality affect the irrigation system adopted in an agroecosystem [4]. These factors have often affected the type of crop, nutrient dynamics and productivity. Let us consider salient features of a few irrigation methods.

35.4.1 FLOOD IRRIGATION

Flood irrigation is an easier and cost effective method whenever water resource is abundant. Crop fields meant for flood irrigation are flat and leveled to avoid stagnation. Slopes, concaving and impediments that generate uneven water flow across field should be avoided. It involves rapid spread of water all across the field. Flood irrigation may generate runoffs, leading to loss of water and dissolved nutrients with it. Sometimes surge flooding is used. Surge flooding is rapid version of conventional surface flooding, but done very quickly to avoid stagnation and runoff. Flood irrigation is well suited to flush fields from undue salt accumulation. Flooding is helpful in areas with saline intolerant crops (e.g., Pigeon pea). Flooding induces anaerobic conditions in the soil horizon, especially the upper layers. It may have impact on soil microbial flora and nutrient transformations mediated by them. Soil anaerobiosis induces N emissions through de-nitrification. Flood irrigation involves low technology and meager investment. Hence, it is adopted by farmers in many of the agroecosystems that thrive in plains. It also suits low input farming adopted in cereal/legume production zones.

35.4.2 FURROW IRRIGATION

Furrow irrigation is actually a variant of flooding system. In this case, water is guided through narrow channels between ridges. Furrows avoid uneven distribution, restrict over flowing and allow better percolation down the profile. Furrows allow water to reach the root zone more accurately. Furrow is common in cropping belts with arable soils and line sown crops.

35.4.3 SPRAY IRRIGATION

Spray irrigation involves complex engineering and gadgetry to regulate water dispensation. Water is channeled through pipes and released to air using spray guns. Water distribution is more accurate and water use efficiency is generally higher. Spray irrigation is amenable in areas that support high input cropping systems. The center-pivot systems and low energy precision application is adapted frequently in intensive cropping zones. For example, spray irrigation systems support large areas in Corn Belt of USA, Maize/soybean in Cerrados of Brazil, Wheat in European Plains, Cereal farms in West Asia (Plate 3).

35.4.4 DRIP IRRIGATION

Drip irrigation is among the most modern irrigation techniques adopted in plantations. Drip irrigation system is relatively expensive, yet remunerative since water distribution is highly efficient and product is priced high. Drip irrigation involves placing perforated pipes or nylon thin tubes that supply water into root zone. Water is channeled drops and slowly, so that water loss from the system is least. Water loss via evaporation is minimized under drip irrigation system. Drip system may require 25% less water resource compared to other systems like flooding. Drip may not be efficient if water has salts in more than permissible levels. To quote a few examples, drip irrigation is common in Citrus Belt of Florida, Grape farming zones in USA, Southern Europe, Russia and India. Drip system is also suited for cash crop production.

35.5 WATER REQUIREMENTS OF CROPS

The water requirement of a crop species or its genotype is an important trait that has immediate effect on growth, nutrient dynamics and productivity of any agroecosystem. The type(s) of cropping expanses possible in a given geographic region is influenced by water requirements of a crop. Crop duration also affects water needs of a crop. Short duration genotypes are preferred in rain fed ecosystems that are prone to drought and short spells of precipitation. Farmers try to maximize precipitation use efficiency by selecting short duration crops and planting them right at start of rainy period. Short duration genotypes with low water needs are also provided with relatively smaller quantities of fertilizers, in order to match low yield goals. Globally, there are large expanses of crops with moderate needs of water ranging from 500–1200 mm per season. There are arable agroecosystems that support an assortment of such crop species/genotypes. Intensive cropping requires relatively large dosages of irrigation. Water requirements increase proportionate to high nutrient inputs and yield goals. In nature, there are crop species that are drought tolerant and genetically endowed to mature with much less water supply. On the other extreme, we have flooded or wetland ecosystems filled with crop species, weeds and natural flora that naturally have high water requirements. For example wetland paddy, sugar cane, bananas require either flooded fields or wet soils throughout most part of vegetative and reproductive phases. Ultimately, water requirements of crop and its supply pattern should match for an agroecosystem to thrive. Since, nutrients are all channeled through water in dissolved state, the nutrient dynamics in an agrocosystem is under stringent influence by water need/crop duration. Following list provides an approximate idea regarding water requirements of various crop species and duration to mature.

Crop duration (days) and Water requirements (mm/season) of a few important crop species:

Cereal Grain Crops

Rice-90–150 d, 450–700 mm; Sorghum-120–130 d, 450–600 mm; Wheat-120–150 d, 450–600 mm; Barley-120–150 d, 450–600 mm; Maize-125–180 d, 500–800 mm; Millet 105–140 d, 400–550 mm; Oats-120–150 d, 450–600 mm

Legumes and Oilseeds

Cajanus 130–160 d, 600–750 mm; Chickpea 120–150 d, 550–700 mm; Cowpea 90–150 d, 550–700 mm; Lentil-150–170 d, 450–600 mm

Canola (Brassicas) 120–150 d, 600–750 mm; Flax-150–220 d, 550–700 mm; Peanut-130–140 d, 500–700 mm; Soybean-135–140 d, 450–700 mm; Sunflower 125–130 d, 600–1000 mm

Vegetables

Beans-75–90 d, 300–500 mm; Brinjal-130–140 d, 600–700 mm; Cabbage-120–140 d, 350–500 mm; Capsicum-120–210 d, 600–900 mm; Carrot-100–150 d, 350–500 mm; Cucumber-105–130 d, 350–500 mm, Lettuce-75–140 d, 400–600 mm; Onion dry-150–210, 350–550 mm; Pea-90–110 d, 350–500 mm; Potato-105–140 d, 500–700 mm; Radish-35–45 d, 300–400 mm; Spinach-60–100 d, 550–750 mm; Squash 95–120 d, 500–650 mm; Tomato-135–180 d, 400–800 mm

Fruits and Cash Crops

Banana-300–365 d, 1200–2200 mm; Citrus-240–265 d, 900–1200 mm; Cotton-180–195 d, 700–1300 mm; Mellon 120–160 d, 400–600 mm; Sugar beet 160–230 d, 550–700 mm; Sugarcane-270–365 d, 1500–2200 mm; Tobacco 130–160 d, 600–700 mm

Natural prairie vegetation with a mixture of grasses, legumes and other species may require 500–1200 mm to complete a season. Now let us compare water needs of crops with that of natural prairies. Among agricultural crops there are many that require almost similar levels of water per season. There are a few crops that need 10–30% less than prairie grasses. On the other hand, there are many crop species, especially major cereals and legumes that need 10–30% more water than natural prairies. Some wet land crops need twice or thrice the levels of water needed by natural prairies. For example, rice needs over 900–2200 mm and sugar cane 1200–1800 mm per season. Crop species has to be matched with precipitation pattern and/or irrigation water channeled.

The basic fact that water is essential in optimum quantities is to be recognized in full, during crop production, anywhere in each continent. Any shortage has its proportionate effects on rooting, nutrient recovery and productivity. Yet, farmers strive hard to search for methods that minimize water usage and try to economize on production costs incurred on fertilizers and water. In many West Asian and North African (WANA) nations, water is a precious resource. The situation holds true even in few other agrarian regions. For example, Dry lands in Peru, Sub-Saharan regions of West Africa, Southern African cropping zones on the fringes of Kalahari, parts of Southern Indian Plains prone to water deficit, North-west China and Dry regions of Australia. Water is also a very costly commodity in some regions. Water is more useful if used for purposes other than crop production. For example, in WANA region, Moutont [6] mentions that, one cubic meter of water channeled to crops earns 15 cents to the farmer. However, if governmental agencies prioritize and supply the same quantity of water to essential industries, it earn 25 US$ per cubic meter. These factors may further force both policy makers and farmers to search for techniques that economize on water usage and still yield same levels of grain/forage. 'Deficit irrigation' is supposedly a technique that allows farmers to economize on supply of water. The idea is to reduce water supply marginally from optimum levels at crop stages that are not crucial to yield formation. Water deficits occur at crop stages that are totally non-sensitive to physiological functions relevant to grain/forage formation. In other words, agroecosystems flourish with slightly lower levels of water usage.

35.6 IRRIGATION AND CROPPING ZONES IN DIFFERENT CONTINENTS: A FEW EXAMPLES

North American cropping zones thrive on variety of irrigation sources. The major riverine belts in the Great Plains and other areas is supplied by mighty Mississippi, Missouri, Arkansas, Ohio and many others (*see* Fig. 1). In addition to rivers, lakes ponds and ground water resources too are used by farmers. Irrigation is a key component in the intensive cropping zones found in Northern Great Plains.

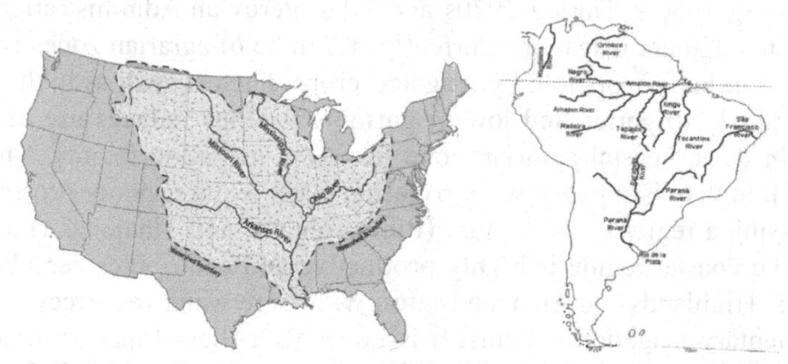

FIGURE 1 Left: Major riverine systems of United States of America. This riverine system irrigates vast stretches of Maize, Wheat, Soybean and Cotton in the Great Plains Region. Right: Major rivers of South America that irrigate Crop Land in the Cerrados, Amazonia and Pampas. *Source:* Several.

Brazilian Agroecosystems have shown marked increase in productivity during recent years. It has been augmented by selection of appropriate crop genotype, nutrient supply, several well time agronomic procedures, investment in farm mechanization and irrigation. Brazilian agricultural enterprise uses about 35% of water resources. Riverine irrigation is the main source. Yet, utilization of water resources and development of irrigated ecosystems is not commensurate. Currently, only 4.5 m ha of cropping expanses are irrigated, which is equivalent to 15% of total cropland of over 60 m ha [7]. Land use estimates indicate that an extra 29 m ha of cropland could be easily brought under irrigation using river and ground water resources. The agroclimate is semiarid in the North-east and Cerrado region of Brazil. Precipitation pattern is insufficient, and hence irrigation is essential. Currently, crops in the Cerrados often receive small dosages of protective irrigation along with that derived from rains. In Northern Brazil, which is humid and endowed with mighty riverine source, irrigation is possible throughout the year. Irrigation is rapidly expanding in Bahia and adjoining regions. Plantations such as coffee, banana, citrus, papaya and other fruit crops have been converted into irrigated farms. The extent of fertilizer supply too has commensurately increased to match revised yield goals. Ground water resources are yet to be exploited in many regions of Brazil [8]. It is believed that Brazil has potential to change the large expanses of dry land crops that are currently rain fed into irrigated agroecosystems. Irrigation could be adopted on crops like cotton, corn, soybean, sorghum and legumes. Adoption of irrigation has improved corn/cotton yield.

Peru is an important agricultural country in South America. Historically, Peruvians had established farms with irrigation facilities. Canal systems belonging to medieval Incas are traced frequently in the agrarian regions of Peru and neighboring countries. At present, cropping ecosystems that thrive on dry lands with scanty rains are prominent in parts of Peru. Arable crops that receive moderate levels of well-distributed rains too are frequently seen in this region. The Peruvian agricultural zone occupies about 4.3% of its territory. About 5.5 m ha of farmland is available for crop production. About 3.75 m ha is rain fed and rest 1.75 is irrigated. Large-scale irrigation projects were started early in 1920. This initiative generated transformation of dry lands to arable, irrigated cropping expanses. Irrigation received greater impetus during 1950 s. During 1970s and 80 s, Peruvian Administration invested large sums to augment irrigation. Currently, 1.7 m ha of agrarian zones is irrigable. Yet only, 1.2 m ha is supported by irrigated crops. Factors such as high costs, lack of profits, timely irrigation and low irrigation efficiency reduces use of irrigation facilities. In Peru, coastal agrarian zone has 67% irrigation facility. The Coastal agrarian belt in Peru is supplied with irrigation water by three major projects namely Majes (Arequipa region), Chira-Piura (Piura Region) and Tinajones (Lambayeque Region). The coastal region is highly productive and contributes cereals and high value crops. Highlands and Amazon region with large water resources still possess only rudimentary irrigation systems. It is generally believed that irrigation has allowed diversification of crops and intensification of cropping belts in Peru. Peruvians use surface irrigation with a network of canals. Crops are often irrigated using furrows. Irrigation projects have generated some problems such as water logging, soil salinization, loss of soil nutrients, pollution of drainage water and channels near farms and loss of natural habitats.

The Argentinian Pampas is a major cropping zone that encompasses both rain fed and irrigated farms. Pampas is a moderately intensive agrarian belt. Pampas produces grain/forage that satisfies its own requirements to feed human and animal population, plus to export to other regions of the world. Irrigation and fertilizer-based nutrient supply are important factors that sustain relatively higher productivity. The agricultural intensification in the Pampas was actually engineered with use of major crops like wheat, soybean, maize and legumes. Irrigation played a crucial role in enhancing crop productivity during 1970 s. Farmers endowed with irrigation facilities overcame droughts and revised their yield goals. The nutrient turnover rates in the fields also increased. Adoption of irrigation also allowed farmers to adopt better cropping systems [9].

According to You et al. [10], currently, irrigation does not play a significant role in the productivity of agroecosystems situated in African continent. Almost entire agricultural landscape is supported by natural precipitation, excepting some areas in Egypt, Libya and Madagascar. The African agricultural expanses experience uncongenial precipitation pattern and periodic droughts. Africa has only 6% of its cultivated land under irrigation compared with 18% worldwide. Riverine surface irrigation is prominent in the vicinity of major rivers such as Nile, Niger, Congo, Zambezi, etc. (see Plates 4 and 5). Agroecosystems in Africa are amply endowed with water resources, yet they have not been used it to advantage. It is said that subsistence farming practices minimizes chances for greater investment on water and nutrient inputs. Groundwater is ample in most agrarian regions within Africa. Ground water is not exploited well except in Southern and North-east Africa. According to researchers at International Food Policy Research Centre, Washington, most African farmers are currently thriving on subsistent level harvest, despite clear knowledge that irrigation is a potent improver of grain/forage yield. Potentially, irrigation could double crop harvest in most regions of Africa. Rather, agroecosystem could be easily intensified using irrigation. Hence, African nations have embarked on rapid development of irrigation resources by 2015. We ought to realize that revising yield goals using improved, high yielding genotypes and irrigation will lead us to higher grain/forage yield. Water and nutrients will have to interact and induce higher productivity of crop genotypes. Aspects like nutrient supply, recycling and grain/forage removal from fields need to match better irrigation levels. Nation wise, more than 95% of cultivated land in Egypt and Djibouti is irrigated. About 30% of cultivated land is irrigated in countries like Sudan, Morocco, Libya, Swaziland and Zimbabwe. In South Africa, for example only 22% of arable land is cropped. In addition, much of it is drought prone, due to erratic rainfall. Irrigated farms extend into mere 1.3 m ha. South Africa needs to revise its irrigation pattern and yield goals, since total agricultural produce is less than sufficient [11]. In countries like Tanzania, Sierra Leone, Senegal, Mali and Algeria about 5–10% of agricultural land is irrigated. In most other countries, irrigation is available for less than 5% of cultivated land [10]. The entire African landscape and agricultural area in particular has been mapped intensively, to decipher irrigation potential [12]. Let us compare a few indicators for status of irrigation in Sub-Saharan Africa and entire Africa. Svendsen et al. [13] have pointed out that currently, only 3.5% of cultivated land is irrigated in Sub-Saharan Africa (SSA). The Sub-Sahara is endowed with optimum levels of water resource, but it needs to be exploited. Potential dam capacity is relatively high at 11.2% and that for renewable ground water is 17.5% of cultivated area. The average increase in irrigation facility of SSA has been low at 2–3.5%, depending on agricultural region. Irrigated area in entire Africa (5.8%) is much below global average (18%). Entire Africa is very well endowed with water resources. The potential for water from dams is 14.6% and that for ground water is 72% of cropped area. Yet these sources have not been exploited to improve nutrient dynamics and productivity of agroecosystems. The average rate of expansion of irrigation facility in both Sub-Sahara and entire Africa is a meager at 2.3% per year. During recent years expansion of irrigation has further slowed down in Africa.

China has large cropping belts that are endowed with irrigation. Irrigated regions are distributed well into intensive wheat/maize and soybean production zones of North-east China. Irrigation is a prominent factor that decides the cropping pattern and productivity in the river valleys of Central and South-eastern China. To a certain extent, irrigation affects the cropping expanses in North-west China. Here, riverine irrigation improves crop productivity, despite low fertility soil and dry land conditions. China is among the top three nations with regard to extent of irrigation in agroecosystems. Yet, high population demands extra food grain production that is possible through irrigated ecosystems. Policy makers suggest that at a rate of 400 kg grain per person/yr, China needs 640 m t grains each year to feed expected 1.6 billion people by the year 2030. To achieve this targeted quantum of food grains, irrigated ecosystem has to expand. Irrigation has to reach 60 m hm^2 by 2030, which is 8.37 m hm^2 more than present level [14, 15]. In China, 75% of irrigation water is used to support the large expanses of cereal/legume crops. Farmers adopt agronomic procedures that improve water-use efficiency. A few important procedures are allocating irrigation resources shrewdly, avoiding crops with high demand for water, encouraging cultivation of genotypes with high water-use efficiency, drought tolerant genotypes, devising irrigation schedules that are efficient in intensive crop production zones, adopting irrigation methods that minimize loss of water, adopting cropping systems with better grain/forage productivity per unit water irrigated etc. Over all, it is clear that irrigation is an important factor that influences expansion, nutrient inputs and yield goals of various agroecosystems. Imparting water use efficiency within each cropping zone is important (Fig. 2).

Large dams across major rivers in South Asia have played a key role in augmenting irrigation to large expanses of crops in countries such as Pakistan, India and Bangladesh. Large irrigation projects have literally converted hitherto dry land regions into agriculturally rich arable cropping belts (Plates 6 and 7) [16].

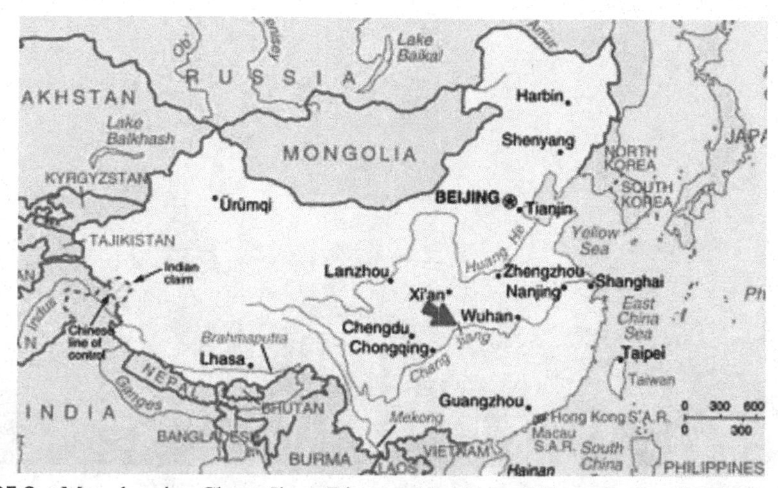

FIGURE 2 Map showing Chang Jiang River and location of Three Gorges dam (see arrow).
Note: Three Gorges dam across river Chang Jiang in the Wuhan region of China is one of the largest dams that man has built. It irrigates large agricultural area in Eastern China.
*Source:*http://www.ccdemo.info/landabee/ChinaTripSlideShows/512x384/16SS-3GorgesDam/slides/002ChinMapDams2. html

PLATE 6 River Indus in Hilly Himalayan Zone.
Note: River Indus irrigates large fertile plains of Pakistan. Its tributaries serve as major source of irrigation to Silty plains of North-west India. Major dams such as Tarbela in Pakistan enhance its irrigation potential. In this region, farmers produce wheat, brassicas and legumes, all through the year. Productivity of cropping belt located in Indus plains is generally higher due to chemical fertilizer, organic manure and irrigation inputs.
Source: International Rivers: People Water, Life. Berkeley, California, USA: http://www.internationalrivers.org/map/15-key-dam-projects-himalayas; www.holyriversofindia; www.dams-info.org/en

PLATE 7 Examples of Southern Indian Water reservoirs and Irrigation sources.
Top: A barrage on river Krishna (Prakasam Barrage) at Vijayawada that serves to supply water to human dwellings and farm land in nearby districts of Eluru, Guntur and Krishna. Bottom: Alamatti Dam on river Krishna near Ilkal, Bagalkot, South India. This dam along with one situated a few kilometers ahead (Narayanpur dam) as the river flows through Vertisols plains, have engineered rapid changes in the nutrient supply, cropping pattern and agroecosystem in general. A famine prone dry land belt has been converted into high input arable cropping zone. The productivity of this rain fed cropping belt improved perceptibly due to water resources.
Source: Krishna, K.R., Bangalore, India.

KEYWORDS

- **Bore well**
- **Drip irrigation**
- **ecosystematic functions**
- **hydathodes**

REFERENCES

1. Doell, V.; Siebert, S. A digital Global Map of Irrigated areas: Documentation. Centre for Environmental Systems Research. University of Kassel, Germany. **1999,** 1–35.
2. CIMMYT. People and Productivity affected by Wheat within each Mega-environment, targeted by Wheat CRP. International Maize and Wheat Research Centre. http://www.cimmytorg **2012,** 1–3 (July 7, 2011).
3. Faures, J. M.; Hoogeveen, J.; Bruinsma, J. The FAO Irrigated Area forecast for 2030, Land and Water Development Division. Food and Agricultural Organization of the United Nations. Rome, Italy. **2010,** 1–14 ftp: //ftp.fao.org/agl/aglw/ docs/fauresetalagadir.pdf.
4. FAO, Choosing an Irrigation Method. Natural Resources Management and Environment Department. Food and Agricultural Organization of United Nations. Rome, Italy. www.fap.org/docrep/S8684E/ s8486e08.htm **2008,** 1–5 (July 11, 2012).
5. FAO, Water lifting Devices: Review of Pumps and Water Lifting devices. http://www.fao.org/docrep/010/ah810e/ AH810E05.htm **2010,** 1–14 (July 17, 2012).
6. Moutont, Yield response factors of Field crops to Deficit Irrigation. FAO corporate Document Repository. Food and Agricultural Organization of the United Nations, Rome, Italy. www.fao.org/docrp/004/yr3655E/ y3655e04.htm **2004,** 1–6 (July 13, 2012).
7. Soybean and Corn Advisor, Only 7% Brazilian crop land is Irrigated.www.soybeanandcorn.com **2011,** 1–3 (July 10, 2012).
8. AgBrazil, Brazils Frontier Regions: Western Bahia. www.agbrazil.com/wesern_bahia_.htm **2012,** 1–6 (July 10, 2012).
9. Reira, C.; Periera, S. G. Irrigated Agricultural Production and Adaptation to climate change in the Argentinian Pampas: An analysis from a Socio-theoretical perspective. International Journal of Sustainability Science and Studies. **2009,** *1,* 35–38.
10. You, L.; Ringler, C.; Nelson. Wood-Sichra, U.; Robertson, R.; Wood, S.; Guo, Z.; Zhu, T.; Sun, T. What is the Irrigation Potential for Africa: A combined Biophysical and Socioeconomic Approach. International Food Policy Research Centre, Washington, D. C. USA. IFPRI Discussion Paper 00993, **2010,** 1–38.
11. SouthAfrica.Info South African Agriculture. www.southafrica.info/business/econnomy/sectors/agricultural-sector.htm **2012,** 1–3 (July 11, 2012).
12. Siebert, S.; Hoogenveen, S.; Frenken, K. Irrigation in Africa, Europe and Latin America: update of Global Digital Map. Frankfurt Hydrology paper 05. Food and Agricultural Organization of the United Nations. Rome, Italy, **2006,** 137.
13. Svendsen, M.; Ewing, M.; Msangi, S. Measuring Irrigation performance in Africa. International Food Policy Research Centre. IFPRI Discussion Paper 00894 43 ifpridp00894.pdf **2009,** (July 20, 2012).
14. Yu Peng, Water-saving irrigation practice in China-Demands, Technical system, Current situation, Development Objective and Counter measures. www. Icid.org/ws1_2001.pdf **2012,** 1–13 (July 18, 2012).

15. Jingsheng, S.; Shaozhong, K. Current Water Resources usage and Measures to develop Water –saving Irrigation in China. Agricultural Engineering Journal. **2000.**
16. Thenkabail, S.; Dheeravath, V.; Biradar, C. M.; Reddy, G. O.; Noojipody, P.; Chandrakanth, G.; Velpuri, M.; Gumma, M.; Li, Y. J. Irrigated area Maps and Statistics for India using Remote Sensing Techniques: A significant Breakthrough. International Water Management Institute (IWMI), Colombo, Srilanka. Internal Report. **2010,** 1–35 http://www.iwmigiam.org/info/GMI-DOC/GIAM-India-Stats.pdf (July 20, 2012).

EXERCISE

1. Mention and explain different methods of irrigation employed in at least five different cropping zones.
2. Mention at least five major riverine irrigation zones of the world.
3. Mention different cropping systems and rotations followed in few precipitation zones.
4. Write about water requirement of different crops grown in dry lands.
5. Describe the pattern of water needs of major cereals and legumes grown in Asia.
6. Define water use efficiency and describe steps required to improve it during water distribution to fields.
7. Write about Irrigation Maps and how it could be used advantageously by farmers.

FURTHER READING

1. Allen, R. G.; Pereira, L. S.; Raes, D.; Smith, F. FAO Irrigation and Drainage Paper: Crop Evapotranspiration-Guidelines for computing Crop Water Requirements. FAO, Water Resources Development and Management Service. Food and Agricultural Organization of the United Nations, Rome, Italy, **2002,** pp. 297.
2. De Fralture, C. Water and Food security in an Insecure World. UNESCO-IHE http://collateral.unescoihe.org/UNESCO-IHE/22/magazine.php?spread=30#/spreadview/12, **2012,** pp. 48.
3. Droagers, P.; Seckler, D.; Martein, I. Estimating the Potential of Rain fed Agriculture. International Water Management Institute. Working Paper, **2001,** *20,* 1–14.
4. FAO (2002 "World Agriculture: Towards 2015/2030, an FAO study. Food and Agricultural Organization of the United Nations. Rome, Italy.
5. Lamm, F. R. Micro irrigation for Changing World: Conserving Resources and Preserving the Environment. Proceedings of the 5th International Conference on Micro-irrigation. American Society of Agricultural Engineers, Wisconsin, USA, **2011**, pp. 97.
6. Lynch, K. Shaping the Future of Water for World Agriculture: A source book of Investment in Agriculture Water Management. International Bank for Reconstruction and Development. Washington D.C. USA, **2005**, pp. 163.
7. Oygard, R., Veldeld, T. and Aune, J. Good Practices in Dry Land Management. The World Bank, Washington D.C., USA, **1999**, pp. 134.
8. Siebert, S.; Hoogenveen, S.; Frenken, K. Irrigation in Africa, Europe and Latin America: Update of Global Digital Map. Frankfurt Hydrology paper 05. Food and Agricultural Organization of the United Nations. Rome, Italy, **2006,** pp. 137.
9. Stewart, B. A.; Nielsen, D. R. Irrigation of Agricultural Crops. American Society of Agronomy, Madison, Wisconsin, USA, **1990,** pp. 1246.

USEFUL WEBSITES

http://www.geo.uni-frankfurt.de/ipg/ag/dl/f_publikationen/1999/doell_siebert_kwws1.pdf (August, 26, 2012)
www.iwmigiam.org/info/main/index.asp (July 12, 2012)
www.fao.org/nr/water/aquastat/main/index.stm (July 12, 2012)
www.fao.org/nr/water/aquastat/irrigationmaps/index50.stm (July 12, 2012)
www.icid.org/imp_data.pdf (July 12, 2012)
http://www.fao.org/ag/agl/aglw/aquastat/irrigationmap/index.stm (July 12, 2012)
ftp://ftp.fao.org/agl/aglw/docs/fauresetalagadir.pdf (July 13, 2012)
www.fao.org/docrep/010/ah810e/Ah810E05.htm (July 16, 2012)
http://www.iwmigiam.org/info/GMI-DOC/GIAM-India-Stats.pdf (July 19, 2012)
http://www.geo.uni-fankfurt.de/ipg/agx.html (July 20, 2012)

CHAPTER 36

CROPS, GENOTYPES AND NUTRIENT DYNAMICS WITHIN AGROECOSYSTEMS

CONTENTS

36.1 CROPS THAT FORM AGROECOSYSTEMS: GENETIC DIVERSITY ISSUES

Agroecosystems described in this volume concern crops that are of immediate utility to human beings in different continents. Agroecosystems could be formed out of large expanses of uniformly single crop species or characteristics that are common to all farming units, for example a single soil type, a uniform agroclimate, precipitation pattern etc. Agroecosystems could harbor a few compatible crop species grown as mixed crops. These situations add diversity to a crop expanse. Further, we should note that, a mono-cropping belt may actually encompass a composite of different genotypes of a crop. For example, vast expanses of wheat in Europe may be an assortment of several genotypes. Factors such as adaptability, economic advantages and farmer's preferences all affect extent to which a genotype(s) spread within an agroecosystem. Now, let us consider aspects of physiological genetics of the crop species and genotypes. Crop genetic diversity has its specific influence on several aspects of nutrient dynamics and productivity of agroecosystems. In fact, to a great extent, modifications in morphogenetic nature of agronomically elite genotypes, genetic enhancement for forage/grain yield and selection of assortment of cultivars has added marked diversity to agroecosystems, to an otherwise large monocropping region. Let us consider a few salient crop genetic features and their impact on agroecosystematic functions.

36.2 CROP PHYSIOLOGICAL ASPECTS THAT AFFECT NUTRIENT DYNAMICS AND PRODUCTIVITY OF AGROECOSYSTEMS

36.2.1 CROP DURATION

Agroecosystems harbor several crop species, grown singly or as intercrops. Crop species vary enormously with regard to duration required to complete various morphogenetic stages and reach maturity. Farmers select crop species based on their ability to adapt to a certain agroclimate and season. Natural factors like shortened day length, frost or drought or end of rainy season may dictate the crop species that fill an agrarian region. Crop species with long duration to maturity need congenial soil nutrients, water and growth conditions for a greater length of time. In case, there is water shortage then irrigation may be needed. Fertilizer inputs should be spaced appropriately. Split application needs to be extended. As a consequence, grain/forage yield could be higher. Let us quote an example. Rice-Wheat-short season cowpea is a highly remunerative crop rotation that is in vogue in Gangetic belt. Here, rice and wheat are fairly long duration crops needing 120–140 days. The third crop encounters a short summer season of 60–70 days before next monsoon starts. Farmers tend to seed the field with extra-short duration genotypes of cowpeas that add to soil-N. Nutrient supply and turnover rates are small but rapid. Here, crop duration affects nutrient dynamics and total grain/forage productivity of an agroecosystem rather stringently.

Now, let us consider genetic differences in crop duration, nutrient recovery pattern and productivity of maize genotypes. For example, maize belt in Southern Indian Plains encompasses at least five different sets of genotypes that are classifiable as extra early short duration (65 days), very early (65–75 days), early (85–90 days), medium

(90–105) and full season maturity (105–130 days). The grain/forage yield potential of early genotypes is proportionately low. It stays in field for short time, absorbs only 36 kg N, 5 kg P and 36 kg K/ha. Hence, nutrient supply is held at low levels of 40:20:30 N: P: K/ha. The grain yield potential is low at 1.8–2.0 t/ha. Such early maturing genotypes are confined to rain fed dry lands. These are preferred in sequences to fill short seasons, say short summer period of 60–80 day before a main crop. Most important, grain harvest and residue recycling occurs in short intervals. Nutrients removed from soil could be effectively replaced. Since productivity is low and nutrient supply schedules are commensurately reduced, nutrient turnover rates too are relatively low. Now let us consider medium or full season genotypes of maize that fill maize belt to a large extent during rainy season in the entire South India. The genetic traits relevant to crop duration, nutrient absorption, photosynthetic efficiency and yield formation exert their influence on agroecosystem productivity. The genetic potential for grain yield is high at 5.5–6 t/ha. The full season genotypes have greater demand for nutrients, since they stay in the field for greater length of time. Farmers usually revize fertilizer schedules to suit the higher grain yield potential. They apply 120 kg N, 60 kg P and 80 kg K. Nutrient recovery is proportionately high at 110 kg N, 22–25 kg P and 110 kg K. Maize genotypes, like that of any other crop species has a set pattern of nutrient recovery from soil. Usually, seedling to silking then until grain formation is the key period, when maximum amount of nutrients are recovered from soil. Therefore, plant breeders try to concentrate and select genotypes that harness soil nutrients best during this period. Therefore, full season genotypes with grain yield potential beyond 5.5 t/ha are provided with higher quantities of nutrients during these stages. The split applications are held at relatively higher levels totally over 120–130 kg/ha in a matter 60–70 days from seedling to grain formation. So, genetic factors related to crop duration/maturity do affect nutrient dynamics and productivity of maize genotypes [1, 2].

36.2.2 PLANTING DENSITY AND CANOPY STRUCTURE

Planting density is an important aspect of crop production procedures. Planting density often depends on soil fertility factors, water resources, plant genetic traits such as tillering, branching pattern, leaf formation, canopy structure, rooting pattern and spread. Planting density has immediate effect on rooting pattern, extent of soil explored, competition for soil moisture and nutrients. Planting density also influences photosynthetic light interception, nutrient partitioning, grain/fruit yield. Let us consider an example involving pearl millet in semi-arids of India and West Africa. In West Africa, dry sandy zones are planted with pearl millet using wide spacing of at least 1 m ′ 1 m. Seeds are planted in hills. This allows efficient rooting and absorption of soil moisture and nutrients from sandy Oxisols (Plate 1). The genetic nature of pearl millet cultivars/composites (e.g., Landraces or CIVT) suits such wide planting and low planting density. The landraces are tall with long panicle and number of tillers per hill is high at 18–32. Pearl millet landrace such as 'Sadorelocal' yields about 900–1400 kg grain/ha. Now compare the planting density and pearl millet morphogenetic traits in Semi-arid India. Pearl millet genotypes are medium in height, slender with short or medium sized panicle of 6–10 cm length. Plant height is not beyond 3 m. Tillering is not profuse. Each plant produces only 4–5 productive tillers. Soil fertility measures and water resources allow genotypes to exert better and yield 2.5–3 t grain/ha. Extraction of soil nutrients is efficient since planting density is high. In nature, we may encounter plasticity for expression of many of the important genetic traits. We have to interpret plant genetic effects carefully.

PLATE 1 Left: Pearl millet hills at Sadore near Niamey in Niger; Right: Pearl millet at Patancheru, Andhra Pradesh India.

Note: Pearl millet genotypes and spacing followed differ enormously based on soil, weather pattern and crop genotype.

In fact, pearl millet genotypes are selected/bred to suit the conditions of Sahelian West Africa and Semi-arid India. Pearl millet genotypes are taller, root profusely, and produce larger biomass per hill in West Africa. In South India, pearl millet genotypes are medium or shorter in height and are spaced very close resulting in high plant population. The crops in West Africa yield 800–1.5 t/ha. On moderately fertile Alfisols, pearl millet yields 2–2.5 t grain plus 4–5 t forage. However, one should realize that genetic potential of most of the genotypes is quite high compared to actual productivity.

Source: Krishna, K.R., Bangalore, India

Now in case of maize, Dobermann et al. [3] opines planting density and grain yield potential has played key role in enhancing productivity. During the past 70 years, planting density has increased from 31,000 to 70,000/ha. It has resulted in increase of grain yield by 4–5 t/ha from previous levels. A few high-density planting farms contain 90,000 seedlings/ha. Maize genotype has to match high planting density procedures, intercept light efficiently and form grains in greater quantity. Genotypes and planting density have a say in the nutrient dynamics that ensue in the agroecosystem and its productivity [4]. This aspect is applicable to all agrarian zones. Over all, tillering, leaf number, canopy structure, panicle traits and grain yield potential have an immense effect on nutrient dynamics and productivity of various agroecosystems.

Let us consider a tree crop. Olive tree plantations differ with regard to genotype, planting density and total productivity. Traditional genotypes of Arabia are sown in low density at 170–230/ha. Nutrient input is commensurately low. Productivity of such genotypes is about 2–5 t/ha. During recent years, Olive orchards are planted with 580–700 trees/ha under high density and up to 1,700/ha under super high-density systems. Olive genotype such as Arbequina, Arbosana and Koreneiki produce 12–17 t fruits/ha. Here, the agroecosystem is intensified using both high yielding genotype and enhanced planting density [5]. Interactive effects of crop morphogenetics and planting density plays a role in grain/forage production of many other crop species.

36.2.3 WATER REQUIREMENTS AND DROUGHT TOLERANCE

Water required by a crop species to germinate, grow, produce tillers/branches, flowers and pods/fruits, i.e., from seedling to maturity is genetically controlled. Crop species vary enormously for total water needed per season. Agroecosystems, even mono-cropping stretches are actually mosaics of different genotypes that have differing water requirements. Individual farmers often try to match water resources with the genetic nature of cultivar with regard to water requirements at different stages of the crop. They would have also studied the critical stages for irrigation. In nature, dry lands often support crops with genetically low water requirements. For example, hardy cereals that with stand drought such as sorghum, pearl millet or finger millet, or legume or horse gram show low water requirements such as cowpea. A low land rice agroeocsystem, supports rice crop that has genetically higher need for water. It needs at least 900–1600 mm per season to grow and mature. Water needs depend on rice variety/hybrid. Clearly, genetic nature of crop species or its genotype dictates the type of agroecosystem that can flourish in an agrarian region. Quite simple! Crops with high demand for water may never make it to dry lands with intermittent drought spells. As stated earlier (*see* Chapter 35), water needs of dry land cereals range from 350–600 mm, those of arable crops grown with irrigation ranges from 550–780 mm and wet land crops like rice or sugarcane need 900–1800 mm per season. These are genetically determined. Of course, we may find genotype within a crop that tolerates drought intensity a bit better than others of the same species. Such genetic traits allow genotypes to survive and still yield marginally in harsh and dry lands.

Through the ages, crop genetic traits that are related to water absorption and its utilization have indeed received their due consideration. Genetic traits relevant to adaptation to rainfall pattern and drought tolerance have been intensely analyzed. Farmers often match season, precipitation pattern and drought tolerance with crop and its cultivar. The pattern of water acquisition and total need per season varies. There are genotypes endowed with genes that tolerate intermittent drought stress. There are others that need relatively lower amounts of water to mature and form grains/forage. Genes relevant to water relation play a key role in deciding the survival perpetuation and productivity of crop species/genotype in a given agroecosystem. Since, nutrient absorption is intricately linked to water. Nutrients are actually acquisitioned in dissolved state along with water, by the crop roots. Therefore, traits such as drought tolerance or those imparting greater WUE also affect nutrient dynamics in the crop belt. Following is an example:

Genotype	Water Use Efficiency g kg^{-1}	Transpiration mm	Harvest Index	Productivity Pods kg ha^{-1}	Nutrient Recovery kg ha^{-1}
Drought Tolerant					
ICR 40	2.52	300	0.33	2210	178
TAG 24	1.36	540	0.31	2285	183
Drought Susceptible Check					
GG2	2.53	265	0.26	1723	138

In a nutshell, genes for water requirements, better water use efficiency, ability to tolerate drought may play crucial role in deciding crop species that dominates a given agroenvironment. It definitely decides the grain/fruit produce that farmer can expect. Nutrient supply schedule, recovery and turnover rates too could be decided based on genetic traits that control water use efficiency and drought tolerance.

36.2.4 NUTRIENT TRANSLOCATION INDEX

Nutrient translocation index (NTI) has direct relevance to amount of each nutrient absorbed and portioned into different plant parts. NTI affects the amount of root biomass, stover and grains/fruits generated within a field or a cropping zone. The amount of nutrients retained below-ground depends firstly on the genetic traits relevant to rooting pattern and its size (biomass). Higher root biomass and nutrients contained in it improves SOM and quality. The root:shoot ratio is an important genetic trait relevant to nutrient dynamics. The ratio of nutrients retained below-ground in the roots and that translocated to shoot system has great effect on nutrient movement from soil to above-ground portion. Crop species or genotypes that retain low amounts of nutrients in soil but translocate more to stem and grains/fruits are profitable. Of course in case of root tubers we need genotypes that translocate less to shoots and retain more to form tubers below-ground. Genotypes of major crops vary enormously with regard to biomass and nutrients partitioned between root system and above-ground portion of crop. Next, crop species (or its genotype) that has better NTI will apportion more of biomass and mineral nutrients in the branches, leaves, twigs and fruits/grains. The ratio of nutrients partitioned into stover and grains is an important trait. It has direct consequences on nutrient removed from field or agroecosystem via grains/fruits, that removed via forage and that recycled *in situ* [9].

36.2.5 TOTAL BIOMASS FORMATION

Total biomass production is an important genetic trait of a crop. It has direct bearing on Organic-C held in the agroecosystem. Crop species or genotypes that accumulate greater quantity of biomass extracts proportionately higher amounts of soil nutrients and fixes C. Soil nutrient depletion and fertilizer schedules are obviously dependent on biomass and nutrient accumulation patterns of crop genotypes. Crop species with better photosynthetic efficiency and light interception traits are preferred, if the aim is to improve biomass. Total biomass affects organic-C and mineral nutrients held in above-ground biomass. In case of field crops, total biomass and HI together decide the extent of grain yield, nutrients removed and that recycled. Total biomass in case of forage crops has impact on amount of soil nutrients removed away from the agro-ecosystem. Mineral nutrient concentration in the succulent tissue of forage decides the nutrients lost from the system. Crop and forage species vary for nutrient accumulated in vegetative tissue. The specific genotype and nutrients contained in its tissue is crucial. It affects nutrient dynamics of the field.

36.2.6 HARVEST INDEX

Harvest Index (HI) is an important genetic trait that affects a series of aspects of nutrient dynamics and grain productivity of agroecosystem. Harvest Index is defined as ratio of biomass, phototsynthates and nutrients partitioned between vegetative tissue and grains. We know that crop genetic potential for high biomass production is of relevance in zones that are low yielding. However, in intensive cropping belts, HI is of great relevance to grain formation. Higher HI means greater grain yield and relatively more nutrients removed via grains. There is no doubt that high biomass/grain formation plus HI requires fertilizer schedules that supply proportionately greater quantity of nutrients. In nature, landraces that were used by farmers generally had low biomass and HI. These landraces also produced more foliage per unit of grain formed. For example, landraces of cereals such as sorghum (HI=0.28), rice (HI=0.32), pearl millet (0.22) did not offer high proportion of grain per t biomass formed. However, consistent genetic improvement for HI resulted in genotypes with better HI. It allowed better ratio of grain harvest. For example, high yielding rice genotypes with HI of 0.35 and higher biomass potential gave over 3 t grains plus forage during 1970s till 1990s. During recent period, rice hybrids with HI of 0.45 and greater biomass potential of 20 t/ha provide 5–8 t grain/ha [10–12, 28]. Similar trends have occurred with wheat, maize, legumes and oil seeds. These genetic changes in HI and biomass potential have firstly affected fertilizer supply into agroecosystems. It has affected planting density and grain yield forecasts. Stover and nutrient recycling trends have also been affected due changes in HI [2, 13].

36.2.7 MINERAL CONTENT OF GRAINS/FRUITS

Cereal grains are known to accumulate different quantities of biochemicals and minerals. In fact, genetic variations for minerals and nutrient content in seeds or fruits have been reported in many other groups of crops, such as legumes (Plate 2), oilseeds, cash crops and plantation species. Minerals accumulated in grains/fruits may vary even between genotypes of a crop species. Harvest index and nutrient translocation index are important characteristics that affect mineral contents of foliage and grains/fruits. Actually, nutrient partitioning between foliage, stem and grains/fruits is affected. The mineral content of grains has an immediate effect on nutrient removals and recycling in a field/agroecosystem. Minerals garnered by foliage/stover could be either removed for farm animals or recycled *in situ*. Mineral accumulated in grains/fruits is of course removed from the field.

PLATE 2 Legume species such as soybean, lentil, field bean, green gram and black eyed bean may vary with regard to mineral content, proteins, fats and other biochemical constituents.
Note: The above legumes may occupy crop belts to different extents depending on planting ratios. The extent of stover and grains produced and minerals removed from field is affected proportionately.
Source: Krishna, K.R., Bangalore, India.

There are several reports about genetic variations in minerals accumulated in fruit crops. Following is an example dealing with minerals in citrus genotypes grown on Spodosols of Central Florida. Genotypes that accumulate higher amounts of minerals in fruits tend to deplete the field, without any provision for recycling. Mineral nutrients are removed away with fruits and transported to farther locations.

Nutrient Accumulated in Citrus Fruits

Citrus Cultivar	N	P	K	Ca	Mg	Fe	Mn	Zn	Cu	Al	Na
		%			%			µg/g			
Hamlin	1.1	0.14	1.16	0.36	0.11	19	03	12	06	19	0.03
Parson Brown	1.0	0.14	1.19	0.44	0.11	26	5 13	5 21	0.03		
Valencia	1.1	0.16	1.19	0.35	0.10	40	4 11	5 16	0.04		
Sunburst	1.2	0.15	1.20	0.29	0.09	33	4 18	6 15	0.05		

Source: Excerpted from Hanlon et al. [14]; Alva et al. [15]; and Alva and Paramisavam [16].

36.2.8 NUTRIENT USE EFFICIENCY

Nutrient use efficiency (NUE) is defined as amount of biomass or grain/fruits produced per unit nutrient held in its tissue. Plants translocate nutrients into grains as they develop and mature. The size, weight and biomass accumulated in each grain/fruit per unit mineral nutrient garnered by it varies. The NUE is an important genetic trait that affects crop produce per unit mineral nutrient channeled into above-ground phase of the agroecosystem. Crops differ enormously for NUE. Further, genotypes within each crop are genetically endowed with low, medium or high nutrient use efficiency. In nature, introduction of each genotype into field or ecosystem induces proportionate changes in nutrient acquisition rates, tolerance to nutrient dearth and nutrient use efficiency. Tolerance to nutrient dearth is a trait useful in areas with low soil fertility. Subsistence farming with low input requires crop genotypes endowed with tolerance to low soil nutrient availability. Crop genotypes that produce deep roots, more of secondary/fine roots active in nutrient acquisition, better biomass formation per unit nutrient translocated to above-ground canopy are mandatory. Tolerance to low nutrient and water resources are almost essential, if a crop genotype has to grow rapidly and produce grains/fruits.

Genetics of nutrient uptake/utilization and set(s) of genes that impart tolerance to low-N or Low-P have been reported [13, 17–19]. Soil N and P are two major nutrients for which deficiency has been perceived worldwide. Agroecosystems that extend into such soils need to be introduced with genotypes that possess genes for low N and P tolerance. We also need genotypes with better N-use efficiency. For example, if 21 kg N needs to be absorbed per ton grain formation in the general course, an N-efficient genotype may require only 16–18 kg N t on grain formation. Hence, for the same amount of N supply/absorption, an N-efficient genotype of cereal may yield 0.25–0.4 t grains more than N-inefficient genotype. Clearly, N dynamics gets affected each time we introduce an N-efficient line. Genotype efficient in N absorption will recover greater amount of N into canopy. Further, one with better N-efficiency will offer farmers more of biomass and grains. Genotypes with greater N-use efficiency are also to be preferred in regions with fertile soil. N uptake is not a constraint in such regions. For example, on Chernozems of Northern Great Plains or European Plains, genes for high N-use efficiency will offer enhanced grain yield/forage. Soil-N status and acquisition rate are not a problem. Genes for greater N-uptake is essential in regions with dearth for soil-N. Many of these arguments applicable for N are equally applicable for P, K and other soil nutrients. We should note that photosynthetic efficiency and C fixation rates of a genotypes coupled with mineral nutrient acquisition/use decides the net productivity of a genotype in a given environment. Let us quote an example. They say in semiarid regions, about 70% of maize cultivation occurs under rain fed conditions on soils low in fertility. Both N and P dearth are easily perceivable. Fertilizer and organic manure supply are often below optimum. Here, low N and P tolerance is required. Maize genotype such as Ageti, Kiran and Low-N tolerant early maturing genotypes such as D741, Pop-49 recover relatively more of soil-N. They explore soil better and yield 1200 kg grain/ha.

36.3 CROP GENETIC IMPROVEMENT AND NUTRIENT DYNAMICS IN AGROECOSYSTEMS: SALIENT EXAMPLES

Crop genetic improvement effected through selection and breeding has immensely affected the agroecosystems world-wide. The genetic nature of crop or its genotype that fills up the agroecosystem needs attention, with regard to its ability to influence nutrient dynamics, ecosystematic functions, grain/fruit productivity and organic matter recycling trends. Cultivars introduced into a cropping belt have often brought about conspicuous changes in biomass, grain yield, nutrient recovery and recycling trends. Genetic advance for traits relevant to morphogenetics, growth, grain yield and forage have occurred rather consistently during past 5–6 decades. We ought to realize that nutrient dynamics and productivity levels achieved currently in different agroecosystems have been achieved gradually, through careful selection of crop genetic traits and its advancement. Let us examine a few crops, regarding the influence of cultivar development programs in various regions and its impact on nutrient recovery, nutrient supply, grain/forage yield and organic matter recycling.

During past 50 years, genetic advance in rice genotypes has meant that, potential yield has improved by 5–6 t grain/ha. Harvest index that influences the grain/forage ratio has changed by 0.14. It means greater quantity of photosynthates and minerals are translocated to grains. Higher grain/forage yield necessitates proportionate increases in nutrient supply via fertilizers and recovery rates by roots (Table 1). Nutrient input increased by 60 kg N, 20 kg P and 40 kg K. We know that 14–21 kg N is required grain/t. Nutrient recovery by rice genotypes also improved enormously during the past 50 years. Nutrient recovery by genotypes increased by 1.8 kg N, 0.5 kg P and 2 kg K/ha each year. This leads to steady increase in nutrient turnover rates in fields. Wheat landraces cultivated in Great Plains, Pampas or in Europe were grown under low input farming systems during early part of 1900s. Wheat landraces derived nutrients from organic residues recycled *in situ*. The genetic potential for grain/forage yield was not high. They yielded about 1–1.2 t grain/ha and 3–5 t forage per season. Advent of fertilizer technology altered wheat grain/forage yield. Wheat genotypes that were responsive to fertilizers, those with high tillering ability and dwarfing genes allowed farmers to harvest high grain yield. Dwarfed wheat genotypes yielded 4–5 t grain/ha. The intensification of major wheat belts were actually achieved partly through better genotypes. Nutrient supply rates jumped by over 80 kg N, 25 kg P and 90 kg K/ha during the past 50 years in the wheat belts. No doubt, nutrient turnover rates markedly improved per season in most wheat belts.

Maize agroecosystem in the Northern Great Plains has been intensified using improved genotypes. Actually, genetic advance for traits such as grain/forage yield has helped the farmers to intensify the belt. Grain yield improved by 6.5 t/ha from 5–5.5t/ha in 1950s to 11.0 t/ha in 2005. Nutrient supply and recovery increased proportionately in order to support higher grain/forage productivity. Nutrient recovery rates of corn genotypes improved remarkably. Nitrogen recovery from soil to above-ground portion improved from 131 kg N/ha in 1950 to 255 kg N/ha in 2000. Similarly, in 50 years, P absorption increased from 39 kg to 78 kg P/ha; and K absorption from 119 to 232 kg K/ha. Maize absorbs 80% of nutrients between 26 and 100th day. Hence, nutrient uptake rate exhibited by genotype during this period is crucial. It regulates nutrient dynamics and productivity to a great extent.

TABLE 1 Rice Genotypes introduced into South Indian Plains and Coasts (1800s to 2005) and its impact on Nutrient Dynamics in the Rice Ecosystem.

YEAR / VARIETIES	Potential Yield (q ha^{-1})	Farmer's Yield (q ha^{-1})	Harvest Index	Nutrient Inputs (kg ha^{-1})	Nutrient Recovery (kg ha^{-1}) N	P_2O_5	K
1800s							
Dodda Butta, Hotay Caimbuti, Arsina Caimbuta, Murargili, Yallic Raja, Caraculla	NA	8-10	0.20-0.25	Crop Residues	21	6.5	33
Early 1900s							
Dodda Butta, Yallic Raja	20-30	8-10	0.20-0.25	FYM, Residues	21	6.5	32
1950s							
ADT 27, Murargil,	30	9-10	0.3	FYM, 20 N	20	6.0	30
2000-2005							
Hybrids: APHR1, APHR2 +120 straw	80-100	65-80	0.44-0.48	FYM, 240 N, 60 P, 120-160 K, FeSO4 25, $ZnSO_4$ 0.2% spray,	145	45	195
Net increase (ha^{-1} $season^{-1}$) due to change from Semi dwarfs to Hybrids Since past 5.5 decades	50-70 + 40 straw	55-70 +30 straw	0.14	60 N, 20 P, 40 K, 5 t FYM	90	28	115

Source: [2].

Note: Nutrient supply was not affected much during the 100 years from 1800 to 1900s. Genetic advance that imparted fertilizer responsive traits and enhanced grain/forage yield occurred markedly during 1950 till date.

Corn Belt of USA	Nutrient Supply kg/ha	Yield (t/ha) Grain	Stover	Nutrient Recovery kg/ha
Landraces (1800s – Early 1900 s)	Subsistence, FYM 2–5 t/ha^{-1}	1.25	3 – 4	20 N, 5 P, 18 K
Hybrids (during 2005–2010)	180 N, 60 P, 200 K, 12 t FYM/ha-1	11.25	14	235 N, 60 P, 240 K
Net Change/Genetic Advance	180 N: 60 P: 200 K, 10 t FYM/ha^{-1}	10.0	10–11	215 N, 55 P, 222 K

Source: [2, 3, 20–23].

Note: Nutrient recovery by Maize genotypes has been computed considering that 22–23 kg N, 5–9 kg P and 22 kg K is needed to produce 1.0 t grains plus proportionate forage based on HI.

Sorghum is a crop well adapted to Southern India Plains. Let us examine effects of introduction of sorghum genotypes on nutrients supplied into the belt and that partitioned and recovered by crop and grain/forage. Records suggest that genotypes grown during 1800s and until mid-1900s were cultivated adopting low input subsistence methods. The genetic potential for yield was markedly low at 1–2 t grains/ha, but soil fertility measures and rainfed production systems allowed farmers to reap only 0.5–0.8 t grain/ha plus forage. During 1950 s, sorghum composites/hybrids were spread all over the plains. They had high grain yield potential of 3–5 t grain/ha. Fertilizer inputs, especially N were practiced, although at low levels of 40 kg/ha. It improved productivity of sorghum belt to 2 t grain/ha forage (see Table 2).

TABLE 2 Sorghum Genotypes cultivated by Southern Indian Farmers, their Productivity and Nutrient Dynamics during past Century

Year / Genotypes	Nutrient Input (kg ha^{-1})	Productivity (t ha^{-1})		Harvest Index	Nutrient Recovery (kg ha^{-1})		
		Grain	Stover		N	P	K
1950 Anakapalle series, Hagari, Bilichigan,	20N, 5 t FYM	0.5	2-3	0.25	10	5.0	17.0
2005	80-120N, 40P, 60K	3.2 -3.5	9-10	0.27-0.32	78	33	118
CSH18	15 t FYM 5-25 $ZnSO_4$,	3.3 -3.5	9-10	0.27-0.30	78	32	118
SPV 1411	2 kg *Azospirillum*	3.5 -3.7	7 -8	0.33-0.35	80	35	122
Net Change in 55 years: Attributable to genotype	60 N, 40 P, 60 K, 10 t FYM	3.0	5.5	0.10	70	30	105

Sources: [2, 24–27];
www.vasat.org/learning_resources/crops/sorghum/sorghum_prodpractices/html/m812/resources/. pp. 1–8.
Note: Major nutrients are shown in kg/ha and Organic manures in t/ha. Nutrient recovery is a computed value derived from the fact that sorghum crop absorbs kg 22–24 N, 9–12 kg P_2O_5 and 34 kg K_2O to produce 1.0 t grain. Potential grain yield of hybrids of CSH series is 3–4 t/ha in Experimental farms and that of varieties like CSV series is 2–2.5 t grain/ha.

During 1970s and 1980s, sorghum belt in India perceived marked changes in crop genotype. Series of hybrids were released into the agroecosystem [28]. Fertilizer and irrigation were augmented better. This resulted in enhanced productivity of sorghum cropping zones. Each change in genotype to greater yield potential meant higher amount of nutrient supply to fields and nutrient recovery by sorghum genotype. Since harvest index was better at 0.35–0.40, nutrients removed via grains were proportionately high. Add to this, greater biomass potential meant soil nutrients were harnessed and recycled better, within the same period of 110–130 days during kharif season. During the past 50 years, net change in aspects like nutrient supply in to agroecosystem, nutrient recovery by crop, and partitioning nutrients into seeds (HI) improved enormously. Developing high yielding sorghum composites/hybrids meant perceptible changes in the sorghum agroecosystem of South India. It was most useful to the human population in the region, since sorghum is among staple cereals consumed. Similar effects on Sorghum agroecosystems in West Africa and Brazil are easily identifiable.

KEYWORDS

- **canopy structure**
- **harvest index**
- **nutrient translocation index**
- **nutrient use efficiency**
- **planting density**

REFERENCES

1. IKISAN, Maize: Varieties. http://www. IKISAN/links/ap_maizeffarieties.shtml **2007,**1–9 (March 12, 2008).
2. Krishna, K. R. Agroecosystems of South India: Nutrient Dynamics, Ecology and Productivity. BrownWalker Press Inc.: Boca Raton, Florida, USA, **2010,** 556.
3. Dobermann, A.; Arkebauer, T.; Cassman, K.; Lindquist, J.; Specht, D.; Walters, L.; Yang, H. Understanding and Managing corn yield potential. Proceedings of the Fertilizer Industry Association Round Table. Charleston, South Carolina, Forest Hill, Maryland, USA. **2002.**
4. Barbieri, A.; Echevaria, H. E.; Sainz Rozas, H. R.; Andrade, F. H. Nitrogen use efficiency in maize as affected by Nitrogen availability and row spacing. Agronomy Journal **2008,** *98,* 1094–1100.
5. Tubeilah, A.; Brugsewen, A.; Turkelboom, F. Growing Olives and other tree species in Marginal Dry Environment with examples: from the Khanessar Valley, in Syria. International Centre for Agricultural Research in Dry Areas, Aleppo, Syria, **2004,** 1–58.
6. Cruickshank, A. W.; Rachupati, N. C.; Wright, G. C.; Nigam, S. N. Breeding of drought-resistant peanuts. In: Proceedings of Collaborative Review Meeting. International Crops Research Institute for the Semi-Arid Tropics. http://www.aciar.gov.au/web.nsf/doc/acia-SSLVWR/$file/PR112.pdf **2002,** 1–102 (April 10, 2004).
7. Basu, M. S.; Mathur, R. K.; Manivel, P. Drought Resistance Research at the National Research for Groundnut. Junaghad, Gujarat, India. Proceedings of Collaborative Review Meeting. International Crops Research Institute for the Semi-Arid Tropics. http://www.aciar.gov.au/web.nsf/doc/acia-SSLVWR/$file/PR112.pdf **2002,** (April 10, 2004) 31–37.
8. Rachupati, N. C.; Wright, G. C. The Physiological basis of selection of peanut genotypes as parents in Breeding for Drought Resistance. In: Proceedings of Collaborative Review Meeting. Cruickshank, A. W.; Rachupati, N. C.; Wright, G. C.; Nigam, S. N. (Eds.). International Crops Research Institute for the Semi-arid Tropics. Hyderabad, India. http://www.aciar.gov.au/web.nsf/doc/acia-SSLVWR/$file/PR112.pdf **2002,** 1–102 (April 10, 2004).
9. Krishna, K. R. Peanut Agroecosystem: Nutrient Dynamics and Productivity. Alpha Science International Inc.; Oxfordshire, England, **2008,** 298.
10. Kush, G. S. Breaking the Yield Frontier of Rice. Geo Journal **1996,** *35,* 329–332.
11. Janaiah, A.; Hybrid Rice in Andhra Pradesh. Findings of a Survey. Economic and Political Weekly, **2003,***38,* 2513–2516.
12. ICAR. History and Development of Rice Varieties in India. http://www.dacnet.nic.in/rice.htm **2006,** 1–48.
13. Krishna, K. R. Maize Agroecosystem: Nutrient dynamics and Productivity. Apple Academic Press Inc.; Toronto and New Jersey, **2013,** 343.

14. Hanlon, E. A.; Obriza, T. A.; Alva, A. K. Tissue and Soil analysis. In: Nutrition of Florida Citrus Trees. Tucker, D. H.; Alva, A. K.; Jackson, L. K.; Wheaton, T. A. (Eds.). University of Florida, Lake Alfred, Florida, USA, **1995,** 13–16.
15. Alva, A. K.; Paramasivam, S.; Graham, G. D. Impact of Nitrogen management practices, nutritional status, and yield of Valencia orange and ground water. Journal of Environmental Quality **1998,** *27,* 904–910.
16. Alva, A. K.; Paramasivam, S. Nitrogen management for high yield and quality. Of citrus in sandy soils. Soil Science Society of America Journal **1998,** *62,* 1335–1342.
17. Krishna, K. R. Crop Improvement towards tolerance to soil related constraints. In: Soil Fertility and Crop Production Krishna, K. R. (Ed.). Science Publishers Inc.; Enfield, New Hampshire, USA, **2002,** 337–370.
18. Banziger, M.; Betran, F. J.; Lafitte, H. R. Efficiency of High-N selection environments for improving maize for Low-N target environments. Crop Science **1997,** *37,* 1103–1109.
19. Beem, J.; Smith, M. E. Variation in Nitrogen Use Efficiency and Root system size in Temperate Maize genotypes. In: Developing Drought and Low-Nitrogen Tolerance in Maize. http://www.cimmyt.org/Research/ Maize/DLNCA/htm/DLNCp2–8.htm **2004,** 6 (October 2007).
20. Duvick, D. N.; Cassman, K. G. Post Green Revolution trends in Yield Potential of Temperate Maize in the North Central USA. Crop Science **1999,** *39,* 1622–1630.
21. Mureithi, J. G. Maize Varieties, Soil Fertility Improvement and Appropriate Agronomic Practices. Kenya Agricultural Research Institute, Kitale, http://www.kari.org/LegumeProject/legume_leaflets/Maizevarieties 020506.pdf **2008,** 1–2 (September 2007).
22. Soliman, M. S. M. Stability and Environment Interactions of some Promising Yellow Maize Genotypes. Research Journal of Agriculture and Biological Sciences **2006,** *2,* 249–255.
23. Abendorth, L.; Elmore, R. Demand for more corn following corn. http://www.agronext.iastate.edu/ corn/corn/production/management/cropping/demand.html. **2007,** 1–2.
24. Buchanan, F. A journey from Madras through the countries of Mysore, Canara and Malabar. W.; Baulmar and Company, Cleveland Row, St James, London **1807,** 3 Volumes, 1–370; 1–540; 1–440.
25. ICRISAT. A review of Fertilizer Use Research on Sorghum in India. International Crops Research Institute for the Semi-Arid Tropics. Research Bulletin **1984,** *8,* 1–59.
26. Venkateswaralu, J. Rain fed Agriculture in India. Indian Council of Agricultural Research, New Delhi, **2004,** 566.
27. UAS. Package of Practices for Karnataka. University of Agricultural Sciences, Bangalore, India, **2006,** 285.
28. ICAR, Hand Book of Agriculture. Indian Council of Agricultural Research, New Delhi, **1997,** 1310.

EXERCISE

1. Mention the genetic traits of cereals that immediately affect nutrients recycled, removed and lost in grains from a field/agroecosystem
2. How does Harvest Index affect nutrient dynamics in a crop field?
3. What is the influence of root:shoot ratio on nutrient partitioning between above and below-ground portion of an agroecosystem?
4. How do total biomass and grain formation accumulation influence nutrient dynamics in a field?
5. What is Genetic Advance for grains, biomass and harvest index. How have they affected agroecosystems in North America and Asia during past five decades-give examples.

FURTHER READING

1. Krishna, K. R. Crop Improvement for Tolerance to Soil Fertility constraints. In: Soil Fertility and Crop Production. Krishna K. R. (Ed.). Science Publishers Inc.: Enfield, New Hampshire, USA, **2002,** pp. 337–369

USEFUL WEBSITES

http://www.cssa.org
http://www.icrisat.org
http://www.icarda.org

CHAPTER 37

CROPPING SYSTEMS AND NUTRIENT DYNAMICS

CONTENTS

37.1 MONOCROPPING

Agroecosystems that support a single species of crop are frequently encountered in certain geographic zones. For example, rice mono-crops grown season after season dominates the landscape in South and South-east Asia. This has been possible both due to precipitation patterns, apt facility to irrigate the crop and supply optimum levels of fertilizer-based nutrients. Monocropping zones of cereals occur in many agrarian areas (Plates 1, 2, 3 and 4). Monocrops of wheat are common in the Northern Great Plains of USA, and Central Plains of Europe. Similarly, monocrops of soybean dominate the landscapes in parts of Matto Grasso and Pampas of South America. Monocrops of sorghum and other coarse grain crops like pearl millet or finger millet are observed in dry land agroecosystems of Africa and Asia. Monocrops of cotton are preferred in large expanses of Vertisol belts of Southern India, Central Plains of USA, etc.

PLATE 1 A field with Wheat Mono crop in Northern Great Plains.
Source: B2 Farm LLC, Brooklyn, Wisconsin, USA.

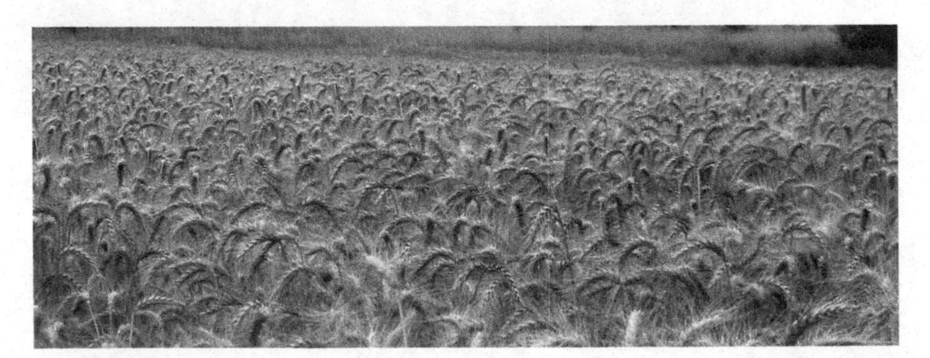

PLATE 2 A monocrop of Wheat in East Africa.
Source: United States of Deparatment of Agriculture, Beltseville, Maryland, USA.

PLATE 3 A Sorghum Monocrop near Lubbock, North Texas. Sorghum expanses that supply grain and forage are common in the Southern Plains of USA. Sorghum cropping zones occur more frequently in Nebraska, Kansas and Texas plains.
Source: Kennedy, 2012, Sorghum Checkoff, Lubbock North Texas, USA. http://www.sorghumcheckoff.com/sites/ default/files/101–0114_IMG_2.jpg

PLATE 4 A Monocrop of Maize grown on Alfisols in South India—A close-up view of Maize canopy. Location: Rural Bangalore, India.
Source: Krishna, K.R., 2012

Some of the prerequisites for adopting monocrops are suitable soil types, weather pattern, genotypes that are productive and adapt effectively to season, disease, insect and drought pressures. Economics of farming enterprises and incessant demand for grains/production seems crucial while opting for monocropping systems. The cropping intensity and productivity of monocrops varies widely depending on soil fertility, fertilizer inputs and demand for grains (Table 1). Monocropping zones of rice are intensively cultivated in South India and Fareast. Wheat belt in European plains is supplied large amounts of fertilizers and irrigated to yield as high as 7–8 ton grains/ha. Maize crop in the 'Corn Belt of Northern USA' is supplied with high quantities of fertilizers and FYM to achieve 8–11 t grain and 15 t forage/ha. Monocrops of cereals, legumes and cotton are moderately intensive in the semiarid zones of Asia. Monocrops are a clear possibility and sometimes inevitable in the arid zones of West Asia and Sahelian region, where precipitation pattern and paucity of fertilizer and other resources occur. These conditions allow only one short cropping season and low productivity levels.

Generally, mono-crops need greater attention and careful mending, since they attract several deleterious effects and long-term mono-crops have led to maladies of soil and other resources. Repeated high supply of fertilizers and soil amendment has indeed induced soil deterioration in many cropping belts. Mono-crops attract and allow build-up of disease/pest inocula. Excessive depletion of soil nutrients and moisture may lead to low productivity. Excessive production through monocrops may lead to market glut and fall in economic value. Obviously, a series of options are weighed out by farmers while deciding on monocrops. Some of these aspects actually regulate the size and intensity of mono-cropping belts.

TABLE 1 A few examples of Mono-cropping belts, their Intensity of Cultivation and Productivity.

Agroecosystem Crop/Region	Fertilizer Supply Inorganic; FYM kg ha^{-1}; t ha^{-1}	Productivity Grain; Forage t ha^{-1}	References
Highly Intensive			
Maize Corn belt of USA	240-280; 12-15	8-11; 10-15	FAOSTAT, 2005; OKSTATE, 2009 Krishna, 2003
Rice, Northern China and Fareast	220-280; 10-12	6-8; 12-14	Janaiah, 2003; ICAR, 2006
Rice in South Indian plains	180-240; 10-12	5-7; 10-12	Raju et al. 2008
Cotton in Vertisol Belt of South India	180-220;	8-10; 5-8	
Moderately Intensive			
Maize in Pampas	120-180; 5-10	4-5; 8-10	Hall et al. 1992; Carcova. et.al. 2000 IPNI, 2008; Krishna, 2010; 2013
Maize in South India	80-140; 5-10	3-5; 8-10	IIRI, 1995; Krishna, 2003
Rice in North Indian plains	120-180; 3-8	3-4; 8-10	Peterson et al. 1996; Oplinger et
Wheat in Great Plains of USA	180-200; 5-10	4-6; 8-12	al. 1998;
	140-180; 3-5	3-4; 5-8	Norwood, 2000; Halvorson et al.
	120-180; 3-5	3-5; 6-8	2001
	80-120; 3-5	2-3; 5-6	

Wheat in Rolling Pampas	140-180; 3-5	3-4; 5-8	Hall, 1992; Gonzalez-Montaner et al. 1997
Wheat in Indo-Gangetic Plains	120-180; 3-5	3-5; 6-8	Velayutham, 1997; Singh and Singh, 2010
Pigeonpea in Vertisol plains of India	80-120; 3-5	2-3; 5-6	Ahlawat and Masood ali,1993; Ganeshmurthy et al. 2006; Krishna, 2010;
Sunflower in Southern India	120-180; 3-5	3-4; 6-8	IKISAN, 2008; Krishna, 2010,
Low Input/Subsistence Farming			
Wheat/Barley in West Asia	40-60; 2-3	1.5-2.5; 5	
Maize in tropical West Africa	nil-40; 2-3	0.7-1.2; 3	Ofori et al.2004
Pearl Millet Sahelian zones	nil-40; nil	0.8-1.8; 3	Mokwunye, 1991; Bationo et al. 1992
Pearl Millet in Arid Northwest Indian	nil-40; nil	0.8-1.5; 5	
Groundnut in Senegal	nil-60; nil	0.8-1.2; 5	Coura Badiane, 2001; Krishna, 2008

Note: Chemical fertilizer (kg/ha) denotes to total major nutrients, namely, Nitrogen, Phosphorus and Potassium supplied in different quantities/proportions to the crop; FYM (t/ha) refers to organic manure of different kinds incorporated into fields. In case of cotton yield refers to bales and forage.

37.2 CROP ROTATIONS

Crop rotations is a term that denotes cultivation of 2, 3 or even more number of crop species one after other on the same piece of land in sequence. Several factors related to geographic conditions, agroclimate, soil fertility and nutrient supply schedules, crop genotype and its physiology, sowing, harvest dates and economic value

need consideration. Farmers often consider the net gain from the entire sequence while deciding on fertilizer supply and irrigation. There are indeed innumerable crop rotations and sequences that are common to different agrarian belts. Crop rotations are also dynamic and keep changing with time. Crop rotations such as maize-soybean or maize cotton in the Great Plains of USA; Soybean-sorghum/maize in the Cerrados of Brazil; Wheat-soybean in Pampas of Argentina; Pearl millet or maize-cowpea in West Africa; Wheat/barley-lentils in Wheat Africa; Rice-Wheat in the Indo-Gangetic plains; and Wheat-soybean in North-east China are some popular rotations that are expansive and dominant. Each crop rotation has its generalized and exclusive advantages to soil fertility, grain productivity and economic benefits [28]. For example, in North China, wheat-maize rotations has intensified the cropping pattern, improved nutrient turnover per unit time and enhanced crop harvest to 4 t/ha of wheat grain and 5–6 t maize grain. In fact, such intensification of crop rotations may help in deriving greater harvests [29].

37.3 INTER-CROPPING AND MIXED CROPS

Intercropping is culture of two or three crop species at different planting densities in the same field. There are different types of intercropping such as mixed intercropping, row intercropping, strip intercropping and relay intercropping. The row ratios and seedling population per unit area may vary depending on wide array of agronomic, soil and crop related factors. Intercrops could be simultaneous or staggered sown at different days. Crop species intercropped are usually compatible and offer lest interference with regard to interception of photosynthetic radiation, root growth, and exploitation water/nutrients. Often intercrops may possess different growth rates, their canopies dominate at different periods with least shallow effects. The rooting depths of intercrops too differ. The deeper roots of legumes exploit water and nutrients from deeper layers. The shallow rooted species, for example, cereals like wheat or finger millet may confine upper horizons of soil and deplete moisture/nutrient from surface layers (Plates 5, 6 and 7).

PLATE 5 Finger millet/Brassica mixed-intercrop in the Alfisol zones of South Indian plains.
Note: Finger millet and Mustard are sown at the onset of monsoon. Finger millet establishes early. It draws moisture and nutrients from the upper horizons of soil. Finger millet is relatively a shallow rooted crop. It matures early by September end. The mustard crop establishes slowly, traps sunlight efficiently and continues to absorb soil moisture and nutrients even after harvest of finger millet. Mustard roots reach deeper into soil exploit nutrients still available unused. Sole crops may yield higher quantity of grains, but intercrops are efficient since they offer two crops whose total harvest is much greater. The land use efficiency in terms of biomass and nutrient is much greater in an intercrop compared to sole crops.
Source: Krishna, K. R., 2011.

PLATE 6 Maize/Finger millet intercrops with 1:5 rows. Location near Kanakpura, Bangalore District, India.
Source: Krishna, K.R. 2012.

PLATE 7 A double legume intercrop of Pigeonpea and Field Bean at 1:1 row ratio. Location: A farmer's field near Kanakpura, Bangalore District. South India.
Source: Krishna, K.R. 2012.

Inter cropping may sometimes involve more than two crop species. Then, it is termed mixed cropping. Crop species meant for mixed cropping is selected carefully considering growth pattern, soil fertility requirements and harvest time (Plates 8, 9, and 10). Intercropping systems also involve two crop species sown in wide strips of several rows. For example, strip cropping of maize and soybean with 6 to 12 rows each is common in the Great Plains of USA. Strips of Agroforestry species such as Sesbania or Leucana with 8–10 rows of maize or sorghum are common eastern Africa. It offers several advantages with regard to agronomic procedures, traction and deployment of farm machinery. At the same time, strip cropping allows farmers to regulate soil fertility. Now-a-days, strip cropping is preferred since it is highly amenable to adoption of GPS-guided precision farming procedures (Plates 11, 12, and 13).

PLATE 8 Mixed Cropping of Pigeonpea, Castor and Maize, along with useful trees at a location near Dodaballapur, near Bangalore in South India (13.17°N latitude and 77.30°E longitude).
Source: Krishna, K. R., 2011.

PLATE 9 A Mixed Cropping Enterprise in Columbian Hills.
Note: The Mixed crop involves culture of annuals, plantains and agroforestry trees in the terraces.
Source: CIAT-CGIAR, Cali, Columbia.

PLATE 10 A Mixed crop of Coffee, Pepper and Fuel wood trees.
Note: Location is in Humid tropics of Western Ghats, Karnataka, State, India.
Source: Krishna, K.R., Bangalore, India.

PLATE 11 A Field being sown with Soybean strip after a Maize strip crop. Soybean seeds are sown at different dates in Strips in a field that supported maize during previous season. Location is Nelson farms, Fort Dodge, IA USA.
Source: David Nelson, Nelson Farms, Iowa, USA.

PLATE 12 A Strip Crop of Forage grass and Mulberry near Kanakpura, Bangalore, South India.
Source: Krishna, K. R., 2012.

PLATE 13 Maize is strip cropped with vegetables in the Alfisol zone of Southern Indian plains.
Source: Krishna, K. R., 2011.

Maize cropping systems that involve intercropping with legume grain or vegetables is popular among small farms in Southern Africa [30]. Soil fertility is maintained using both inorganic and organic manures. Organic residue recycling is important. Intercrops with maize as staple cereal and a short duration legume to serve the protein needs are predominant in the agrarian regions of South and Central Africa. The farmers intend to intensify this combination of intercrop using integrated nutrient management [31]. The aim is to improve productivity of both intercrops by 30% over current levels using nutrients and supplemental irrigation. Maize is sown early immediately after onset of rains. It allows rapid development of seedlings, matches with precipitation pattern that ends by late September. The legume crop sown in between cereal rows continues to grow and fix-N for few days more. Such an intercrop is said to restore soil fertility plus profitability. Reports by CIAT-CGIAR [32] suggest that bean-based intercrops traced in agricultural belts of African continent, play a vital role in maintaining soil fertility and supplying carbohydrates plus proteins. Bean is intercropped with crops such as cassava, sweet potato, sorghum, pearl millet, banana or maize. The bean-based intercropping is mostly regions with Ferralsols and Nitisols. It is said that in Southern Africa, between 15–38% of beans is derived from intercrops. Bean and intercrop is usually sown simultaneously in March/April and harvested by October. This system coincides with rainfall pattern. It offers best precipitation and nutrient use efficiency. Farm holding in the bean-intercropping region is small with 1.2 ha as average size. Hence, N derived via BNF is important to sustain soil-N fertility.

Relay cropping is adopted mainly to use land resources efficiently. Relay cropping involves seeding of two different crops with a lapse of time. Staggering and delaying sowing of second crop is mainly aimed at using soil and water resources efficiently. There should be enough time to grow both the crops to maturity. Also, there must be appropriate fertilizer inputs and water resources that allow both crops to maturity. Extended rainfall or use of stored moisture to support the second crop is necessary. Relay cropping is practiced by farmers in different agrarian regions, most often to attain greater harvests compared with mono-crops. Let us consider a few examples:

In the Canadian Plains, intercropping is effectively employed to produce an annual wheat crop and a green manure legume. Relay cropping allows harvest of legume well ahead of frost. The relay crop of legume is seeded in between rows of already established cereal. Cereal is harvested first but the legume continues to fix N throughout the season and until harvest. In a double crop, where in both cereal and alfalfa is seeded simultaneously, the legume fixes for relatively shorter duration [33]. Relay cropping actually aims at improving N fixation, biomass formation by green manure legume and recycling of organic matter.

Maize farms in North-east USA, adopt relay-cropping procedures rather routinely. There are specially designed interrow seeders that dibble seeds of second crop without disturbing the first crop i.e., maize. They say relay crop captures nutrient in soil efficiently, reduces N surplus in the soil profile, reduces runoff and percolation loss of nutrients. Relay crop protects soil from erosion via water or wind. Relay crop, if recycled entirely acts as a good organic manure. Relay crop could be a good forage supplier too. Relay intercropping is quite popular in Nebraska. It essentially involves sowing of wheat a few weeks ahead of the second crop-soybean. This system allows both wheat and soybean to share the same field for a length of time, use photosynthetic radiation and heat units efficiently. It also improves exploitation of soil nutrient accumulated in both upper and lower horizons of soil profile. At the time of harvest, wheat is generally harvested first without disturbing the soybean rows [34]. Relay cropping is said to reduce loss of soil-N via leaching, since it is effectively trapped and used by roots of both crops. Soybean extracts greater quantities of all three major nutrients N, P and K, since it accumulates these minerals at higher concentrations.

Let us consider a few more examples. Wheat is often intercropped with soybean in the Northern Great Plains. At Wooster in Ohio, USA, soybean is preferred as the second crop to be sown. Soybean is a photosensitive crop that matures in response to day length. Planting dates for relay crop, i.e., soybean could be flexible within limits. This allows efficient use of irrigation resources and/or stored moisture. Nutrient inputs are scavenged better for an extended period and by different crops. Fertilizer inputs and irrigation requirements of relay intercrop is generally more than sole crops. Regarding grain harvest, a sole crop of wheat produces 71 Bu/ac and that of soybean gives 48 Bu/ac. However, a relay crop of wheat-soybean in the same area produces 65 Bu/ac wheat grain and 36 Bu/ac soybean grain [35]. The soil nutrient exploitation is greater in an intercrop. Relay cropping is a preferred system when crop season is short for two consecutive sole crops. Nutrient recovery and recycling per unit time is better under intercrop than in a mono-cropping system.

Farmers in Meso-America adopt several variations of intercropping. Mixed intercrops involving, perennial trees, maize and bean plus cucumbers are an ancient cropping system that continues to dominate the tropical landscapes. Maize/bean intercrop is the predominant system covering 3–5% cropping area in Meso- and Latin America. Other intercropping systems encountered in Meso-America are coastal plantation, cereal plus bean. Temperate mixed intercropping systems such as wheat/soybean in the Pampas of Argentina, maize-sorghum and wheat /cotton are other intercropping pairs traced in Pampas. Mixed cropping involving maize/sorghum/soybean is common to Cerrados and Llanos regions of Brazil. Each of these intercropping procedures has several advantages to farmers. Efficient utilization of land and soil fertility resources, water, fertilizers and pesticides are advantages. Intercrops also allow appropriate residue recycling and C-sequestration [36]. The reduction in fertilizer usage, soil erosion control and better harvest are major advantages derived by farmers. Relay intercropping is also common in Latin American agrarian zones. For example, relay cropping of sunflower/maize/ soybean in the large farms of Pampas allows for zero or conservation tillage, restricts soil erosion to minimum, reduces fertilizer requirements since residue recycling procedures become efficient and total productivity of crops per unit land improves significantly. The three crops namely sunflower, maize and soybean are sown in February, March and April, respectively, in relay in-between rows. Later, they are harvested by September, October and November, respectively.

In the Sahelian West Africa, soils are sandy and low in nutrient content/availability. Fertilizer input, if any is small but it has to be efficiently used, in time without loss to lower horizon or via emissions. The precipitation pattern too is uncongenial for extended rainy season crops. Therefore, cropping systems have to aim at higher efficiency with regard to precipitation and nutrient use. Sahelian farmers maximize fertilizer use efficiency by intercropping millet/sorghum with a legume, such as cowpea. The planting pattern and density in the Sahel is peculiar. The pearl millet is spaced 1 m apart. This leaves sufficiently wide space for cowpea intercrop to flourish. The photosynthetic rates are not affected much. Usually pearl millet is grown with first rains. This allows the cereal to use early rain and soil fertility. The cowpea sown a couple of weeks later continues in the field for a month beyond the harvest of pearl millet. This type of intercropping allows efficient utilization of crop season, precipitation and soil fertility. Cowpea, being a legume adds to soil N fertility (Plate 8).

37.5 STRIP CROPPING

Strip cropping is common in large farms of Great Plains of North America. Major cereals such as wheat, maize and legumes such as soybean are grown in wide strips that are amenable to farm operations (Plate 11). Similarly, Strip cropping is in vogue in farms that cultivate soybean, cotton or sorghum in the Brazilian Cerrados. Strip cropping is common in European plains. Here, it is easy to allocate different crops to specific strips and control most of the operations through GPS guided vehicles. Strip cropping is common in tropics of Africa. Here, farmers tend to grow large strips of Agroforestry tree and alternate it with strips of maize or cowpea or vegetables. Strip cropping allows farmers to regulate various operations during cultivation of cereals, mulberry and forage. Strip cropping is adopted in the dry land of South India (Plates 12 and 13). Farmers grow finger millet and groundnut in strips. Evaluations suggest that sole crop of groundnut produces 905 kg pods/ha, but under strip cropping conditions groundnut produces slightly less than sole crop at 685 kg pod/ha plus 375 kg finger millet grain/ha. Nutrient recovery is significantly higher in the strip crops compared to sole crops. Residue recycling is much less in a sole crop compared to strip intercrops. The productivity of strip crops in terms of biomass and pods/grains is much higher [37].

37.6 FALLOWS

Agroecosystems are a matrix of cropped fields and those kept under fallow. We may encounter large areas or even expanses becoming fallows for a season. For example, weather pattern and natural resources in Sub-Sahara allow only one season of pearl millet/cowpea intercrop. Almost all fields become fallows immediately after the rainy season ends and first crop is harvested. The situation is similar in agrarian regions of northern latitudes. Fields go into fallows after the harvest of temperate cereal such as barley or wheat. Onset of frost and cold decides against cropping. Dry lands in South Asia that depend entirely on rainfall pattern also support one crop of cereal or cereal/legume intercrop. In contrast, within intensive cropping belts, farmers tend to grow 2, 3 or even 5 crops before making a field fallow. There are several variations of crop rotations and sequences that include fallow (Plates 14 and 15).

PLATE 14 A Field under Fallow after a Crop of Cereal in the Dry Land region of South India.
Note: This field is left to fallow without crop. Stray grasses grow naturally.

PLATE 15 A Finger millet field left to fallow with stubbles. Location-Hebbal Regional Experiment Station, University of Agricultural Sciences, Bangalore, India.
Source: [27].

Why do we practice fallows? There are indeed many reasons that necessitate fallows. Insufficient soil fertility status, lack of continued irrigation and economic aspects are major reasons for unplanted fallows. Farmers tend to maintain fallows in different ways. Fields are left without a crop. Such fallows may be prone to erosion and loss of top soil. Fields are given deep plow with disks; soils are turned and left to nature. There are fallows that support a short season crop or a catch crop. This type of fallow improves soil-N fertility, if the catch crop is a legume. Catch crops upon incorporation into field will improve SOM and nutrients. Fallows prone to loss of stored soil moisture and residual nutrients are used to produce a short season legume. There are clear suggestions that planting an N-fixing legume during the fallow period enormously improves soil fertility, especially soil-N. Crop residue recycling enhances SOM status [38]. Crop species useful to fill the fallows are chickpeas, field pea, lentil, soybean, mung bean, short season maize or sorghum.

37.7 COVER CROPS

Cover crops are cultivated in most agrarian zones mainly to conserve soil fertility, recycle nutrients efficiently, add to soil-N status, avoid loss of top soil in-between rows of plantation crops and lessen weed infestation. Cover crops also improve land-use efficiency. Cover crops can play a vital role in regulating nutrient dynamics, with in a small field or an agroecosystem at large, if the procedure gets popular in the entire agrarian zone. Adoption of leguminous cover during off-season in an otherwise cereal-farming zone is a useful proposition. In most cereal belts legumes are rotated as major crops. Otherwise, short season legume is interjected in the sequence. Such legumes add to soil-N status when they are harvested early in succulent stage and incorporated into fields. For example, in the corn cultivating zones of North America, cover crops like rye (*Secale cereal*) and vetch (*Vicia villosa*) is practiced to scavenge nutrients accumulated in soil, which otherwise will be lost due to environmental vagaries in the off season. Planting a rye/vetch mixture may improve nutrient recovery by the main cereal crop [39]. A vetch crop may accumulate large amounts of N in the biomass, much of it derived from biological nitrogen fixation. A slightly warm climate is enough to decompose the N rich succulent tissue of vetch and release nutrients to soil. A legume such as vetch has a C:N ratio of 8:1 to 15:1. Reports from North America indicate that on an average a rye/vetch mixture adds about 150 kg N/ha for succeeding maize crop to thrive. Maize yield increased by 2.0–2.2 t grain/ha due to incorporation of a cover crop such as vetch. Wheat and Sorghum production in the Central Plains, especially Kansas, Oklahoma or Tennessee involves cultivation of cover crops during fallow. The crops may or not improve grain yield of subsequent cereal. The advantages to count are improvement in soil structure, i.e., aggregate stability, moderation in soil temperature during winter, reduction in soil compaction, improvement in SOM content upon incorporation and increase of total soil-N.

Farmers in the Canadian prairies cultivate catch crops mainly to sequester C and N in the soil. In the absence of a cover crop, soil-N that gets accumulated rapidly is clearly prone to loss via leaching and emissions. Cover crops such as oats, annual rye grass, radish, peas and red clover are popular in Southern Canada. They do affect nutrient dynamics, especially C and N perceptibly and usefully. On an average, a cover crop grown in southern Canadian Prairies may recover 80–90 kg N/ha [40]. A legume crop may fix about 40 kg N/ha through its symbiotic relationship with nodule bacteria. Reasonably, such a cover crop releases about 25–40 kg N/ha into soil upon incorporation.

Cover crops also lessen loss of soil-N that occurs due erosion and emissions. They also reduce growth of weeds.

Researchers in Florida believe that cover crops are useful in reducing inorganic-N fertilizer requirements. Further, improved water and nutrient retention associated with cover crops like Rhizoma peanut may actually enhance productivity of citrus or cereal crops by 10% over control fields that never had cover crops. Methods to seed and cultivate rhizome peanuts as cover crops are available (Rouse et al. 2012). Cover crops grown in citrus groves of Florida reduced weed growth effectively by 30–40% compared to control grooves without cover crops (Plate 16).

PLATE 16 Rhizoma groundnut (*Arachis glabrata*) is a popular cover crop or intercrop in the Citrus groves of Florida and elsewhere in the cereal and cotton producing zones of South and South-eastern United States of America. Rhizoma groundnut avoids loss of nutrients via soil erosion. It uses excess fertilizer applied to citrus rather efficiently and preserves it for recycling. Rhizoma groundnut being a legume improves soil-N fertility. Rhizoma groundnut is also known to thwart ground water deterioration in Citrus belt of Florida. It is known to trap fertilizer-N applied effectively within the upper horizons thus avoiding percolation of N to vadose zone and groundwater.
Source: IFAS, Citrus Research and Educational Centre, Lake Alfred, University of Florida, Gainesville, Florida, USA.

Reports from parts of West African farming zones suggest that soil deterioration, loss of fertility and progressive reduction of crop yield are common. In some pockets, grain yield from cereals has never reached even 1.0 t/ha. Most of researchers attribute it to improper soil management procedures, absence of nutrient replenishment schedules and conservation practices. Lack of cover crops during fallow makes surface soils vulnerable to loss of nutrients. In the tropical West Africa, mainly in Ivory and Gold coast region, crops such as maize, cowpea, pigeonpea, groundnut are cultivated in sequence that is interjected with a short season, nonfood legume such as *Macuna, Crotalaria* or *Leucana.* Cultivation of legume cover crop such as *Macuna* or *Cajanus* during off-season reduces N requirement of the main cereal by 30%. Reports from Northern Ghana suggest that cover crops generate biomass equivalent to 115–136 kg N/ha during off-season and this could be recycled effectively. Analyzes of N budgets indicate that in two years a legume crop may cause a gain of 200 kg N/ha. Such N gain occurs both due to biological N-fixation and through reduction of loss of N from the system. It is said that fields under cover crops often experienced only marginal loss of soil-N, which is <20% of initial measurements. Macuna and Cajanus cover crops do recycle perceptible amounts of soil-P which otherwise is prone to loss via leaching or soil chemical fixation process [41]. Cultivation of hardy legumes such as horse gram (Plate 17), cowpea or grass species as a cover crop is routinely practiced in Southern Indian Plains. Fallows with cover crops add to SOM and soil-N if it is a legume species.

PLATE 17 Horse gram grown as a Cover crop in the Dry tracts of South Indian plains.
Note: The Alfisol fields in this region are mostly gravely, low/moderate in soil fertility and need periodic fallowing to refurbish soil fertility. The above field is planted (broadcasted) sparsely with a hardy legume-Horse gram (*Macrotyloma utilissima*). Horse gram negotiates soil fertility variations effectively. Horse gram is a legume and it forms symbiotic association with Bradyrhizobium that has the ability to fix atmospheric-N. The extent of soil-N fertility improvement achieved depends on horse gram genotype, bradyrhizobial strain and environmental parameters. Generally, a cover crop of horse gram adds 20–40 kg N/ha upon incorporation.
Source: Krishna, K. R., 2011.

KEYWORDS

- ***Macrotyloma utilissima***
- **Rhizoma groundnut**
- **rye**
- **strip cropping**
- **vetch**

REFERENCES

1. FAOSTAT, Maize statistics. Food and Agricultural Organization of the United Nations, Rome, Italy, www.faostat.fao.org. **2005.**
2. OKSTATE, World Wheat and Maize Production. http://nue.okstate.edu/Crop_information/World_Wheat_ Producion.htm **2009,** 1–11.

3. Krishna, K. R. Agrosphere: Nutrient Dynamics, Ecology and Productivity. Science Publishers Inc.: Enfield, New Hampshire, USA, **2003,** 343.
4. Janaiah, A. Hybrid Rice in Andhra Pradesh: Findings of a survey. Economic and Political Weekly **2003,** *38,* 2513–2516.
5. ICAR. History and Development of Rice varieties in India. http://dacnet.nic.in/rice.htm **2006,** 1–48.
6. Raju, A.; Reddy, A. R.; Bharambe, R.; Khadi, B. M. Agro-economic analysis of Cotton-based cropping systems. Indian Journal of Fertilizers **2008,** *4,*19–57.
7. Hall, A. J.; Vivella, F.; Trapani, N.; Chimenti, C. A. The Effects of Water stress and Genotype on the dynamics of Pollen shedding and Silking in Maize. Field Crops Research **1992,** *5,* 349–363.
8. Carcova, J.; Maddonni, G. A.; Ghersa, C. M. Long-term cropping effect on Maize productivity. Agronomy Journal **2000,** *92,* 1256–1265.
9. IPNI, Maize Planting and Production in the World. International Plant Nutrition Institute, Norcross, Georgia, USA, http://www.ipni.net/ppiweb /nchina.nsf/$webindex/3FA09D72EC945–883482573BB0030EB2A **2008,** 1–8.
10. Krishna, K. R. Agroecosystems of South India: Nutrient Dynamics, Ecology and Productivity. Brown Walker Press Inc.; Boca Raton, Florida, USA, **2010,** 240–278.
11. Krishna, K. R. Maize Agroecosystem: Nutrient Dynamics and Productivity. Apple Academic Press Inc.; Toronto, Canada, **2013,** 342.
12. IRRI, Rice Almanac. International Rice Research Institute. Manila, Philippines, **1995,** 175.
13. Peterson, G. A.; Schlegel, A. J.; Tanaka, D. K.; Jones, O. R. Precipitation use efficiency as affected by cropping and tillage. Journal of Production Agriculture **1996,** *9,* 180–186.
14. Oplinger, E. S.; Wiersma, D. W.; Gran, D. R.; Kolling, K. A. Intensive wheat management. University of Wisconsin-Extension, Madison, Wisconsin, USA, **1998,** 1–18.
15. Norwood, C. A. Dry land winter wheat as affected by previous crop. Agronomy Journal **2000,** 121–127.
16. Halvorson, A. D.; Weinhold, B. J.; Black, A. L. Tillage and Nitrogen fertilization influences on grain and soil nitrogen in a spring wheat-fallow system. Agronomy Journal **2001,** *93,* 1130–1135.
17. Gonzalez-Mongtaner, J. H.; Madonni, G. A.; DiNaploi, M. R. Modelling grain yield and grain yield responses to nitrogen in Spring wheat crops in the Argentinian Southern Pampas. Field Crops Research **1997,** *51,* 241–252.
18. Velayutham, M. Sustainable Productivity under Rice-Wheat cropping system-Issues and Imperatives in Research. In: Sustainable Soil Productivity under Rice Wheat system. Biswas, T. D.; Narayanaswamy, G. (Ed.). Indian Society of Soil Science Bulletin **1997,** *18,* 1–6.
19. Singh, B.; Singh, Y. Significance of Nutrient ratios in NPK fertilization in Wheat-Rice cropping system in Indo-Gangetic Plains. Indian Journal of Fertilizers **2010,** *6,* 44–50.
20. Ahlawat, I. S.; Masood Ali, Fertilizer Management of Pulse crops. In: Fertilizer Management in Food crops. Tandon, H. L. S. (Ed.) FDCO, New Delhi, **1993,** 114–138.
21. Ganeshamurthy, A. N.; Ali, M.; Srinivasarao, Ch. Role of *Pulses* in sustaining soil health and crop production. *Indian Journal of Fertilizers*, **2006,** *1(3),* 29–40.
22. IKISAN, Sunflower: Hybrid Sunflower. http://www.IKISAN.com/links /ap_sunflowerHybrid %20Sunflower.shtml **2008,** 1–6.
23. Ofori, E.; Kyei-Baffour, N.; Agodzo, S. K. Developing effective climate information for managing rained crop production in some selected farming centers in Ghana. Proceedings of the School of Engineering Research. Accra, Ghana **2004,** 1–18.

24. Mokwunye, V. Alleviating soil fertility related constraints to increased Crop Productivity. Kluwer Academic Publishers, Dordrecht, Netherlands, **1991,** 342.
25. Bationo, A.; Christianson, C. B.; Baethgen, W. A.; Mokwunye, A. A farm level evaluation of Nitrogen and Phosphorus fertilizer use and planting density for pearl millet production in Niger. Fertilizer Research **1992,** *312,* 175–184.
26. Coura Badiane, TED case studies number 646. http://www.Senegalcase study.htm. **2001,** 1–32.
27. Krishna K. R. Peanut Agroecosystem: Nutrient Dynamics and Productivity. Alpha Science International Inc.: Oxfordshire, United Kingdom, **2008,** 74–117.
28. FAO, Sustainable Crop Rotations. Food and Agricultural Organization of the United Nations. http://www.fao.org/docrep/V9926E/v9926e06.htm **2006,** 1–23 (August 16, 2012).
29. Brown, L. Rescuing a Planet under stress and a cultivation trouble. Earth Policy Institute, Washington, USA, **2003,** 1–3.
30. Kumwenda, J. D. T.; Waddington, S. R.; Snapp, S. S.; Jones, R. B.; Blackie, M. J. Soil Fertility Management Research for the maize cropping system of small holders in Southern Africa. NRG paper 96–02, International Maize and Wheat Centre (CYMMIT), Mexico, **1996,** 1–3.
31. Mekuria, M. Sustainable intensification of Maize-Legume Cropping systems for food security in Eastern and Southern Africa. Australian Centre for International Agricultural Research, Canberra, Australia. aciar.gov.au/project/ CSE/2009/024 **2009,** 1–3 (August 16, 2012).
32. CIAT-CGIAR Cropping Systems. Webapp.ciat.cgiar.org/Africa/pdf/atlas_bean_africa_cropping_systems.df **2012,** 1–11.
33. Grant, C. Crop Management to Reduce N Fertilizer Use. http://www.umannitoba.ca/afs/agronomists_conf/proceedings/ 2005/grant_crop_management.htm **2005,** 1–8 (August 3, 2012).
34. UNL Extension Services relay cropping offers Economic benefits and reduces Nitrate leaching from soils. Crop Watch: Wheat. University of Nebraska-Lincoln. http://ianrhome.unl.eduhome **2003,** 1–5 (August 18, 2012).
35. Beurlein, J. Relay cropping Wheat and Soybean. Ohio State University Fact Sheet. Ohioline.osu.edu/agf-fact/0106.html **2011,** 1–5 (August 16, 2012).
36. FAO, Farming Systems: Latin America and Caribbean. Food and Agricultural Organization of the United Nations. http://www.fao.org/decrep/003/yr1860E/yr1860e09.htm **2009,** 1–30 (August 17, 2012).
37. Hegde, G.; Reddy, R.; Balamatti, A. Strip crop-a ray of hope for dry land farmers. Agriculture, Man, Ecology. Bangalore, India. India Development Gateway.htm **2010,** 1–3 (August 23, 2010).
38. DFID, Promotion of Rain fed Rabi cropping in Rice Fallows of Eastern India. Natural Resources Department, University of Wales, Bangor, United Kingdom. http://www.dfid.gov.uk/R4D/Project/3592/Default.aspx **2006,** 1–3 (August 21, 2012).
39. Zotarelli, L.; Avila, L.; Scholberg, J. M. S.; Alves, B. J. R. Benefits of Vetch and Rye Cover crops to Sweet Corn under No-tillage. Agronomy Journal **2009,** *101,* 252–260.
40. Stewart, G. Greenhouse gas mitigation. Ontario Ministry of Agriculture and Food. Canada, **2005,** 1–3.
41. Fosu, M.; Kuhne, R. F.; Vlek, P. L. G. Improving Maize yield in the Guinea Savanna Zone of Ghana with Legume crops and P, K fertilizers. Journal of Agronomy **2004,** *3,* 115–121.
42. DeSoussa, M. G.; Lobato, E.; Rein, T. A. Uso de gosso agricolo nos solos do Cerrados. Planaltina, D. F.; EMBRAPA-CPAC, Circular Technical No. 32, **1995,** 1–20.

43. Diop, A. SAFE – world project/initiative summary: Senegal. http://www2.essex.ac.uk/ces/reseachprograms/SusAg/ SAFEAfrica-projects/senegal/rodaleregenerating Agricenter.pdf **2000,** 1–3.

EXERCISE

1. Mention various cropping systems andsdiscuss them with examples.
2. What are main advantages of Mixed Cropping?
3. Compare productivity of three different cereal monocrops with mixed crops.

FURTHER READING

1. Shreshta, A.4Cropping Systems: Trends and Advances. CRC Pres : Boca Raton, Florida, USA, **2004**, pp. 720.

USEFUL WEBSITES

http:/dacnet.nic.in/rice; http://www.indiancommodity.com/statistic/sunflowcr.htm
http://www.fao.org/docrep/V9926E/v9926e03.htm

INDEX

A

B

D

E

O

R

X

Y

Z

Printed in the United States
by Baker & Taylor Publisher Services

Printed in the United States
by Baker & Taylor Publisher Services